Advances in Wastewater Treatment

Advances in Wastewater Treatment

Giorgio Mannina, George Ekama,
Hallvard Ødegaard and Gustaf Olsson

PUBLISHING

Published by **IWA Publishing**
Alliance House
12 Caxton Street
London SW1H 0QS, UK
Telephone: +44 (0)20 7654 5500
Fax: +44 (0)20 7654 5555
Email: publications@iwap.co.uk
Web: www.iwapublishing.com

First published 2018

British Library Cataloguing in Publication Data
A CIP catalogue record for this book is available from the British Library

ISBN: 9781780409702 (Paperback)
ISBN: 9781780409719 (eBook)

Contents

List of Abbreviations

AAO – anaerobic ammonium oxidizers
AGS – aerobic granules sludge
AOB – ammonia-oxidizing bacteria
BOD_5 – biochemical oxygen demand
CAS – conventional activated sludge
COD – chemical oxygen demand
DO – dissolved oxygen
DAMOA – denitrifying anaerobic methane oxidizing archaea
DAMOB – denitrifying anaerobic methane oxidizing bacteria
DOHO – denitrifying OHO
DGAO – denitrifying GAO
DPAO – denitrifying PAO
EPS – extracellular polymeric substances
F/M – food to microorganism ratio g-COD/g-VSS·d
FISH – fluorescence in situ hybridization
GAO – glycogen-accumulating organisms
HRT – hydraulic retention time
MBBR – moving bed biofilm reactor
NOB – nitrite oxidizing bacteria
OHO – ordinary heterotrophic organisms
OLR – organic loading rate (kg-COD/$m^3 \cdot d$)
ORP – oxidation-reduction potential
OUR – oxygen uptake rate (mg $O_2/L \cdot h$)
PAO – phosphate-accumulating organisms
PHA – polyhydroxyalkoanates
PPCP – pharmaceutical and personal care products
SBR – sequencing batch reactor
SND – simultaneous nitrification-denitrification

SRT – solids retention time
SVI – sludge volume index (mL/g)
TN – total nitrogen
TP – total phosphorus
TSS – total suspended solids
VER – volume exchange ratio
VFA – volatile fatty acid
VSS – volatile suspended solids
WWTP – wastewater treatment plant

Preface

Many new wastewater treatment technologies have been developed over the last decades, to be applied in combination with or as an alternative to traditional wastewater treatment systems, such as conventional activated sludge. In parallel with technology development there has been significant progress in mathematical modelling, reflecting an increasing understanding of the various process mechanisms as well as the complex couplings between basic processes.

The book is an extension of an *Advanced Course on innovative wastewater treatment and modelling* supported by the International Water Association (IWA) and organized by Palermo University (Palermo, Italy). The course has been presented by the editors of the book, while the book is extended with contributions from several other experts in process technology and modelling.

The book presents advances in wastewater treatment processes, including modelling topics and also describes "lessons learned for the future".

The aim of the book is to present both fundamentals and practices of innovative wastewater treatment processes, such as activated sludge processes in membrane bioreactors, moving bed biofilm processes and granular processes. Greenhouse gas formation in treatment processes is getting increasing attention, and micropollutant control is becoming a critical challenge. Furthermore, the book highlights the essential role of models in process design, operation and control.

The book is organized in eleven chapters, covering:

- Primary treatment: Particle separation by rotating belt sieves;
- Biological nutrient removal activated sludge systems with membranes;
- MBBR and IFAS Systems;
- Aerobic granular sludge: state of the art, applications, and new perspectives;

Editors: Giorgio Mannina, George Ekama, Hallvard Ødegaard and Gustaf Olsson
doi: 10.2166/9781780409719_00xv

- Membrane based processes;
- Micropollutant control;
- Anaerobic processes;
- Greenhouse gas emissions from membrane bioreactors;
- Mixing – new insights and opportunities through computational fluid dynamics;
- Making water operations smarter;
- Global sensitivity analysis in wastewater treatment modelling.

It is our hope that the reader will find the book a source of inspiration for both research and practicing engineering.

Giorgio Mannina
George Ekama
Hallvard Ødegaard and
Gustaf Olsson

Chapter 1

Primary treatment: Particle separation by rotating belt sieves

B. Rusten[1] and H. Ødegaard[2]

[1]Aquateam COWI, Oslo, Norway

[2]Hallvard Ødegaard, Scandinavian Environmental Technology and NTNU, Trondheim, Norway (SET AS, Hellemsveien 421, N 7165 Oksvoll, Norway)

ABSTRACT

Fine mesh rotating belt sieves (RBSs) offer a compact solution for removal of particles in primary treatment of wastewater. This chapter shows examples from lab-scale, pilot-scale and full-scale testing of primary treatment and chemically enhanced primary treatment (CEPT). In a Norwegian full-scale survey, the use of a 350 μm belt showed more than 40% removal of total suspended solids (TSS) and 30% removal of chemical oxygen demand (COD) at sieve rates as high as 160 $m^3 \cdot m^{-2} \cdot h^{-1}$. Maximum sieve rate tested was 288 $m^3 \cdot m^{-2} \cdot h^{-1}$ and maximum particle load was 80 kg TSS $\cdot m^{-2} \cdot h^{-1}$. When the sludge mat on the belt increased from 10 to 55 g TSS $\cdot m^{-2}$, the removal efficiency for TSS increased from about 35 to 60%. CEPT is a simple and effective way of increasing the removal efficiency of an RBS. When adding about 0.7–1.0 g $\cdot m^{-3}$ of cationic polymer and using about 2 min of flocculation time, the removal of TSS typically increased from 40–50% without polymer to 60–75% with polymer. The particulate organic matter that was removed in the RBS had little or no effect on the denitrifying capacity of the wastewater. The high volatile solids (VS) content of the RBS sludge as compared to primary clarifier sludge, gave a higher methane potential in anaerobic digesters for the RBS sludge. The high caloric value of RBS sludge makes it attractive for incineration.

Keywords: Municipal wastewater, primary treatment, rotating belt sieves

Editors: Giorgio Mannina, George Ekama, Hallvard Ødegaard and Gustaf Olsson
doi: 10.2166/9781780409719_0001

1.1 INTRODUCTION

Traditionally primary treatment has been synonymous with settling in clarifiers. Primary settling may be used as the only treatment when discharging to the ocean, or as pretreatment in order to lower the load and sludge production of subsequent more advanced treatment processes. Depending on the raw water characteristics, removal efficiencies of primary settling typically range from 40–60% with respect to total suspended solids (TSS) and 15–30% with respect to biochemical oxygen demand (BOD_5).

In order to improve the separation efficiency of primary clarifiers, chemically enhanced primary treatment (CEPT) may also be used. This implies addition of coagulants and introduction of flocculation tanks ahead of the clarifier so that colloidal, non-settleable particles may be removed in addition to the suspended, settleable ones. If an inorganic coagulant is used (Al or Fe), phosphate removal is achieved as well, and in many cases this is the main purpose of using CEPT. Lately, however, focus has been on the enhanced removal of organic matter as pretreatment before de-ammonification processes (Water Environment Research Foundation [WERF], 2014; Ødegaard, 2016). Again depending on the raw water characteristics but also design and operation of coagulant mixing and flocculation, removal efficiencies of CEPT range from 70–90% with respect to TSS and 60–80% with respect to BOD_5 (Ødegaard, 1992, 2016).

Clarifiers require quite a lot of space and, since low foot-print of wastewater treatment plants has become a goal in many instances, fine mesh sieves have been introduced as an alternative to settling for primary treatment. Particularly rotating belt sieves (RBSs), also referred to as rotating belt filters (RBFs), have gained popularity. In this chapter we shall focus on the development and the experiences of rotating belt sieves for primary treatment. An overview of the RBS technology will be given, as well as examples of test results from different types of applications. These are mainly based on a R&D programme on primary treatment that was carried out in 2005 (Ødegaard, 2005; Rusten & Ødegaard, 2006) and on recent experiments at municipal wastewater treatment plants in Norway (Paulsrud *et al.,* 2014; Rusten *et al.,* 2017; Sahu *et al.,* 2017). A bench-scale procedure for choosing mesh size and evaluating potential treatment results is also presented (Rusten & Ødegaard, 2006).

1.1.1 The Norwegian primary treatment evaluation programme

Due to the European Union (EU) requirements for wastewater treatment, the Norwegian State Pollution Control Agency (SFT) took an initiative to evaluate and test several different technologies for primary treatment. This R&D programme was carried out with contributions from research institutions (NTNU, SINTEF and NIVA), consulting companies (Asplan Viak, Rambøll and Aquateam), the cities of Bergen and Tromsø, and the regional water and wastewater utility company IVAR (Ødegaard, 2005).

The goal was to find dependable and cost-efficient technologies that fulfilled the EU criteria for primary treatment, i.e. at least 20% removal of organic matter (measured as BOD_5) and 50% removal of total suspended solids (TSS). For treatment plants >10,000 pe (population equivalent), with 24 control samples per year, at least 21 samples must fulfil the requirements. This is a lot stricter than looking at average removal efficiencies and the R&D programme showed that an average TSS-removal of about 65% was necessary for enough samples to pass the 50% removal requirement.

Several types of primary treatment options, such as clarifiers, fine mesh sieves, large septic tanks, dissolved air flotation (DAF) and deep bed filtration were initially evaluated. Of the technologies that were considered fully developed, clarifiers and different types of fine mesh sieves were found most suitable for primary treatment. These technologies were then tested in full scale at several treatment plants, for both primary treatment and chemically enhanced primary treatment.

Historically primary treatment has been synonymous with sedimentation in clarifiers. The R&D programme revealed, however, that most of the primary clarifiers tested, were unable to achieve primary treatment according to the EU requirements. Surveying of data from five treatment plants showed that only one plant fulfilled the EU requirements for primary treatment, and this plant had an average surface area overflow rate (v_f) of only 0.36 m · h^{-1} (Misund *et al.*, 2004).

Often the particle size distribution of the wastewater was such that the required primary treatment removal efficiencies could not be achieved by sedimentation, regardless of how low the surface area overflow was. Therefore, chemically enhanced primary treatment (CEPT) would normally be required when using primary clarifiers to meet the standard for primary treatment (>50% TSS-removal and 20% BOD_5-removal). CEPT has commonly been used in Scandinavia, by dosing precipitation coagulants and/or polymers to the wastewater followed by flocculation and clarification. In some upgraded primary plants, only aerated sand and grit traps have been used for flocculation. The main goal has been to remove phosphorus, but removal efficiencies have also been high for both BOD (>70%) and TSS (>90%) (Ødegaard, 1992, 2016).

The R&D programme also revealed the great need for a better understanding of how to design and operate fine mesh sieves in order to optimize primary treatment performance. This became the main focus of the R&D programme and will be discussed below.

A number of fine mesh sieves are on the market and were tested in the R&D programme (Ødegaard, 2005). They include stationary sieves, rotating drum sieves, rotating disc sieves and rotating belt sieves. Full-scale tests were carried out at nine treatment plants with predominantly municipal wastewater. Six of these plants used rotating belt sieves (from two different manufacturers), two plants used rotating disc sieves and one plant had both a stationary sieve and a rotating drum sieve. Mesh sizes ranged from 80 to 850 μm. Of all the sieves tested on predominantly municipal wastewater, only rotating belt sieves (RBSs) fulfilled the EU primary treatment requirements and this type of sieve will be focused on here.

RBS technology offers a very compact solution and has been used successfully for primary treatment of municipal wastewater (Rusten & Ødegaard, 2006; Sutton *et al.*, 2008). Advantages of RBSs include compact footprint, reduced civil engineering site work, and modular construction. The latter allowing for reduced design work, faster installation, and ease of plant expansion (Franchi & Santoro, 2015).

1.2 ROTATING BELT SIEVE (RBS) TECHNOLOGY

Rotating belt sieves (RBSs) are also referred to as rotating belt filters (RBFs) in some of the literature, but the term RBS will be used in this chapter. The experiences and data reported here, from commercial and prototype RBS systems, are based on RBSs manufactured by Salsnes Filter (Namsos, Norway). Commercial systems have a submerged belt area from about 0.1 m² for the smallest unit (SF500) to 2.5 m² for the largest unit (SFK600). For primary treatment of municipal wastewater, the most common mesh size is 350 μm, but mesh sizes from 30 μm to 2 mm are available. The SF systems, shown on the left-hand side in Figure 1.1, are free-standing and enclosed, while the SFK systems, shown on the right-hand side in Figure 1.1, are open for installation in concrete channels. The sieves are modular and multiple units are used to accommodate large flows.

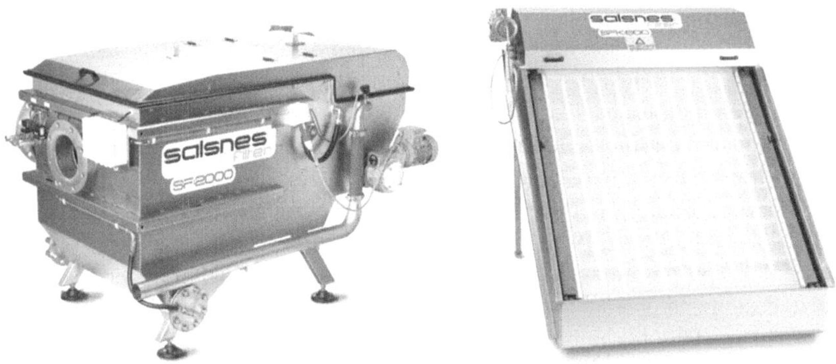

Figure 1.1 Salsnes Filter RBS systems. SF systems (left) are free-standing and enclosed, while SFK systems (right) are open for installation in concrete channels. (Courtesy of Salsnes Filter AS.)

The operating principle for the Salsnes Filter RBS systems is described in Figure 1.2. Particles larger than the mesh openings are collected on the belt and gradually these particles will create a filter mat and remove particles significantly smaller than the mesh openings. This will reduce the flow through the belt and the belt needs to rotate so that it can be cleaned and thereby sustain the necessary hydraulic capacity. The RBS can be operated with either a fixed belt speed and a variable water level, or a fixed water level and variable belt speed. The latter is most common, and the belt speed will then depend on the water flow and the amount of suspended solids in the water.

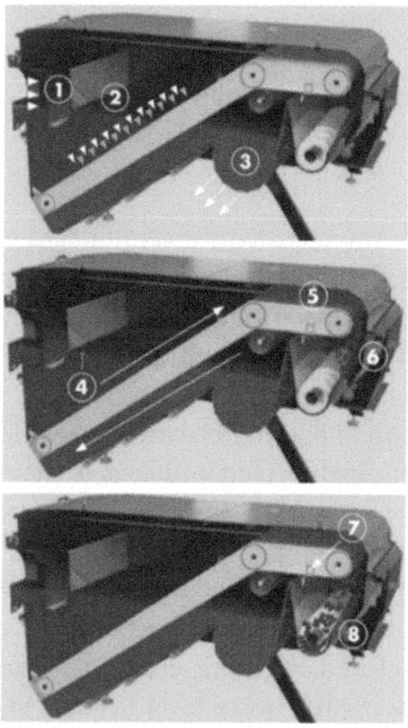

1) Wastewater enters the inlet chamber
2) Solids above the filter mesh create a "filter mat", that enhances separation performance as particle build-up on the mesh creates progressively smaller holes.
3) Outlet for water filtered through the mesh.
4) WW influent rises to a certain level and the filter mesh starts to rotate like a conveyer belt, transporting sludge and enabling the thickening process.
5) Gravity thickens sludge to 3 – 8 % TS.
6) Sludge drops into the collection area.
7) Using air (not water) the Air Knife automatic cleaning system removes any remaining sludge from the filter mesh into the sludge collection area.
8) A screw press typically dewaters the sludge to 25 – 30 % DS before it exits the unit.

Figure 1.2 Operating principle for Salsnes Filter RBS systems. (Courtesy of Salsnes Filter AS.)

A patented air knife is normally used to blow the sludge off the belt, but scrapers and intermittent water spray have also been used.

1.2.1 Characterization of wastewater through screening tests

Characterization of the wastewater is very important in order to predict the removal efficiencies and hydraulic capacities that can be expected for a given sieve. A simple screening test apparatus and procedure (Rusten, 2004) was developed and used in the R&D programme. A sketch of the equipment and photos of a 350 μm sieve cloth prior to testing and after development of a filter mat are shown in Figure 1.3.

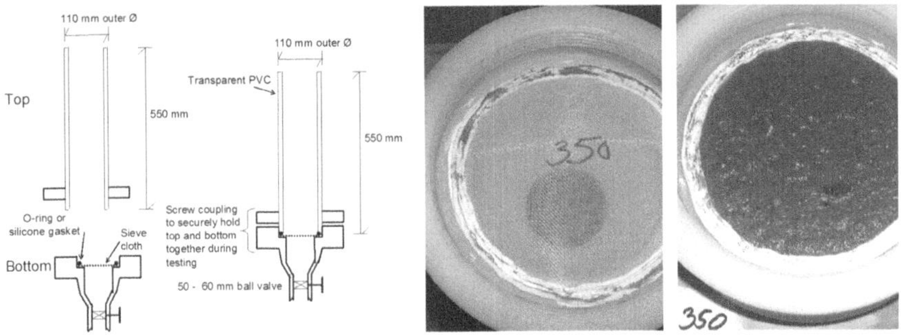

Figure 1.3 Sketch of screening test apparatus and photos of clean sieve cloth and sieve cloth with filter mat (Rusten & Ødegaard, 2006).

A batch of wastewater with sufficient volume to run tests with several different sieve cloths was placed in a large tank. To ensure a homogenous distribution of particulate material the batch would be vigorously stirred prior to taking wastewater out of the tank for analysis or to put through the test apparatus. Samples of the wastewater filtered through the sieve cloths were taken of the first litre of wastewater filtered, when the sieve cloth was clean. Then more wastewater was added until a build-up of particles on the sieve cloth had formed a filter mat. Tests with a filter mat simulated operation of a fine mesh sieve with a significant pressure drop over the sieve cloth and a low hydraulic load.

The transparent PVC tube of the apparatus had marks at 200 mm and 300 mm above the surface of the sieve cloth. After the first litre was filtered through the sieve cloth, the valve at the bottom of the apparatus was closed and more wastewater was added. Then the valve was partially opened, allowing the water level in the PVC tube to drop at a rate of 3 to 4 $cm \cdot s^{-1}$. When a proper filter mat had formed on the sieve cloth, the valve was opened all the way and filtered wastewater was collected while the water level dropped from the 300 mm mark to the 200 mm mark. The time it took for the water level to drop from 300 mm to 200 mm was also recorded. For most test runs this procedure was done repeatedly after more wastewater had been added and a gradually thicker filter mat had developed, resulting in a longer period of time for the water level to drop from the 300 mm to the 200 mm mark.

1.3 RESULTS AND EXPERIENCES FROM RBS OPERATION IN THE NORWEGIAN R&D PROGRAMME ON PRIMARY TREATMENT

1.3.1 Screening test results

The screening tests showed that required primary treatment removal efficiencies could be achieved with all tested wastewaters (from 11 different treatment plants) if the proper mesh size was used and a sufficiently thick filter mat was allowed to develop. However, use of sieves would not always be recommendable due to the low hydraulic loads necessary to achieve sufficiently high removal efficiencies with some of the wastewaters. Sieves that could not be operated with a significant filter mat would likely fail to meet primary treatment requirements, even with mesh sizes in the 50 to 100 μm range.

To be considered suitable for primary treatment with fine mesh sieves, the screening tests indicated that at least 20% of the TSS in the wastewater should consist of particles larger than 350 μm and the ratio between filtered (Whatman GF/C) chemical oxygen demand (FCOD) and total COD (TCOD) should be <0.4. Once a filter mat was formed on the sieves, there were practically no differences in the performances of sieve cloths with different mesh sizes, with regard to both % TSS removal and filtration rate. This would normally favour the use of larger mesh sizes, like the 350 μm sieve cloth. However, if the wastewater has a very small amount of larger particles there may not be enough particles present to form a filter mat, and a smaller mesh size would be recommended to initiate the formation of the necessary filter mat.

Examples of screening test results are shown in Figure 1.4. As mentioned above, the mesh sizes of the sieve cloths had very little influence on the results, within the ranges tested. With a given wastewater, the removal of TSS was mainly a function of the hydraulic flow through the sieve cloth, referred to as sieve rate, which again was a function of the development of a filter mat on the sieve. At a low sieve rate of only 20 m^3 per m^2 submerged sieve cloth area per hour ($m^3 \cdot m^{-2} \cdot h^{-1}$) more than 70% removal of TSS was achieved with wastewater from the Nordre Follo wastewater treatment plant (WWTP). When the sieve rate was increased to 100 $m^3 \cdot m^{-2} \cdot h^{-1}$ the removal efficiencies dropped to about 60%. The example from the Tiendeholmen WWTP shows a concentrated wastewater that was very well suited for fine mesh sieve treatment. Even at a sieve rate of 224 $m^3 \cdot m^{-2} \cdot h^{-1}$ the removal of TSS was 69% with a 350 μm sieve cloth.

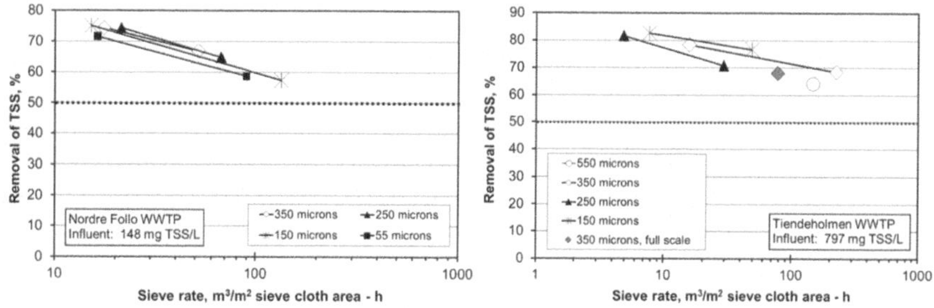

Figure 1.4 Examples of screening test results with different sieve cloths, showing removal of TSS versus sieve rate. For the Tiendeholmen WWTP the result from the full-scale plant at the time of the screening test is also shown (Rusten & Ødegaard, 2006).

1.3.2 Full-scale results

Very good agreement was found between screening tests and full-scale tests. An example of this is shown in Figure 1.4 where the full-scale result for the Tiendeholmen WWTP is shown together with the screening test results. Figure 1.5 shows a photo of the rotating belt sieves at the Tiendeholmen WWTP. Full-scale testing demonstrated the importance of gentle handling of the particles to prevent them from breaking and then going through the sieve openings. It also verified the need for a filter mat (Figure 1.6). Only rotating belt sieves had the ability to control filter mat development in our tests.

Figure 1.5 Tiendeholmen WWTP with three rotating belt sieves (Salsnes Filter SF6000) and 350 μm mesh size (Rusten & Ødegaard, 2006).

Of all the sieves tested, on predominantly municipal wastewater, only rotating belt sieves fulfilled the EU primary treatment requirements. The results from Breivika WWTP in Tromsø, summarized in Figure 1.7, show that every single sample fulfilled the EU primary treatment requirements. Average influent concentrations for 19 samples were 331 g TSS $\cdot$ m^{-3} and 176 g BOD_5 $\cdot$ m^{-3}, while average effluent concentrations were 34 g TSS $\cdot$ m^{-3} and 36 g BOD_5 $\cdot$ m^{-3}. This corresponds to average removal efficiencies of 90% for TSS and 80% for BOD_5. These extraordinary good results can be explained by the fact that the sieves were operated with a very thick filter mat (Figure 1.6) and at a sieve rate of only 25 $m^3 \cdot m^{-2} \cdot h^{-1}$.

At higher sieve rates the removal efficiencies will normally be lower. However, for wastewater with a very favourable particle composition very high sieve rates may be used. In cases where the focus is on high capacity and not high removal efficiencies, sieves can be operated without a filter mat, and hydraulic capacities may be as high as 300 $m^3 \cdot m^{-2} \cdot h^{-1}$, depending on wastewater composition and sieve cloth properties (Rusten & Ødegaard, 2006).

Figure 1.6 Thick filter mat on a rotating belt sieve with a mesh size of 350 μm at the Breivika WWTP (Rusten & Ødegaard, 2006).

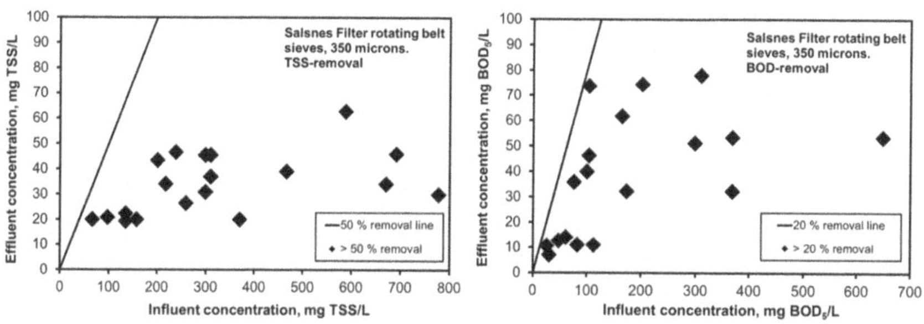

Figure 1.7 Results from primary treatment at the Breivika WWTP using three rotating belt sieves (Salsnes Filter SF4000) with 350 μm mesh size (Rusten & Ødegaard, 2006).

A broad general observation from these full-scale plants was that the highest possible removal efficiencies were achieved if the plants were operated in such a way that they treated the least amount of water over the longest possible time. This means that fine mesh sieves with pumped influent should have frequency-controlled pumps to avoid on/off operation. At plants with several sieves in parallel, all sieves should be running even at low water flows. This will enable operation with thick filter mats and high removal efficiencies (Rusten & Ødegaard, 2006). Section 1.4.1

contains a further discussion of how different operational parameters influence the particle removal in RBS systems.

1.3.3 Chemically enhanced primary treatment

Screening tests and full-scale tests at the Bangsund WWTP showed the wastewater to be unfavourable for primary treatment with conventional fine mesh sieves. Initial full-scale tests, with two rotating belt sieves in series and coagulant/flocculant addition and flocculation between the two sieves, showed that prior removal of particles smaller than 850 μm had a detrimental effect on the flocculation. Use of cationic polymer alone worked better than different combinations of metal salts and polymer, and the best results were obtained when the wastewater bypassed the first sieve. However, excellent results were achieved with a mesh size of 850 μm on the first sieve, addition of about 1 g · m^{-3} of a cationic polymer (Pemcat 163, medium charge density, high molecular weight), flocculation in a static flocculator and solids separation on a rotating belt sieve with a mesh size of 250 μm (Rusten & Lundar, 2004). Results achieved with a small dose of cationic polymer easily fulfilled the EU criteria for primary treatment and they are shown in Figure 1.8. For 5 of the 23 data points in Figure 1.8 the first sieve was bypassed. Average TSS-removal was 66% at an average sieve rate of 25 $m^3 \cdot m^{-2} \cdot h^{-1}$ (Rusten & Ødegaard, 2006).

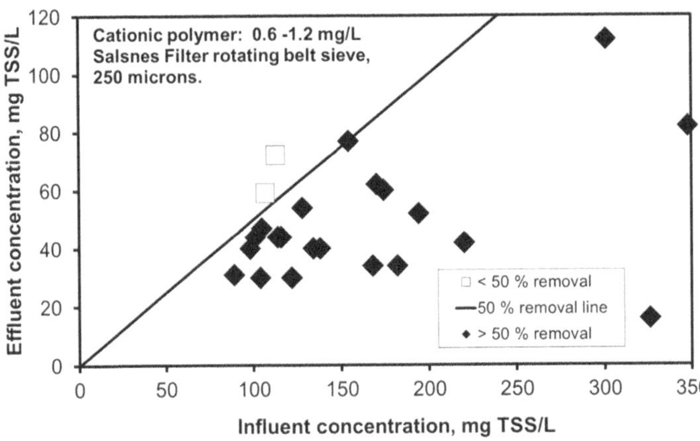

Figure 1.8 Results from the Bangsund WWTP using cationic polymer and a rotating belt sieve with a mesh size of 250 μm (Rusten & Ødegaard, 2006).

1.3.4 Sludge dewatering

All the different rotating belt sieves had simple screw presses for sludge dewatering, either integrated or as separate units. Dewatered primary sludge had total solids (TS) concentrations from 17 to 37%, with an average of 27%. There was no significant difference between the different types of sieves or between plants with or without chemically enhanced primary treatment. The volatile solids (VS) fraction was very high in all the sludge samples and averaged 90% (Paulsrud, 2005).

1.3.5 Cost comparison

A cost comparison of primary treatment, including sludge dewatering, was carried out for rotating belt sieves and clarifiers. The cost comparison was for a dry weather flow of 200 $m^3 \cdot h^{-1}$ and an influent concentration of 250 g TSS $\cdot m^{-3}$. The maximum wet weather flow was 400 $m^3 \cdot h^{-1}$. The clarifier overflow rate was 1.2 $m \cdot h^{-1}$ at dry weather flow and 2.4 $m \cdot h^{-1}$ at maximum flow. The sieve rate was 100 $m^3 \cdot m^{-2} \cdot h^{-1}$ at dry weather flow and 200 $m^3 \cdot m^{-2} \cdot h^{-1}$ at maximum flow. The cost of land was set at zero and the clarifiers were uncovered. A 7% annual interest rate and 15 years depreciation was used to calculate annual capital costs.

For the above conditions, savings were found to be substantial when using rotating belt sieves for primary treatment. Both investment costs and total annual costs (annual capital costs plus operation & maintenance costs) for the rotating belt sieves were about 50% of the costs for the primary clarifiers (Ødegaard, 2005).

1.4 RESULTS AND EXPERIENCES FROM RECENT STUDIES OF RBS

1.4.1 Primary treatment

At the Nedre Romerike WWTP (Strømmen, Norway) a Salsnes Filter SF2000 RBS, with a 350 µm belt and 0.5 m^2 submerged belt area, was installed and tested under a variety of operating conditions and hydraulic loads (Rusten *et al.*, 2017). The plant is in a rock cavern and had primary clarifiers followed by a MBBR (moving bed biofilm reactor) process for nitrogen removal. The objectives were to (1) see how much space can be saved by RBS primary treatment, (2) produce a primary effluent that is optimum for the downstream pre-denitrification MBBR process, and (3) produce a primary sludge with a higher methane gas potential (Paulsrud *et al.*, 2014) for future anaerobic sludge digestion.

A total of 40 test runs were performed over a period of 3 months. Wastewater was pumped to the SF2000 from the influent channel, immediately after the sand traps. Wastewater temperatures were normally between 8 and 9°C. Only one run had a temperature below 7°C.

In addition to wastewater characteristics being very important, removal rates and hydraulic capacity were influenced by water level and belt speed. The amount of particles deposited on the belt, prior to being blown off by the air knife, varied from below 10 to above 100 g TSS $\cdot m^{-2}$ and is shown in Figure 1.9. These particles created a filter mat that was important for the performance. Wet sludge blown off the belt had a total solids (TS) concentration from 4 to 10%, depending on the belt speed. After dewatering in an integrated screw press the median concentration was 25% TS.

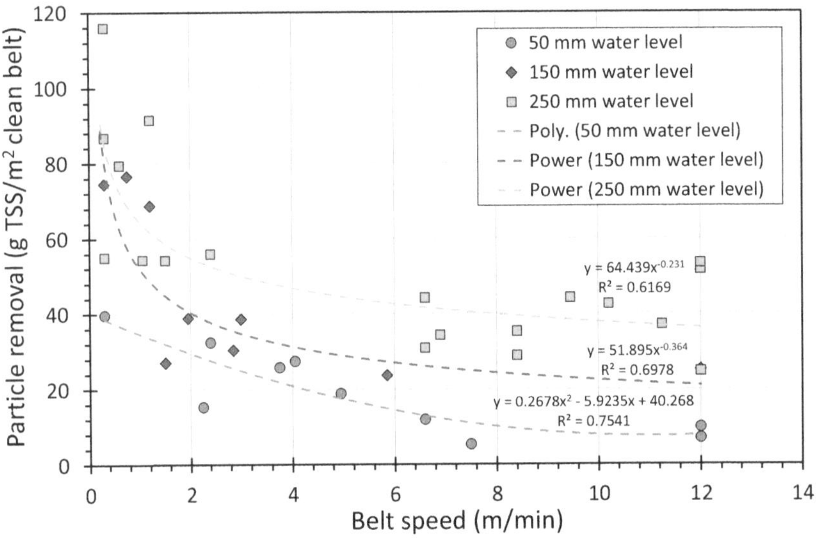

Figure 1.9 SF2000 primary treatment at Nedre Romerike WWTP. Amount of particles deposited on belt prior to removal by air knife, as a function of belt speed and water level on the inlet side of the sieve (Rusten *et al.*, 2017).

Figure 1.10 shows the TSS removal efficiency as a function of the sieve rate. Data are plotted for four different influent TSS concentration ranges as well as the three different water levels on the inlet side of the sieve. High influent TSS concentrations resulted in higher % removal of TSS than low influent TSS concentrations. More than 40% removal of TSS and more than 30% removal of COD were achieved for all test conditions at sieve rates between 140 and 160 $m^3 \cdot m^{-2}$ submerged belt area $\cdot$ h^{-1}.

The maximum sieve rate tested was 288 $m^3 \cdot m^{-2} \cdot h^{-1}$ and the maximum particle load was 80 kg TSS $\cdot$ $m^{-2} \cdot h^{-1}$. For all sieve rates above approximately 170 $m^3 \cdot m^{-2} \cdot h^{-1}$, the maximum water level of about 250 mm was needed to push the water through the sieve.

Table 1.1 shows key data for the test runs marked A, B and C in Figure 1.10. They are used to demonstrate the importance of the filter mat. Run A is an example of a very thin filter mat with visually very few particles on the belt. The water level on the inlet side of the sieve was only 50 mm and the belt speed was as high as 7.5 m $\cdot$ min^{-1}, even though the sieve rate was only 63 $m^3 \cdot m^{-2} \cdot h^{-1}$. This resulted in a filter mat of only 5.3 g TSS $\cdot$ m^{-2} belt area and the removal of TSS was only 22.4%.

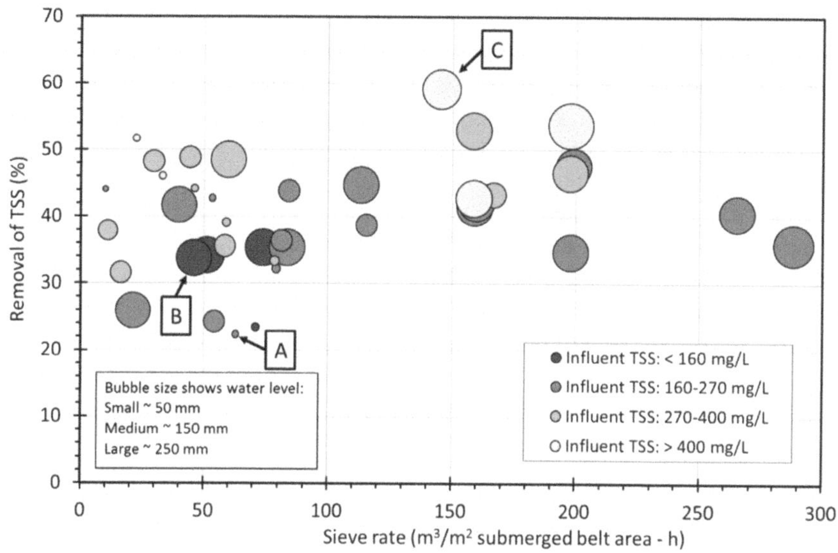

Figure 1.10 SF2000 primary treatment at Nedre Romerike WWTP. Shows removal of TSS as a function of sieve rate, influent TSS concentration, and water level on the inlet side of the sieve. Further data for the three selected test runs marked A, B and C are shown in Table 1.1 (Rusten *et al.*, 2017).

Table 1.1 SF2000 primary treatment at Nedre Romerike WWTP. Key data for test runs marked A, B and C in Figure 1.10 (Rusten *et al.*, 2017).

	Run A	Run B	Run C
Sieve rate, $m^3 \cdot m^{-2}$ submerged belt area $\cdot h^{-1}$	63	46	146
Water level on inlet side of sieve, mm	50	248	271
Belt speed, $m \cdot min^{-1}$	7.5	0.3	12.0
Particles deposited on belt (filter mat), g TSS $\cdot m^{-2}$	5.3	86.7	53.4
Influent concentration, g TSS $\cdot m^{-3}$	204	121	534
Removal efficiency for TSS, %	22.4	33.7	59.2
Removal efficiency for total COD, %	18.7	23.6	39.0

Run B is an example of a very thick filter mat. The water level on the inlet side was 248 mm and the belt moved very slowly, at a belt speed of only 0.3 $m \cdot min^{-1}$. This resulted in a filter mat of 87 g TSS $\cdot m^{-2}$ belt area. Together with a low sieve rate of 46 $m^3 \cdot m^{-2} \cdot h^{-1}$ very good removal of TSS should be expected (Rusten & Ødegaard, 2006), but the removal efficiency was only 33.7%. There are two reasons for this, firstly the very low influent concentration of 121 g TSS $\cdot m^{-3}$ (the lowest of all the 40 test runs), and secondly the low water flow resulting in a very high proportion of the water going through the lower part of the belt where there is no or only a very thin filter mat.

Run C is an example of a relatively thick filter mat, and similar filter mats can be seen in the far-right photo in Figure 1.3 and in Figure 1.6. The water level on the inlet side was 271 mm and the sieve rate was as high as 146 $m^3 \cdot m^{-2} \cdot h^{-1}$. In spite of the belt moving at the maximum speed of 12 $m \cdot min^{-1}$ the filter mat was found to be as high as 53 g TSS $\cdot m^{-2}$ belt area. This was possibly due to the combination of a high flow rate and a very high influent concentration of 534 g TSS $\cdot m^{-3}$, the highest of all the 40 test runs. The removal efficiency for TSS was 59.2%. This was partly due to the high influent concentration, and partly due to the high flow rate that forced a significant fraction of the water to pass through the upper part of the sieve cloth (with a thicker filter mat) because the lower part of the sieve cloth rapidly reached the maximum hydraulic capacity for a clean cloth.

The achieved removal efficiencies, and the observations of how influent concentrations and RBS operational parameters influenced the removal efficiencies, were in agreement with the results found by Franchi *et al.* (2012) during a demonstration-scale operation of an RBS unit at a WWTP in California, USA.

The Nedre Romerike WWTP has decided to replace existing primary clarifiers with RBS primary treatment. Salsnes Filter type SFK600 sieves are being installed for a maximum design flow of 5040 $m^3 \cdot h^{-1}$. The goal is to operate the sieves to achieve 40–50% removal of TSS, and this can be done by the proper combination of belt speed and number of sieves in operation.

1.4.2 Chemically enhanced primary treatment in RBS

Chemically enhanced primary treatment (CEPT) can be used to increase the removal of TSS and particulate COD, either where the wastewater is not suitable for primary treatment with fine mesh sieves (Rusten & Ødegaard, 2006) or where the goal is to maximize the particle removal. When chemical precipitation of phosphorus is not an objective, the simplest form of CEPT is to add a small amount of polymer directly upstream of the RBS. This has successfully been done at full-scale primary treatment RBS plants to improve the removal efficiencies. However, a systematic approach has been taken in a R&D project, using a pilot-plant with controlled chemical dosing and mixing.

1.4.2.1 CEPT testing at the Nordre Follo WWTP

Experiments were conducted at the Nordre Follo WWTP (Ås, Norway) (Rusten *et al.*, 2017). Wastewater was pumped from the influent channel, just downstream of the 3 mm screens, and to the pilot plant. Tests were performed as a worst-case scenario, since a grinder pump was used to feed the pilot system and thus reduced the particle sizes in the wastewater.

The tank flocculator had a wet volume of 170 L and a variable speed stirrer with 3 blades. For all the tests reported here, the stirrer speed was 60 rpm. The SF500 had a belt angle of 20° and the control system was set to run the sieve with a water depth on the inlet side of 88 mm and a maximum belt speed of 4.3 m · min^{-1}. The submerged belt area was 0.09 m^2. Collected solids were removed by a scraper, followed by an air knife to blow off any residual particles.

Belts with 250-micron and 350-micron openings were tested. Hydraulic retention times in the flocculator varied from 1.6 to 2.8 minutes. The hydraulic loads on the RBS varied from 40 to 68 m^3 · m^{-2} submerged belt area · h^{-1} and were limited by the maximum capacity of the grinder pump.

Some results are shown in Figure 1.11. Typically, a small amount of polymer (low cationic charge, very high molecular weight polyacrylamide) increased the removal efficiency by about 20 percentage points, from 40–50% TSS-removal without polymer to 60–70% TSS-removal with polymer. This increased removal efficiency is very important when the RBS is used as the only treatment step and the plant has to meet the EU requirements for primary treatment. It will be very reassuring for the plant owners that if need be, they can significantly increase the plant performance by adding a small amount of polymer.

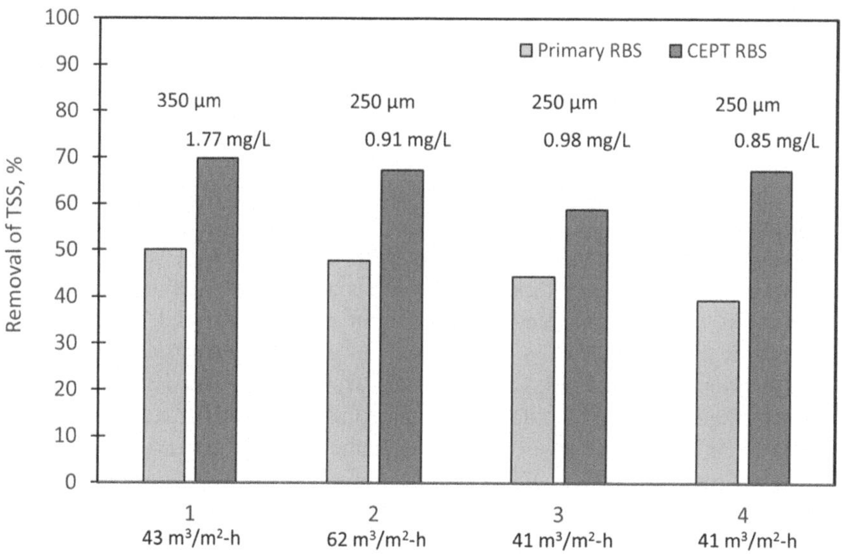

Figure 1.11 Results from pilot-scale CEPT testing with SF500 RBS at the Nordre Follo WWTP. Numbers above columns show openings in belt and polymer dose. Numbers below show test run and hydraulic load on the submerged belt area (Rusten *et al.*, 2017).

The average removal efficiency (66%) for TSS shown for the CEPT process in Figure 1.11 was identical to the removal efficiency for the CEPT process at the Bangsund RBS plant (Rusten & Ødegaard, 2006). Polymer doses were also similar. However, the Bangsund RBS plant was operated at a sieve rate of 25 $m^3 \cdot m^{-2} \cdot h^{-1}$ while the runs in Figure 1.11 were performed at sieve rates from 41 to 62 $m^3 \cdot m^{-2} \cdot h^{-1}$.

The particles scraped off the SF500 belt had a concentration between 5 and 7% TS in most of the test runs. A methane potential test (Bioprocess Control AMPTS, Lund, Sweden) was performed with sieve sludge from primary treatment and CEPT, respectively. The test was run at the mesophilic temperature of 37°C and with sludge collected from run 4 (see Figure 1.11). This test showed that the primary sieve sludge produced 317 NmL $CH_4 \cdot g\ VS^{-1}$, while the CEPT sieve sludge produced 483 NmL $CH_4 \cdot g\ VS^{-1}$. If the sludge goes to an anaerobic digester, this will be a great advantage for the CEPT sludge. It is not clear why the CEPT sludge had so much higher methane potential, but one hypothesis is that the smaller particles captured by the polymer are easier to degrade and contribute more biogas than larger particles.

1.4.2.2 CEPT testing at the Sandefjord and Namsos WWTPs

A portable pilot-plant, mounted inside a trailer, was used at the Sandefjord and Namsos WWTPs. This pilot-plant was equipped with a control panel, auto-samplers, coagulant and polymer dosing system, coagulation and flocculation tanks, and a SF500 RBS. Different mesh sizes were investigated. The belt angle was maintained at 23° with a submerged belt area of 0.06 m^2. The RBS was also equipped with an air knife (AK) connected to a blower, a water knife and a mechanical scraper (SC) to scrape the sludge cake off the mesh into the sludge compartment.

Testing at Sandefjord and initial testing at Namsos were done with 250 μm and 350 μm belts, comparing addition of cationic polymer to no addition of polymer. Results were basically independent of the mesh size, and also independent of the different sieve rates tested (54–94 $m^3 \cdot m^{-2} \cdot h^{-1}$). Flocculation time varied from 3 to 5 minutes. As an example, average results from testing with a 250 μm belt are shown in Figure 1.12. At Sandefjord the air knife was used for removing sludge from the belt, and addition of 0.7 $g \cdot m^{-3}$ cationic polymer increased the removal of TSS by 23% (from 45% to 68%). Removal of total COD was 51% with polymer addition. The Namsos wastewater was periodically heavily polluted with discharge from a local dairy and this resulted in very large variations in RBS performance when no polymer was added. However, as seen in Figure 1.12, very good results (72–73% TSS removal) were achieved with 0.85 $g \cdot m^{-3}$ cationic polymer, both when the air knife and when the scraper was used for removing the sludge cake (Sahu *et al.*, 2017).

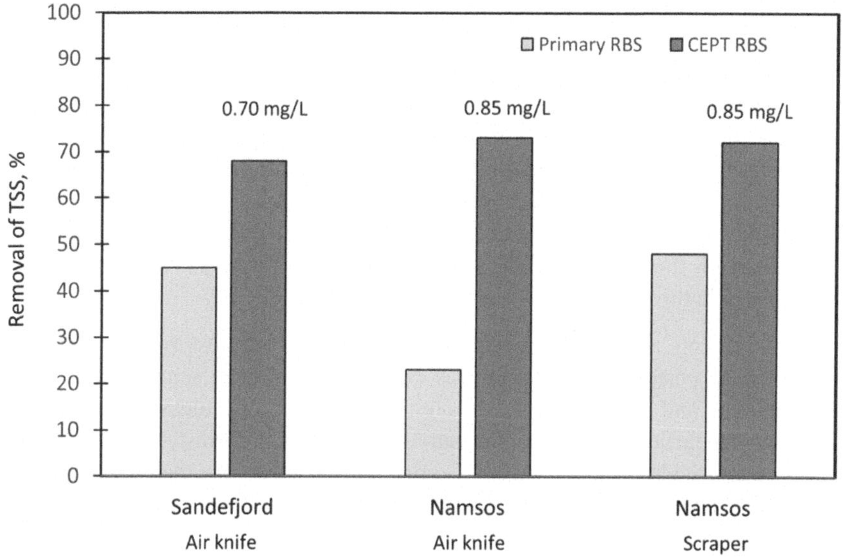

Figure 1.12 Percentage removal of TSS using SF500 RBS with 250 μm mesh, without and with addition of polymer, at Sandefjord and Namsos WWTPs. Primary RBS has no addition of polymer, for CEPT RBS the cationic polymer dose is shown above the column. Air knife or mechanical scraper shows how the sludge cake was removed from the belt. Data from Sahu *et al.* (2017).

At the Sandefjord WWTP increasing the polymer dose from 0.7 to 1.5 $g \cdot m^{-3}$ had no effect on the removal efficiencies (Rathnaweera, 2017). However, when the polymer dose was increased from 0.85 to 1.75 $g \cdot m^{-3}$ at the Namsos WWTP, the removal efficiencies for both TSS and TCOD dropped and the belt speed increased in order to handle the same flow as with the lower polymer dose (Sahu *et al.*, 2018). This indicates that optimum polymer dosing will normally be $<1\ g \cdot m^{-3}$.

A Malvern Mastersizer was used at the Sandefjord WWTP to measure the particle size distribution (PSD), as well as the total particle volume, of influent wastewater, flocculated wastewater (after addition of cationic polymer) and RBS effluent (Rathnaweera, 2017). Area based diameter (ABD) was used to describe the particle size. These measurements were taken for two different runs with 250 μm mesh size at a sieve rate of 90 $m^3 \cdot m^{-2} \cdot h^{-1}$. About 70–75% of the influent particle volumes were smaller than 250 μm. After flocculation with cationic polymer, grab samples showed that 75 to 80% of particle volumes were larger than 250 μm. This explains the significant increase in removal of TSS when adding polymer. However, the PSD measurements also showed that 25–35% of the particle volumes in the

RBS effluent samples were larger than the sieve size of 250 μm. Three different phenomena may have contributed to this (Rathnaweera, 2017):

(a) Remaining particles and polymer residuals continue the flocculation process in effluent water.
(b) Wastewater particles are not spherical and particles that are larger than sieve openings based on ABD can have a dimension smaller than the sieve openings (rods etc.). These particles can easily pass through the filter.
(c) Fluffy particles (and also other particles) may be flexible and can go through the sieve, especially when there is a hydraulic force pressing the particle through the sieve.

Measurements showed 68% removal of TSS for both test runs where PSD analysis were performed, and this was close to the particle removal calculated from influent and effluent particle volumes. Polymer flocculation significantly increased the particle volume, as measured by the Malvern Mastersizer, without a similar increase in particle dry mass. So the density, as mg dry matter per ml, was significantly lower for the flocculated particles than for the particles in the influent wastewater (Rathnaweera, 2017).

During the second period of testing at the Namsos WWTP coagulants (Al or Fe based) were added, either alone or in combination with an anionic polymer, to see how much phosphorus can be removed. PO_4-P is easily precipitated, but this adds more particles to the wastewater entering the RBS and a very high removal of TSS is needed in order to get good removal of total phosphorus (TP). Belts with mesh sizes from 250 to 40 μm were tested. Despite large variations in wastewater quality, performance tests with 90 μm mesh size indicated reliable removal of 60–65% TP and 70–75% TSS at reasonable coagulant doses (12 g $Al \cdot m^{-3}$, 2.1 mol Al/mol P) and high hydraulic load (90 $m^3 \cdot m^{-2} \cdot h^{-1}$) based on submerged belt area. Reported removal efficiencies are based on influent concentrations prior to coagulant dosing, so actual particle removal over the RBS will be higher. Sludge cake removal mechanisms, air knife or mechanical scraper, did not have a clear influence on the particle removal efficiency (Sahu *et al.*, 2018).

1.4.3 Sludge from rotating belt sieves

Paulsrud *et al.* (2014) carried out a survey of the sludges from primary treatment plants in Western and Northern Norway, where the discharge requirement follows the EU primary treatment criteria. As the basis for sludge characterization, grab samples of sludge were collected from 19 full-scale primary treatment plants employing fine mesh sieves (all rotating belt sieves of the Salsnes Filter type, see Figure 1.1). At those plants the separated sludge is dewatered in an integrated screw press, and all the samples were taken after dewatering. Only five of those plants had grit chambers installed ahead of the RBS and none had fine screens ahead of the sieves. This means that the sieve sludge samples contained more debris than what would be normal in a more advanced treatment plant with fine screening and a grit chamber ahead of the RBS, where the RBS would be a substitute for primary clarifiers.

The results revealed a TS content of 13.6–36.9% TS (mean = 27.3% TS). The mean VS content of the dewatered sieve sludges was 91.6% of TS, significantly higher than the VS content (80.8% of TS) of the sludge from the primary clarifier step of full-scale wastewater treatment plants that were also analysed.

The high VS content of the RBS sludge as compared to the primary clarifier sludge, would indicate a higher methane potential in anaerobic digesters for the RBS sludge. This was supported by bio-methane potential (BMP) tests showing a mean value of 345 NmL $CH_4 \cdot g\ VS^{-1}$ for RBS sludge and 287 NmL $CH_4 \cdot g\ VS^{-1}$ for primary clarifier sludge. These benefits for RBS sludge, over sludge from primary clarifiers, were confirmed in a recent study by Ho *et al.* (2016).

When incinerating sludge or using sludge as a fuel in cement kilns, the caloric value (CV) is of great importance. The CV-analysis of the RBS sludge demonstrated CVs at a high level (about 19 $MJ \cdot kg\ TS^{-1}$), indicating that the RBS sludge would be attractive for this purpose. The reader is referred to Paulsrud *et al.* (2014) for more details about this survey.

Sludge recovery from primary wastewater using coagulation/flocculation by polymer dosage ahead of RBS technology was investigated at three full-scale WWTPs in Norway (Sahu *et al.*, 2017; Sahu & Rathnaweera, 2017). The objective was to test the BMP of the sludge obtained from RBS with and without polymer addition to the municipal wastewater at the Nordre Follo, Sandefjord and Namsos WWTPs. For the tests at Nordre Follo WWTP, a tank flocculator was tested in combination with the pilot scale SF500 filter. A belt with 250-micron openings was investigated for these tests. A well-equipped pilot unit with a polymer-dosing station, a flocculator and a SF500 was used at Sandefjord and Tiendeholmen WWTPs. Belts with 250 μm and 350 μm openings were used in both tests.

A summary of the specific BMP results at these three test sites is given in Table 1.2. Addition of polymer resulted in 1.5 to 2.9 times more biogas than for tests without polymer addition. The increased biogas production at the Nordre Follo WWTP was partly due to a higher specific BMP, and partly due to the higher removal of TSS. For the Namsos WWTP the increased biogas production was primarily due to the higher TSS removal when polymer was added.

Most samples in Table 1.2 had a BMP between 330 and 400 NmL $CH_4 \cdot g\ VS^{-1}$, even though there was a tendency to be in the upper BMP range for sludge samples obtained when polymer was used. However, the biggest advantage with regard to biogas production is that CEPT significantly increased the removal of organic particles and thus increased the total amount of primary VS available for gas production.

Table 1.2 Comparison of bio-methane potential (BMP) at different WWTPs using sludge from Salsnes Filter RBS. Sludge cake removal with either air knife (AK) or scraper (SC). Data from Sahu & Rathnaweera (2017).

WWTP	Scenario	BMP NmL $CH_4 \cdot g\ VS^{-1}$	TSS Removal %	Ratio for net CH_4 Production (polymer/no polymer)
Nordre Follo	No Polymer – AK	317	40	2.5
	0.85 $g \cdot m^{-3}$ polymer – AK	483	67	
Sandefjord	0.75 $g \cdot m^{-3}$ polymer – AK	391	68	–
Namsos	No Polymer – AK	334	23	2.9
	0.85 $g \cdot m^{-3}$ polymer – AK	333	73	
	No Polymer – SC	350	48	1.5
	0.85 $g \cdot m^{-3}$ polymer – SC	389	72	

A simple economic calculation showed the cost of polymer for RBS CEPT to be insignificant compared to the benefits obtained in higher solids removal, more biogas generation, less particles in downstream processes and reduced oxygen consumption in downstream bioreactors (Sahu *et al.*, 2017). The sludge cake blown off or scraped from the belt had 8–10% TS at the Sandefjord WWTP and 7–12% TS at the Namsos WWTP (Rathnaweera, 2017; Sahu *et al.*, 2017). This is close to ideal for use in anaerobic sludge digesters, and there is no need for thickening of this sludge.

1.5 IMPACT OF RBS PRIMARY TREATMENT ON NITROGEN REMOVAL

It has been argued that RBSs will remove too much of the carbon that might be available as carbon source in pre-denitrification processes. However, relatively large particles that are removed by RBS are slowly biodegradable, so the question is how far down in particle size is the optimum particle removal before any negative effects are seen in biological N- and P-removal processes. Based on a literature study, Newcombe *et al.* (2011) expected the optimum particle size cut-off in front of standard biological nutrient removal processes to be in the 15–20 μm range.

A three-year R&D project was initiated in 2012 to find the optimum particle size cut-off for particle removal in front of biological nitrogen removal processes with pre-denitrification. Removal of too many particles may reduce the carbon-to-nitrogen (C/N) ratio to the point where the nitrogen removal is affected due to reduced denitrification, while not removing enough particles may negatively affect the nitrification process. Some initial short-term tests were performed using wastewater from two municipal wastewater treatment plants that were tested separately in both small-scale activated sludge reactors and small-scale biofilm reactors (Razafimanantsoa *et al.,* 2014a). This was followed by longer-term tests (Razafimanantsoa *et al.,* 2014b) with activated sludge sequencing batch reactors (SBRs). For the wastewaters tested, the results showed that sieves with 33 μm mesh size provided the optimum primary treatment (with respect to nitrogen removal) when operated without a filter mat on the sieve cloth.

The next investigation step consisted of side-by-side continuous flow testing of nitrogen removal, where one train received screened wastewater and the other train received wastewater that had passed through a 33 μm fine mesh RBS (Salsnes Filter). Tests were completed with both MBRs (membrane bio-reactors) and MBBRs (moving bed biofilm reactors). Only the tests with MBBR (Rusten *et al.,* 2016) have been published so far, but a paper with results from the MBR tests is in preparation (Razafimanantsoa *et al.,* 2018).

The two MBBR pilot-plants and the two MBR pilot-plants were located at the Nordre Follo WWTP, south of Oslo, Norway. The Nordre Follo WWTP is a large MBBR plant with combined pre- and post-denitrification. Coarse screened wastewater was pumped from just downstream of the screens at the full-scale plant and through either a 2 mm screen or a Salsnes Filter RBS with 33 μm sieve cloth to storage tanks from where it was fed to the four pilot-scale plants. The RBS was operated without a filter mat in order to prepare wastewater for the tests, and it removed on average 41% of TSS, 31% of total COD, 12% of total N and 14% of total P. Fresh batches of wastewater were prepared three times per week and stored in tanks with mechanical mixers to prevent settling of particulate matter.

1.5.1 Impact on MBBR

The MBBR lab-scale plants for nitrogen removal were operated in parallel where each train had four reactors in series, where reactors 1 and 2 were anoxic, and reactors 3 and 4 were aerobic. Anoxic reactors had mechanical mixers and aerobic reactors had diffusers for aeration at the bottom of the tanks. All reactors had 60% fill of the Kaldnes K1 biofilm carriers, resulting in a protected biofilm surface area of 300 $m^{-2} \cdot m^{-3}$ of wet reactor volume. All anoxic reactors had a wet volume of 4.0 L and aerobic reactors had a wet volume of 6.0 L, resulting in 40% anoxic volume and 60% aerobic volume. Nitrified effluent was recycled from reactor 4 to reactor 1 at approximately two times the influent flow.

The overall results are shown in Table 1.3, with the 2 mm screened influent wastewater, the 33 μm sieved wastewater and the effluents from reactors 4 in Train A and Train B, respectively. This means biological treatment for Train A, and primary plus biological treatment for Train B. The MBBR plants did not have final solids separation stages, so when analysing overall removal of organic matter and nitrogen, effluent concentrations measured on filtered (GF/C) samples were used. Based on this, average removal for total COD was 91% for both Train A and Train B. Average removal of total N was 68% for Train A (2 mm screen) and 66% for Train B (33 μm sieve). Average removal of total P was 70% for both trains. P was removed both as particulate P and as P assimilated in to the produced biomass.

Table 1.3 Overall results for the two pilot-scale MBBR plants. Shows averages and standard deviations. Data from Rusten *et al.* (2016).

Parameter	2 mm Screen, Train A		33 μm Sieve, Train B	
	Influent ($g \cdot m^{-3}$)	Effluent ($g \cdot m^{-3}$)	Influent ($g \cdot m^{-3}$)	Effluent ($g \cdot m^{-3}$)
TSS	281 ± 96	204 ± 85	165 ± 70	137 ± 76
Total COD	521 ± 129	302 ± 106	355 ± 93	209 ± 90
Filtered COD	168 ± 53	45 ± 9	143 ± 42	43 ± 10
BSCOD*	118 ± 45	–	98 ± 37	–
Total N	44.0 ± 11.9	23.2 ± 5.2	38.9 ± 11.1	23.0 ± 6.4
Filtered TN	33.6 ± 9.9	15.1 ± 6.3	32.2 ± 9.6	16.1 ± 5.8
NH_4-N	31.7 ± 10.7	1.3 ± 2.3	30.4 ± 10.3	1.1 ± 2.6
NO_2-N	0.03 ± 0.02	0.21 ± 0.19	0.04 ± 0.05	0.19 ± 0.26
NO_3-N	0.41 ± 0.2	10.9 ± 4.2	0.41 ± 0.2	12.5 ± 4.1
Total P	4.1 ± 1.4	3.3 ± 1.5	3.5 ± 1.1	3.4 ± 0.99
Filtered TP	1.7 ± 0.6	1.1 ± 0.6	1.3 ± 0.6	1.1 ± 0.6

*Biodegradable soluble COD.

Table 1.3 confirmed that, for the wastewater characteristics at the test plant, Salsnes Filter primary treatment with a 33 μm RBS and no filter mat, produced a primary effluent that was close to the optimum with respect to how much organic matter that is removed. Removal of organic matter with the 33 μm sieve had no significant, negative effect on the denitrification process. Nitrification rates improved by 10–15% in the train with 33 μm RBS primary treatment. Mass balance calculations showed that without RBS primary treatment, the oxygen demand in the biological system was 36% higher.

The average sludge production for the two treatment trains were 199 $g\ TSS \cdot m^{-3}$ for Train A (2 mm screen) and 244 $g\ TSS \cdot m^{-3}$ (114 $g\ TSS \cdot m^{-3}$ primary + 130 $g\ TSS \cdot m^{-3}$ biological) for Train B (33 μm sieve). The specific sludge yield for biological sludge was 0.45 $g\ TSS \cdot g\ TCOD^{-1}$ removed for both trains. Using SF fine mesh sieves increased the total sludge production by about 20–25%. However, almost half of the sludge was primary sludge, which can dewater to a high solids concentration and has a significantly higher methane gas potential (Gavala *et al.*, 2003) than biological sludge.

1.5.2 Impact on MBR

The two pilot-scale MBRs were operated in parallel. Each MBR train was composed of two anoxic reactors of 10 L, equipped with a mechanical mixer rotating at about 215 rpm, one aerobic reactor of 25 L, and a submerged hollow fibre membrane ZeeWeed-10 (ZW10, Zenon Environmental Systems Inc., Oakville, Ontario, Canada) with a 40 nm nominal pore size. Nitrified activated sludge was recycled from Reactor 3 to Reactor 1 at twice the flow of the influent wastewater. The membrane ZW10 was operated at normal forward flow for 9.5 minutes and backwashed for 0.5 minute at twice the permeate flow. All pumps (feed, recycle and permeate) were controlled by a programmable logic controller (PLC). A pressure transmitter and a dissolved oxygen probe were also connected to the PLC to record continuously the transmembrane pressure and dissolved oxygen in the aerobic reactors.

The MBRs were operated at an influent flow rate of about $5\ L \cdot h^{-1}$ and a recycle flow rate of $10\ L \cdot h^{-1}$. The hydraulic retention time was 9 h. The mixed liquor suspended solids (MLSS) in the anoxic reactors was about $5\ kg \cdot m^{-3}$, while the MLSS in the aerobic reactors were maintained at around $7\ kg \cdot m^{-3}$. The operating temperature varied between 16 and 21°C and the pH was maintained around 7. The dissolved oxygen (DO) concentration in the aerobic zones was around $4\ kg \cdot m^{-3}$. The membrane permeate flux was $6.3–6.4\ L \cdot m^{-2} \cdot h^{-1}$ in both trains. Over the 96 days of active testing the transmembrane pressure (TMP) was very low, with 46 ± 9 mbar in Train A (2 mm screen) and 26 ± 7 mbar in Train B (33 μm sieve). Even though the TMP was higher in the train treating 2 mm screened wastewater, no chemical cleaning of the membrane was performed throughout the experiment as the TMPs were well below the 300 mbar maximum limit recommended by the membrane supplier.

The overall results are summarized in Table 1.4, with the 2 mm screened influent wastewater, the 33 μm sieved wastewater and the effluents (membrane permeate) from Train A and Train B, respectively. This means biological treatment for Train A, and primary plus biological treatment for Train B.

Final effluent concentrations were very similar for the two trains. With a 40 nm membrane effluent TSS should per definition be zero, and was well below the detection limit. Based on these averages removal for total COD was 94% and removal of total N was 74% for both Train A and Train B. Average removal of total P was 81% for Train A (2 mm screen) and 83% for Train B (33 μm sieve). P was removed both as particulate P and as P assimilated in to the produced biomass. The results confirmed that the selective removal of particulate organic matter with a 33 μm sieve with no filter mat was close to the optimum, and did not negatively affect the performance for nitrogen removal in the pilot-scale MBR (Razafimanatsoa *et al.*, 2018).

Table 1.4 Overall results for the two pilot-scale MBR plants. Shows averages and standard deviations. Data from Razafimanatsoa *et al.* (2018).

Parameter	2 mm Screen, Train A		33 μm Sieve, Train B	
	Influent ($g \cdot m^{-3}$)	Effluent ($g \cdot m^{-3}$)	Influent ($g \cdot m^{-3}$)	Effluent ($g \cdot m^{-3}$)
TSS	275 ± 99	$\ll 1$	161 ± 69	$\ll 1$
Total COD	522 ± 134	32.4 ± 7.2	349 ± 88	32.1 ± 7.5
Filtered COD	168 ± 47	29.7 ± 7.3	145 ± 35	29.1 ± 7.7
Total N	43.2 ± 12.0	11.3 ± 3.5	38.7 ± 11.8	11.3 ± 2.5
NH_4-N	31.5 ± 10.5	0.24 ± 0.74	30.1 ± 9.9	0.32 ± 1.37
NO_2-N	0.03 ± 0.01	0.14 ± 0.22	0.03 ± 0.05	0.13 ± 0.19
NO_3-N	0.41 ± 0.15	8.63 ± 3.0	0.40 ± 0.15	10.0 ± 3.2
Total P	4.26 ± 1.51	0.82 ± 0.62	3.70 ± 1.25	0.74 ± 0.55
PO_4-P	1.51 ± 0.60	0.57 ± 0.44	1.09 ± 0.65	0.55 ± 0.52

In Train A (2 mm screen), the overall sludge produced from the system was only composed of biosludge. In Train B (33 μm sieve), the total sludge production was a combination of RBS sludge and biosludge. The biosludge production in Train B was only about half of the biosludge produced in Train A. However, the total amount of sludge produced in Train B was about 16% higher compared to that of Train A because of the primary sludge from the RBS. The specific biological sludge yield was slightly lower in Train B (0.31 g TSS $\cdot$ g COD^{-1}) than in Train A (0.34 g TSS $\cdot$ g COD^{-1}), which was probably due to the longer solids retention time of about 16.8 d in Train B compared to 13.7 d in Train A. The reduction of organic matter prior to biological treatment also allowed a decrease of the oxygen demand by about 30%. Therefore, removal of particles with the 33 μm sieve was beneficial for the biological process (Razafimanantsoa *et al.*, 2018).

1.5.3 Operation of RBS in front of biological nitrogen removal process

Lab- and pilot-scale testing has demonstrated that the optimum particle removal prior to biological nitrogen removal is to use a 33 μm mesh size sieve with no filter mat on the sieve (Razafimanantsoa *et al.*, 2014a,b, 2018; Rusten *et al.*, 2016). However, this is not practical for operation of RBSs at large WWTPs. Typically mesh sizes of 250 μm or 350 μm are used at municipal WWTPs, and the sieves are operated with a sludge cake on the belt that will remove a large fraction of the particles that are smaller than the openings in the mesh. This larger mesh will be much stronger, easier to clean and have a larger hydraulic capacity than a 33 μm mesh.

The question is if it is possible to operate an RBS with a filter mat on a 250 μm or 350 μm belt in such a way that we get a removal of particles that is similar to using a 33 μm belt without a filter mat. The TSS removal efficiency and effluent particle size distribution (PSD) from an RBS with filter mat on the belt can be manipulated by changing operational parameters like flow, belt speed, water level and belt cleaning procedure. A preliminary test was performed at the Nordre Follo WWTP, using a SF1000 RBS with 33 μm and 350 μm belts. A Malvern Mastersizer was used to measure the PSD of the effluent (Rusten *et al.*, 2014).

Figure 1.13 shows an example of PSD curves for three test runs, where the PSD from a run with a 33 μm mesh and no filter mat is compared to two runs with 350 μm mesh operated at different belt speeds. The belt speed is proportional to the Hz setting for the belt motor. It can be seen that at 30 Hz the shape of the PSD curve was very similar to the shape of the PSD curve for the 33 μm belt with no filter mat, but slightly less TSS was removed with the 350 μm belt. At 20 Hz belt speed the filter mat was thicker, and slightly more TSS was removed with the 350 μm belt than with the 33 μm belt and no filter mat.

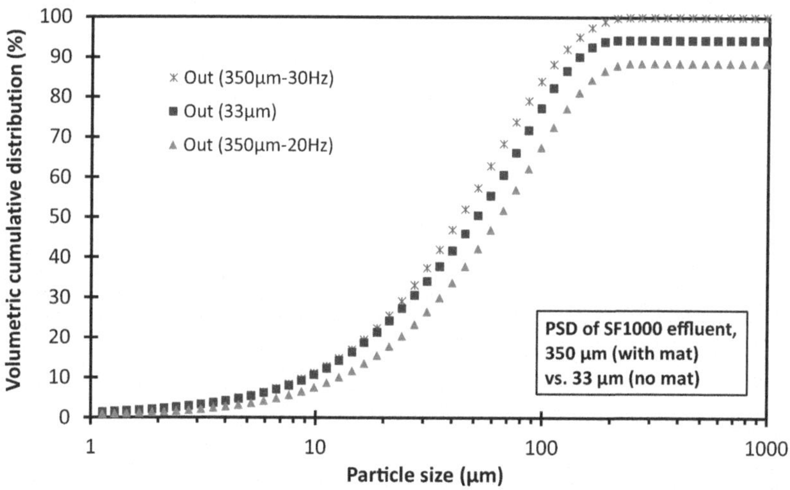

Figure 1.13 Example of PSD curves for three test runs with SF1000 RBS. A 33 μm belt operated with no filter mat and 350 μm belt operated at belt motor speeds of 20 Hz and 30 Hz.

The preliminary results in Figure 1.13 indicate that it will be possible to operate RBSs with filter mats in such a way that they produce a primary effluent that will be close to optimum for downstream biological nitrogen removal. A belt speed of 25 Hz would most likely have produced a PSD curve very close to the curve for the 33 μm belt with no filter mat. However, since influent wastewater characteristics constantly change it is a challenge to come up with good control algorithms for these sieves.

1.6 CONCLUSIONS

Rotating belt sieves (RBSs) may effectively be used for compact primary treatment of wastewater. Most importantly they will remove relatively large, slowly biodegradable particles. Normally the separating efficiency is such that the defined EU removal efficiencies for primary treatment (50% removal of TSS and 20% removal of BOD_5) is achieved, if the RBSs are properly built and equipped, properly designed, and operated with a filter mat. This is achieved by operating the belt at low speed or discontinuously.

Design sieve rates should be established by screening tests (as described above) and will range from 20 $m^3 \cdot m^{-2} \cdot h^{-1}$ to about 300 $m^3 \cdot m^{-2} \cdot h^{-1}$, depending on wastewater characteristics and required removal efficiencies. To meet the EU primary treatment requirements, sieve rates will normally be below 200 $m^3 \cdot m^{-2} \cdot h^{-1}$.

To be considered suitable for primary treatment with fine mesh sieves the screening tests indicated that at least 20% of the TSS in the wastewater should consist of particles larger than 350 µm. However, cationic polymer in combination with rotating belt sieves has successfully treated wastewater that was originally classified as unfavourable for fine mesh sieves.

A sieve opening in the range of 250–500 microns will normally be the appropriate choice for typical municipal wastewater, but this should be determined after screening tests. Once a filter mat is formed on the sieve, there is practically no difference in the performance of sieve cloths within this size range, with regard to both % TSS removal and filtration rate. A mesh size of 350 µm is most commonly used for primary treatment. Removal efficiencies and hydraulic capacities have been demonstrated to be dependent on influent concentrations, water level on the inlet side of the sieve and belt speed. Creation of a filter mat on the belt is important for high removal efficiencies. Test results showed up to 60% TSS removal with a good filter mat, and less than 25% TSS removal with a very thin filter mat. Best results were achieved with maximum water level on the inlet side of the sieve and sieve rates in the range of 140 to 160 $m^3 \cdot m^{-2} \cdot h^{-1}$.

It has been argued that the use of RBS reduces the capacity to denitrify because of the removal of organic matter. However, pilot studies demonstrated that even with a sieve as fine as 33 µm (operated without a sludge mat), the removal of organic matter caused by the sieve had no negative effect on the denitrification process and overall N-removal.

Chemically enhanced primary treatment (CEPT) is a simple and effective way of increasing the removal efficiency of an RBS. Pilot-scale SF500 RBS tests with 250 or 350 µm belts and sieve rates from 40 to 94 $m^3 \cdot m^{-2} \cdot h^{-1}$, demonstrated that the removal of TSS typically increased from 40–50% without polymer to 60–75%

with polymer (with 0.7–1 g · m^{-3} of polymer and ~2 min of flocculation time). The total bio-methane potential (BMP) of the recovered sludge was significantly higher with polymer than with no polymer. A simple economic calculation shows the cost of polymer with RBS to be insignificant compared to the benefits obtained in higher solids removal, more biogas generation, and less particles in downstream processes.

Sludge scraped or blown off the belt has typically 5 to 7% TS, but has been as high as 12% TS. The most commonly used RBS (Salsnes Filter) is, however, equipped with a screw press for the sludge that typically dewaters the sludge to about 25–30% TS.

The high VS content of RBS sludge as compared to primary clarifier sludge indicates a higher methane potential in anaerobic digesters for the RBS sludge and this was supported by batch bio-methane potential (BMP) showing a mean value of 345 NmL CH_4 · g VS^{-1} for RBS sludge and 287 NmL CH_4 · g VS^{-1} for primary clarifier sludge.

When incinerating sludge or using sludge as a fuel in cement kilns, the caloric value (CV) is of great importance. The CV-analysis of the RBS sludge demonstrated CVs at a high level (about 19 MJ · kg TS^{-1}), indicating that the RBS sludge would be attractive for this purpose.

1.7 REFERENCES

Franchi A. and Santoro D. (2015). Current status of the rotating belt filtration (RBF) technology for municipal wastewater treatment. *Water Practice and Technology*, **10**(2), 319–327.

Franchi A., Stedman K. and Mitchell C. (2012). Demonstration-scale of primary solids harvesting for energy recovery at the UC Davis Wastewater Treatment Plant. *Presented in Session 78 at WEFTEC 2012*, September 29–October 3, New Orleans, LA. Published in proceedings, 5699–5714.

Gavala H. N., Yenal U., Skiadas I. V., Westermann P. and Ahring B. K. (2003). Mesophilic and thermophilic anaerobic digestion of primary and secondary sludge. Effect of pre-treatment at elevated temperature. *Water Research*, **37**, 4561–4572.

Ho D., Gupta M., Sarathy S., Batstone D., Nakhla G. and Santoro D. (2016). Primary treatment to optimize secondary biological processes and anaerobic digestion. *Proceedings of the IWA World Water Congress & Exhibition*, October 9–14, Brisbane, Australia.

Misund A. K., Fjorden S. and Jacobsen J. (2004). Experiences From Primary Treatment Plants Using Sedimentation. (In Norwegian). Report from Asplan Viak Sør, December 10.

Newcombe R., Lopp M., Higby T., Martini R. and Rusten B. (2011). Particle size separation implications on COD removal before BNR: a case study. *Proceedings from WEF-IWA Specialized Conference "Nutrient Recovery and Management 2011"*, Miami, Florida, pp. 574–581.

Ødegaard H. (1992). Norwegian experiences with chemical treatment of raw wastewater. *Water Science and Technology*, **25**(12), 255–264.

Ødegaard H. (2005). Primary Treatment. (In Norwegian). Norwegian State Pollution Control Agency, SFT TA-2088/2005.

Ødegaard H. (2016). A road-map for energy-neutral wastewater treatment plants of the future based on compact technologies (including MBBR). *Frontiers of Environmental Science & Engineering* **10**(4), 2. DOI 10.1007/s11783-0835-0.

Paulsrud B. (2005). Handling of Sludge from Primary Treatment Plants. (In Norwegian). Aquateam report no. 04-047, January.

Paulsrud B., Rusten B. and Aas B. (2014). Increasing the sludge energy potential of wastewater treatment plants by introducing fine mesh sieves for primary treatment. *Water Science and Technology*, **69**(3), 60–65.

Rathnaweera S. (2017). Pilot Scale Tests on Flocculation and Particle Separation Using SF500 for Chemically Enhanced Primary Treatment (CEPT) of Municipal Wastewater - Activity 8: Sandefjord Wastewater Treatment Plant. Aquateam COWI report 16-006.

Razafimanantsoa V. A., Ydstebø L., Bilstad T., Sahu A. K. and Rusten B. (2014a). Effect of selective organic fractions on denitrification rates using Salsnes filter as primary treatment. *Water Science and Technology*, **69**(9), 1942–1948.

Razafimanantsoa V. A., Ydstebø L., Bilstad T., Sahu A. K. and Rusten B. (2014b). Impact of selective size distribution of influent suspended solids on downstream biological processes. *Proceedings, A069. IWA Conference on Pretreatment of Water and Wastewater*, 18–21 May, Shanghai, China.

Razafimanantsoa V. A., Adyasari D., Sahu A. K., Rusten B., Bilstad T., Ydstebø L. (2018). Pilot-Scale Study to Investigate the Impact of Rotating Belt Filter Upstream of a MBR for Nitrogen Removal. Paper in preparation.

Rusten B. (2004). Development of a Simple Test Apparatus and Procedure to Predict the Full Scale Treatment Performance of Salsnes Filter Fine Mesh Sieves – Preliminary Testing. Aquateam report no. 04-002, January.

Rusten B. and Lundar A. (2004). Primary Treatment Project 10: Use of Chemical Conditioning Prior to Fine Mesh Sieve Treatment. (In Norwegian). Aquateam report 04–048, December.

Rusten B. and Ødegaard H. (2006). Evaluation and testing of fine mesh sieve technologies for primary treatment of municipal wastewater. *Water Science and Technology*, **54**(10), 31–38.

Rusten B., Paulsrud B. and Razafimanantsoa V. A. (2014). Physical processes: micro-sieves and micro-filters for carbon control. *Presented at Workshop W02: Knowledge Development Forum: Carbon Removal and Recovery Technologies for Sustainable, Energy-Efficient Water Resources Recovery Facilities*, September 27, 2014, WEFTEC, New Orleans.

Rusten B., Razafimanantsoa V. A., Andriamiarinjaka M. A., Otis C. L., Sahu A. K. and Bilstad T. (2016). Impact of fine mesh sieve primary treatment on nitrogen removal in moving bed biofilm reactors. *Water Science and Technology*, **73**(2), 337–344.

Rusten B., Rathnaweera S. S., Rismyhr E., Sahu A. K. and Ntiako, J. (2017). Rotating belt sieves for primary treatment, chemically enhanced primary treatment and secondary solids separation. *Water Science and Technology*, **75**(11), 2598–2606.

Sahu A. K. and Rathnaweera S. S. (2017). Biogas Potential from Sludge Produced Using SF500 for Chemically Enhanced Primary Treatment (CEPT) of Municipal Wastewater – Activity 10. Aquateam COWI report 17-010.

Sahu A. K., Rathnaweera S. S., Szatkowska B. and Rusten, B. (2017). Enhanced filtration with rotating belt filter for higher methane potential from primary wastewater sludge. *Proc. IWA Specialist Conference on Sludge Management, SludgeTech 2017, Theme 5, Imperial College*, July 9–13, London, UK.

Sahu A. K., Rathnaweera S., Lyng T. O. and Rusten B. (2018). Chemically enhanced primary treatment of combined dairy and municipal wastewaters using rotating belt filter. *Extended abstract submitted to 4th IWA Specialized International Conference – Ecotechnologies for Wastewater Treatment*, June 25–27, London, Ontario, Canada.

Sutton P. M., Rusten B., Ghanam A., Dawson R. and Kelly H. (2008). Rotating belt screens: an attractive alternative for primary treatment of municipal wastewater. *Presented in Session 24 at WEFTEC 2008*, October 18–22, Chicago, IL. Published in proceedings, 1671–1687.

WERF (2014). Nutrient Challenge. Deammonification. In: J. B. Neethling (ed.), Water Environment Research Foundation (WERF) 2014, http://www.werf.org.

Chapter 2

Biological nutrient removal activated sludge systems with membranes

V. Parco[1], G. J. G. du Toit[2] and G. A. Ekama[2]

[1]Hydraulic and Environmental Engineering Department, University of Palermo, 90100, Italy

[2]Water Research Group, Department of Civil Engineering, University of Cape Town, Rondebosch, Cape Town, South Africa (george.ekama@uct.ac.za)

ABSTRACT

The impact of including membranes for solid-liquid separation and high volatile suspended solids (VSS) concentration (3–12 gVSS/ℓ) on the kinetics of biological nitrogen and phosphorus removal was investigated. To achieve this, a membrane bioreactor (MBR) biological nutrient removal activated sludge (BNRAS) system was operated for 450 days in parallel with a conventional BNR system with a secondary settling tank (CAS, SST). The influence of high VSS concentration (up to 12 gVSS/ℓ) in the MBR system on the system performance and kinetic rates was measured in 85 batch tests on sludge harvested from the MBR and CAS systems.

The maximum specific growth rate (μ_{A},/d) of the nitrifiers in the MBR system was observed to decrease by 25% with increase in VSS concentration from 2–3 gVSS/ℓ to 13–14 gVSS/ℓ. The OHOVSS specific denitrification rate [K_{2OHO}, $mgNO_3$-N/(mgOHOVSS · d)] obtained at different MBR VSS concentrations indicated no effect of VSS concentration – the average rate in the MBR system was similar to that in the CAS system [~0.25 $mgNO_3$-N/(mgOHOVSS · d)] and also to those measured in previous studies with CAS systems. Also, no difference in PAOVSS specific rates of anaerobic acetate uptake and P release and aerobic P uptake were observed at different VSS concentrations.

Editors: Giorgio Mannina, George Ekama, Hallvard Ødegaard and Gustaf Olsson
doi: 10.2166/9781780409719_0031

The results allow the BNRAS steady state and kinetic models and their default kinetic rates developed for low VSS concentration BNRAS systems with SST to be applied with reasonable confidence to predict the performance of high VSS concentration MBR BNRAS systems, except for the maximum specific growth rate of the nitrifiers. Equations for converting between anaerobic, anoxic and aerobic zone sludge mass and volume fractions are given for various biological nutrient removal configurations.

Keywords: Biological phosphorus removal, denitrification, kinetics, membrane, nitrification, reactor sizing, settling tanks

2.1 INTRODUCTION

For conventional (with settling tanks) activated sludge (CAS) systems for biological nutrient removal (BNR), considerable knowledge has been accumulated on their performance, design and operation. Design procedures and performance simulation models have been developed based on well structured and researched stoichiometric and kinetic principles of the underlying fundamental biologically mediated processes. It is not certain whether this knowledge developed for CAS BNR systems can be applied directly to membrane bioreactor (MBR) BNR systems, given the significant differences that may arise when membranes are included such as (i) floc structure (Zhang *et al.*, 1997; Cicek *et al.*, 1999; Huang *et al.*, 2001; Yamamoto, 2001; Gao *et al.*, 2004; Manser *et al.*, 2005a), (ii) bacterial communities (Ghyoot *et al.*, 1999; Luxmy *et al.*, 2000; Liebig *et al.*, 2001; Smith *et al.*, 2002; Manser *et al.*, 2005b), (iii) metabolic activities (Rosenberger *et al.*, 2002; Witzig *et al.*, 2002; Lee *et al.*, 2003; Han *et al.*, 2005; Li *et al.*, 2005; Sperandio *et al.*, 2005), and (iv) sludge production (Cicek *et al.*, 1999; Smith *et al.*, 2002; Holbrook *et al.*, 2005; Monti *et al.*, 2005).

Ramphao *et al.* (2005) concluded that incorporating membranes in BNR systems makes a profound difference not only to the design of the BNR system itself, but also to the approach to design of the whole wastewater treatment plant. This paper presents research that investigates whether the steady state and kinetic models developed for CAS BNR systems can be applied also with reasonable accuracy to model MBR BNR systems operating at high TSS.

Accordingly, the kinetic rates of nitrification, denitrification, anaerobic acetate uptake and P release, anoxic P release/uptake and aerobic P uptake were measured in batch tests over a range of volatile suspended solids (VSS) concentrations (3–12 gVSS/ℓ) on sludge harvested from an MBR BNR system and compared with the corresponding rates measured in a parallel CAS BNR system at 3 gVSS/ℓ. Also, the influence of the limitation of substrate (ammonia, oxygen, nitrate, phosphorus, acetic acid) concentrations on the kinetic rates was investigated in batch tests. To provide additional information on the anoxic behaviour of phosphate accumulating organisms (PAO), the ability of the sludge in the MBR BNR system to denitrify under anoxic conditions with simultaneous phosphate uptake was investigated and quantified.

2.2 MATERIAL AND METHODS

Two parallel lab-scale MBR and CAS BNR systems were operated for 450 days at 20°C allowing their behaviour to be monitored and their performance to be compared. Both systems were University of Cape Town (UCT) configurations (Figure 2.1, Table 2.1) so that denitrification and enhanced biological phosphorus removal (EBPR) could function independently, provided the recycles did not overload the anoxic reactor with nitrate. System design and operational parameters such as zone mass fractions, inter-reactor recycles and sludge ages were kept the same in both systems (Table 2.1). Five A4 size Kubota® membrane panels submerged in the aerobic reactor of the MBR system replaced the function of the secondary settling tank (SST).

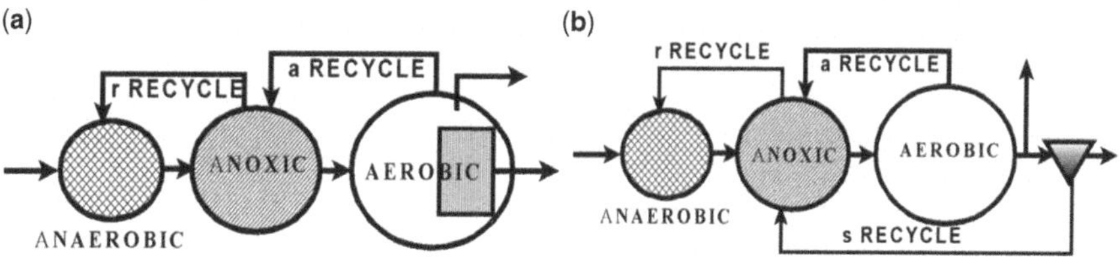

Figure 2.1 Schematic layouts of the MBR (a) and CAS (b) UCT systems with mixed liquor "a" from aerobic to anoxic, underflow "s" from secondary settling tank (SST) to anoxic and mixed liquor "r" from anoxic to anaerobic recycle ratios.

The systems were fed screened (1 mm mesh) raw unsettled municipal wastewater from the Mitchell's Plain Wastewater Treatment Plant (Cape Town, South Africa), augmented with sodium acetate (200 mg chemical oxygen demand [COD]/ℓ to accentuate EBPR), ammonia (20 mgN/ℓ to increase total Kjeldahl nitrogen [TKN]/COD), phosphorus (to ensure >5 mgP/ℓ in effluent) and sodium bicarbonate (for pH buffering). The wastewater was collected in 2 m^3 batches, macerated and stored in stainless steel tanks at 4°C and served as feed for both systems for 15 to 20 d. Daily, after thorough mixing, the required volume of wastewater for feeding the two systems was withdrawn from the stainless tanks and diluted with tap water to the target COD concentration (800 mgCOD/ℓ). After adding the supplements, a sample was taken and the feed volume for 1 day transferred into the systems' refrigerated (8°C) feed drums. The feed drums were gently stirred (1–2 rpm) to keep settleable solids in suspension and covered with a floating lid to minimize oxygen entrainment. The influent was pumped into each system with a multi-channel peristaltic pump, which also pumped the recycle flows. The influent tube was passed through a water bath at 20°C to avoid temperature decrease in the anaerobic reactor, in particular in the MBR system with the very short hydraulic retention time.

Table 2.1 MBR and CAS UCT systems' design and operating parameters.

System Parameters	MBR UCT	CAS UCT
Sludge age (d)	20	20
Anaerobic (R1) mass fraction (%), Vol (ℓ), Vol fraction (%)	12.6[1], 19, 25[1]	12.6[1], 5.6, 22[1]
Anoxic (R2) mass fraction (%), Vol (ℓ), Vol fraction (%)	27.9[1], 21, 28[1]	27.9[1], 6.2, 25[1]
Aerobic (R3) mass fraction (%), Vol (ℓ), Vol fraction (%)	59.5[1], 35, 47[1]	59.5[1], 13.2, 53[1]
a-recycle (R3 to R2)	3:1	2:1
r-recycle (R2 to R1)	1:1	1:1
s-sludge Return Recycle (SST to R2)	–	1:1
Hydraulic retention time (d)	0.53	1.67
MLVSS[2] concentration (mg/ℓ)	12500	3600
MLTSS[2] concentration (mg/ℓ)	18000	5000
Influent flow (l/d)	140	15
Feed COD concentration (mg/ℓ)	1000	1000
Membrane flux ($m^3/m^2/d$)	0.239	–

[1]For the given a- and r-recycle ratios.
[2]MLVSS and MLTSS are mixed liquor volatile and total suspended solids respectively.

The two systems were monitored daily via the parameters listed in Table 2.2. Taking due account of the mass of sludge taken for samples, the required volume of sludge to establish 20 d sludge age was removed daily from the aerobic reactor of the two systems. Additionally, recycle flow rates and trans-membrane pressure (TMP) at the constant flux of 11 $\ell/(m^2 \cdot h)$ were monitored daily. Once monthly, mixed liquor samples were analysed by a microbiologist for filament identification and floc morphology. Also, mixed liquor samples were sent fortnightly for fluorescent in situ hybridization (FISH) analysis (Maharaj *et al.*, 2007). The influent readily biodegradable organics (RBO) COD concentration (before supplement addition) was measured daily in a fully aerobic square wave fed (12 h feed on, 12 h feed off) system at 2.5 days sludge age according to Ekama *et al.* (1986).

For each wastewater batch (which was accepted to represent a steady-state period), the daily results were averaged (after analysis for outliers). The averages of the influent and effluent concentrations are given in Table 2.3. These steady-state averages were used to assess the performance of the systems and the following process characteristics were calculated: System COD and N mass balances; influent unbiodegradable soluble and particulate COD fractions ($f_{S'us}$ and $f_{S'up}$ respectively, Ekama & Wentzel, 1999); mixed liquor VSS/TSS, COD/VSS and TKN/VSS ratios; nitrate and P mass changes across each reactor, sludge production and the influent readily biodegradable (RB) COD concentration from the OUR measured in the square-wave fed 2.5 d sludge age activated sludge (AS) system (du Toit *et al.*, 2007).

Table 2.2 Sampling position and parameter measurement.

Test	COD	TKN	FSA	NO_3	NO_2	TP	TSS	VSS	OUR	DSVI	pH
Influent	F; UF	UF	F			UF					
Anaerobic				F	F	F	UF	UF			
Anoxic				F	F	F	UF	UF			
Aerobic	UF	UF		F	F	F	UF	UF	D	D[1]	D
Final effluent	F; UF	F; UF	F	F	F	F; UF					

F = 0.45 μm filtered; UF = unfiltered samples; D = direct measurement taken. COD; TKN; FSA (free and saline ammonia); TP (total phosphorus); TSS (total suspended solids); VSS (Standard Methods, 1985). DSVI = dilute sludge volume index; (Ekama & Marais, 1984); OUR = oxygen utilization rate (Randall et al., 1991).

[1]For the MBR system, the unfiltered COD was measured at the 800 ml mark of the 1000 ml measuring cylinder after 30 min settling.

Table 2.3 Average influent and effluent concentrations and system COD, N and P removals of both UCT systems.

Parameter		Influent	MBR UCT		CAS UCT	
			Effluent	Efficiency	Effluent	Efficiency
COD	mgCOD/ℓ	951	41 (139[1], 69[2])	95.7% (88%[1], 92.7%[2])	73.4[1] (50.7[2])	92.3%[1] (94.7%[2])
TKN	mgN/ℓ	106.5	1.52	98.60%	3.34[1] (2.24[2])	96.9%[1] (97.97%[2])
FSA	mgN/ℓ	81.5	0.74	99.10%	0.99	98.8%
NO_3	mgNO_3-N/ℓ	0	16.53	–	16.8	–
TN	mgN/ℓ	106.5	18.05	83.10%	20.1	81.10%
TP	mgP/ℓ	30.7	8.4	22.5	12.06	17.4
TSS	mgTSS/ℓ	N/A	0	–	19.3	–
E. coli	CFU/100 mℓ	N/A	0	–	2250	–

[1]unfiltered sample (for MBR at 800 mℓ level in 1ℓ measuring cylinder after 30 min settling;
[2]0.45 filtered sample; N/A = value not available.

To determine the kinetics rates, 85 batch tests on the mixed liquor harvested from the two BNR systems were conducted (Parco, 2006; Parco *et al.*, 2006, 2007). Particularly on the MBR system, the influence of the VSS concentration and of the limitation of ammonia, oxygen, nitrate and acetate concentrations on the kinetic rates was examined. Moreover, to provide additional information on the anoxic behaviour of phosphate accumulating organisms (PAO), the ability of AS in MBR and CAS systems to denitrify under anoxic conditions with simultaneous phosphate uptake was investigated and quantified to check the extent of anoxic P uptake EBPR in the systems. This is important to accurately separate ordinary heterotrophic organism (OHO) and PAO denitrification behaviour. Details of the entire investigation are given by du Toit *et al.* (2010).

2.3 OVERALL MBR AND CAS UCT SYSTEM PERFORMANCE

The overall (wastewater batch) average removals and effluent quality of both the MBR and CAS systems are summarized in Table 2.3. The MBR system exhibited removals that were equivalent or superior to that produced by the CAS system.

2.3.1 Organics (COD) removal

The COD removal efficiency of the MBR system (96%) was superior to that of the CAS system (92% unfiltered, 95% 0.45 μm filtered). However, the MBR system produced an "effluent" 0.45 μm filtered COD concentration (measured on the supernatant of the MBR aerobic reactor mixed liquor) that was consistently higher than the effluent filtered (0.45 μm) COD concentration from the CAS system (Table 2.3). Similarly, the MBR unfiltered "effluent" COD values (measured on the unfiltered supernatant at the 800 mℓ mark in the 1ℓ DSVI measuring cylinder test on aerobic reactor sludge after 30 minutes settling) were consistently higher (139 mgCOD/ℓ) than those from the CAS system (73 mgCOD/ℓ).

From these observations (Table 2.3), the following aspects can be highlighted:

(1) the membranes, because of smaller pore size, retain organics that would be considered "soluble" (< 0.45 μm) in a CAS system with SSTs;
(2) even though the total COD removal of the MBR appeared to show a better performance (lower effluent concentration) due to the filtration effect of the membranes, in terms of biological removal the CAS system seems to have better performance. The FISH analysis results confirmed this observation because the samples of the MBR system on average indicated that only 50% of total DAPI (total DNA) were eubacteria (EUB probe) compared with 80% in the CAS system (Maharaj *et al.*, 2007);
(3) although the nominal pore size of the Kubota® membranes used in this study was 0.4 μm, the considerably lower MBR effluent COD (41 mgCOD/ℓ) than

the 0.45 μm filtered COD (69 mgCOD/ℓ) indicates that a dynamic gel layer formed on the membranes which reduced their effective pore size.

The differences between the MBR and CAS effluent COD can be accommodated in the steady-state and dynamic models as differences in the soluble unbiodegradable COD fraction ($f_{S'us}$) which were 0.045 and 0.066 respectively for the two systems. As expected, the MBR system produced a solids-free effluent, whereas from the CAS system there was a continual loss of solids with the effluent (19 mgTSS/ℓ), quantified as the difference between unfiltered and filtered effluent COD concentrations (Table 2.3). This confirms that the MBR system effluent quality is independent of the floc characteristics of the mixed liquor. The MBR system's sludge settleability did not exceed 120 mℓ/g DSVI, but a significant non-settling particulate fraction remained suspended after 30 min, i.e. an average of 139-41 (MBR effluent) = 98 mgCOD/ℓ. The DSVI of the CAS system varied between 80 and 180 mℓ/g over the investigation.

2.3.2 Pathogen (faecal coliform) removal

The microbiological quality (faecal coliforms) of the MBR system effluent was superior. The results indicated pathogen counts were not detectable for the MBR system whereas for the CAS system, the average pathogen count was 2250 CFU/100mℓ (Table 2.3). While these absolute CFU concentrations are not reflective of effluents from full-scale wastewater treatment plants (WWTPs), from the relative removals by the MBR and CAS systems, the membrane effluent has a higher quality for reuse purposes.

2.3.3 Trans-membrane pressure (TMP)

Due to the constant flow and load conditions, the membrane flux was constant and the TMP was controlled to maintain the flux [0.26 $m^3/(m^2 \cdot d)$]. Because the volume of the MBR system aerobic reactor needed to be constant to maintain its VSS concentration, the external U-bend over which the effluent exited the aerobic reactor was lowered below the reactor mixed liquor (ML) level to increase the TMP as required. From day 50 to 350, the TMP increased linearly from around 20 to 80 mm (0.2 mm/d). While the membranes were rinsed under running water once every month, no chemical cleans were applied. From day 350 to 430 the TMP increased linearly to 200 mm (1.5 mm/d) caused by inorganic colloidal solids from construction work at the WWTP from which the wastewater feed was sourced, which also increased the influent and reactor inorganic suspended solids (ISS) concentrations. A standard chemical clean recommended by the membrane suppliers on day 433 restored the TMP to its earlier low value of 55 mm.

2.3.4 N and COD mass balances

Nitrogen mass balances were calculated from the wastewater batch average results by comparing the exiting N via the effluent, waste sludge stream and nitrate denitrified (from a nitrate balance over the anoxic and anaerobic reactors) with the N entering the systems via the influent TKN. N balances for the MBR system ranged between 75 and 120% with an overall average of 96%. Similarly, the COD balance was calculated by comparing the exiting COD via the effluent, waste sludge stream and oxygen utilized in the aerobic reactor (corrected for nitrification) with the COD entering the systems via the influent COD. COD balances for the MBR system ranged between 89 and 107% with an overall average of 103%. These mass balances are good and validate that the measured results are reliable. In fact, the COD balances over the MBR system are among the best achieved in nitrification denitrification enhanced biological phosphorus removal (NDEBPR) systems at UCT (Ekama & Wentzel, 1999), possibly as a result of the membranes.

2.3.5 Biological nitrogen removal

The TKN removal efficiency of the MBR system was marginally better than that of the CAS system (Table 2.3). This is again attributed to the retention of solids by the membranes compared with those lost with the effluent from the CAS system. In both systems, nitrification was virtually complete, as indicated by the low residual free and saline ammonia (FSA) concentrations. The two systems achieved similar total N removals indicated by the similar effluent nitrate concentrations (Table 2.3).

2.3.6 Biological phosphorus removal

In both systems, total phosphorus (TP) was dosed in excess of the system removal capacity in order to observe the maximum EBPR possible. Thus, P removal performance is represented by total P removals not as %. System average P removals of 22.5 mgP/ℓ and 17.4 mgP/ℓ were achieved in the MBR and CAS systems showing that the P removal performance of the CAS was inferior to that of the MBR system. Reasons for this are: (1) Anoxic P uptake was higher in the CAS system at 22% of total P uptake in the anoxic reactor compared with the MBR system at 8%. This was also evident in the P removal anoxic batch tests results. With anoxic P uptake by denitrifying PAOs, significantly reduced EBPR has been reported (Ekama & Wentzel, 1999; Hu *et al.*, 2002) probably due to less efficient utilization of stored polyhydroxyalkanoates (PHA) by PAOs when nitrate serves as an electron acceptor (Hu *et al.*, 2002, 2007a,b). (2) The nitrate load on the anoxic reactor of the MBR system was lower than the reactor's denitrification potential identified by zero nitrate concentration in this reactor. In contrast, denitrification was not complete in the anoxic reactor of the CAS system, identified by > 1mgNO_3-N/ℓ nitrate concentration in this reactor. This is one of the main factors that stimulates anoxic P uptake (Ekama &

Wentzel, 1999; Hu *et al.*, 2002). For the CAS system, the recycle of nitrate to the anaerobic reactor from the anoxic reactor reduced the RBO available for PAOs and additionally reduced the P removal. The main reason for the difference in denitrification capacity of the MBR and CAS anoxic reactors was the dissolved oxygen (DO) in the recycles, which had a negligible effect on denitrification at the very high VSS concentration in the MBR system.

The observations above indicate that the nitrate load in the CAS should have been reduced by reducing the aerobic to anoxic a-recycle ratio to maximize EBPR. While reducing the a-recycle ratio does not change the anoxic mass fraction and denitrification potential, the N removal performance will decrease, i.e. more nitrate out with the effluent rather than recycled to the anaerobic reactor via the anoxic reactor. In contrast, in the MBR system, from the design procedures of Ramphao *et al.* (2005), reducing the a-recycle ratio causes a corresponding reduction in the anoxic mass fraction and nitrate load and hence produces a greater decrease in N removal performance. This highlights the conflicting requirements between N and P removal in biological nutrient removal activated sludge (BNRAS) systems at elevated influent TKN/COD ratios, which is by no means new but operates differently in the MBR BNR systems than in the CAS BNR systems. The balance between optimizing N and P removal will be influenced by a number of factors, such as influent TKN/COD ratio, N and P removal requirements and influent RBO fraction.

At present anoxic P uptake EBPR is not explicitly incorporated in the steady-state design procedures for NDEBPR systems (Wentzel *et al.*, 1990; Henze *et al.*, 2008), as quantitative relationships linking the extent of anoxic P uptake to the system design and operating parameters have not been definitively established. Several dynamic models attempt to include it (ASM2d, Henze *et al.*, 1999; UCTPHO+, Hu *et al.*, 2007a,b; Cosenza *et al.*, 2013; Mannina *et al.*, 2018a) with varied success because the stimuli that promote it and the changes in kinetic behaviour it causes are not well defined yet.

2.3.7 System stability

With regard to COD, N and P removal over the whole investigation, the MBR system showed a greater stability than the CAS system – the MBR system performance recovered more quickly after power failures or pump stoppages and was generally less sensitive to disruptions such as sludge withdrawal for batch tests and monthly system cleaning (draining all the sludge from the reactors for about 2–3 h).

2.3.8 Sludge production

The MBR system had a greater sludge production per COD load (0.31 kgVSS wasted/kgCOD load) than the CAS system (0.20 kgVSS wasted/kgCOD load). This increased sludge production can be explained only in part by

(1) the retention of solids that would normally escape from CAS systems as effluent TSS and organics that would normally would be considered "soluble" (< 0.45 μm) and (2) a higher P removal resulting in a larger PAO population which produces more VSS and ISS per unit influent COD than OHOs due to their lower endogenous respiration rate (Wentzel *et al.*, 1990) and higher ISS content (Ekama & Wentzel, 2004). In the literature, previous studies comparing CAS and MBR BNR systems run under the same operating conditions (sludge age, influent wastewater) have indicated similar results as those outlined above (Cicek *et al.*, 1999; Smith *et al.*, 2002; Holbrook *et al.*, 2005). More data at different sludge ages are required to determine if this increase is consistent (predictable) and to identify the underlying causes.

2.4 CALCULATING THE BIOPROCESS SPECIFIC KINETIC RATES

In the steady-state design procedures and dynamic models, the increased sludge production in MBR systems can be accommodated by increasing the influent unbiodegradable particulate COD fraction ($f_{S'up}$). This was done in this investigation. Fixing the unbiodegradable soluble COD fraction ($f_{S'us}$) for the MBR and CAS systems at the values found above, i.e. 0.045 and 0.066 respectively, the $f_{S'up}$ fraction for the MBR and CAS systems were calculated to be 0.241 and 0.084 mgCOD/mgCOD respectively to match the measured average mass of VSS in the systems (Ekama & Wentzel, 1999). Noting that the model takes account of the different masses of PAOs in the two systems, it is a concern that for two systems with the same design and operating parameters fed the same wastewater, different $f_{S'up}$ fractions are obtained – if $f_{S'up}$ is really a wastewater characteristic, it should be the same for both systems. The problem of obtaining different $f_{S'up}$ fractions for the MBR and CAS systems, is that they result in different OHO (f_{avOHO}) and PAO (f_{avPAO}) biomass fractions of the VSS in the systems, where $f_{avOHO} = X_{BH}/X_v$ and $f_{avPAO} = X_{BG}/X_v$ and X_{BH}, X_{BG} and X_v are the OHO, PAO and total VSS concentrations respectively. The method of calculating $f_{S'up}$ by matching the calculated mass of VSS in the system with that measured has been applied in the past to determine the f_{avOHO} and f_{avPAO} active fractions and the OHO and PAO specific kinetic rates (van Haandel *et al.*, 1981; Wentzel *et al.*, 1990; Clayton *et al.*, 1991; Ekama & Wentzel, 1999) and these specific rates have been adopted as default values in the ASM1 and ASM2 kinetic models. Because this is the usual way of determining biomass specific kinetic rates from experimental systems fed real wastewater, the uncertainty that different $f_{S'up}$ fractions will have on the kinetic rates, while not ideal, has to be accepted as it has been in the past (Ekama & Wentzel, 1999) – expressing kinetic rates in terms of VSS makes the rates incomparable between different BNR systems and wastewater characteristics. The steady state models (Water Research Commission [WRC], 1984; Wentzel *et al.*, 1990; Henze *et al.*, 2008) aligned with and based on the same but simplified principles as the dynamic kinetic models (such ASM1 or ASM2, Henze *et al.*, 1987, 1995), are a useful interface between experimental systems and the dynamic kinetic models.

Because the kinetic rates determined from the batch tests results were assigned to the biomass population mediating the particular bioprocess, and the steady state NDEBPR model (Wentzel *et al.*, 1990) was used to determine the OHO (f_{avOHO}) and PAO (f_{avPAO}) active fractions from the measured data on the MBR and CAS systems, it was important for the OHO specific denitrification rate and the PAO specific P release and P uptake rates determined from the batch test results that the observed and predicted P removal of the MBR and CAS source sludge systems matched well. This ensured that the OHO and PAO specific kinetic rates were consistent with estimates of the OHO (f_{avOHO}) and PAO (f_{avOHO}) active fractions determined in the past. Figure 2.2 shows the wastewater batch average observed and steady state model predicted P removal of the MBR system, where the latter is based on the known system operating parameters, dosed acetate (200 mg/ℓ) and measured wastewater RBO concentration (model inputs). The nitrification batch tests, for which a close correlation between predicted and measured P removal was not important, were conducted at the beginning of the investigation when the predicted and measured P removal did not match well. The denitrification (anoxic) and P release and P uptake (anaerobic-anoxic/aerobic) batch tests were conducted during wastewater batches 10 to 25, when the predicted and measured P removal did match well. The measured kinetic rates in the MBR and CAS systems can therefore be legitimately compared with rates measured in previous investigations.

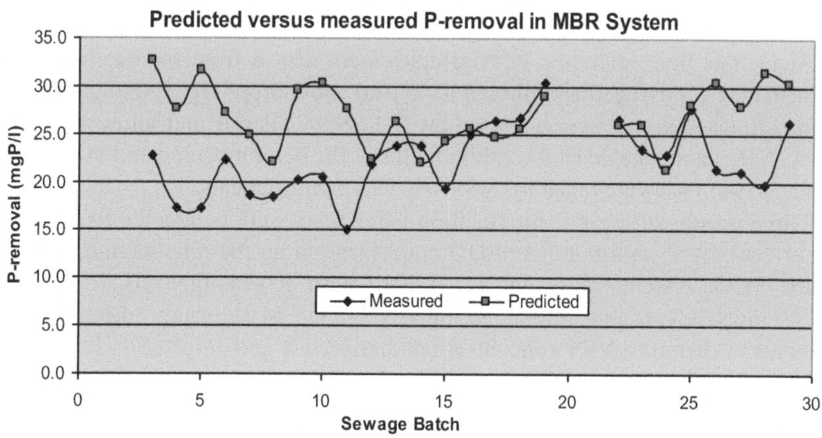

Figure 2.2 Sewage (wastewater) Batch average measured and predicted P removal for the MBR system. The denitrification, P release and P uptake batch tests were conducted between Sewage Batch 10 to 25 when the predicted and measured P removal matched well.

2.5 NITRIFICATION KINETICS – AEROBIC BATCH TESTS

2.5.1 Test and calculation procedures

Nitrification rates were measured with aerobic batch tests conducted on sludge harvested from the aerobic reactors of the MBR and CAS systems. The batch tests were conducted in pairs in two separate 4 ℓ mixed reactors constructed of 120 mm diameter clear cylindrical Perspex aerated through a diffuser with low pressure humidified air.

At the start of an aerobic batch test, 3 ℓ aerobic reactor sludge was transferred to the batch reactor. Because endogenous conditions are required at the start of the test, initially the sludge was aerated overnight (at least 12 hours) in the absence of substrate. This ensured complete degradation of all residual biodegradable organics and reduced OHO activity to endogenous respiration only. Thereafter, ammonia was added directly to the reactor as a small volume (65–175 mℓ) of concentrated ammonium chloride solution (5 $gNH_4Cl/ℓ$).

During the batch test, the pH of the mixed liquor was controlled at around 7.5 by adding small volumes of a sodium bicarbonate solution as required. The DO concentration, and concomitantly oxygen utilization rate (OUR), were measured continuously and automatically using a YSI 5739 DO probe connected to a HiTech Microsystem DO controller/OUR meter (Randall *et al.*, 1991). The OUR was recorded until endogenous conditions were re-established. The difference between the total OUR and the OUR associated with endogenous respiration was accepted as the OUR due to ammonia (NH_4-N) nitrification (OUR_{Nit}). The oxygen utilized (OU) for nitrification (area under the OUR_{Nit} – time curve) matched stoichiometrically to the concentration of ammonia added to the batch test through OU/4.57 = ammonia concentration added.

At regular time intervals, grab samples were drawn from the batch reactor, immediately centrifuged and filtered (0.45 μm) and 2 drops $HgCl_2$ (8.6 g/ℓ) added to the filtrate. Filtrates were analysed for NO_3, NO_2, TP, free and saline ammonia (FSA, NH_4^+) and soluble COD. Additionally, at the beginning and end of the tests duplicate samples were taken for VSS and TSS measurements.

Three groups of aerobic nitrification batch tests were conducted to evaluate the effect of VSS, ammonia and DO concentration on the nitrification kinetics in the MBR system: 29 Group (1), i.e. 10 with 10–20 mgN-NH_4/ℓ, 12 with 30–40 mgN-NH_4/ℓ and 7 with 50 mgN-NH_4/ℓ on MBR sludge diluted (with effluent) to different VSS concentrations between 2 and 14 gVSS/ℓ, i.e. 8 with 2–3 gVSS/ℓ, 2 with ~ 4 gVSS/ℓ, 6 with ~5–6 gVSS/ℓ, 5 with 7–10 gVSS/ℓ, 3 with 10–11 gVSS/ℓ and 5 with 12–14 gVSS/ℓ on MBR sludge, 2 Group (2), i.e. 2 on MBR sludge at the same VSS concentration (~9 gVSS/ℓ) but at different DO concentrations 2–5 and 10–15 mgO/ℓ) and 6 Group (3), i.e. in parallel, 3 on each of MBR and CAS sludge with MBR sludge diluted to the same low VSS concentration as that from the CAS system (2–3 gVSS/ℓ) to determine the effect of the membranes.

For all the batch tests, the nitrification, or ammonia utilization rate [AUR, mgNH$_4$-N/(ℓ·h)] was calculated from the OUR profile (Figure 2.3) as the average maximum value of OUR_{Nit} and from the slope of the FSA concentration – time profile (Figure 2.4). The AURs determined from the two methods were found to be in close agreement for all the nitrification batch tests. The VSS specific nitrification or ammonia utilization rate [SAUR, mgNH$_4$-N/(gVSS·h)] was defined as AUR/X_v where X_v was the measured VSS concentration in the batch test.

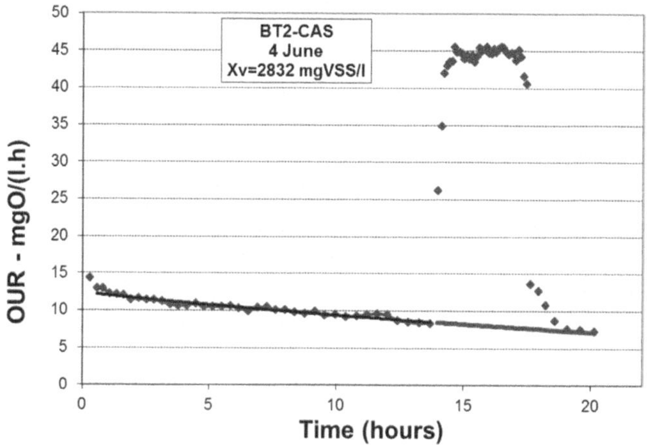

Figure 2.3 Oxygen utilization rate (OUR, mgO/[l.h]) time profile for aerobic batch test on mixed liquor from the CAS system at 2832 mgVSS/l and 30 mgNH$_4$-N/l dosed ammonia.

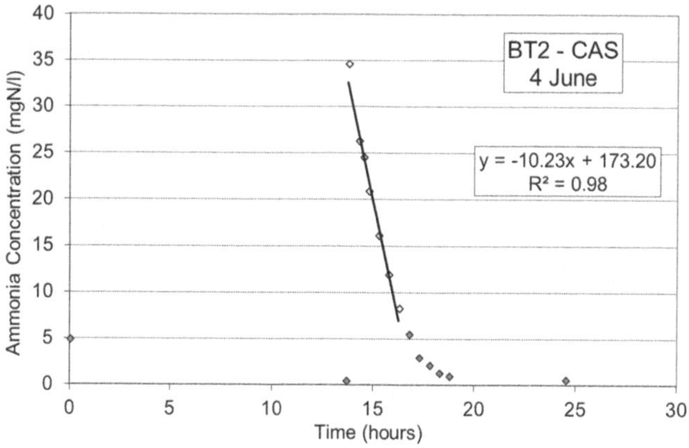

Figure 2.4 Free and saline ammonia (FSA) concentration-time profiles for aerobic batch test on mixed liquor from conventional (CAS) UCT system.

To quantify the maximum specific growth rate of autotrophic nitrifying organisms (ANOs, μ_A), nitrification was considered as a single composite bioprocess, where NH_4-N to NO_2-N oxidation limits the overall transformation of NH_4-N to NO_3-N. Consequently, the nitrification kinetics were expressed as:

$$\frac{[NH_4 - N]}{dt} = \frac{\mu_A}{Y_A} \frac{[NH_4 - N]}{K_{NH4-N} + [NH_4 - N]} X_{BA} \tag{2.1}$$

where

$\frac{[NH_4 - N]}{dt}$ = ammonia utilization rate AUR – mgNH_4-N/(ℓ·h)
μ_A = maximum specific growth rate of ANOs – mg ANOVSS/(mgANOVSS.d)
Y_A = yield coefficient of ANOs – mgANOVSS/mgNH_4-N
[NH_4-N] = ammonia concentration (mgN/ℓ)
K_{NH4_N} = ANO Monod half saturation coefficient for ammonia (mgNH_4-N/ℓ);
X_{BA} = ANO biomass concentration (mgANOVSS/ℓ)

Stoichiometrically the oxygen requirement for the conversion of ammonia to nitrate is 4.57 mgO/mgNH_4-N nitrified. Hence, ignoring the ANO biomass formed, which is negligible, the oxygen consumption rate for nitrification is:

$$\frac{dO}{dt} = \frac{4.57\mu_A}{Y_A} \frac{[NH_4 - N]}{K_{NH4-N} + [NH_4 - N]} X_{BA} \tag{2.2}$$

where

dO/dt = oxygen utilization rate (OUR) for nitrification [mgO/(ℓ · h)].

From Eq. 2.1, to determine μ_A from the batch test data, the concentration of ANO in the sludge (X_{BA}) needs to be known. From the nitrogen mass balances over the parent systems, the fluxes of nitrate produced in the MBR and CAS systems were calculated. From these nitrate fluxes, the masses of nitrifiers in the systems were calculated (WRC, 1984; Henze *et al.*, 2008), which requires the yield coefficient (Y_A) and endogenous respiration rate (b_A) of the ANOs to be defined. The values accepted were Y_A = 0.1 mgANOVSS/mgNH_4-N and b_A = 0.04/d (Henze *et al.*, 2008). The mass of ANOs divided by the measured mass of VSS in the systems gives the ANO VSS biomass fractions (f_{ANOVSS}), which were 0.0127 and 0.0187 mgANOVSS/mgVSS for the MBR and CAS systems respectively. These fractions were nearly the same for different sewage batches due to the approximately constant influent TKN/COD concentration ratio during the investigation. With X_{BA} known from $f_{ANOVSS}X_v$, the μ_A values were calculated. Two methods were applied: (1) the slope of the linear portion of the NH_4-N concentration-time profile (Figure 2.4), i.e. AUR, was determined by linear regression. Then because the NH_4-N concentration is much larger than the K_{NH4-N} for most of the ammonia-time profile, $\mu_A \approx Y_A$ AUR/($f_{ANOVSS} \cdot X_v$), where X_v

is the measured VSS in the batch test. (2) From the nitrification OUR (OUR_{Nit}) – time profile, the AUR versus time was calculated from $OUR_{Nit}/4.57 = dNH_4\text{-}N/dt = AUR_t$. This AUR_t was plotted versus the calculated NH_4-N concentration at time t, which was calculated from $[NH_4\text{-}N]_t = [NH_4\text{-}N]_{t-1} - AUR_t \times \Delta t$, using the initial ammonia concentration as the starting ammonia concentration. Using Eq. 2.1, the values for the Monod μ_A and $K_{NH4\text{-}N}$ that gave the least squares best fit to the AUR_t – time plot (Figure 2.5) were accepted. A probability plot of the $K_{NH4\text{-}N}$ values obtained with this method from the different batch tests is given in Figure 2.6. The average $K_{NH4\text{-}N}$ value was 1.9 $mgNH_4$-N/ℓ. This $K_{NH4\text{-}N}$ value is nearly double that usually accepted for CAS systems, i.e. $K_{NH4\text{-}N}$ = 1.0 $mgNH_4$-N/ℓ (WRC, 1984; ASM1, Henze *et al.*, 1987). The μ_A values obtained from the two methods were almost the same for all batch tests and are listed in Table 2.4, grouped in VSS concentration and initial ammonia concentration ranges.

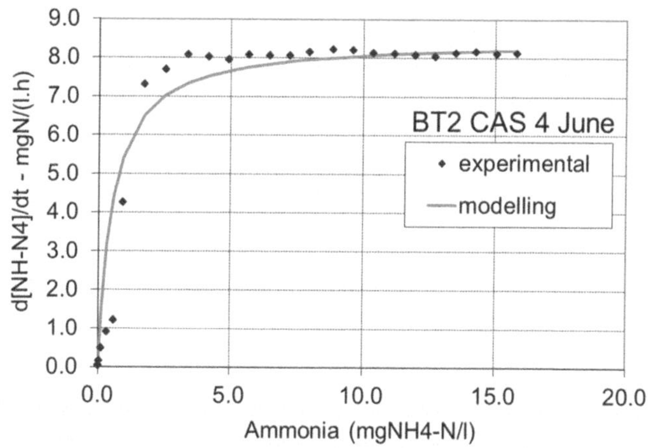

Figure 2.5 A typical ammonia utilization rate (AUR) versus the NH_4-N concentration for aerobic batch test on mixed liquor from the CAS system.

2.5.2 Nitrification – results and discussion

The VSS specific ammonia utilization rate [SAUR, $mgNH_4$-N/(gVSS·h)] and maximum specific growth rate (μ_A) of the ANOs were calculated as described above for the 29 nitrification batch tests. The SAUR and μ_A results are shown in Figures 2.7 and 2.8.

From Figures 2.7 and 2.8, as the VSS concentration increases, the SAUR and μ_A decrease. It was hypothesized that this decrease arises either from oxygen or ammonia diffusion limitations. This was tested in the two Group (2) batch tests – one was done using pure oxygen for aeration to elevate the DO concentration throughout the test to 10–15 mgO/ℓ and the other was done in the parallel test with air for aeration

(DO 2–5 mgO/ℓ). The results showed that the VSS concentration effect in Figures 2.7 and 2.8 was not related to DO transfer limitations because the SAUR and μ_A were both lower at the high DO than at the low DO. From the increase in SAUR and μ_A with increase in initial ammonia concentration evident in Figures 2.7 and 2.8, it was concluded that the effect was possibly due to ammonia diffusion limitation into the floc at increasing VSS concentration.

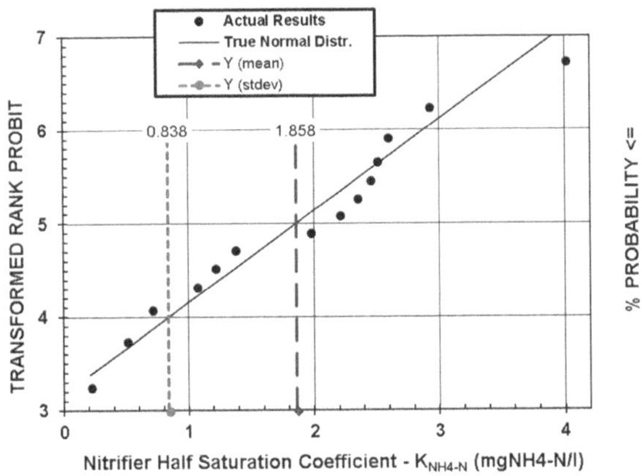

Figure 2.6 Probability plot of Monod half saturation concentration ($K_{NH4\text{-}N}$, mgN/ℓ) determined from the nitrification OUR measured in the aerobic nitrification batch tests.

Table 2.4 MBR system μ_A (/d) ranges at different VSS and initial ammonia concentration ranges.

Batch Test VSS (mgVSS/ℓ)	Ammonia Added (mgN/ℓ)			
	10–20	**30–40**	**50**	**70**
2000–3000	0.34	0.38–0.47	0.53–0.57	
3000–5000	0.25–0.31		0.48	
5000–7000	0.20–0.25	0.31-0.39	0.41–0.45	
>7000	0.18–0.20	0.25–0.37	0.33–0.43	0.58–0.52

Comparing the OUR profiles in the nitrification batch tests at different VSS concentrations clearly demonstrates a change in nitrification behaviour with change in VSS concentration (Figures 2.9 and 2.10). At low VSS concentration (Figure 2.9), on addition of ammonia, the OUR increases sharply to a plateau, remains at this plateau for a period, and then drops precipitously to the endogenous level. In contrast, at high

VSS concentration (Figure 2.10), the increase and decrease in OUR on ammonia addition and depletion is slow. Whether this effect of increasing VSS concentration on the SAUR and μ_A is related to ammonia transport limitation is difficult to tell from the results. The batch tests at increasing VSS concentration have increasing OUR, e.g. 35 mgO/(ℓ·h) at 2.3 gVSS/ℓ (Figure 2.9) and 90 mgO/(ℓ·h) at 8.7 gVSS/ℓ (Figure 2.10). Stenstrom and Song (1991) show that DO diffusion resistance into the flocs for nitrification increases with increasing nitrification OUR.

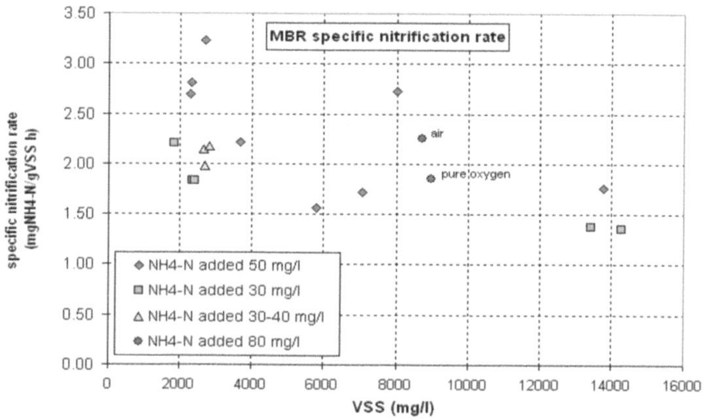

Figure 2.7 VSS specific nitrification rates (SAUR) for the batch tests on MBR system mixed liquor at different VSS concentrations.

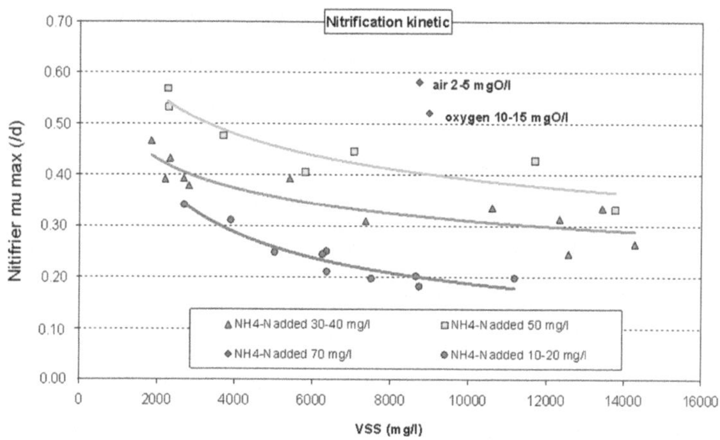

Figure 2.8 Maximum specific growth rate of nitrifiers (μ_A) for the batch tests on MBR system mixed liquor at different VSS concentrations.

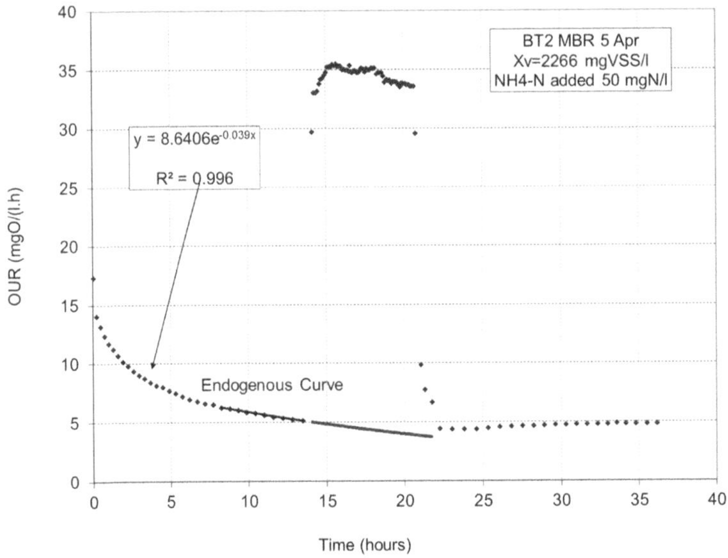

Figure 2.9 Oxygen Uptake Rate (OUR) profile for the batch test on MBR system mixed liquor at low VSS concentration (2.2 gVSS/ℓ).

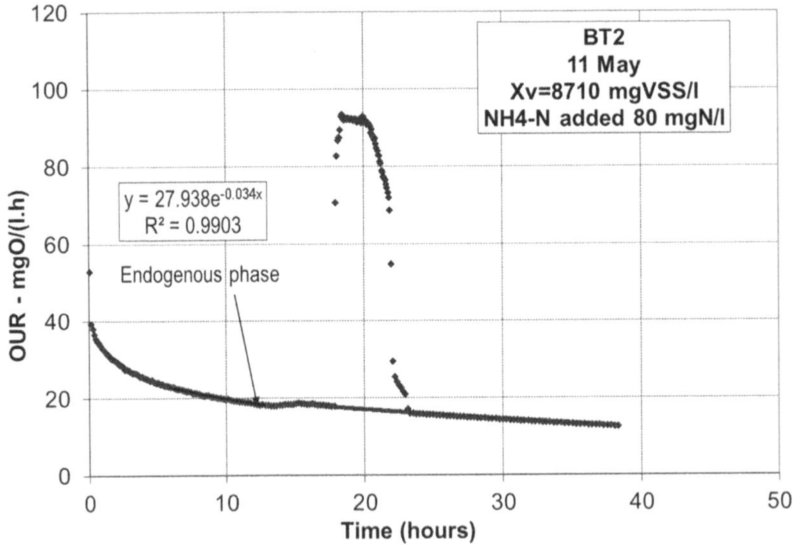

Figure 2.10 Oxygen Uptake Rate (OUR) – time profile for the batch test on MBR system mixed liquor at high VSS concentration (8.8 gVSS/ℓ).

The six Group (3) parallel nitrification batch tests on the CAS and MBR systems' sludges demonstrate that the SAUR and μ_A for the CAS system are approximately 1.8 and 1.2 times higher than those for the MBR system at the same low VSS concentration (see Table 2.5). The μ_A is less different than the SAUR because the sludge production in the MBR system is higher than in the CAS system, so the ANOs are diluted in a greater VSS mass at the same sludge age in the MBR and CAS systems.

Table 2.5 Group (3) nitrification batch results for the MBR and CAS system sludge at the same low VSS concentration (2–3 gVSS/ℓ).

Parameter	MBR System			CAS System		
	2305	3105	0406	2305	3105	0406
Ammonia added (mgNH_4-N/ℓ)	42	33	33	42	33	33
VSS (mgVSS/ℓ)	2692	2670	2824	2710	2196	2832
Ammonia/VSS load (mg NH_4-N/gVSS)	15.75	12.26	11.59	15.65	14.91	11.56
AUR [mgNH_4-N/(ℓ·h)]	5.32	5.60	6.13	10.87	7.79	10.23
SAUR [mgNH_4-N/(gVSS·h)]	1.98	2.10	2.17	4.01	3.55	3.61
ANO max specific growth rate (μ_A,/d)	0.374	0.396	0.410	0.514	0.455	0.463
ANO/VSS fraction (f_{avANO})	0.0127	0.0127	0.0127	0.0187	0.0187	0.0187

From all the batch tests, it is evident that the ANOs in the MBR and CAS systems exhibit a different activity, apparently induced by different environments under which the ANOs develop. The possible reasons for this are: (1) In CAS systems with SSTs, organism loss via the effluent occurs including ANOs. Therefore, CAS systems may select ANOs with higher maximum specific growth rates (μ_A) than MBR systems. In the MBR system all the ANOs are retained, including slow growing ones. (2) At the high VSS concentrations in the MBR system, ammonia transport limitations decrease the observed SAUR and μ_A.

In summary for nitrification:

(1) At the same VSS concentration, the MBR system exhibits lower VSS specific ammonia utilization rate (SAUR) and ANO maximum specific growth rates (μ_A) than the parallel CAS system, apparently due to different selection pressures imposed by membranes and SSTs for solid-liquid separation.
(2) In the MBR system, as the VSS concentration increases, the SAUR and μ_A decrease, apparently due to ammonia transfer limitation.
(3) In the MBR system, as the initial ammonia concentration is increased for similar VSS concentrations, the SAUR and μ_A increase, indicating possible ammonia transport limitation at increasing VSS concentration.

2.6 DENITRIFICATION KINETICS – ANOXIC BATCH TESTS

2.6.1 Batch test and calculation procedures

In nitrification denitrification (ND) systems, the kinetics of biological denitrification have been extensively investigated and a simplified general kinetic model has been developed (van Haandel *et al.*, 1981; Ekama & Wentzel, 1999) as:

$$\frac{[NO_3-N]}{dt} = KX_v = K_{2OHO} f_{avOHO} X_{OHO} \tag{2.3}$$

where

$d[NO_3\text{-}N]/dt$ = nitrate utilization rate NUR – $mgNO_3$-N/(ℓ·d)

*K_{2OHO} = OHO specific denitrification rate constant utilizing slowly biodegradable particulate organics (BPO) in the primary anoxic reactor of NDEBPR systems – $mgNO_3$-N/(mgOHOVSS·d)

$[NO_3\text{-}N]$ = nitrate concentration ($mgNO_3$-N/ℓ)

f_{avOHO} = OHO active fraction of VSS

X_v = measured VSS concentration (mgVSS/ℓ)

*Note that K_{2OHO} here refers specifically to the denitrification rate by OHO in the primary anoxic reactor utilizing BPO in the NDEBPR system. This K_{2OHO} rate is 2.5 times faster than the equivalent rate in ND systems for reasons not yet known (Clayton *et al.*, 1991; Ekama & Wentzel, 1999). This difference between the K_{2OHO} rates in ND and NDEBPR systems is taken into account in ASM1 and ASM2 models by different η_G, the reduction factor relative to aerobic conditions for the hydrolysis rate of the BPO under anoxic conditions – 0.33 in ASM1 and 0.60 in ASM2.

In the NDEBPR system, an initial rapid rate of denitrification by influent readily biodegradable organics (RBO), K_{1OHO}, is usually absent because these organics have been converted to volatile fatty acids (VFAs) and taken up by the PAOs in the anaerobic reactor. Therefore, provided the PAOs do not participate in the denitrification, the denitrification proceeds as a single slow rate by the OHOs (K_{2OHO}) utilizing the BPO (Clayton *et al.*, 1991; Ekama & Wentzel, 1999). To determine the OHO specific denitrification rate utilizing BPO K_{2OHO} [$mgNO_3$-N/(mgOHOVSS·d)] with Eq. 2.3, nitrate concentration versus time profiles (nitrate utilization rate, NUR) in anoxic batch tests were measured. The anoxic batch tests were conducted in the same batch reactors as the nitrification batch tests, except the air supply was replaced with nitrogen gas during the non-aerated (anoxic) part of the batch tests.

At the start of an anoxic batch test, the required volumes of sludge were transferred from the anaerobic and aerobic reactors of the MBR and CAS systems to the batch reactors. Because the VSS concentration in the anaerobic reactor is lower than that in the aerobic reactor, to achieve high VSS concentrations in the batch tests (8–12 gVSS/ℓ), the sludge needed to be concentrated. The required volume of mixed aerobic and anaerobic reactor sludge was gently centrifuged (500 rpm for 3 min) and half the supernatant discarded. The substrate available for denitrification in the batch tests is the BPO from the wastewater and endogenous process enmeshed in the anaerobic and aerobic reactor mixed liquor. Because anoxic conditions are required, the mixed liquor was de-oxygenated to remove the DO from the aerobic reactor sludge by the nitrogen gas flow. When the DO concentration was zero, nitrate was added as a small volume (120–240 mℓ) of concentrated potassium nitrate solution (7.2 $gKNO_3$/ℓ), sufficient to ensure that nitrate was present for the entire test period (5–6 hours). At regular intervals, grab samples were drawn from the batch reactor, immediately centrifuged and filtered (0.45 μm) and 2 drops $HgCl_2$ (8.6 g/ℓ) added to the filtrate. Filtrates were analysed for NO_3, NO_2, TP, NH_4-N and soluble COD. Additionally, at the beginning and end of the tests duplicate samples were taken for VSS and TSS measurements. During the test the pH of the mixed liquor was controlled at around 7.5 by adding small volumes of 20% HCl solution. After the anoxic period, aeration was commenced for 18–20 h which was controlled with the DO controller/OUR meter used in the nitrification batch tests. Sampling was continued during the aerobic period to observe the ANO and PAO behaviour.

Five groups of anoxic batch tests were conducted, viz. Group (1): On MBR sludge at different VSS concentrations between 2.5 and 12 gVSS/ℓ with sludge from the anaerobic and aerobic reactors mixed in proportion to the recycles entering the anoxic reactor; Group (2): Like Group (1), but at different nitrate concentrations; Group (3) which subdivides into three subsets: Like Groups (1) and (2) but with different proportions of anaerobic and aerobic reactor sludge, i.e. Set I - 50/50 by VSS mass, Set II - 100% anaerobic and Set III - 100% aerobic (from which the post denitrification rate, K_{3OHO} was determined); Group (4) on MBR and CAS sludge in parallel with the MBR sludge diluted to the same low VSS concentration as that from the CAS system (2–3 gVSS/ℓ) and with sludge from the anaerobic and aerobic reactors mixed in proportion to the recycles entering the anoxic reactor; and Group (5): like Group (4) but with wastewater added.

2.6.2 Denitrification kinetics – results and discussion

From the 14 Group (1) and (2) anoxic batch tests, a general denitrification behaviour could be defined (Figure 2.11), i.e. nitrate reduction took place via a single linear rate, in agreement with observations on BNR sludge by others (Clayton *et al.*, 1991; Ekama & Wentzel, 1999). Thus denitrification with RBO via the K_1 rate was absent (Figure 2.11) which confirms that all RBO was removed in the anaerobic reactor of the MBR system. The P concentration remained essentially constant (Figure 2.11), indicating that apparently, the PAOs did not take part in the denitrification process, even though PAOs were known to be present in the mixed liquor from the high P removal exhibited in the MBR system. This implied that the observed NUR

(Eq. 2.3) could be ascribed solely to the OHOs, i.e. zero contribution by PAOs. So the OHO specific denitrification rate K_{2OHO} was calculated from Eq. 2.3, i.e. $K_{2OHO} = NUR/(f_{avOHO}\ X_v)$ mgNO$_3$-N/(mgOHOVSS·d).

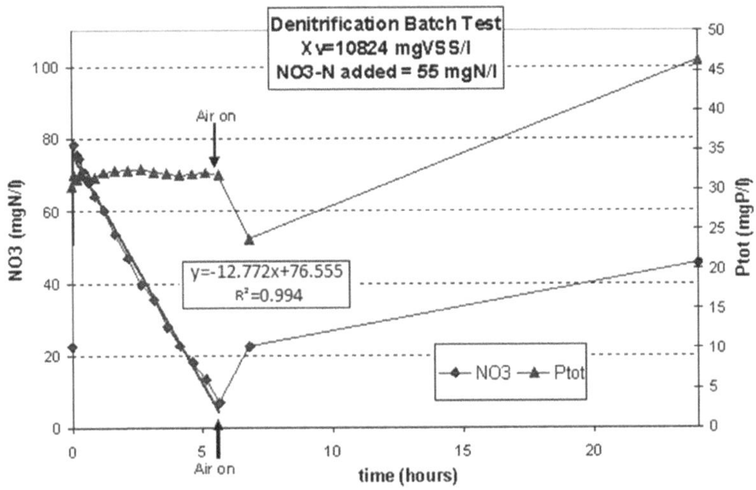

Figure 2.11 Nitrate (NO_3) and total soluble phosphorus (Ptot) concentration time profiles.

The K_{2OHO} rates obtained at different MBR system VSS concentrations (2.5–12 gVSS/ℓ) and different initial nitrate concentrations ranging from 30 to 90 mgN/ℓ (Groups (1) and (2)) are shown in Figure 2.12. From the nitrate concentration-time profiles, the NUR was zero order with respect to nitrate concentration, in agreement with past work (van Haandel *et al.*, 1981; Clayton *et al.*, 1991; Ekama & Wentzel, 1999). The K_{2OHO} rates observed at different VSS concentrations indicate no effect of VSS on the rate (Figure 2.12), in contrast to the observed effect of VSS on the nitrification rate. From all the anoxic batch tests, an average $K_{2OHO} = 0.264$ mgNO$_3$-N/(mgOHOVSS d) was obtained at 20°C. This is very close to the average K_{2OHO} rate reported in the literature for conventional (low VSS) BNR systems with SSTs, i.e. 0.255 at 20°C from Ekama and Wentzel (1999).

For the one Group (3) batch test with no anaerobic reactor sludge (Set III), the nitrate (NO$_3$-N) and filtered total P (P_{tot}) concentration time profiles were the same as those of Groups (1) and (2), i.e. denitrification took place at a single linear rate [slope = NUR, mgNO$_3$-N/(ℓ·h)] and the filtered total P concentration remained essentially constant (Figure 2.13). Because the PHA of the PAOs in the aerobic reactor is low, most likely only the OHOs mediate this "post" denitrification (K_{3OHO}), where the energy source is derived from the BPO generated by organism death and cell lysis. So the K_{3OHO} rate was calculated from $K_{3OHO} = 24{\cdot}NUR/(f_{avOHO}X_v)$.

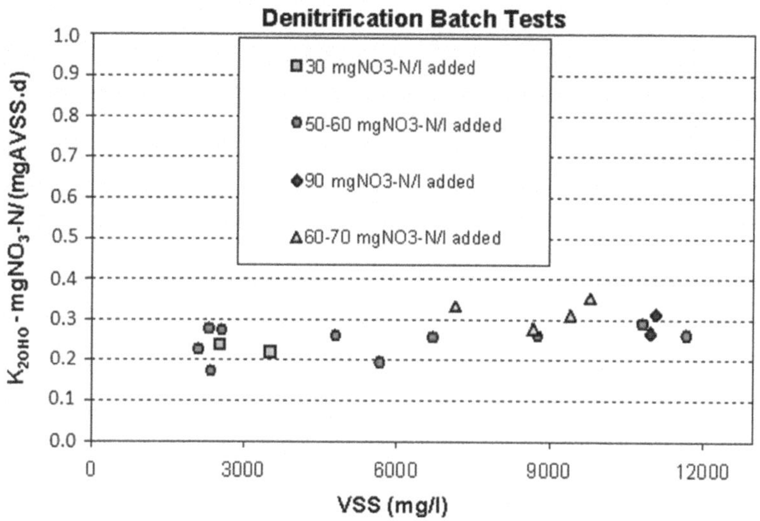

Figure 2.12 Specific denitrification rate constants (K_{2OHO}) at different VSS and NO_3 concentrations.

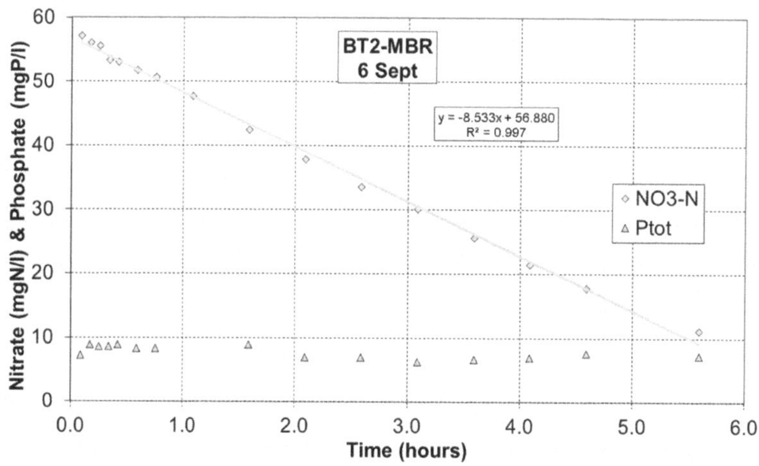

Figure 2.13 Nitrate ($mgNO_3$-N/ℓ) and total soluble P (mgTP/ℓ) concentration-time profiles for the anoxic batch tests on MBR sludge from the aerobic reactor only.

For the three Group (3) batch tests with a high proportion of anaerobic reactor ML (two Set I with 50/50 anaerobic and aerobic reactor sludge and one Set II with 100% anaerobic reactor sludge), the nitrate (NO_3-N) and filtered total P (P_{tot}) concentration time profiles (Figure 2.14) were different to those with low or no anaerobic reactor sludge. In these three batch tests, denitrification took place via two sequential linear rates – an initial fast rate for less than an hour followed by a second slower rate. Because P uptake terminated at about the same time as the change in denitrification rate, it seemed that PAOs also denitrified during the initial fast rate but stopped within 1 h. Then the OHOs continued to denitrify during the slower rate for the remainder of the anoxic period while nitrate was present. A similar behaviour was observed in the P removal batch tests (see below). It seems that the high proportion of anaerobic sludge in these anoxic batch tests resulted in a high concentration of PHA, which made denitrification by PAOs become visible. In contrast, in the batch tests with a low (or zero, as in Set III) proportion of anaerobic sludge, the PHA concentration is diluted by the aerobic reactor sludge with low PHA so that denitrification by the PAOs is not detectable. The PHA profiles measured in the P removal batch tests confirmed this (see below). To determine the OHO (K_{2OHO}) and PAO (K_{PAO}) specific denitrification rates, the NUR of the second slower rate [slope = NUR_{OHO}, mgNO_3-N/(ℓ·h)] was deducted from the NUR of the first faster rate (slope = $NUR_{PAO+OHO}$) to obtain NUR_{PAO}. The OHO and PAO specific denitrification rates were then calculated from $K_{2OHO} = 24 \cdot NUR_{OHO}/(f_{avOHO}X_v)$ and $K_{PAO} = 24 \cdot NUR_{PAO}/(f_{avPAO}X_v)$, where X_v is the measured VSS concentration in the batch test in mgVSS/ℓ.

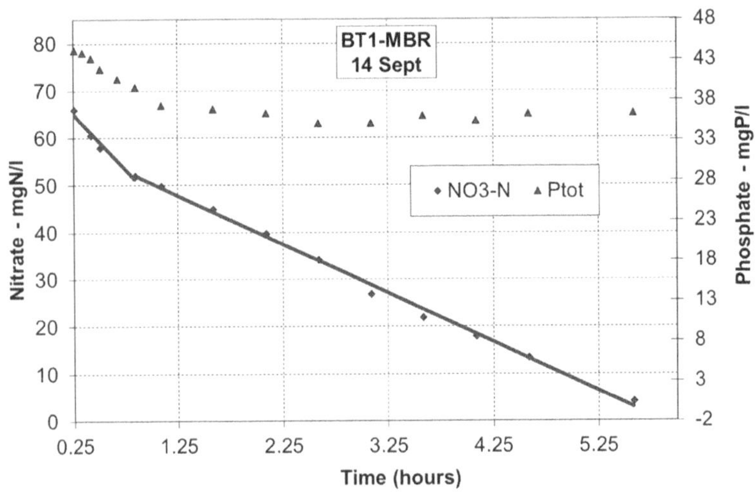

Figure 2.14 Typical nitrate and total soluble phosphorus concentration time profiles for anoxic batch tests on MBR system mixed liquor with high proportion of anaerobic reactor sludge.

The calculated specific denitrification rates (K_{2OHO}, K_{PAO}, K_{3OHO}) from the Group (3) anoxic batch tests were added to Figure 2.12 and are plotted in Figure 2.15.

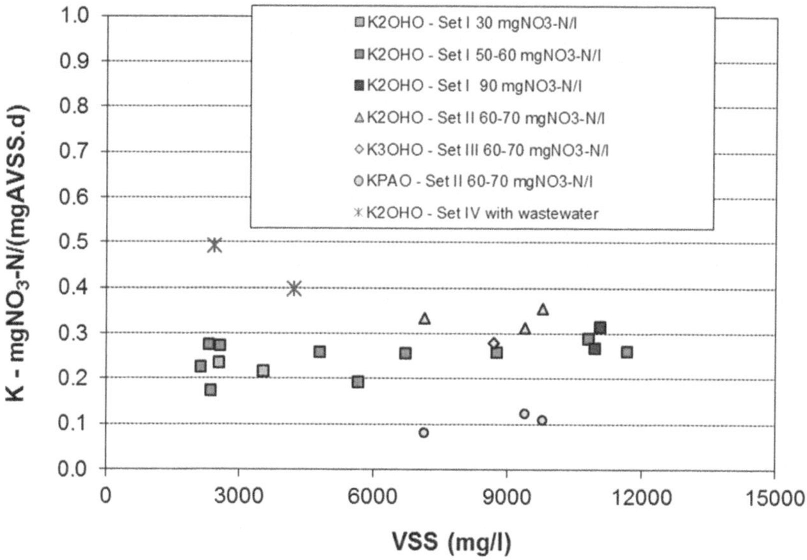

Figure 2.15 OHO and PAO specific denitrification rate constants (K_{2OHO}, K_{3OHO}, K_{PAO}) on MBR sludge at different VSS and initial nitrate concentrations and different proportions of anaerobic and aerobic reactor sludge (Set I - 50/50 by VSS mass, Set II - 100% anaerobic and Set III - 100% aerobic).

From Figure 2.15 the following can be observed:

The K_{2OHO} rates fall within the same band of 0.2 to 0.3 mgNO$_3$-N/(mgOHOVSS·d) with an average of 0.25. Also, no effect of VSS or initial nitrate concentration was apparent.

At a mean of 0.104 mgNO$_3$-N/(mgPAOVSS·d), the K_{PAO} is 2.5 times lower than K_{2OHO}. This conforms to the observation of Hu *et al.*, (2002) that the PAOs denitrify with PHA significantly slower than the OHOs with BPO. This is quantified further with the anaerobic-anoxic/aerobic batch tests discussed below.

At a rate of 0.278 mgNO$_3$-N/(mgOHOVSS·d) from the single Group (3) Set III batch test, the post denitrification K_{3OHO} rate was in the same range as the K_{2OHO} rates. From the K_{3OHO} rate of 0.10 mgNO$_3$-N/(mgOHOVSS·d) reported by Ekama and Wentzel (1999), this was unexpectedly high. The reason for this high post denitrification rate is difficult to establish so this result from the single batch test was rejected.

The results of the Group (4) batch tests on MBR and CAS sludge at low VSS concentration (with sludge from the aerobic and anaerobic reactors mixed in proportion to the flows entering the anoxic reactor) are listed in Table 2.6. In conformity with Groups (1) and (2), the denitrification took place at a single linear rate [slope = NUR,

$mgNO_3$-N/(ℓ·h)] and the filtered total P concentration remained essentially constant. The average OHO specific denitrification rate for the MBR sludge was 0.242 $mgNO_3$-N/(mgOHOVSS·d) and that for the CAS sludge was 0.206 $mgNO_3$-N/(mgOHOVSS·d). Although the MBR system average rate is 20% higher than that of the CAS system, the two rates are not statistically different due to the low number of tests (three on each system) and the high variability in the measured rates.

Table 2.6 Specific OHO denitrification rates (K_{2OHO}) for parallel batch tests on MBR and CAS system mixed liquor at the same low VSS concentration (2.5–3.5 gVSS/ℓ).

Batch Test	System	NUR	VSS	VSS Specific Denit. Rate	OHO Active Fraction	OHO Specific Denit. Rate K_{2OHO}
		$mgNO_3$-N/(ℓ·h)	mgVSS/ℓ	$mgNO_3$-N/(gVSS·h)	f_{avOHO}	$mgNO_3$-N/(mgOHOVSS·d)
1706C	CAS	2.86	2294	1.25	0.120	0.249
0607C	CAS	3.16	2830	1.12	0.156	0.172
1209C	CAS	2.39	2302	1.04	0.126	0.198
1706M	MBR	2.37	2550	0.93	0.095	0.235
0607M	MBR	3.37	3534	0.95	0.106	0.216
1209M	MBR	2.27	2324	0.97	0.085	0.275

In the Group (5) batch tests, which were the same as Group (4) except with added wastewater comprising RBO and SBO but not acetate, the denitrification took place in two phases (Figures 2.16a MBR and 2.16b CAS), i.e. an initial rapid NUR phase due to the utilization of influent RBO and BPO followed by a second slower NUR phase due to the utilization of the only the BPO. During both phases the filtered total P concentration remained essentially constant, indicating that the PAOs did not take part in the denitrification.

To determine the OHO specific denitrification rates utilizing RBO (K_{1OHO}) and BPO (K_{2OHO}), the NUR of the second slower rate [slope NUR_{2OHO}, $mgNO_3$-N/(ℓ·h)] was deducted from the NUR of the first rate (slope $NUR_{1OHO+2OHO}$). The OHO specific denitrification rates were then calculated from $K_{2OHO} = 24 \cdot NUR_{2OHO}/(f_{avOHO}X_v)$ and $K_{1OHO} = 24 \cdot NUR_{1OHO}/(f_{avOHO}X_v)$, where X_v is the measured VSS concentration in the batch test in mgVSS/ℓ (Table 2.7). The average K_{1OHO} and K_{2OHO} rates are 0.256 and 0.413 $mgNO_3$-N/(mgOHOVSS·d) respectively. Usually for ND systems, the K_{1OHO} rate is significantly faster than the K_{2OHO} rate (van Haandel *et al.*, 1981; Ekama & Wentzel, 1999), but this is not the case here. Also, the K_{1OHO} is also slower than 0.70 $mgNO_3$-N/(mgOHOVSS·d) observed by Clayton *et al.* (1991) in their NDEBPR systems. This is probably because the OHOs rarely encounter RBO in the anoxic reactor of the NDEBPR system – most of it is converted to VFA in the anaerobic reactor and taken up by PAO and so is rarely utilized by OHOs for denitrification. Consequently, the K_{1OHO} is of little consequence. The K_{2OHO} rate is higher than the 0.25 mgNO$_3$-N/(mgOHOVSS·d) observed in Group (1) and (2) batch tests because in the Group (5) batch tests the BPO/VSS ratio is much higher from the added wastewater BPO. This increase in specific denitrification rate is accommodated in dynamic kinetic models like ASM1 and ASM2 via the saturation kinetics for utilization of BPO, which is essentially a Monod type equation dependent on the BPO/OHO concentration ratio.

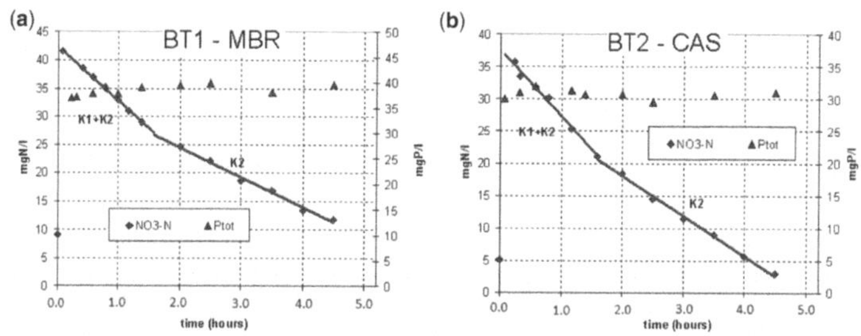

Figure 2.16 Nitrate and filtered total soluble phosphorus concentration time profiles for the batch tests on MBR (BT1, a) and CAS (BT2, b) sludges with added influent wastewater. K1 is the specific OHO denitrification rate (K_{1OHO}) on readily biodegradable organics (RBO) and K_2 is the specific OHO denitrification rate (K_{2OHO}) on slowly biodegradable organics (BPO).

Table 2.7 Specific OHO denitrification rates (K_{1OHO} and K_{2OHO}) for parallel batch tests on MBR and CAS system mixed liquor at the same low VSS concentration (2.5–3.5 gVSS/ℓ).

Batch Test	System	NUR	VSS	VSS Specific Denit. Rate	OHO Active Fraction	OHO Specific Denit. Rates	
		mgNO$_3$-N/(ℓ·h)	mgVSS/ℓ	mgNO$_3$-N/(gVSS·h)	f_{avOHO}	mgNO$_3$-N/(mgOHOVSS·d)	
2306C1	CAS	3.78	2994	1.26	0.156	0.194	K_{1OHO}
2906C1	CAS	3.85	2786	1.38	0.156	0.212	
2306M1	MBR	3.14	4210	0.75	0.106	0.169	
2906M1	MBR	4.82	2424	1.99	0.106	0.450	
2306C2	CAS	7.97	2994	2.66	0.156	0.410	K_{2OHO}
2906C2	CAS	6.39	2786	2.29	0.156	0.353	
2306M2	MBR	7.37	4210	1.75	0.106	0.396	
2906M2	MBR	5.26	2424	2.17	0.106	0.491	

2.7 BIOLOGICAL P REMOVAL KINETICS – ANAEROBIC-ANOXIC/AEROBIC BATCH TESTS

2.7.1 Batch test and calculation procedures

The sludge withdrawn from the MBR anoxic reactor (in which nitrate was <1 $mgNO_3$-N/ℓ) was subjected to sequential anaerobic and anoxic and/or aerobic phases. To achieve high VSS concentrations in the batch tests (8–12 gVSS/ℓ), the anoxic reactor sludge was concentrated by slow centrifugation (500 rpm for 3 min) and the required volume of supernatant decanted. At the start of the batch test anaerobic phase, defined masses of concentrated sodium acetate and potassium phosphate or real wastewater (without acetate) were added. The anaerobic phase was maintained for 4 to 5 h, which was sufficient for complete acetate uptake and P release. Then the batch sludge volume was halved – one half was aerated with online OUR measurement and nitrate was dosed to the other half without aeration to establish anoxic conditions. The pH in the batch tests was maintained between 7.0 and 7.2 by dosing concentrated HCl or NaOH as required throughout the batch test.

During the batch tests, grab samples were taken at specific time intervals, immediately centrifuged (3500 rpm for 10 min), filtered (0.45 μm) and analysed for HAc, NO_3-N, NO_2-N, PO_4-P, FSA and soluble COD. VSS and TSS were measured at the start, phase change and end of the batch tests. The solids fraction of all the samples was analysed for polyhydroxybutyrate (PHB) using the method of Satoh *et al.* (1996).

From the batch test results the following slopes were calculated from the concentration-time profiles: Anaerobic maximum P release and acetate uptake rates, the P release to acetate uptake (P/HAc) ratio and the aerobic and anoxic maximum P uptake rates. Also from the anoxic NUR, the denitrification rates and from the aerobic AUR, the nitrification rates were determined.

From a direct comparison of the P uptake rate under aerobic and anoxic conditions, the proportion of denitrifying PAOs (X_{DNPAO}/X_{PAO}) was estimated using the method of Meinhold *et al.* (1999), i.e.

$$X_{PAO} = X_{DNPAO} + X_{APAO} \quad \text{and} \quad \frac{q_{anx}}{\eta_{NO_3}} \frac{1}{q_{aer}} = \frac{X_{DNPAO}}{X_{PAO}} \tag{2.4}$$

where

X_{PAO} = concentration of PAOs = f_{avPAO} X_v (mgPAOVSS/ℓ)
X_{DNPAO} = concentration of denitrifying PAOs (mgDNPAOVSS/ℓ)
X_{APAO} = concentration of aerobic PAOs (mgAPAOVSS/ℓ)
q_{anx} = specific PAOVSS anoxic P uptake rate - mgP/(gPAOVSS·h)
q_{aer} = specific PAOVSS aerobic P uptake rate - mgP/(gPAOVSS·h)
η_{NO_3} = reduction factor for anoxic conditions relative to aerobic conditions.

The initial slope of the anoxic or aerobic P uptake profiles, i.e. where the P uptake proceeded without PHA limitation, was used to determine the P uptake rates (q_{anx}, q_{acr}). The η_{NO_3} value reflects the observation that under anoxic conditions nitrate is less efficient than oxygen as an electron acceptor (Payne, 1981; Kuba *et al.*, 1993; Casey *et al.*, 1994, 1999; Orhon *et al.*, 1996; Meinhold *et al.*, 1999; Sperandio *et al.*, 1999; Muller *et al.*, 2003). The OHO and PAO active fractions (f_{avOHO} and f_{avPAO}) were calculated from the MBR system results as described above. Unfortunately, the MBR anaerobic batch tests were not comparable with those from the CAS system due to the differing P removal behaviour in the CAS system mentioned earlier.

2.7.2 Anaerobic P release and anoxic/aerobic P uptake behaviour

Fifteen anaerobic batch tests were conducted, 13 (BTs 1 to 13) with low to moderate acetate dosages varying from 0.009 to 0.043 mgHAcCOD/mgVSS and VSS concentrations ranging from 2.7 to 11.2 gVSS/ℓ, one (BT 14) with excess acetate addition at 0.166 gHAcCOD/gVSS at 6.37 gVSS/ℓ and one (BT 15) with wastewater addition at 5.52 gVSS/ℓ. The results are listed in Tables 2.8a and 2.8b.

Typical anaerobic-aerobic/anoxic batch test behaviour with low to moderate acetate dose (BTs 1 to 13) is shown in Figure 2.17 (BT 8). The P release took place at a high rate while acetate was present. After the acetate was all taken up, a slow P release took place while conditions were still anaerobic, presumably due to some PAO maintenance processes. After the anaerobic phase, P uptake took place in the aerobic and anoxic phases. The aerobic P uptake rate was considerably faster than the anoxic P uptake rate (Figure 2.17). In the anoxic phase, the P uptake gradually diminished and stopped after ~2 h. In the tests where an aerobic phase followed the anoxic phase, P uptake started again under aerobic conditions and the final P concentration at 24 h reached a low value but not as low as the half of the batch volume that was aerated immediately after the anaerobic phase (Figure 2.17).

During the anoxic phase, nitrate reduction took place via two linear rates with a reduction in slope (rate) at about the same time as the anoxic P uptake stopped (Figure 2.18), like the Group (3) denitrification batch test with a high proportion of anaerobic sludge. The reason for this behaviour is the same – PAO denitrification due to high PHA in the batch test. In the anaerobic batch tests, a high PHA was stimulated by the acetate dose.

The observations from Figures 2.17 and 2.18 may be explained by dividing PAOs into two groups – one capable of utilizing nitrate and oxygen as electron acceptors (DNPAO) and one group capable of utilizing only oxygen (APAO) as an electron acceptor (Meinhold *et al.*, 1999). When P uptake ceases under

Table 2.8a Anaerobic – anoxic/aerobic batch tests results (Batch Tests 1 to 8).

Date			**1810**	**2111**	**2811**	**0412**	**0602**	**0902**	**1302**	**2903**
Batch Test No			**1**	**2**	**3**	**4**	**5**	**6**	**7**	**8**
VSS concentration	mg/ℓ	1	8530	3357	2697	3001	3896	8643	6360	8837
Organic Load	mgHAC/mgVSS	2	0.034	0.04	0.043	0.042	0.035	0.034	0.039	0.04
PAOActive fraction (f_{avPAO})	mgPAOVSS/mgVSS	3	0.237	0.247	0.243	0.243	0.282	0.282	0.282	0.276
OHOActive fraction (f_{avOHO})	mgOHOVSS/mgVSS	4	0.125	0.073	0.136	0.136	0.148	0.148	0.148	0.127
Prelease/Acetate uptake ratio	mgP/mgHAc	5	0.51	0.546	0.638	0.583	0.571	0.509	0.642	0.472
Prelease/Acetate uptake ratio	mol/mol	6	0.5	0.53	0.619	0.565	0.554	0.494	0.622	0.458
Anaerobic P Release rate	mgP/(gVSS·h)	7	12.05	23.69	21.08	23.07	19.36	16.9	20.33	19.63
Anaerobic P Release rate	mgP/(gPAOVSS·h)	8	50.8	95.9	86.7	94.9	68.7	59.9	72.1	71.1
Anaerobic acetate uptake rate	mgHAc/(gVSS·h)	9	19.41	29.19	31.79	36.99	31.6	27.93	31.65	36.39
Anaerobic acetate uptake rate	mgHAc/ (gPAOVSS·h)	10	81.9	118.2	130.8	152.2	112.1	99.0	112.2	131.8
Aerobic P Uptake rate	mgP/(gVSS·h)	11	11.12	11.1	5.48	9.5	13.19	11.07	11.55	10.38
Aerobic P Uptake rate	mgP/(gPAOVSS·h)	12	46.92	44.94	22.55	39.09	46.77	39.26	40.96	37.61
Anoxic P Uptake rate	mgP/(gVSS·h)	13	4.53	3.85	2.68	3.2	3.22	2.16	3.24	3.43
Anoxic P Uptake rate	mgP/(gPAOVSS·h)	14	19.11	15.59	11.03	13.17	11.42	7.66	11.49	12.43
DNPAO/APAO ratio	DNPAO/APAO, %	15	51	43	61	42	31	24	35	41
Anx/Aer P uptake rate ratio	q_{anx}/q_{aer}, %	16	41	34	49	34	25	19	28	33
OHO Denitrification rate	$mgNO_3$-N/(gVSS·h)	17	1.74	1.33	1.25	1.14	1.76	1.12	1.52	1.41
PAO Denitrification rate	$mgNO_3$-N/(gVSS·h)	18	1.57	20.6	2.87	2.63	1.14	2.04	1.9	1.8
% PAO Denitrification	%	19	12.04	24.61	21.56	32.51	17.69	52.28	18.9	21.01
NO_2 formation	Anoxic $mgNO_2$-N/ℓ	20	0.85	4.46	0.24	0	1.81	0	2.83	12.14
Nitrification Rate	mgFSA-N/(gVSS·h)	21	0.71		1.82		1.65	1.09	1.34	0.82

Table 2.8b Anaerobic – anoxic/aerobic batch tests results (Batch Tests 9 to 15).

Date			**0305**	**0811**	**1511**	**0112**	**1804**	**0404**	**1502**
Batch Test No			**9**	**10**	**11**	**12**	**13**	**14**	**15**
VSS concentration	mg/ℓ	1	8721	6258	5026	11182	7503	6369	5522
Organic Load	mgHAC/mgVSS	2	0.034	0.02	0.02	0.01	0.02	0.166	0.009
PAOActive fraction (f_{avPAO})	mgPAOVSS/mgVSS	3	0.264	0.215	0.247	0.243	0.288	0.276	0.282
OHOActive fraction (f_{avOHO})	mgOHOVSS/mgVSS	4	0.095	0.105	0.073	0.136	0.159	0.127	0.148
Prelease/Acetate uptake ratio	mgP/mgHAc	5	0.508	0.515	0.606	0.588	0.753	0.57	0.59
Prelease/Acetate uptake ratio	mol/mol	6	0.493	0.499	0.587	0.57	0.73	0.56	0.57
Anaerobic P Release rate	mgP/(gVSS·h)	7	18.24	17.47	17.68	16.43	14.88	19.49	6.43
Anaerobic P Release rate	mgP/(gPAOVSS·h)	8	69.1	81.3	71.6	67.6	51.7	70.6	22.8
Anaerobic acetate uptake rate	mgHAc/(gVSS·h)	9	31.39	34.92	28.99	27.87	19.36	29.29	22.4
Anaerobic acetate uptake rate	mgHAc/(gPAOVSS·h)	10	118.9	162.4	117.4	114.7	67.2	106.1	79.4
Aerobic P Uptake rate	mgP/(gVSS·h)	11	9.77	8.62	13.37	8.28	10.49	5.74	6.91
Aerobic P Uptake rate	mgP/(gPAOVSS·h)	12	37.01	40.09	54.13	34.07	36.42		24.50
Anoxic P Uptake rate	mgP/(gVSS·h)	13	0.51	1.59	4.67	3.12	2.4		1.88
Anoxic P Uptake rate	mgP/(gPAOVSS·h)	14	1.93	7.40	18.91	12.84	8.33		6.67
DNPAO/APAO ratio	DNPAO/APAO	15	7	23	44	47	28		34
Anx/Aer P uptake rate ratio	q_{anx}/q_{aer}	16	6	18	35	38	22		27
OHO Denitrification rate	$mgNO_3$-N/(gVSS·h)	17	1.38	1.99	1.19	1.43	1.63		1.7
PAO Denitrification rate	$mgNO_3$-N/(gVSS·h)	18	0.72		1.12	1.21	1.16		2.03
% PAO Denitrification	%	19	12		13.52	25.27	15.74		20.67
Anoxic NO_2 formation	$mgNO_2$-N/ℓ	20	10.35	7.96	4.95	2.14	16.91	6.74	9.48
Nitrification Rate	mgFSA-N/(gVSS·h)	21	0.97	1.15	1.5	1.01	1.05	1.15	1.02

anoxic conditions, the DNPAO have utilized their stored PHB. Then only the APAO still have PHB, so when aeration starts, only the APAO take up P. In the anaerobic-aerobic tests, both groups of PAO have stored PHB and consequently both groups take up P. This was confirmed by the PHB profiles, which mirror the P release and P uptake profiles (Figure 2.19). In the first part of the anoxic and aerobic phases, where there is no PHB limitation for both PAO groups, the maximum anoxic P uptake rate is slower than aerobic P uptake, which according to Eq. 2.4 is due to a lower DNPAO VSS concentration and a lower energy capture when the nitrate is utilized as an electron acceptor compared with oxygen, which reduces the yield coefficient under anoxic conditions compared to aerobic conditions (Kuba *et al.*, 1996; Beun *et al.*, 2000; Muller *et al.*, 2003) and is taken into account via the η_{NO_3} factor.

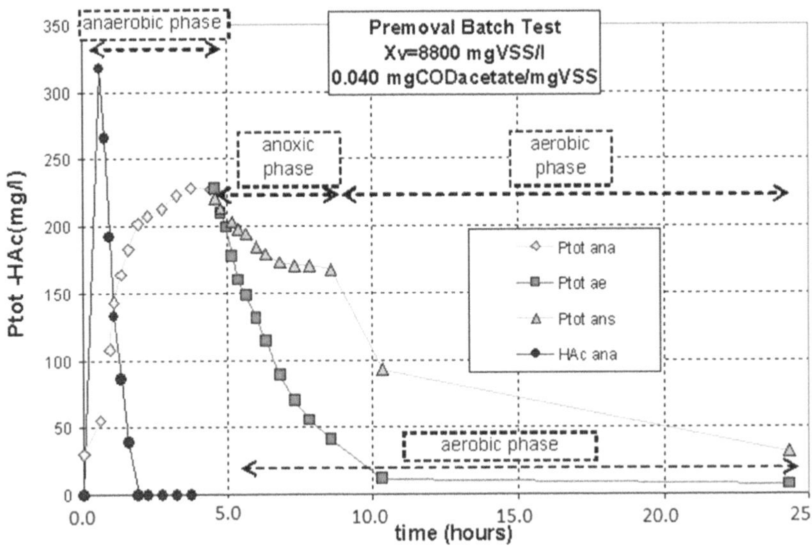

Figure 2.17 Acetate (HAc) and total soluble phosphorus (Ptot) concentration-time profiles.

From the batch tests and the literature information, it seemed reasonable to accept that both PAOs and OHOs participate in the denitrification process ($K_{2OHO} + K_{PAO}$) in the first part (first slope) of the anoxic phase while only the OHO continue to denitrify (second slope) for the remainder of the anoxic phase while nitrate is present (Figure 2.18).

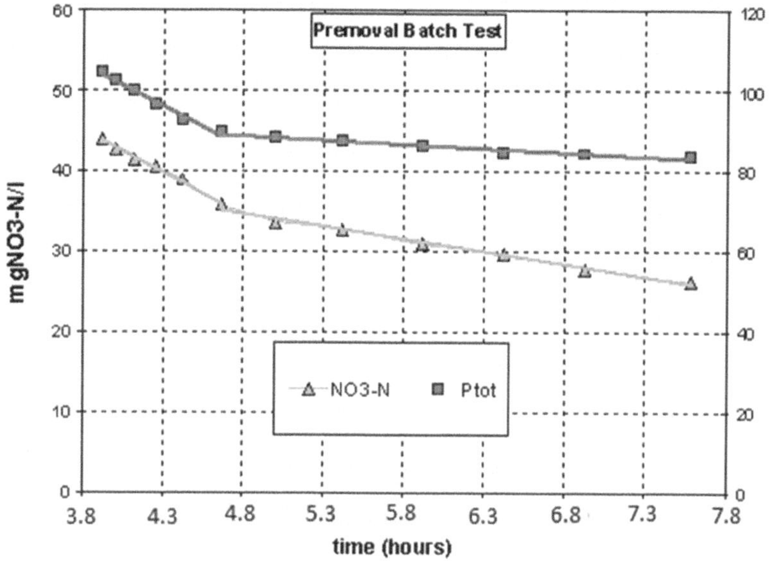

Figure 2.18 Nitrate (NO_3), (left axis) and total soluble phosphorus (Ptot, right axis) concentration-time profiles.

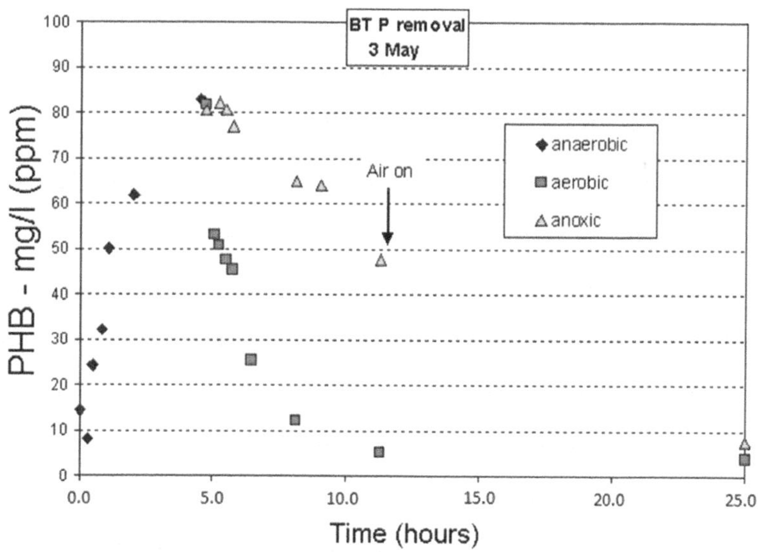

Figure 2.19 Typical PHB concentration-time profiles in mg/l observed in anaerobic and aerobic/anoxic batch tests on MBR sludge.

2.7.3 Anaerobic acetate uptake and P release kinetics

The specific VSS and specific PAOVSS anaerobic acetate (as COD) uptake and P release rates obtained at the different VSS concentrations for the 15 batch tests are shown in Figures 2.20 and 2.21, respectively. For Batch Tests 1 to 13, the acetate uptake and P release rates vary over quite a wide range from 19 to 37 mgCOD/(gVSS·h) or 67 to 162 mgCOD/(gPAOVSS·h) and 12 to 24 mgP/(gVSS·h) or 51 to 96 mgP/(gPAOVSS·h), respectively. No effect of VSS or initial acetate concentration on the rates is apparent from the results (Figures 2.20 and 2.21). The results obtained with different concentrations of acetate added showed that the HAc uptake rates were zero order with respect to HAc concentration, in agreement with the literature studies (Wentzel *et al.*, 1985, 1989), and in conformity with the observations in BT 14 with excess HAc addition where similar anaerobic HAc uptake and P release rates were observed as in the low to moderate acetate dose BTs 1 to 13. The average (and standard deviation) acetate uptake and P release rates obtained over the VSS concentration range (2.7–11.2 gVSS/ℓ) for the 14 batch tests with acetate dose including the excess acetate dose BT 14 are 29.8 ± 5.3 mgCOD/(gVSS·h) or 116.1 ± 24.7 mgCOD/(gPAOVSS·h) and 18.6 ± 3.1 mgP/(gVSS·h) or 72.3 ± 13.7 mgP/(gPAOVSS·h), respectively. The P release to acetate uptake ratio in the 14 batch tests varied between 0.47 and 0.75 with an average of 0.57 ± 0.07 mgP/mgHAc and also shows no effect with acetate dose and VSS concentration.

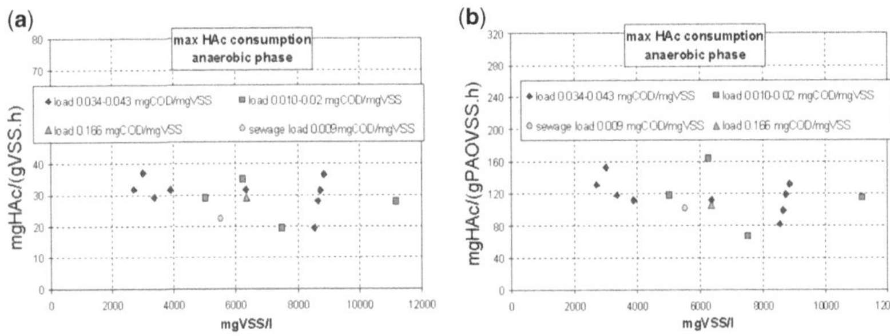

Figure 2.20 Anaerobic VSS specific maximum anaerobic acetate uptake rate mgHAc/(gVSS·h) (a, left) and PAOVSS specific maximum anaerobic acetate uptake rate (mgHAc/(gPAOVSS·h) (b, right).

2.7.4 Aerobic and anoxic P uptake rates

The specific VSS and specific PAOVSS aerobic and anoxic P uptake rates obtained in the 15 batch tests over the VSS concentration range from 2.7 to 11.2 gVSS/ℓ are shown in Figures 2.22a and b and Figures 2.23a and b respectively. For Batch Tests 1 to 14, aerobic and anoxic P uptake rates vary over quite a wide range from 5.5 to 13.4 mgP/(gVSS·h) or 22.6 to 54.1 mgP/(gPAOVSS·h) and 0.5 to 4.7 mgP/(gVSS·h) or 1.9 to 19.1 mgP/(gPAOVSS·h), respectively. Again no effect of VSS or initial acetate concentration on these rates is apparent. The average (and standard deviation) aerobic and anoxic P uptake rates obtained over the VSS concentration range (2.7–11.2 gVSS/ℓ) for the 13 batch tests with acetate dose (excluding the excess acetate dose BT 14 in which anoxic P release was observed) are 10.0 ± 2.3 mgP/(gVSS·h) or 40.0 ± 7.6 mgP/(gPAOVSS·h) and 3.0 ± 1.1 mgP/(gPAOVSS·h) or 11.6 ± 4.7 mgP/(gPAOVSS·h) respectively.

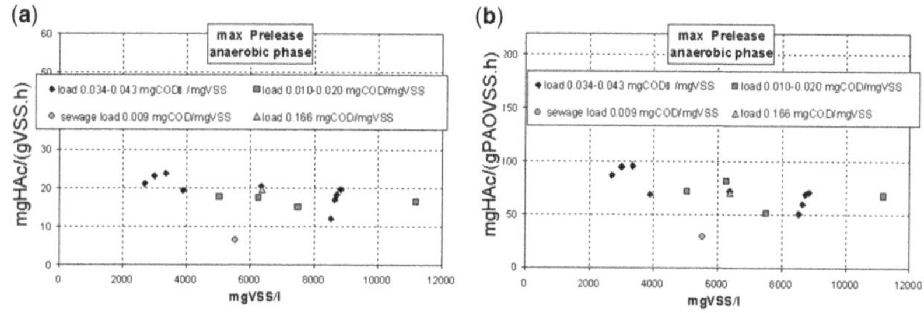

Figure 2.21 Anaerobic VSS specific maximum anaerobic P release rate mgP/(gVSS·h) (a, left) and PAOVSS specific maximum anaerobic P release rate mgP/(gPAOVSS·h) (b, right).

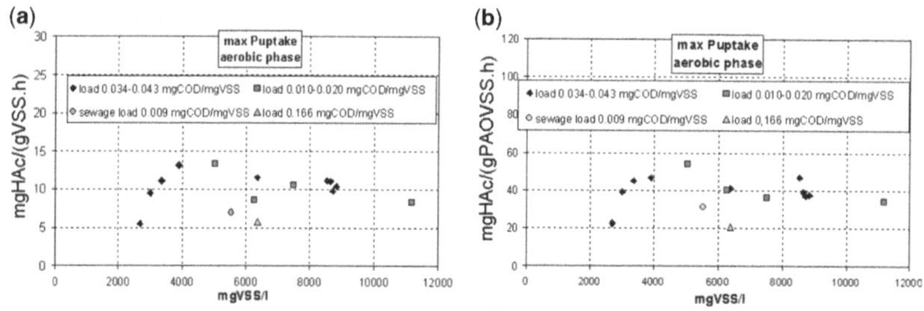

Figure 2.22 Aerobic VSS specific maximum P uptake rate mgP/(gVSS·h) (a, left) and PAOVSS specific maximum P uptake rate mgP/(gPAOVSS·h) (b, right).

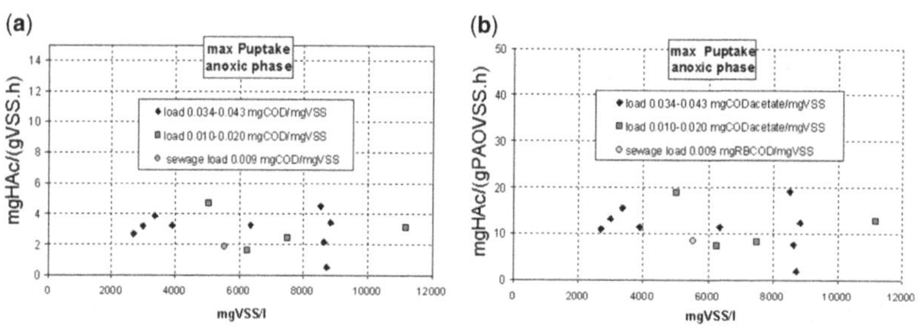

Figure 2.23 Anoxic VSS specific maximum P uptake rate mgP/(gVSS·h) (a, left) and PAOVSS specific maximum P uptake rate mgP/(gPAOVSS·h) (b, right).

2.7.5 Fermentation of readily biodegradable organics (RBO)

For comparison, also shown in Figures 2.20 and 2.21 (as o) are the specific VSS and specific PAOVSS anaerobic soluble COD uptake and P release rates for BT 15 with wastewater addition. Although shown in Figures 2.20 and 2.21, it is strictly not correct to compare these rates of BT 15 with those of BTs 1 to 14. This is because the wastewater soluble RBO are first hydrolysed and fermented to VFAs by the OHOs in the anaerobic reactor before the VFAs can be taken up by the PAOs with P release (Wentzel *et al.*, 1985). So the decrease in soluble (0.45 μm filtered) COD concentration is associated with the hydrolysis/fermentation of RBO to VFAs by OHOs. Similarly, the specific VSS or specific PAOVSS anaerobic P release rate of BT 15 (Figure 2.21) also is not comparable to those of BTs 1 to 14 because this rate is constrained by the VFAs fermentation rate by OHOs. It can be seen that the initial P release rate with wastewater RBO is much slower (only ~1/3 of the rate) than with acetate (Figure 2.21). This is because the hydrolysis rate of RBO by OHOs is much slower than the acetate uptake rate by PAOs so in BT 15 the P release rate is constrained by the production rate of VFAs by OHOs. However, the P release per COD uptake ratio is comparable because the soluble COD reduction rate is linked to the P release rate via the RBO fermentation and VFA uptake rates. The P release per COD uptake ratio for BT 15 was 0.59, which is close to the average of 0.57 obtained for BTs 1 to 14.

The hydrolysis/fermentation rate of RBO to acetate mediated by the OHOs, and the associated P release by the PAOs from the acetate uptake, is modelled as a first order rate with respect to the RBO COD concentration (Wentzel *et al.*, 1985, 1992; Henze *et al.*, 1995), viz.

$$dS_{bs}/dt = K_{\text{ferm}} S_{bs} f_{\text{avOHO}} X_v \qquad \text{mgCOD}/(\ell.\text{d}) \tag{2.5a}$$

$$dP/dt = C_{sp} K_{\text{ferm}} S_{bs} f_{\text{avOHO}} X_v \qquad \text{mgP}/(\ell.\text{d}) \tag{2.5b}$$

where

K_{ferm} = OHO hydrolysis/fermentation rate of RBO - ℓ/(mgOHOVSS·d)
C_{sp} = P release per acetate taken up ratio (mgP/mgCOD)

The average K_{ferm} rate and C_{sp} ratio measured by Wentzel *et al.* (1985) in CAS BNR systems fed real wastewater were 0.06 ℓ/(mgOHOVSS·d) at 20°C and 0.5 mgP/mgCOD respectively.

Applying the Wenzel *et al.* (1985) fermentation equation to the results of BT 15 yielded K_{ferm} = 0.036 ℓ/(mgOHOVSS·d) when based on the measured soluble COD-time profile and 0.027 ℓ/(mgOHOVSS·d) when based on the P release-time profile. This rate is considerably lower than the 0.06 rate measured by Wentzel *et al.* (1985). Considering the rather wide variation in the other kinetic rates measured in this investigation, and in the individual K_{ferm} rates measured by Wentzel *et al.* (1985), it is possible that the K_{ferm} rate measured in this one BT 15 with wastewater is a low value in a population with an average around 0.06 ℓ/(mgOHOVSS·d). In hindsight, fewer batch tests with acetate addition and more with wastewater addition should have been done so the effect of high VSS concentration on the fermentation rate could be established.

Considering that…

(1) the variation in the individual K_{ferm} rate measured by Wentzel *et al.* (1985) of which 0.06 ℓ/(mgOHOVSS·d) was the average,
(2) the VSS concentration in the anaerobic reactor of MBR BNR systems is not much higher than in CAS BNR systems: from Ramphao *et al.* (2005) the VSS concentration in the anaerobic reactor will not be greater than half that in the aerobic reactor due to the diluting effect of the mixed liquor cycles in the MBR system (see Reactor Sizing Considerations below) – if the aerobic reactor VSS concentration is say 10 gVSS/ℓ, then the anaerobic reactor VSS concentration will be less than 5 gVSS/ℓ, which is not much greater than the VSS concentration in the anaerobic reactor of CAS systems;
(3) the acetate uptake and P release rates by PAOs showed no VSS concentration effect up to 11 gVSS/ℓ,
(4) the OHOs, which mediate the RBO fermentation to acetate, showed no decrease in denitrification rate with VSS concentration,

….it is likely that the fermentation rate of OHO (K_{ferm}) also is not affected by high VSS concentrations so that the rate observed in CAS systems also can be accepted for MBR systems.

2.7.6 Comparing kinetic rates with those of other investigations

The acetate uptake, P release and P uptake kinetic rates determined in this investigation are compared in Table 2.9 with rates determined in previous investigations with batch tests on sludge harvested from different BNRAS systems at conventional BNR system VSS concentrations, such as multi-reactor UCT systems and sequencing batch reactors (SBR) fed real wastewater (mixed OHO and PAO cultures) or acetate only (enhanced PAO cultures).

The VSS (mixed culture) specific rates from this investigation are within the rather wide range of VSS specific values in the literature for CAS BNR systems. The literature rates vary widely because the PAO active fraction (f_{avPAO}) differs widely depending on the P removal achieved in the systems relative to the influent COD concentration – the higher the P removal, the higher the f_{avPAO} of the measured VSS and therefore the higher the VSS specific rate. The PAOVSS specific rates measured in this investigation are also within the range of PAOVSS specific rates measured on enhanced PAO cultures (Wentzel *et al.*, 1989; Kuba *et al.*, 1994;

Kuba *et al.*, 1997; Wachtmeister *et al.*, 1997; Hu *et al.*, 2003). The range of literature PAOVSS specific rates is wide and is due to some enhanced PAO culture systems being influent P limited rather than influent COD limited. Accepting that in this investigation (in which the two systems were not P limited), the f_{avPAO} was estimated reasonably accurately, the conformity of the acetate uptake, P release and P uptake kinetic rates to literature rates from CAS BNR systems, confirms that these PAO kinetic rates are not affected by high VSS concentrations in MBR systems. The investigation also highlights the importance of specifying kinetic rates of bioprocesses with respect to the actual biomass concentration mediating the particular bioprocess. Rates with respect to a lumped parameter like VSS are not comparable between different BNR systems and not useful for kinetic models.

Table 2.9 Average P removal kinetic rates and stoichiometry measured in this investigation compared with those from other investigations.

Parameter		This Investigation	Other (Refs)
Anaerobic P release rate with acetate	mgP/(gVSS·h)	18.6 ± 3.1	5–19 (6,7,9)
	mgP/(gPAOVSS·h)	72.3 ± 13.7	40–106 (1,3,9)
Anaerobic HAc uptake rate	mgHAcCOD/(gVSS·h)	29.8 ± 5.3	7–50 (6)
	mgHAcCOD/(gPAOVSS·h)	116.1 ± 24.7	8–236 (1,3,9)
Aerobic P uptake	mgP/(gVSS·h)	10.0 ± 2.3	4–30 (4,5,6,9)
	mgP/(gPAOVSS·h)	40.0 ± 7.6	40–72 (3,5,8,9)
Anoxic P uptake	mgP/(gVSS·h)	3.0 ± 1.1	1.2–13 (4,5,6,7)
	mgP/(gPAOVSS·h)	11.6 ± 4.7	10–40 (1,3,5,9)
$P_{release}/HAc_{uptake}$	mgP/mgCOD	0.57 ± 0.07	0.15–0.60 (1,2,3,6,8,9)

(1) Kuba *et al.* (1993), (2) Kerrn-Jespersen and Henze (1993), (3) Kuba *et al.* (1994), (4) Sorm *et al.* (1996), (5) Wachtmeister *et al.* (1997), (6) Kuba *et al.* (1997), (7) Sorm *et al.* (1997), (8) Wentzel *et al.* (1989), (9) Hu *et al.* (2003).

2.7.7 Comparing PAO and OHO denitrification behaviour in this and with other investigations

The anoxic P uptake and denitrification by PAOs can vary significantly with time in a particular system (Hu *et al.*, 2002). This has been observed in this investigation and others (Kuba *et al.*, 1997; Ekama & Wentzel, 1999; Hu *et al.*, 2002). From these observations, it appears the magnitude of anoxic P uptake relative to the total (anoxic and aerobic) P uptake depends on mainly two factors for significant anoxic P uptake to take place, viz. (1) the duration of exposure of the activated sludge to aerobic and anoxic conditions, i.e. high anoxic and low aerobic mass fractions; and

(2) the nitrate load on the anoxic reactor in relation to its denitrification potential,. i.e. high nitrate load which exceeds the denitrification potential so that nitrate is present in the anoxic reactor outflow. Different methods have been developed in order to determine the fraction of denitrifying PAOs, X_{DNPAO}/X_{PAO} (Meinhold *et al.*, 1999; Hu *et al.*, 2003), the PAO contribution to denitrification (Hu *et al.*, 2002) or the level of PAO activity in the anoxic zone (Wachmeister *et al.*, 1997). With these methods, the relative magnitude of PAO denitrification can be determined for a specific sludge drawn from an EBPR system.

In this investigation, the PAO denitrification (K_{PAO}) and anoxic P uptake rates were calculated from the anaerobic-anoxic-aerobic batch test results, where the first slope corresponded to $K_{PAO} + K_{2OHO}$ and the second flatter slope to the K_{2OHO} only (Figure 2.21). The PAO denitrification in the MBR system was calculated from the system P uptake and denitrification performance according to the method of Hu *et al.* (2002). This method assumes that the nitrate concentration denitrified by PAOs is proportional to the % PHB utilized by the DNPAOs in the anoxic reactor which in turn is proportional to the % anoxic P uptake relative to the total anoxic and aerobic P uptake (calculated from P mass balances around the anoxic and aerobic reactors). So the nitrate concentration denitrified by the PAOs is the catabolic fraction [$(1\text{-}f_{cv}Y_{G,ANX})/2.86 = 6.2$ mgCOD/mgNO$_3$-N] of the influent RBO taken up by PAOs in the anaerobic reactor (ΔS_{bsi} = PHBCOD) multiplied by the % anoxic P uptake. The nitrate concentration denitrified by OHOs is the difference between the observed nitrate concentration denitrified in the anoxic reactor and the nitrate concentration denitrified by PAOs. Once the concentration of nitrate denitrified by the OHOs and PAOs was known, the VSS specific and PAOVSS and OHOVSS specific denitrification rates were calculated from the PAO and OHO active fractions (f_{avPAO}, f_{avOHO}) as described above.

Comparing the % anoxic P uptake determined for the MBR system with that measured in the batch tests over the same period in which the batch tests were conducted (Table 2.10a and 2.10b), it can be seen that this was much lower in the MBR system than in the batch tests. This was probably due to the low nitrate load on the anoxic reactor of the MBR system relative to its denitrification potential (<1mgN/ℓ, Table 2.10a and 2.10b), which indicates the anoxic reactor was under loaded with nitrate. Consequently, the PAOs had little opportunity to use the limited nitrate available in the anoxic reactor of the MBR system – they had to compete with the faster OHO for entering nitrate. In contrast, in the batch tests with excess acetate dosing during the anaerobic phase leading to high PHB and excess nitrate during the anoxic phase (both conditions not prevalent in the MBR system), the DNPAOs can utilize their PHB with nitrate without competition from OHOs. So the conditions in the batch tests are not the same as the conditions in the MBR system. This is the reason for the large difference between the PAO denitrification rate [K_{PAO}, mgNO$_3$-N/(mgPAOVSS·d)] measured in the batch tests compared with those calculated from the MBR system performance results (Table 2.7). For the batch tests, the K_{PAO} varied between 0.10 and 0.28 mgNO$_3$-N/(mgPAOVSS.d)

Table 2.10a Comparison of batch test 1 to 8 and MBR system OHO and PAO VSS specific denitrification rates.

Batch Test Date			**1801**	**2111**	**2811**	**0412**	**0602**	**0902**	**1302**	**2903**
Batch Test No			**1**	**2**	**3**	**4**	**5**	**6**	**7**	**8**
OHOActive fraction (f_{avOHO})	mgOHOVSS/mgVSS	1	0.125	0.073	0.136	0.136	0.148	0.148	0.148	0.127
PAOActive fraction (f_{avPAO})	mgPAOVSS/mgVSS	2	0.237	0.247	0.243	0.243	0.282	0.282	0.282	0.276
OHO Denitrification rate	mgNO_3-N/(gVSS·h)	3	1.74	1.33	1.25	1.14	1.76	1.12	1.52	1.41
OHO Denitrification rate	mgNO_3-N/(mgOHOVSS·d)	4	0.334	0.437	0.221	0.201	0.285	0.182	0.246	0.266
PAO Denitrification rate	mgNO_3-N/(gVSS·h)	5	1.57	2.06	2.87	2.63	1.14	2.04	1.9	1.8
PAO Denitrification rate	mgNO_3-N/(mgPAOVSS·d)	6	0.159	0.200	0.283	0.260	0.097	0.174	0.162	0.157
MBR System										
Wastewater Batch No		1	15	17	18	18	22	22	22	26
Batch Period		2	1310	1111	1211	2911	3101a	1402	3101b	2403
Anoxic P uptake	%	3	18.7	11.2	11.2	13.1	6.1	6.1	6.1	17.5
Aerobic P uptake	%	4	81.3	88.8	88.8	86.9	93.9	93.9	93.9	82.5
Anx/Aer P uptake ratio	%	5	28.8	15.8	15.8	18.8	8.1	8.1	8.1	26.5
OHO denitrification rate	mgNO_3-N/(mgOHOVSS·d)	6	0.239	0.314	0.158	0.173	0.271	0.271	0.271	0.297
PAO denitrification rate	mgNO_3-N/(mgPAOVSS·d)	7	0.033	0.023	0.022	0.022	0.01	0.01	0.01	0.029
Anoxic reactor nitrate conc	mgNO_3-N/ℓ	8	1.33	1.32	0.14	0.14	0.27	0.27	0.27	0.39

Table 2.10b Comparison of batch test 9 to 15 and MBR system OHO and PAO VSS specific denitrification rates.

Date		**0305**	**0811**	**1511**	**0112**	**1804**	**0404**	**1502**
Batch Test No		**9**	**10**	**11**	**12**	**13**	**14**	**15**
OHOActive fraction (f_{avOHO})	mgOHOVSS/mgVSS	0.095	0.105	0.073	0.136	0.159	0.127	0.148
PAOActive fraction (f_{avPAO})	mgPAOVSS/mgVSS	0.264	0.215	0.247	0.243	0.288	0.276	0.282
OHO denitrification rate	$mgNO_3$-N/(gVSS·h)	1.38	1.99	1.19	1.43	1.63		1.7
OHO denitrification rate	$mgNO_3$-N/(mgOHOVSS·d)	0.349	0.455	0.391	0.252	0.246		0.276
PAO denitrification rate	$mgNO_3$-N/(gVSS·h)	0.72		1.12	1.21	1.16		2.03
PAO denitrification rate	$mgNO_3$-N/(mgPAOVSS·d)	0.065		0.109	0.120	0.097		0.173
MBR System								
Wastewater Batch No		28	16	17	18	27		23
Batch Period		2304	0810	1111	1211	0804		1502
Anoxic P uptake	%	15.2	4.6	11.2	13.1	14.8		4.3
Aerobic P uptake	%	84.8	95.4	88.8	86.9	85.2		95.7
Anx/Aer P uptake ratio	% = Anx/(Aer)·(1/η_{NO_3})	22.4	6.0	15.8	18.8	21.7		5.6
OHO denitrification rate	$mgNO_3$-N/(mgOHOVSS·d)	0.202	0.202	0.314	0.158	0.25		0.228
PAO denitrification rate	$mgNO_3$-N/(mgPAOVSS·d)	0.028	0.008	0.023	0.022	0.023		0.007
Anoxic reactor nitrate conc	$mgNO_3$-N/ℓ	0.04	0.33	1.32	0.14	0.23		0.01

with a mean of 0.16 whereas in the MBR system the K_{PAO} varied between 0.01 and 0.03 mgNO_3-N/(mgPAOVSS.d) with a mean of 0.02. In contrast, the OHO denitrification rates [K_{2OHO}, mgNO_3-N/(mgOHOVSS·d) measured in the batch tests were similar to those calculated from the MBR system performance results (Table 2.7). In the batch tests, the K_{2OHO} varied between 0.18 and 0.45 mgNO_3-N/(mgOHOVSS·d) with a mean of 0.33 and in the MBR system, it varied between 0.16 and 0.31 mgNO_3-N/(mgPAOVSS·d) with a mean of 0.27. The batch test and MBR system K_{2OHO} rates are neither statistically different from each other nor to those measured in the anoxic-aerobic batch tests described earlier.

In a previous investigation comparing conventional BNRAS systems with external nitrification (EN) BNRAS systems, Hu *et al.* (2002) observed a similar large difference between the K_{PAO} and K_{2OHO} rates as that in the MBR system of this investigation. In the conventional BNRAS systems K_{2OHO} varied from 0.05 to 0.32 mgNO_3-N/(mgOHOVSS·d) with a mean of 0.14 while K_{PAO} varied from 0.005 to 0.07 mgNO_3-N/(mgPAOVSS·d) with a mean of 0.028. In the ENBNRAS systems, which promoted anoxic P uptake EBPR due to its high anoxic mass fraction and overloaded anoxic reactor, the K_{PAO} varied from 0.03 to 0.07 mgNO_3-N/(mgPAOVSS·d) with a mean of 0.05. The K_{2OHO} varied from 0.12 to 0.18 mgNO_3-N/(mgOHOVSS·d) with a mean of 0.15. So the specific PAOVSS denitrification rate of the PAOs on internally stored PHB was 3 to 5 times lower than that of the OHOs on slowly biodegradable particulate organics (BPO) in conventional and EN BNR systems. In this investigation, the specific PAOVSS denitrification rate of the PAOs on internally stored PHB was 14 times lower than that of the OHOs on slowly biodegradable particulate organics (BPO) in the MBR system.

During the investigation, in the MBR system, the PAOs achieved an average % anoxic P uptake of 9% and contributed 11% to the total denitrification. In the batch tests, the PAOs achieved an average % anoxic P uptake of 37% and contributed 22% to the total denitrification. The relationship between the proportion of DNPAOs to total PAOs (X_{DNPAO}/X_{PAO}) and the PAO contribution to the total denitrification is shown plotted in Figure 2.24. As expected, the greater the proportion of DNPAOs, the greater the % denitrification by PAOs. For the results of this investigation, the X_{DNPAO}/X_{PAO} ratio varied from 8 to 61% while the contribution by the PAOs to denitrification varied from 12 to 25%. This indicates, as found by Hu *et al.* (2002), that the contribution by the PAOs to denitrification was small compared with that of the OHOs. The large reduction in P removal (20 to 30%) with significant anoxic P uptake EBPR compared with aerobic P uptake EBPR appears to be a high price to pay for the low additional denitrification.

Finally, with regard to the nitrification rates obtained during the aerated part of the anaerobic batch tests, VSS specific nitrification rates [mgFSA-N/(gVSS·h)] with VSS concentration are shown in Figure 2.25. As found in the aerobic nitrification tests above, as the VSS concentration in the batch tests increases, the specific nitrification rate decreases. This confirms the VSS concentration effect on the VSS specific nitrification rate.

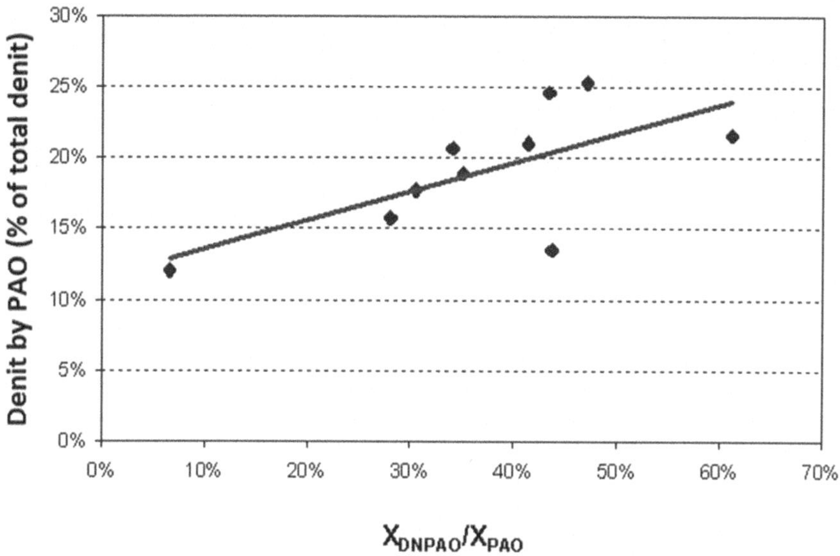

Figure 2.24 Denitrification by PAOs (as a percentage of the total denitrification) versus proportion of denitrifying PAO (as a percentage of the total PAOs, X_{DNPAO}/X_{PAO}).

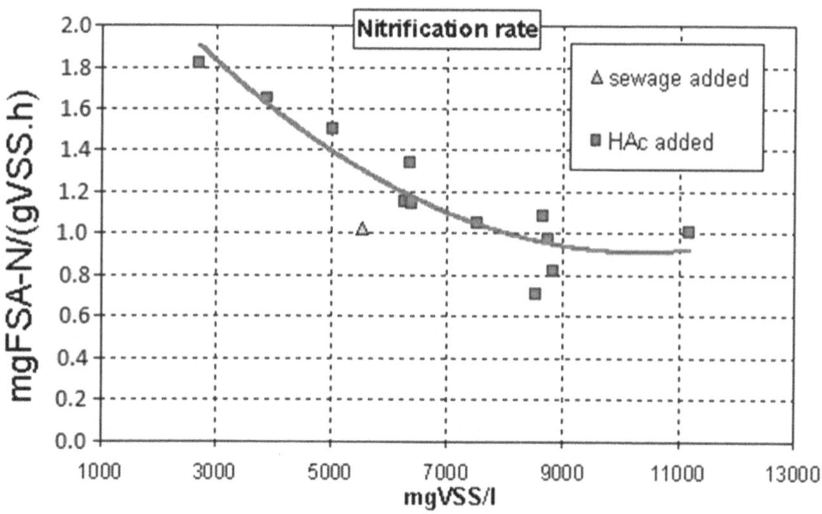

Figure 2.25 VSS specific ammonia utilization rate (AUR) measured in the aerobic phase of the anaerobic-anoxic-aerobic batch tests at different VSS concentrations.

2.8 MEMBRANE NDEBPR SYSTEM REACTOR SIZING CONSIDERATIONS

2.8.1 Converting between sludge mass fractions and volume fractions – general considerations

Steady state BNR models are commonly used for sizing BNR system reactor volumes such as the Metcalf & Eddy (2014), Deutsche Vereinigung für Wasserwirtschaft, Abwasser and Abfall (DWA, German Association for Water, Wastewater and Waste) (2016) or the University of Cape Town (UCT, WRC, 1984; Henze *et al.*, 2008; Ekama, 2011) steady state models. One outcome of these steady state models is the mass of TSS (MX_t, kgTSS) in the biological reactor at the selected solids retention time (SRT, sludge age) and the sludge mass fractions of the various zones of the reactor such as anaerobic (f_{xana}), anoxic (f_{xanx}) and aerobic (f_{xaer}). These zone sludge mass fractions need to be transformed to volume fractions to determine the volumes of the different zones of the reactor. Similarly, for existing NDEBPR systems, the zone volume fractions of the different zones of the whole reactor may need to be converted to sludge mass fractions. Equations for transforming sludge mass fractions to volume fractions and *vice versa* for various BNR configurations with SSTs or membranes for solid-liquid separation are given in this section.

A total reactor volume V_R, i.e. a non-configuration specific volume, can be calculated from a selected average reactor TSS concentration ($\overline{X_t}$) for the reactor as a whole, i.e.

$$V_R = \frac{MX_t}{\overline{X_t}} \quad \text{m}^3 \tag{2.6}$$

where

MX_t = mass of TSS in the reactor (kgTSS)

$\overline{X_t}$ = zone volume weighed average TSS concentration in the whole BNR reactor (kgTSS/m^3).

The volume fraction (f_v) of a zone is the volume of the zone divided by the whole reactor volume (V_R). The sludge mass fraction (f_x) of a zone is the mass of sludge in it (kgTSS) divided by the total mass of sludge (MX_t, kgTSS) in the whole reactor (V_R, m^3). For example, a UCT system with membranes in the aerobic reactor (Figure 2.1a) has anaerobic, anoxic and aerobic zone volumes of 1500, 3500 and 5000 m^3. Therefore the whole reactor volume (V_R) is 10000 m^3 and the anaerobic, anoxic and aerobic volume fractions are f_{vana} = 1500/10000 = 0.150, f_{vanx} = 3500/10000 = 0.350 and f_{vaer} = 5000/10000 = 0.500 respectively. If the aerobic zone TSS concentration (X_{taer}) is 10 kgTSS/m^3, and the mixed liquor *a*-recycle ratio from the aerobic to the anoxic zone is 4:1 and the mixed liquor *r*-recycle ratio from the anoxic to the anaerobic zone is 1.5:1, then the anaerobic (X_{tana}) and anoxic (X_{tanx}) zone TSS concentrations are 4.8 kgTSS/m^3 and 8.0 kgTSS/m^3. These zone concentrations are determined by the TSS mass balance over each zone, for which the wastage sludge flow rate can be ignored to keep the equations simple. This is demonstrated below in Eqs. 2.7 to 2.36. From the zone TSS concentrations, the sludge mass in each zone and whole reactor is $1500 \times 4.8 + 3500 \times 8.0 + 5000 \times 10.0 = 7200 + 28000 + 50000 = 85200$ kgTSS. Hence the sludge mass fraction of each zone is: anaerobic (f_{xana}) = 7200/85200 = 0.0845; anoxic (f_{xanx}) = 28000/85200 = 0.3286 and aerobic (f_{xaer}) = 50000/85200 = 0.5869 and the volume weighted average TSS concentration ($\overline{X_t}$) is 85200/10000 = 8.52 kgTSS/m^3. The sum of the volume fractions is 1 and the sum of the mass fractions also is 1. All the sludge mass fraction to volume fractions (and reverse) equations listed in the tables below were derived following the same principles set out below. The principles followed in these derivations can be applied to any BNR system with membranes or SST for solid-liquid separation and was the procedure followed to derive the volume to sludge mass fraction (and reverse) equations for the MBR BNR pilot plant operated by Mannina *et al.* (2018b), which had a different configuration to those considered in this chapter.

In sizing the SST (Ekama *et al.*, 1997) or MBR solid-liquid separation units, it is the TSS concentration feed to the SST or in the MBR that is usually size defining and so is selected for sizing of the BNR system volume. For BNR systems with SSTs or membranes in the aerobic zone (e.g. submerged panel membranes), this size defining selected TSS concentration is the aerobic zone TSS concentration (X_{taer}). For BNR systems with membranes in a separate membrane reactor following the aerobic zone, the size defining TSS concentration is the membrane zone TSS concentration (X_{tmbr}). So with the equations below, which link the configuration zone volume fractions to the design sludge mass fractions, equations for the aerobic or membrane zone TSS concentration in terms of the average system TSS concentration $\overline{X_t}$ are given. These equations allow the average system TSS concentration $\overline{X_t}$ to be calculated from a selected aerobic (X_{tear}) or membrane (X_{tmbr}) zone TSS concentration. This average system TSS concentration $\overline{X_t}$ is then used in the sludge mass fraction to volume fraction conversion equations. For existing systems, the aerobic or membrane zone TSS concentration is the measured TSS concentration in these zones and the recycle ratios can be calculated (or checked) from the measured long-term average TSS concentrations in the different zones of the system.

2.8.2 Derivation of the sludge mass – volume fraction equations

The procedure for deriving the sludge mass fraction – volume fraction equations is demonstrated for the UCT system with membranes in the aerobic zone (Figure 2.1a).

From a TSS mass balance over the anaerobic zone....

$$0\,x1 + rX_{\text{tanx}} = (1+r)X_{\text{tana}} \text{ from which } X_{\text{tana}} = \frac{r}{(r+1)}X_{\text{tanx}} \tag{2.7}$$

where

X_t = TSS concentration and subscripts *ana*, *anx* and *aer* refer to the anaerobic, anoxic and aerobic zone TSS concentrations

r = anoxic to anaerobic zone recycle ratio.

From a TSS mass balance over the anoxic zone....

$$(1+r)X_{\text{tana}} + aX_{\text{taer}} = (a+r+1)X_{\text{tanx}}$$

where

a = aerobic to anoxic zone recycle ratio.

Substituting Eq. 2.7 for X_{tana} yields $X_{\text{taer}} = \frac{a+1}{a}X_{\text{tanx}}$ (2.8)

From a TSS mass balance over the aerobic zone....

$$(a+1)X_{\text{tanx}} = aX_{\text{taer}}$$

Substituting Eq. 2.7 for X_{tanx} yields $X_{\text{taer}} = \frac{(a+1)(r+1)}{ar}X_{\text{tana}}$ (2.9)

The total TSS sludge mass in the whole reactor is

$$MX_t = V_R\overline{X_t} = V_{\text{ana}}X_{\text{tana}} + V_{\text{anx}}X_{\text{tanx}} + V_{\text{aer}}X_{\text{tae}}$$

where

V = volume and subscripts *ana*, *anx* and *aer* refer to the anaerobic, anoxic and aerobic zones.

Dividing through by V_R yields

$$\overline{X_t} = \frac{V_{\text{ana}}}{V_R}X_{\text{tana}} + \frac{V_{\text{anx}}}{V_R}X_{\text{tanx}} + \frac{V_{\text{aer}}}{V_R}X_{\text{taer}}$$
$$\overline{X_t} = f_{\text{vana}}X_{\text{tana}} + f_{\text{vanx}}X_{\text{tanx}} + f_{\text{vaer}}X_{\text{taer}} \tag{2.10}$$

where

f_v = volume fraction and subscripts *ana*, *anx* and *aer* refer to the anaerobic, anoxic and aerobic zones.

With Eqs. 2.7 to 2.9, substitute out X_{tanx} and X_{taer} in Eq. 2.10 to give Eq. 2.10 in terms of X_{tana} only, i.e.,

$$\overline{X_t} = f_{\text{vana}} X_{\text{tana}} + f_{\text{vanx}} X_{\text{tana}} \frac{(r+1)}{r} + f_{\text{vaer}} X_{\text{tana}} \frac{(r+1)(a+1)}{ar} \tag{2.11}$$

Dividing Eq. 2.11 through by X_{tana} yields

$$\frac{\overline{X_t}}{X_{\text{tana}}} = f_{\text{vana}} + f_{\text{vanx}} \frac{(r+1)}{r} + f_{\text{vaer}} \frac{(r+1)(a+1)}{ar} \tag{2.12}$$

The mass of TSS sludge in the anaerobic zone is given by

$$\overline{X_t} V_R f_{\text{xana}} = V_{\text{ana}} X_{\text{tana}} \text{ from which } \frac{\overline{X_t}}{X_{\text{tana}}} = \frac{f_{\text{vana}}}{f_{\text{xana}}} \tag{2.13}$$

So from Eqs. 2.12 and 2.13,

$$\frac{f_{\text{vana}}}{f_{\text{xana}}} = f_{\text{vana}} + f_{\text{vanx}} \frac{(r+1)}{r} + f_{\text{vaer}} \frac{(r+1)(a+1)}{ar} \tag{2.14}$$

Similarly, with Eqs. 2.7 to 2.9, substitute out X_{tana} and X_{taer} in Eq. 2.10 to give Eq. 2.10 in terms of X_{tanx} only, i.e.,

$$\overline{X_t} = f_{\text{vana}} X_{\text{tanx}} \frac{r}{(r+1)} + f_{\text{vanx}} X_{\text{tanx}} + f_{\text{vaer}} X_{\text{tanx}} \frac{(a+1)}{a} \tag{2.15}$$

Dividing Eq. 2.15 through by X_{tanx} yields

$$\frac{\overline{X_t}}{X_{\text{tanx}}} = f_{\text{vana}} \frac{r}{r+1} + f_{\text{vanx}} + f_{\text{vaer}} \frac{(a+1)}{a} \tag{2.16}$$

The mass of TSS sludge in the anoxic zone is given by

$$\overline{X_t} V_R f_{\text{xanx}} = V_{\text{anx}} X_{\text{tanx}} \text{ from which } \frac{\overline{X_t}}{X_{\text{tanx}}} = \frac{f_{\text{vanx}}}{f_{\text{xanx}}} \tag{2.17}$$

So from Eqs. 2.16 and 2.17,

$$\frac{f_{\text{vanx}}}{f_{\text{xanx}}} = f_{\text{vana}} \frac{r}{r+1} + f_{\text{vanx}} + f_{\text{vaer}} \frac{(a+1)}{a} \tag{2.18}$$

Similarly, with Eqs. 2.7 to 2.9, substitute out X_{tana} and X_{tanx} in Eq. 2.10 to give Eq. 2.10 in terms of X_{taer} only,

$$\overline{X_t} = f_{vana}X_{taer}\frac{ar}{(a+1)(r+1)} + f_{vanx}X_{taer}\frac{a}{a+1} + f_{vaer}X_{taer} \tag{2.19}$$

Dividing this equation through by X_{taer} yields

$$\frac{\overline{X_t}}{X_{taer}} = f_{vana}\frac{ar}{(a+1)(r+1)} + f_{vanx}\frac{a}{(a+1)} + f_{vaer} \tag{2.20}$$

The mass of TSS sludge in the aerobic zone is given by

$$\overline{X_t}V_R f_{xaer} = V_{aer}X_{taer} \text{ from which } \frac{\overline{X_t}}{X_{taer}} = \frac{f_{vaer}}{f_{xaer}} \tag{2.21}$$

So from Eqs. 2.20 and 2.21,

$$\frac{f_{vaer}}{f_{xaer}} = f_{vana}\frac{ar}{(a+1)(r+1)} + f_{vanx}\frac{a}{a+1} + f_{vaer} \tag{2.22}$$

The sum of the volume fractions is 1. Hence $f_{vanx} = 1 - f_{vana} - f_{vaer}$. Substituting $1 - f_{vana} - f_{vaer}$ for f_{vanx} into Eq. 2.18 for the anoxic reactor yields

$$\frac{f_{vanx}}{f_{xanx}} = f_{vana}\frac{r}{r+1} + (1 - f_{vana} - f_{vaer}) + f_{vaer}\frac{(a+1)}{a} \tag{2.23}$$

Simplifying the right hand side of Eq. 2.23 yields

$$\frac{f_{vanx}}{f_{xanx}} = \left[1 - \frac{f_{vana}}{r+1} + \frac{f_{vaer}}{a}\right] \tag{2.24}$$

So from Eq. 2.24, the anoxic sludge mass fraction in terms of the anaerobic and aerobic volume fractions is given by

$$f_{xanx} = f_{vanx}G \tag{2.25}$$

$$\text{where } G = \frac{1}{\left[1 - \frac{f_{vana}}{r+1} + \frac{f_{vaer}}{a}\right]} \tag{2.26}$$

Similarly, substituting $1 - f_{vana} - f_{vaer}$ for f_{vanx} into Eq. 2.14 for the anaerobic reactor yields

$$\frac{f_{vana}}{f_{xana}} = f_{vana} + (1 - f_{vana} - f_{vaer})\frac{(r+1)}{r} + f_{vaer}\frac{(r+1)(a+1)}{ar} \tag{2.27}$$

Simplifying the right hand side of Eq. 2.27 yields

$$\frac{f_{vana}}{f_{xana}} = \frac{(r+1)}{r}\left[1 - \frac{f_{vana}}{r+1} + \frac{f_{vaer}}{a}\right] \tag{2.28}$$

So from Eq. 2.28, the anaerobic sludge mass fraction in terms of the anaerobic and aerobic volume fractions is given by

$$f_{xana} = \frac{rf_{vana}}{(r+1)\left[1 - \frac{f_{vana}}{r+1} + \frac{f_{vaer}}{a}\right]} = \frac{rf_{vana}G}{(r+1)} \tag{2.29}$$

where G is given by Eq. 2.26.

Again substituting $1 - f_{vana} - f_{vaer}$ for f_{vanx} into Eq. 2.22 for the aerobic reactor yields

$$\frac{f_{vaer}}{f_{xaer}} = f_{vana}\frac{ar}{(a+1)(r+1)} + (1 - f_{vana} - f_{vaer})\frac{a}{a+1} + f_{vaer} \tag{2.30}$$

Simplifying the right hand side of Eq. 2.30 yields

$$\frac{f_{vaer}}{f_{xaer}} = \frac{a}{a+1}\left[1 - \frac{f_{vana}}{r+1} + \frac{f_{vaer}}{a}\right] \tag{2.31}$$

So from Eq. 2.31, the aerobic sludge mass fraction in terms of the anaerobic and aerobic volume fractions is given by

$$f_{xaer} = \frac{(a+1)f_{vaer}G}{a\left[1 - \frac{f_{vana}}{r+1} + \frac{f_{vaer}}{a}\right]} = \frac{(a+1)}{a}f_{vaer}G \tag{2.32}$$

where G is given by Eq. 2.26.

To convert sludge mass fractions to volume fractions requires G to be expressed in terms of the anaerobic and aerobic mass fractions. This is done by making the anaerobic and aerobic volume fractions the subjects of

Eqs. 2.29 and 2.32, substituting these in to Eq. 2.26 for G and solving for G. This yields,

$$G = 1 + \frac{f_{\text{xana}}}{r} - \frac{f_{\text{xaer}}}{a+1} \tag{2.33}$$

To find the anaerobic, anoxic and aerobic volume fractions in terms of sludge mass fractions, the anaerobic, anoxic and aerobic volume fractions are made the subject of Eqs. 2.29, 2.25 and 2.32. This yields,

$$f_{\text{vana}} = \frac{(1+r)f_{\text{xana}}}{rG} \tag{2.34}$$

$$f_{\text{vanx}} = \frac{f_{\text{xanx}}}{G} \tag{2.35}$$

$$f_{\text{vaer}} = \frac{af_{\text{xaer}}}{(a+1)G} \tag{2.36}$$

The equations derived above are listed in Table 2.15. All the other equations in Tables 2.11 to 2.17 were derived following the same principles. The principles followed in these derivations can be applied to any BNR system with membranes or SST for solid-liquid separation and is the procedure followed to derive the volume to sludge mass fraction (and reverse) equations for the MBR BNR pilot plant operated by Mannina *et al.* (2018b), which had a different configuration to those considered in this chapter.

For the derivation of the equations for other configurations, Eq. 2.21 can be extended to all BNR system configurations with secondary settling tanks (SSTs) or with membranes (MBR) for solid-liquid separation, at steady state, average dry weather flow (ADWF) conditions and negligibly small waste flow rate; the concentrations of TSS in the pre-anoxic, anaerobic (if present) and anoxic, aerobic and membrane zones (X_{tpax}, X_{tana}, X_{tanx}, X_{taer}, X_{tmbr}), as fractions of the average system TSS concentration $\overline{X}_t$ of the reactor are equal to the ratio of the sludge mass fraction and volume fraction of the zones, as Eqs. 2.13, 2.17 and 2.21 above show, i.e.

$$\frac{X_{\text{tana}}}{\overline{X}_t} = \frac{f_{\text{xana}}}{f_{\text{vana}}};\quad \frac{X_{\text{tanx}}}{\overline{X}_t} = \frac{f_{\text{xanx}}}{f_{\text{vanx}}};\quad \frac{X_{\text{taer}}}{\overline{X}_t} = \frac{f_{\text{xaer}}}{f_{\text{vaer}}};\quad \frac{X_{\text{tmbr}}}{\overline{X}_t} = \frac{f_{\text{xmbr}}}{f_{\text{vmbr}}};\quad \frac{X_{\text{tpax}}}{\overline{X}_t} = \frac{f_{\text{xpax}}}{f_{\text{vpax}}} \tag{2.37}$$

where

f_x, f_v = zone sludge mass and volume fractions respectively, and
subscripts *ana*, *anx*, *anxp*, *aer*, *anxs*, *mbr* and *pax* refer to the anaerobic, anoxic, primary anoxic, aerobic, secondary anoxic, membrane and pre-anoxic zones respectively.

Table 2.11 Equations to convert sludge mass fractions to volume fractions and vice versa for the UCT and JHB systems with SSTs.

Anoxic	Anaerobic	Aerobic	Pre-anoxic
UCT system with SST (Figure 2.1b)			
From sludge mass fraction to volume fraction: $B = \left(1 + \frac{f_{xana}}{r}\right)$ (2.38)			
$f_{vana} = \frac{f_{xana}(1+r)}{rB}$ (2.39)	$f_{vanx} = \frac{f_{xanx}}{B}$ (2.40)	$f_{vaer} = \frac{f_{xaer}}{B}$ (2.41)	—
$X_{tana} = \overline{X_t}\frac{rB}{(1+r)}$ (2.42)	$X_{tanx} = \overline{X_t}B$ (2.43)	$X_{tanx} = \overline{X_t}B$ (2.44)	—
From volume fraction to sludge mass fraction: $B = \frac{1}{\left(1 - \frac{f_{vana}}{1+r}\right)}$ (2.45); TSS conc (X_t) Eqs. above apply			
$f_{xana} = \frac{f_{vana}rB}{r+1}$ (2.46)	$f_{xanx} = f_{vanx}B$ (2.47)	$f_{xaer} = f_{vaer}B$ (2.48)	—
JHB system with SST (Figure 2.30)			
From mass fraction to volume fraction: $C = \left(1 - \frac{f_{xpax}}{(1+s)}\right)$ (2.49)			
$f_{vana} = \frac{f_{xana}}{C}$ (2.50)	$f_{vanx} = \frac{f_{xanx}}{C}$ (2.51)	$f_{vaer} = \frac{f_{xaer}}{C}$ (2.52)	$f_{vpax} = \frac{sf_{xpax}}{(s+1)C}$ (2.53)
$X_{tana} = \overline{X_t}C$ (2.54)	$X_{tanx} = \overline{X_t}C$ (2.55)	$X_{taer} = \overline{X_t}C$ (2.56)	$X_{tmbr} = \overline{X_t}\frac{(s+1)}{s}$ (2.57)
From volume fraction to sludge mass fraction: $C = \frac{1}{\left(1 + \frac{f_{xpax}}{s}\right)}$ (2.58); TSS conc (X_t) Eqs. above apply			
$f_{xana} = f_{vana}C$ (2.59)	$f_{xanx} = f_{vanx}C$ (2.60)	$f_{xaer} = f_{vaer}C$ (2.61)	$f_{xpax} = \frac{(s+1)Cf_{vmbr}}{s}$ (2.62)

Table 2.12 Equations to convert sludge mass fractions to volume fractions for the MLE system.

Anoxic Zone	Aerobic Zone	Membrane Zone
MLE with membranes in the aerobic zone (Figure 2.31a)		
From sludge mass fraction to volume fraction: $D = \left(1 + \frac{f_{xanx}}{a}\right)$ (2.63)		
$f_{vanx} = \frac{f_{xanx}(1+a)}{aD}$ (2.64)	$f_{vaer} = \frac{f_{xaer}}{D}$ (2.65)	—
$X_{tanx} = \overline{X_t}\frac{aD}{(a+1)}$ (2.66)	$X_{taer} = \overline{X_t}D$ (2.67)	—
From volume fraction to sludge mass fraction: $D = \frac{(a+1)}{(a+1-f_{vanx})}$ (2.68); TSS conc (X_t) Eqs. above apply		
$f_{xanx} = \frac{f_{vanx}aD}{(a+1)}$ (2.69)	$f_{xaer} = f_{vaer}D$ (2.70)	—
MLE with membranes in membrane zone following the aerobic zone (Figure 2.31b)		
From mass fraction to volume fraction: $E = \left(1 + \frac{(1-f_{xmbr})}{a}\right)$ (2.71)		
$f_{vanx} = \frac{f_{xanx}(1+a)}{aE}$ (2.72)	$f_{vaer} = \frac{f_{xaer}(1+a)}{aE}$ (2.73)	$f_{vmbr} = \frac{f_{xmbr}}{E}$ (2.74)
$X_{tanx} = \overline{X_t}\frac{aE}{(a+1)}$ (2.75)	$X_{taer} = \overline{X_t}\frac{aE}{(a+1)}$ (2.76)	$X_{tmbr} = \overline{X_t}E$ (2.77)
From volume fraction to sludge mass fraction: $E = \frac{(a+1)}{(a+fvmbr)}$ (2.78); TSS conc (X_t) Eqs. above apply		
$f_{xanx} = \frac{f_{vanx}aE}{(a+1)}$ (2.79)	$f_{xaer} = \frac{f_{vaer}aE}{(a+1)}$ (2.80)	$f_{xmbr} = f_{vmbr}E$ (2.81)

Table 2.13 Equations to convert sludge mass fractions to volume fractions and vice versa for the 4 Stage Bardenpho system.

Primary Anoxic	Aerobic	Secondary Anoxic	Membrane/ Re-aeration
4 Stage Bardenpho with membranes in the aerobic zone (Figure 2.32)			
From sludge mass fraction to volume fraction: $F = \left(1 + \frac{(1 - f_{xmbr})}{s}\right)$ (2.82)			
$f_{vanxp} = \frac{f_{xanxp}(1+s)}{sF}$ (2.83)	$f_{vaer} = \frac{f_{xaer}(1+s)}{sF}$ (2.84)	$f_{vanxs} = \frac{f_{xanxs}(1+s)}{sF}$ (2.85)	$f_{vmbr} = \frac{f_{xmbr}}{F}$ (2.86)
$X_{tanxp} = \overline{X_t}\frac{sF}{(s+1)}$ (2.87)	$X_{taer} = \overline{X_t}\frac{sF}{(s+1)}$ (2.88)	$X_{tanxs} = \overline{X_t}\frac{sF}{(s+1)}$ (2.89)	$X_{tmbr} = \overline{X_t}F$ (2.90)
From volume fraction to sludge mass fraction: $F = \frac{1+s}{s + f_{vmbr}}$ (2.91); TSS (X_t) concentrations above apply			
$f_{xanxp} = \frac{f_{vanxp}sF}{(s+1)}$ (2.92)	$f_{xaer} = \frac{f_{vaer}sF}{(s+1)}$ (2.93)	$f_{xanxs} = \frac{f_{vanxs}sF}{(s+1)}$ (2.94)	$f_{xmbr} = Ff_{vmbr}$ (2.95)

Table 2.14 Equations to convert sludge mass fractions to volume fractions and vice versa for the UCT system.

Anaerobic	Anoxic	Aerobic	Membrane/Re-aeration
UCT system with membranes in the aerobic zone (Figure 2.33a)			
From sludge mass fraction to volume fraction: $G = \left(1 + \frac{f_{xana}}{r} - \frac{f_{xaer}}{(a+1)}\right)$ (2.96)			
$f_{vana} = \frac{f_{xana}(1+r)}{rG}$ (2.97)	$f_{vanx} = \frac{f_{xanx}}{G}$ (2.98)	$f_{vaer} = \frac{af_{xaer}}{(a+1)G}$ (2.99)	—
$X_{tana} = \overline{X_t}\frac{rG}{(1+r)}$ (2.100)	$X_{tanx} = \overline{X_t}G$ (2.101)	$X_{taer} = \overline{X_t}\frac{(a+1)}{a}G$ (2.102)	—
From volume fraction to sludge mass fraction: $G = 1 \Big/ \left(1 - \frac{f_{vana}}{r+1} + \frac{f_{vaer}}{a}\right)$ (2.103); TSS conc (X_t) Eqs. above apply			
$f_{xana} = \frac{f_{vana}rG}{r+1}$ (2.104)	$f_{xanx} = f_{vanx}G$ (2.105)	$f_{xaer} = \frac{(a+1)Gf_{vaer}}{a}$ (2.106)	—
UCT system with membranes in membrane zone following the aerobic zone (Figure 2.33b)			
From mass fraction to volume fraction: $H = \left(1 + \frac{f_{xana}}{r} - \frac{f_{xmbr}}{(a+1)}\right)$ (2.107)			
$f_{vana} = \frac{f_{xana}(1+r)}{rH}$ (2.108)	$f_{vanx} = \frac{f_{xanx}}{H}$ (2.109)	$f_{vaer} = \frac{f_{xaer}}{H}$ (2.110)	$f_{vmbr} = \frac{af_{xmbr}}{(a+1)H}$ (2.111)
$X_{tana} = \overline{X_t}\frac{rH}{(1+r)}$ (2.112)	$X_{tanx} = \overline{X_t}H$ (2.113)	$X_{taer} = \overline{X_t}H$ (2.114)	$X_{tmbr} = \overline{X_t}\frac{(a+1)}{a}H$ (2.115)
From volume fraction to sludge mass fraction: $H = \frac{1}{\left(1 - \frac{f_{vana}}{r+1} + \frac{f_{vmbr}}{a}\right)}$ (2.116); TSS conc (X_t) Eqs. above apply			
$f_{xana} = \frac{f_{vana}rH}{r+1}$ (2.117)	$f_{xanx} = f_{vanx}H$ (2.118)	$f_{xaer} = f_{vaer}H$ (2.119)	$f_{xmbr} = \frac{(a+1)Hf_{vmbr}}{a}$ (2.120)

Table 2.15 Equations to convert sludge mass fractions to volume fractions and vice versa for the JHB system.

Anaerobic	Anoxic	Aerobic	Membrane	Pre-anoxic
JHB system with membranes in the aerobic zone (Figure 2.34a)				
From sludge mass fraction to volume fraction: $J = 1 + \frac{f_{xana}}{s} + \frac{f_{xanx}}{(a+s)}$ (2.121)				
$f_{vana} = \frac{f_{xana}(1+s)}{sJ}$ (2.122)	$f_{vanx} = \frac{f_{xanx}(1+a+s)}{(a+s)J}$ (2.123)	$f_{vaer} = \frac{f_{xaer}}{J}$ (2.124)	—	$f_{vpax} = \frac{f_{xpax}}{J}$ (2.125)
$X_{tana} = \overline{X_t}\frac{sJ}{(1+s)}$ (2.126)	$X_{tanx} = \overline{X_t}\frac{(a+s)J}{(a+s+1)}$ (2.127)	$X_{taer} = \overline{X_t}J$ (2.128)	—	$X_{tpax} = \overline{X_t}J$ (2.129)
From volume fraction to sludge mass fraction: $J = \frac{1}{\left(1 - \frac{f_{vana}}{s} - \frac{f_{vanx}}{(a+s+1)}\right)}$ (2.130); TSS conc (X_t) Eqs. above apply				
$f_{xana} = \frac{f_{vana}sJ}{s+1}$ (2.131)	$f_{xanx} = \frac{f_{vanx}(a+s)J}{(a+s+1)}$ (2.132)	$f_{xaer} = f_{vaer}J$ (2.133)	—	$f_{xpax} = f_{vpax}J$ (2.134)
JHB system with membranes in membrane zone following the aerobic zone (Figure 2.34b)				
From sludge mass fraction to volume fraction: $L = \left(1 + \frac{f_{xana}}{s} + \frac{(f_{xanx} + f_{xaer})}{(a+s)}\right)$ (2.135)				
$f_{vana} = \frac{f_{xana}(1+s)}{sL}$ (2.136)	$f_{vanx} = \frac{f_{xanx}(1+a+s)}{(a+s)L}$ (2.137)	$f_{vaer} = \frac{f_{xaer}(1+a+s)}{(a+s)L}$ (2.138)	$f_{vmbr} = \frac{f_{xmbr}}{L}$ (2.139)	$f_{vpax} = \frac{f_{xpax}}{L}$ (2.140)
$X_{tana} = \overline{X_t}\frac{sL}{(1+s)}$ (2.141)	$X_{tanx} = \overline{X_t}\frac{(a+s)L}{(a+s+1)}$ (2.142)	$X_{taer} = \overline{X_t}\frac{(a+s)L}{(a+s+1)}$ (2.143)	$X_{tmbr} = \overline{X_t}L$ (2.144)	$X_{tpax} = X_tL$ (2.145)
From volume fraction to sludge mass fraction: $L = \frac{1}{\left(1 - \frac{f_{xana}}{s+1} - \frac{(f_{xanx} + f_{xaer})}{(a+s+1)}\right)}$ (2.146); TSS conc (X_t) Eqs. above apply				
$f_{xana} = \frac{f_{vana}sL}{s+1}$ (2.147)	$f_{xanx} = \frac{f_{vanx}(a+s)L}{(a+s+1)}$ (2.148)	$f_{xaer} = \frac{f_{vaer}(a+s)L}{(a+s+1)}$ (2.149)	$f_{xmbr} = f_{vmbr}L$ (2.150)	$f_{xpax} = f_{vpax}L$ (2.151)

Table 2.16 Equations to convert sludge mass fractions to volume fractions and vice versa for the 3 Stage Bardenpho system.

Anaerobic	Anoxic	Aerobic	Membrane/Re-aeration
3 Stage Bardenpho system with membranes in the aerobic zone (Figure 2.35a)			
From sludge mass fraction to volume fraction: $K = \left(1 + \frac{f_{xana}}{s} + \frac{f_{xanx}}{(a+s)}\right)$ (2.152)			
$f_{vana} = \frac{f_{xana}(1+s)}{sK}$ (2.153)	$f_{vanx} = \frac{f_{xanx}(1+a+s)}{(a+s)K}$ (2.154)	$f_{vaer} = \frac{f_{xaer}}{K}$ (2.155)	—
$X_{tana} = \overline{X_t}\frac{sK}{(1+s)}$ (2.156)	$X_{tanx} = \overline{X_t}\frac{(a+s)K}{(a+s+1)}$ (2.157)	$X_{taer} = \overline{X_t}K$ (2.158)	—
From volume fraction to sludge mass fraction: $K = \frac{1}{\left(1 - \frac{f_{vana}}{s+1} - \frac{f_{vanx}}{(a+s+1)}\right)}$ (2.159); TSS (X_t) concentrations above apply			
$f_{xana} = \frac{f_{vana}sK}{s+1}$ (2.160)	$f_{xanx} = \frac{f_{vanx}(a+s)K}{(a+s+1)}$ (2.161)	$f_{xaer} = f_{vaer}K$ (2.162)	—
3 Stage Bardenpho system with membranes in membrane zone following the aerobic zone (Figure 2.35b)			
From mass fraction to volume fraction: $M = \frac{(a+s+1)}{(a+s)} + \frac{af_{xana}}{s(a+s)} - \frac{f_{xmbr}}{(a+s)}$ (2.163)			
$f_{vana} = \frac{f_{xana}(1+s)}{sM}$ (2.164)	$f_{vanx} = \frac{f_{xanx}(1+a+s)}{(a+s)M}$ (2.165)	$f_{vaer} = \frac{f_{xaer}(1+a+s)}{(a+s)M}$ (2.166)	$f_{vmbr} = \frac{f_{xmbr}}{M}$ (2.167)
$X_{tana} = \overline{X_t}\frac{sM}{(1+s)}$ (2.168)	$X_{tanx} = \overline{X_t}\frac{(a+s)M}{(a+s+1)}$ (2.169)	$X_{taer} = \overline{X_t}\frac{(a+s)M}{(a+s+1)}$ (2.170)	$X_{tmbr} = \overline{X_t}M$ (2.171)
From volume fraction to sludge mass fraction; $M = \frac{(a+s+1)}{a+s-\frac{af_{vana}}{(s+1)}+f_{vmbr}}$ (2.172); TSS (X_t) concentrations above apply			
$f_{xana} = \frac{f_{vana}sM}{s+1}$ (2.173)	$f_{xanx} = \frac{f_{vanx}(a+s)M}{(a+s+1)}$ (2.174)	$f_{xaer} = \frac{f_{vaer}(a+s)M}{(a+s+1)}$ (2.175)	$f_{xmbr} = f_{vmbr}M$ (2.176)

Table 2.17 Equations to convert sludge mass fractions to volume fractions and vice versa for the 5 Stage Bardenpho with membranes in the re-aeration zone.

Anaerobic	Primary Anoxic	Aerobic	Secondary Anoxic	Membrane/ Re-aeration
5 Stage Bardenpho system with membranes in the re-aeration aerobic zone (Figure 2.36)				
From sludge mass fraction to volume fraction: $N = \left(1 + \frac{(1 - f_{xmbr})}{s}\right)$ (2.177)				
$f_{vana} = \frac{f_{xana}(1+s)}{sN}$ (2.178)	$f_{vanxp} = \frac{f_{xanxp}(1+s)}{sN}$ (2.179)	$f_{vaer} = \frac{f_{xaer}(1+s)}{sN}$ (2.180)	$f_{vanxs} = \frac{f_{xanxs}(1+s)}{sN}$ (2.1811)	$f_{vmbr} = \frac{f_{xmbr}}{N}$ (2.182)
$X_{tana} = \overline{X_t}\frac{sN}{(1+s)}$ (2.183)	$X_{tanxp} = \overline{X_t}\frac{sN}{(s+1)}$ (2.184)	$X_{taer} = \overline{X_t}\frac{sN}{(s+1)}$ (2.185)	$X_{tanxs} = \overline{X_t}\frac{sN}{(s+1)}$ (2.186)	$X_{tmbr} = \overline{X_t}N$ (2.187)
From volume fraction to sludge mass fraction: $N = \left(\frac{s+1}{s + f_{vmbr}}\right)$ (2.188) TSS (X_t) concentrations above apply				
$f_{vana} = \frac{f_{xana}sN}{s+1}$ (2.189)	$f_{xanxp} = \frac{f_{vanxp}sN}{s+1}$ (2.190)	$f_{xaer} = \frac{f_{vaer}sN}{s+1}$ (2.191)	$f_{xanxs} = \frac{f_{vanxs}sN}{s+1}$ (2.192)	$f_{xmbr} = f_{xmbr}N$ (2.193)

2.8.3 BNR systems with secondary settling tanks for solid-liquid separation

For BNR systems with SSTs in which the sludge mass is uniformly distributed and the TSS concentrations are the same in the anaerobic (if present), anoxic and aerobic zones of the reactor, the sludge mass and volume fractions are equal. This is the case for the Modified Ludzack-Ettinger (MLE, Figure 2.26) and 4 stage Bardenpho systems (Figure 2.27) for N removal and in the 3 and 5 stage Bardenpho systems (Figures 2.28 and 2.29) for N and P removal. For example, if a 3 stage Bardenpho system (Figure 2.28) requires anaerobic, anoxic and aerobic mass fractions (f_{xana}, f_{xanx}, f_{xaer}) of 0.15, 0.35 and 0.50 respectively, the corresponding volume fractions of these zones (f_{vana},f_{vanx},f_{vaer}) with respect to the reactor volume (V_R) will also be 0.15, 0.35 and 0.50 respectively. This is because the influent flow dilutes the SST return sludge concentration in the first zone by the same amount as the SST concentrates it after the last zone. This equality of sludge mass and volume fractions does not apply to any multi-zone BNR system with membrane solid-liquid separation, either in the aerobic zone itself (immersed panel membranes) or with membranes in a separate membrane zone/reactor following the aerobic (last) zone. This is because the aerobic or membrane (MBR) zone TSS concentrations are the equivalent of the thickened return sludge concentration from the SST (if there were SSTs).

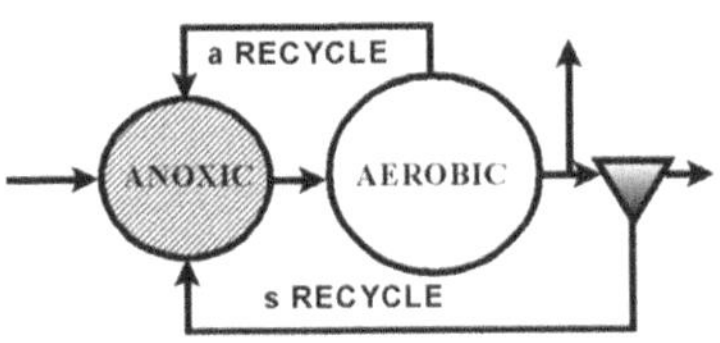

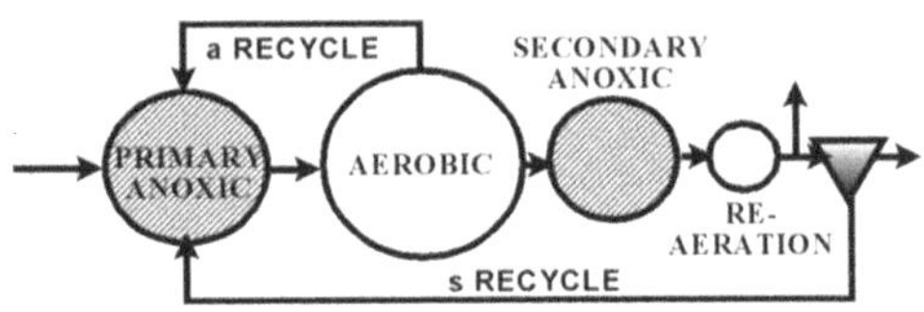

Figure 2.26 (left) and Figure 2.27 (right) The Modified Ludzack-Ettinger (MLE) and 4 Stage Bardenpho systems for biological N removal with SST for solid-liquid separation.

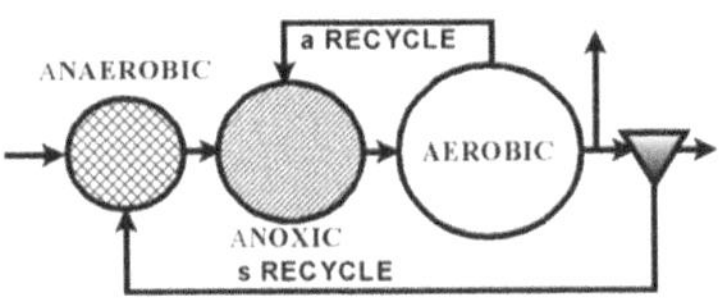

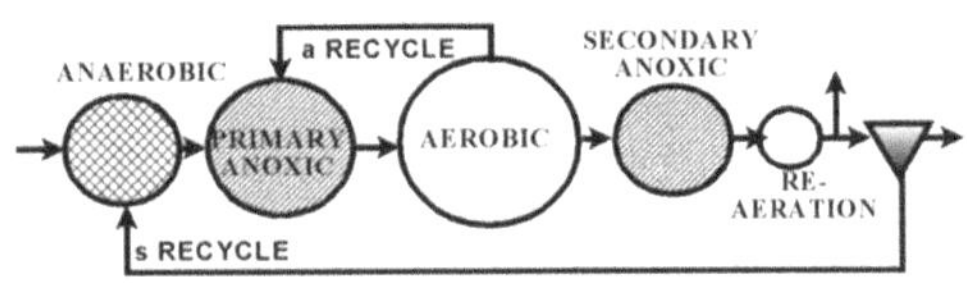

Figure 2.28 (left) and Figure 2.29 (right) The 3 Stage and 5 Stage Bardenpho systems for biological N and P removal with SST for solid-liquid separation.

For BNR systems *with* SSTs, in which the TSS concentrations are not the same in the pre-anoxic, anaerobic, anoxic or aerobic zones, e.g. in the University of Cape Town (UCT, Figure 2.1b) or Johannesburg (JHB, Figure 2.30) systems, the volume and mass fractions are not equal. The equations to convert between sludge mass fractions and volume fractions are given in Table 2.11 as Eqs. 2.38 to 2.62.

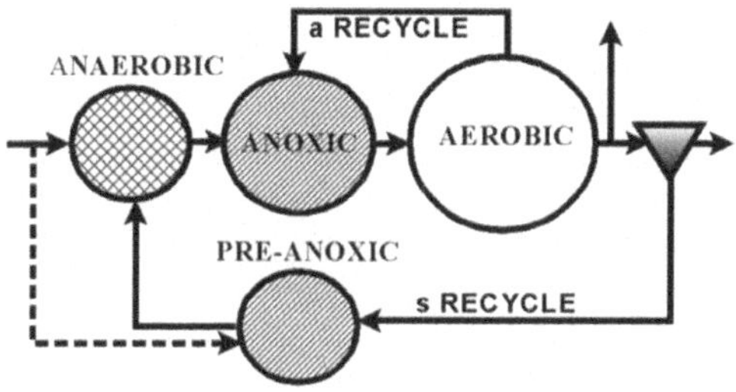

Figure 2.30 The Johannesburg system for biological N and P removal with SST for solid-liquid separation.

2.8.4 BNR systems with membranes for solid-liquid separation

In BNR systems with *membrane solid-liquid separation* in the aerobic zone such as immersed panel membranes, the sludge mass distributes itself differently in the different zones of the system compared with systems with SSTs. This is because the effluent is withdrawn via the membranes from the aerobic zone which concentrates the sludge in this zone relative to that in the other zones. However, in recycling this concentrated aerobic zone sludge to an upstream zone, it is diluted by the less concentrated incoming sludge stream from the upstream zones. The higher the recycle ratios from downstream zones to upstream zones, the more uniformly the sludge mass is distributed around the system and the closer the TSS concentrations in the different zones are to the system average TSS concentration $\overline{X_t}$ and to one another.

2.8.4.1 The MLE ND system

For the MLE system with membranes in the aerobic zone (e.g. immersed panel membranes, Figure 2.31a) or with membranes in a dedicated membrane reactor (Figure 2.31b), the equations to convert between sludge mass fractions and volume fractions are given in Table 2.12 as Eqs. 2.63 to 2.81.

2.8.4.2 The 4 stage Bardenpho ND system

In the 4 stage Bardenpho system for N removal, membranes cannot be placed in the aerobic zone because the secondary anoxic zone follows it. However, membranes can be placed in the re-aeration zone, which then becomes the membrane zone (Figure 2.32). The equations for converting sludge mass fractions to volume fractions are Eqs. 2.82 to 2.95 in Table 2.13.

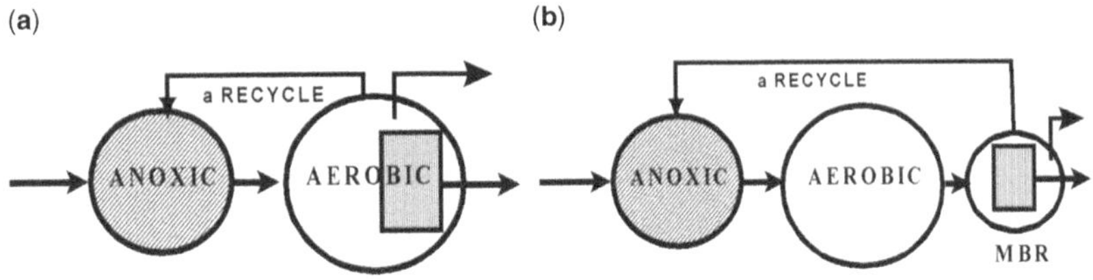

Figure 2.31 The Modified Ludzack-Ettinger (MLE) for biological N removal with membranes for solid-liquid separation in the aerobic reactor (a) or with a membrane tank following the aerobic zone (b).

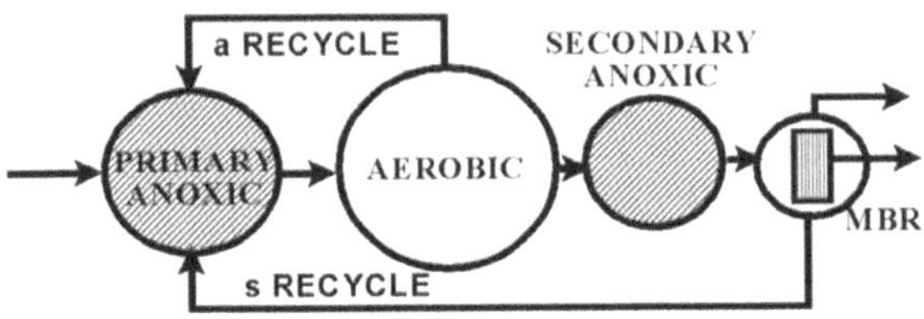

Figure 2.32 The 4 stage Bardenpho for biological N removal with membranes for solid-liquid separation in a membrane tank following the secondary anoxic zone.

2.8.4.3 The University of Cape Town (UCT) NDEBPR system

For the UCT system with membranes in the aerobic zone (Figure 2.33a) or with membranes in a dedicated membrane reactor (Figure 2.33b), the equations for converting between sludge mass fractions and volume fractions are Eqs. 2.96 to 2.120 in Table 2.14.

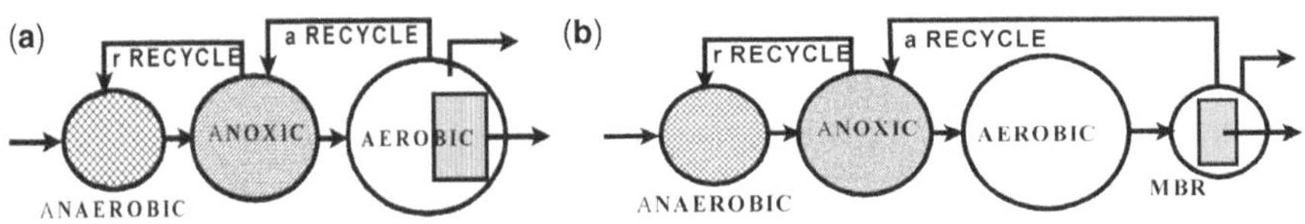

Figure 2.33 The UCT system for biological N and P removal with membranes for solid-liquid separation in the aerobic reactor (a) or with a membrane tank following the aerobic zone (b).

2.8.4.4 The Johannesburg (JHB) NDEBPR system

For the JHB system with membranes in the aerobic zone, e.g. immersed panel membranes (Figure 2.34a), or with membranes in a dedicated membrane reactor

(Figure 2.34b), the equations for converting between sludge mass fractions and volume fractions are Eqs. 2.121 to 2.151 in Table 2.15.

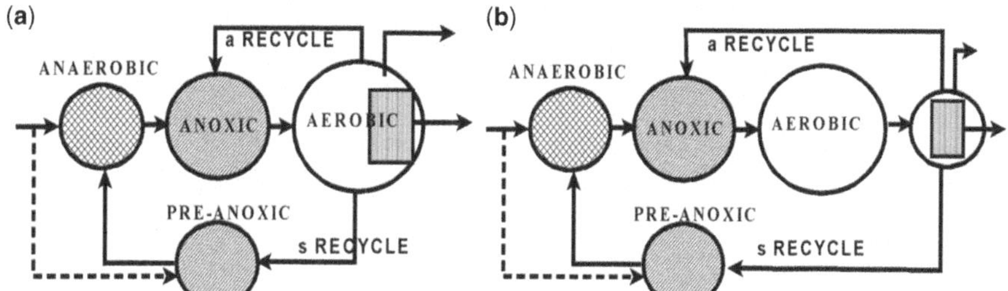

Figure 2.34 The JHB system for biological N and P removal with membranes for solid-liquid separation in the aerobic reactor (a) or with a membrane tank following the aerobic zone (b).

2.8.4.5 The 3 stage NDEBPR Bardenpho system

For the 3 stage Bardenpho system with membranes in the aerobic zone (Figure 2.35a) or with membranes in a dedicated membrane reactor (Figure 2.35b), the equations for converting between sludge mass fractions and volume fractions are Eqs. 2.152 to 2.176 in Table 2.16.

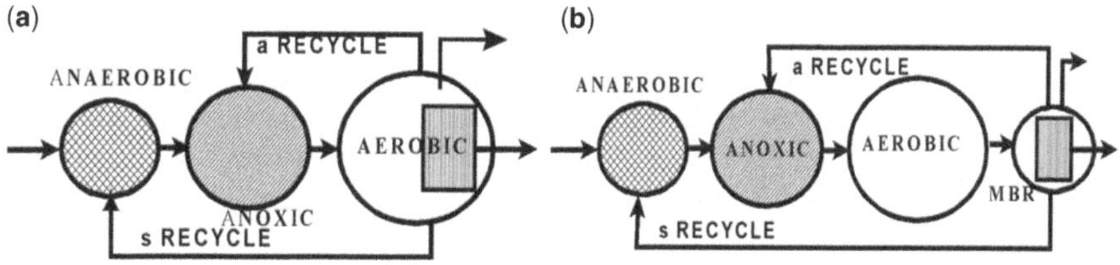

Figure 2.35 The 3 stage Bardenpho system for biological N and P removal with membranes for solid-liquid separation in the aerobic reactor (a) or with a membrane tank following the aerobic zone (b).

2.8.4.6 The 5 stage NDEBPR Bardenpho system

Like for the 4 stage Bardenpho system, in the 5 stage Bardenpho system for N and P removal, membranes cannot be placed in the aerobic zone because the secondary anoxic zone follows it. However, membranes can be placed in the re-aeration zone, which then becomes the membrane zone (Figure 2.36). The equations for converting between sludge mass fractions and volume fractions are Eqs. 2.177 to 2.193 in Table 2.17.

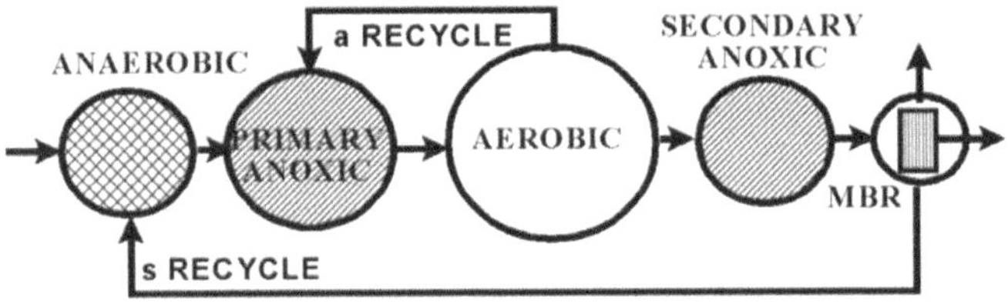

Figure 2.36 The 5 stage Bardenpho system for biological N and P removal with membranes for solid-liquid separation in the re-aeration tank following the secondary anoxic zone.

2.8.5 Mass fraction flexibility in MBR BNR systems

In multi-zone BNR systems with membranes in the aerobic reactor and fixed volumes for the anaerobic, anoxic and aerobic zones (i.e. fixed volume fractions), the mass fractions can be varied (within a range) by varying the inter-reactor recycle ratios. For example, in a UCT system with anaerobic, anoxic and aerobic zone volume fractions of 0.25, 0.35 and 0.40 and an *r*-recycle ratio from the anoxic to the anaerobic zones of 1:1, the anaerobic, anoxic and aerobic zone mass fractions can be varied from 0 to 0.131, 0 to 0.366 and 1 to 0.503 respectively by changing the *a*-recycle ratio from 0:1 to 5:1. Increasing the *a*-recycle ratio concomitantly increases the nitrate load on the anoxic reactor thereby increasing the denitrification and N removal as the anoxic mass fraction increases. Increasing the *r*-recycle ratio increases the anaerobic sludge mass fraction (at the expense of the other two zone sludge mass fractions) and increases (not proportionally) the P removal. This zone sludge mass fraction flexibility is a significant advantage of membrane BNR systems over conventional BNR system with SSTs because it allows changing of the sludge mass fractions to optimize biological N and P removal in conformity with influent wastewater characteristics and the effluent N and P concentrations required.

2.8.6 Modelling MBR BNR systems

Finally, because the kinetic rates of the BNR bioprocesses do not appear to be influenced by high VSS concentration (3–12 gVSS/ℓ), the performance of membrane BNR systems can be simulated with current BNR activated sludge simulation models by returning the SST underflow with a point separator back into the zone from which the SST feed flow exits. However, such simulations require *a priori* information on the reactor and zone volumes and recycle flows. These are best determined with steady state calculation procedures such as those set out by Metcalf & Eddy (2014); DWA (2016) or the University of Cape Town (UCT, WRC, 1984; Henze *et al.*, 2008; Ekama, 2009, 2011). However, it should be noted that these steady state models do not yield the same results for the sludge age (SRT), mass of sludge in the reactor, sludge mass (or volume) fractions, oxygen demand and effluent N and P concentrations. The best steady state model to use would be the one that best aligns with the dynamic simulation models ASM1 and ASM2, which is the last one mentioned.

2.9 CONCLUSIONS

To assess the impact of high VSS concentration in membrane bioreactor (MBR) biological nutrient removal (BNR) activated sludge (AS) systems on the bioprocess kinetic rates that mediate biological N and P removal, two identical (except for the hydraulic retention time) parallel laboratory scale University of Cape Town (UCT) nitrification denitrification (ND) enhanced biological phosphorus removal (EBPR) systems fed the same real wastewater were operated for 450 days. One was operated at a low VSS concentration (3 gVSS/ℓ) with a secondary settling tank (SST) for solid-liquid separation (CAS system), the other was operated at a high VSS concentration (13 gVSS/ℓ) with submerged panel membranes (MBR system) for solid-liquid separation. From the BNR performance of these two systems and from the 85 aerobic, anoxic-aerobic and anaerobic-anoxic-aerobic batch tests on sludge harvested from the two systems, the following conclusions were drawn.

The MBR system achieved a higher COD removal (effluent COD 41 mgCOD/ℓ) compared with the CAS system (unfiltered 74 mgCOD/ℓ, 0.45 μm filtered 51 mgCOD/ℓ). Both systems achieved similar N removals (MBR 83%, CAS 81%). Nitrification was complete in both systems (<1.0 mgFSA-N/ℓ). Denitrification was better in the MBR system (effluent nitrate MBR 18.0 mgNO_3-N/ℓ and CAS 20.0 mgNO_3-N/ℓ) due to the negligible impact of the dissolved oxygen in the recycle to the anoxic reactor at the high VSS concentration of the MBR system. The P removal in the MBR system (22.5 mgP/ℓ) was higher than that in the CAS system (17.4 mgP/ℓ). This was due to the recycle of nitrate from the anoxic reactor to the anaerobic reactor and greater anoxic P uptake in the CAS system due to the non-zero nitrate concentration in the anoxic reactor. The sludge production by the MBR system was higher [0.31 (gVSS/d)/(gCOD/d)] than by the CAS system [0.20 (gVSS/d)/(gCOD/d)].

From the 37 aerobic nitrification batch tests: (1) At the same low VSS concentration, the MBR system exhibited lower VSS specific ammonia utilization rate (SAUR) and autotrophic nitrifier organism (ANO) maximum specific growth rates (μ_A) than the parallel CAS system. (2) For the MBR system, as the VSS concentration increased, the SAUR and μ_A decreased, apparently due to ammonia transfer limitation. (3) For the MBR system at its high VSS concentration, as the initial ammonia concentration increased, the SAUR and μ_A increased, indicating possible ammonia transport limitation at increasing VSS concentration. Possible reasons for these differences are: (1) In CAS systems with SSTs, organism loss via the effluent occurs including ANOs. Therefore CAS systems may select ANOs with higher maximum specific growth rates (μ_A) than MBR systems. In the MBR system all the ANOs are retained, including slow growing ones. (2) At the high VSS concentrations in the MBR system, ammonia transport limitation decreased the observed SAUR and μ_A.

From the 33 anoxic-aerobic batch tests, the OHOVSS specific denitrification rate by OHOs (K_{2OHO}) utilizing slowly biodegradable particulate organics (BPO) obtained at different MBR system VSS concentrations (2.5–12 gVSS/ℓ) and different initial nitrate concentrations ranging from 30 to 90 mgN/ℓ showed no effect of initial nitrate concentration, in agreement with past work (van Haandel *et al.*, 1981; Clayton *et al.*, 1991; Ekama & Wentzel, 1999) and no effect of VSS concentration. The average K_{2OHO} was 0.264 $mgNO_3$-N/(mgOHOVSS·d), which is very close to the average K_{2OHO} rate reported in the literature for conventional (low VSS) BNR systems with SSTs, i.e. 0.255 from Ekama and Wentzel (1999).

From the 15 anaerobic-anoxic-aerobic batch tests, the specific VSS and specific PAOVSS anaerobic acetate (as COD) uptake and P release rates showed no effect of VSS or initial acetate concentration. Also, the results obtained with different concentrations of acetate added showed the acetate uptake rate to be zero order with respect to acetate concentration, which is in agreement with literature studies (Wentzel *et al.*, 1985, 1989). The P release to acetate uptake ratio also showed no effect of acetate dose and VSS concentration. The specific VSS and specific PAOVSS aerobic and anoxic P uptake rates also showed no effect of VSS concentration. The average PAOVSS specific anaerobic acetate uptake and P release rates and the aerobic P uptake rate obtained over the VSS concentration range were within the range of literature rates observed on enhanced PAO culture systems, confirming that within experimental variation, high VSS concentration did not affect the rates.

In the 13 anaerobic-anoxic/aerobic batch tests with acetate uptake, the PAOs showed significantly higher anoxic P uptake and denitrification rates than in the MBR system itself, where high acetate and excess nitrate did not occur. In the former, the PAOs denitrified 22% of the nitrate whereas in the MBR system this value was only 11%. The OHOVSS specific denitrification rates were within the same 0.2 to 0.3 $mgNO_3$-N/(mgOHOVSS·d) range in all the batch tests with an anoxic phase. While the PAOVSS specific denitrification rate in the anaerobic-anoxic/aerobic batch tests was about half of the OHOVSS specific denitrification rate, in the MBR system, the PAOVSS specific denitrification rate was only 1/14 of the OHOVSS specific denitrification rate because the conditions in the anaerobic-anoxic/aerobic batch tests (high acetate and nitrate) were not prevalent in continuous flow BNR systems fed real wastewater. The large reduction in P removal (up to a third) resulting from significant anoxic P uptake EBPR seems counter-productive for the very small PAO contribution to denitrification.

Specific denitrification rates are zero order with respect to nitrate concentration and HAc uptake rates are zero order with respect to HAc concentration in agreement with previous observations on conventional BNR systems. Anoxic P uptake has been consistently observed and the existence of two groups of PAO bacteria has been demonstrated. Anoxic P uptake is detrimental to the EBPR performance in a BNR system resulting in a reduction in P removal by up to a third. However, definitive quantitative links between design and operational parameters and the extent of anoxic P uptake that is obtained have not been established. This has hindered incorporation of anoxic P uptake in the design and simulation models for BNR systems, with or without membranes, and requires resolution.

The results from this investigation show that the BNRAS steady state and kinetic models developed for low VSS concentration BNRAS systems with secondary settling tanks can be applied with reasonable confidence to predict the performance of high VSS concentration (up to 13 gVSS/ℓ) BNRAS systems with membranes, except for the maximum specific growth rate of the nitrifiers, which was observed to be significantly lower in the MBR system at high VSS concentration.

For design and evaluation of BNR systems with membranes for solid-liquid separation, equations for converting between anaerobic, anoxic, aerobic and membrane zone sludge mass fractions and zone volume fractions were derived for the MLE and 4 stage Bardenpho ND systems and the UCT, JHB, and 3 and 5 Stage Bardenpho NDEBPR systems. Because the kinetic rates of the BNR bioprocesses do not appear to be influenced by high VSS concentration (3–12 gVSS/ℓ), the performance of membrane BNR systems can be simulated with current BNR activated sludge simulation models by returning the SST underflow with a point separator back into the zone from which the SST feed flow exits. However, such simulations require *a priori* information on the reactor and zone volumes and recycle flows. These are best determined with steady state calculation procedures such as those set out by Metcalf & Eddy (2014); DWA (2016) or University of Cape Town (UCT, WRC, 1984; Henze *et al.*, 2008; Ekama, 2009, 2011). However, it should be noted that these steady state models do not yield the same results for the sludge age (SRT), mass of sludge in the reactor, sludge mass (or volume) fractions, oxygen demand and effluent N and P concentrations. The best steady state model to use would be the one that best aligns with the dynamic simulation models ASM1 and ASM2, which is the last one mentioned.

2.10 ACKNOWLEDGEMENTS

Gratitude is expressed to Mr Taliep Lakay and Mr Hector Mafungwa for assistance with operating and testing the MBR and CAS BNR systems. This research was conducted in the Water Research Laboratory in the Department of Civil Engineering at the University of Cape Town (UCT) by Geoff du Toit, Masters student at UCT, and Valentina Parco, PhD student at University of Palermo. The research was financially supported by the National Research Foundation, Water Research Commission and University of Cape Town and is published with their permission.

2.11 REFERENCES

Beun J., Verho E. V., van Loosdrecht M. C. M. and Heijnen J. J. (2000). Stoichiometry and kinetics of poly-B-hydroxybutyrate metabolism under denitrifying conditions in activated sludge cultures. *Biotechnology and Bioengineering*, **68**, 496–507.

Casey T. G., Wentzel M. C. and Ekama G. A. (1999). Filamentous organism bulking in nutrient removal activated sludge systems. Paper 10: Metabolic behaviour of heterotrophic facultative aerobic organisms under aerated/unaerated conditions. *Water SA* **25**(4), 425–442.

Casey T. G., Wentzel M. C., Ekama G. A., Loewenthal R. E. and Marais G. v. R. (1994). An hypothesis for the cause and control of anoxic-aerobic (AA) filament bulking in nutrient removal activated sludge systems. *Water Science and Technology*, **29**(7), 203–212.

Cicek N., Franco J. P., Suidan M. T., Urbain V. and Manem J. (1999). Characterization and comparison of membrane bioreactor and a conventional activated sludge system in the treatment of wastewater containing high molecular weight compounds. *Water Environment Research*, **71**(1), 64–70.

Clayton J. A., Ekama G. A., Wentzel M. C. and Marais G. v. R. (1991). Denitrification kinetics in biological N and P removal activated sludge systems treating municipal wastewaters. *Water Science and Technology*, **23**, 1025–1035.

Cosenza A., Mannina G., Neumann M. B., Viviani G., Vanrolleghem P. A. (2013). Biological nitrogen and phosphorus removal in membrane bioreactors: model development and parameter estimation. *Bioprocess and Biosystems Engineering*, **36**(4), 499–514.

du Toit G. J. G., Parco V., Ramphao M. C., Wentzel M. C. and Ekama G. A. (2007). Design and performance of BNR activated sludge systems with flat sheet membranes for solid liquid separation. *Water Science and Technology*, **56**(6), 105–113.

du Toit G. J. G., Parco V., Ramphao M. C., Wentzel M. C., Lakay M. T., Mafungwa H. Z. and Ekama G. A. (2010). The Performance and Kinetics of Biological Nitrogen and Phosphorus Removal with Ultra-Filtration Membranes for Solid-Liquid Separation. Final WRC Report for Projects K8/814 and K5/1537, Water Research Commission, Private Bag X03, Gezina 0031, Pretoria, South Africa.

DWA (2016). DWA Regelwerk, Arbeitsblatt DWA – A 131, Bemessung von einstufigen Belebungsanlagen. DWA Deutsche Vereinigung fur Wasserwirtschaft, Abwasser und Abfall e. V., Theodor-Heuss-Allee 17, 53773 Hennef, Deutschland. ISBN: 978-3-88721-331-2 (Print), 978-3-88721-332-9 (E-Book).

Ekama G. A. (2009). Using bio-process stoichiometry to build a steady state plant wide wastewater treatment plant model. *Water Research*, **43**(8), 2101–2120.

Ekama G. A. (2011). Biological nutrient removal, Chapter 14. In: Volume 4 of Treatise on Water Science, P. Wilderer (ed.). Academic Press, Elsevier, Amsterdam, Oxford, pp. 409–528. Hardcover ISBN: 9780444531933, eBook ISBN: 9780444531995.

Ekama G. A. and Marais G. v. R. (1984). Two improved activated sludge settleability parameters. *IMIESA*, **9**(6), 20–27.

Ekama G. A. and Wentzel M. C. (1999). Denitrification kinetics in biological N and P removal activated sludge systems treating municipal wastewaters. *Water Science and Technology*, **39**(6), 69–77.

Ekama G. A. and Wentzel M. C. (2004). A predictive model for the reactor inorganic suspended solids concentration in activated sludge systems. *Water Research*, **38**(19), 4093–4106.

Ekama G. A., Dold P. L. and Marais G. v. R. (1986). Procedures for determining influent COD fractions and the maximum specific growth of the heterotrophs in activated sludge systems. *Water Science and Technology*, **18**(6), 91–114.

Ekama G. A., Barnard J. L., Günthert F. W., Krebs P., McCorquodale J. A., Parker D. S. and Wahlberg E. J. (1997). Secondary Settling Tanks: Theory, Modelling Design and Operation. Scientific and Technical Report (STR) No 6, IAWQ (now IWA), London, ISBN 1 900222 03 5.

Gao M., Yang M., Li H., Yang H. and Zhang Y. (2004). Comparison between a submerged bioreactor and a conventional activated sludge on treating ammonia-bearing inorganic wastewater. *Journal of Biotechnology*, **108**, 265–268.

Ghyoot W., Vandale S. and Verstraete W. (1999). Nitrogen removal from sludge reject water with a membrane assisted bioreactor. *Water Research*, **33**(1), 23–32.

Han S. S., Bae t. H., Jang G. G. and Tak T. M. (2005). Influence of sludge retention time on membrane fouling and bioactivities in membrane bioreactor system. *Process Biochemistry*, **40**, 2393–2400.

Henze M., Grady C. P. L. Jr., Gujer W., Marais G. v. R. and Matsuo T. (1987). Activated Sludge Model no.1. IAWQ Sci. & Tech. Report No.1. IAWQ, London, UK.

Henze M., Gujer W., Mino T., Matsuo T., Wentzel M. C. and Marais G. v. R. (1995). Activated Sludge Model No.2. IAWQ Sci. & Tech. Report No 3. IAWQ, London. ISSN 1025-0913, 32 pp.

Henze M., Gujer W., Mino T., Matsuo T., Wentzel M. C. and Marais G. v. R. (1999). Activated sludge Model No.2d, ASM2d. *Water Science and Technology*, **39**(1), 165–182.

Henze M., van Loosdrecht M. C. M., Ekama G. A. and Brdjanovic D. (2008). Biological Wastewater Treatment: Principles, Modelling and Design. IWA Publishing, London. 528 p. ISBN - 10:1843391880; 13:9781843391883.

Holbrook R. D., Massie K. A. and Novak J. T. (2005). A comparison of membrane bioreactor and a conventional activated sludge mixed liquor and biosolids characteristics. *Water Environment Research*, **77**(4), 323–360.

Hu Z.-R., Wentzel M. C. and Ekama G. A. (2002). The significance of denitrifying polyphosphate accumulating organisms in biological nutrient removal activated sludge systems. *Water Science and Technology*, **46**(1/2), 129–138.

Hu J.-H., Ong S. L., Lu N. F. and Fan X. J. (2003). A new method of characterizing denitrifying phosphorus removing bacteria by using three different types of electron acceptors. *Water Research*, **37**, 3463–3471.

Hu Z.-R., Wentzel M. C. and Ekama G. A. (2007a). A general model for biological nutrient removal activated sludge systems - model development. *Biotechnology and Bioengineering*, **98**(6), 1242–1258.

Hu Z.-R., Wentzel M. C. and Ekama G. A. (2007b). A general model for biological nutrient removal activated sludge systems - model evaluation. *Biotechnology and Bioengineering*, **98**(6), 1259–1275.

Huang X., Gui P. and Qiuan Y. (2001). Influence of sludge retention time on microbial behaviour in a submerged membrane bioreactor. *Process Biochemistry*, **36**(10), 1001–1006.

Kerrn-Jespersen J. P. and Henze M. (1993). Biological phosphorus uptake under anoxic and aerobic conditions. *Water Research*, **27**(4), 617–624.

Kuba T., Smolders G., van Loosdrecht M. C. M. and Heijen J. J. (1993). Biological phosphorus removal from wastewater by anaerobic-anoxic sequencing batch reactor. *Water Science and Technology*, **27**(5–6), 241–252.

Kuba T., Wachmeister A., van Loosdrecht M. C. M. and Heijen J. J. (1994). Effect of nitrate on phosphorus release in biological phosphorus removal systems. *Water Science and Technology*, **30**, 263–269.

Kuba T., Murnleitenr E., van Loosdrecht M. C. M. and Heijen J. J. (1996). A metabolic model for biological phosphorus removal by denitrifying organisms. *Biotechnology and Bioengineering*, **52**(6), 685–695.

Kuba T., van Loosdrecht M. C. M., Brandse F. A. and Heijen J. J. (1997). Occurrence of denitrifying phosphorus removing bacteria in modified UCT-type wastewater treatments plants. *Water Science and Technology*, **31**(4), 777–786.

Lee W., Kang S. and Shin H. (2003). Sludge characteristics and their contribution to microfiltration in submerged membrane bioreactors. *Journal of Membrane Science*, **216**, 217–227.

Li H., Gao M., Yang H., Zhang Y. and Kamagata Y. (2005). Comparison of nitrification performance and microbial community between a submerged bioreactor and a conventional activated sludge. *Water Science and Technology*, **51**(6–7), 193–200.

Liebig T., Wagner M., Bjerrum L. and Denecke M. (2001). Nitrification performance and nitrifier community composition of a cheostat and a membrane-assisted bioreactor for the nitrification of sludge reject water. *Bioprocess and Biosystems Engineering*, **24**, 203–210.

Luxmy B. S., Nakajima F. and Yamamoto K. (2000). Analysis of bacterial community in membrane separation bioreactors by fluorescent in situ hybridization and DGGE techniques. *Water Science and Technology*, **41**(10–11), 259–268.

Maharaj S., du Toit G. J. G., Wentzel M. C. and Bux F. (2007). Molecular approaches to study the dynamics of nitrifying bacteria in a conventional activated sludge system and membrane bioreactor. *4th International Water Association Leading-Edge Conference and Exhibition on Water & Wastewater Technology*, 3–6 June, Singapore (Poster).

Mannina G., Capodici M., Cosenza A., di Trapani D. and Ekama G. A. (2018a). The effect of the solids and hydraulic retention time on moving bed membrane bioreactor performance. *Journal of Cleaner Production*, **170**, 1305–1315.

Mannina, G., Cosenza, A., Ekama, G. A. (2018b). A comprehensive integrated membrane bioreactor model for greenhouse gas emissions. *Chemical Engineering Journal*, **334**, 1563–1572.

Manser J., Gujer W. and Siegrist H. (2005a). Consequence of mass transfer effects on the kinetics of nitrifiers. *Water Research*, **39**, 4633–4642.

Manser J., Gujer W. and Siegrist H. (2005b). Membrane bioreactor versus conventional activated sludge system: population dynamics of nitriefiers. *Water Science and Technology*, **52**(10–11), 417–425.

Meinhold J., Filipe-Carlos C. D. M., Daigger G. T. and Isaacs S. (1999). Characterization of the denitrifying fraction of phosphorus accumulating organisms in biological phosphate removal. *Water Science and Techonology*, **39**(1), 31–42.

Metcalf & Eddy (2014). Wastewater Engineering: Treatment and Reuse, 4th edn, Revised by Tchobanoglous G., Burton F. L. and Stensel H. D. McGraw Hill, New York.

Monti A., Hall E. R., Dawsin R. N., Husain H. and Kelly H. G. (2005). Comparative study of biological nutrient removal (BNR) processes with sedimentation and membrane based separation. *Biotechnology and Bioengineering*, **94**(4), 740–752.

Muller A., Wentzel M. C., Ekama G. A. and Loewenthal R. E. (2003). Heterotroph anoxic yield in anoxic aerobic activated sludge systems treating municipal wastewater. *Water Research*, **37**(10), 2435–2441.

Orhon D., Sozen S. and Artan N. (1996). The effect of heterotrophic yield on the assessment of the correction factor for anoxic growth. *Water Science and Technology*, **34**(5–6), 67–74.

Parco V. (2006). Bioreattori a membrana per la rimozione biologica dei nutrienti: cinetiche di processo ed efficienze (Membrane bioreactors for the biological removal of nutrients: kinetics and efficiency). PhD thesis, University of Palermo, Palermo, Sicily (in Italian).

Parco V., Wentzel M. C. and Ekama G. A. (2006). Kinetics of nitrogen removal in a MBR nutrient removal activated sludge system. *Desalination*, **199**(1–3), 89–91.

Parco V., du Toit G. J. G., Wentzel M. C. and Ekama G. A. (2007). Biological nutrient removal in membrane bioreactors: denitrification and phosphorus removal kinetics. *Water Science and Technology*, **56**(6), 125–134.

Payne W. J. (1981). Denitrification. John Wiley and Sons, New York.

Ramphao M. C., Wentzel M. C., Ekama G. A. and Alexander W. V. (2005). The impact of membrane solid-liquid separation on the design of biological nutrient removal activated sludge systems. *Biotechnology and Bioengineering*, **89**(6), 630–646.

Randall E. W., Wilkinson A. and Ekama G. A. (1991). An instrument for the direct determination of oxygen utilization rate. *Water SA*, **17**(1), 11–18.

Rosenberger S., Kuger U., Witzig R., Manz U, Szewzyk M. and Kraume M. (2002). Performance of a bioreactor with submerged membranes for aerobic treatment of municipal wastewater. *Water Research*, **36**, 413–420.

Satoh H., Ramey W., Koch F. A., Oldham W. K., Mino T. and Matsuo T. (1996). Anaerobic substrate uptake by the enhanced biological phosphorus removal activated sludge treating real sewage. *Water Science and Technology*, **34**(1–6), 9–16.

Smith S., Jefferson B. and Judd S. J. (2002). Membrane Bioreactors: Hybrid Activated Sludge or a New Process? CHISA, Prague.

Sorm R., Bortone G., Saltarelli R., Jenicek P., Wanner J. and Tilche A. (1996). Phosphate uptake under anoxic conditions and fixed-film nitrification in a nutrient removal activated sludge system. *Water Research*, **30**(7), 1573–1584.

Sorm R., Wanner J., Saltarelli R., Bortone G. and Tilche A. (1997). Verification of anoxic phosphate as the main biochemical mechanism of the DEPHANOX process. *Water Science and Techonolgy*, **35**(10), 87–94.

Sperandio M., Urbain V., Audic J. M. and Paul E. (1999). Use of carbon dioxide evolution rate for denitrification heterotrophic yield and characterizing denitrifying biomass. *Water Science and Technology*, **39**(1), 139–146.

Spérandio M, Masse A, Espinosa-Bouchot MC, Cabassud C. (2005). Characterization of sludge structure and activity on submerged membrane bioreactor. *Water Science and Technology*, **52**(10–11), 401–408.

Standard Methods (1985). Standard Methods for the Examination of Water and Wastewater. 16th edn, APHA, WEF, AWWA, Washington, DC, USA.

Stenstrom M. K. and Song S. S. (1991). Effect of oxygen transport limitation on nitrification in the activated sludge process. *Journal WPCF*, **63**(3), 208–219.

Van Haandel A. C., Ekama G. A. and Marais G. v. R. (1981). The activated sludge process Part 3 - Single sludge denitrification. *Water Research*, **15**(10), 1135–1152.

Wachmeister A., Kuba T., van Loosdrecht M. C. M. and Heijen J. J. (1997). A sludge characterization assay for aerobic and denitrifying phosphorus removing sludge. *Water Research*, **31**(3), 471–478.

Wentzel M. C., Dold P. L., Ekama G. A. and Marais G. v. R. (1985). Kinetics of biological phosphorus release. *Water Science and Technology*, **17**, 57–71.

Wentzel M. C., Dold P. L., Ekama G. A. and Marais G. v. R. (1989). Enhanced polyphosphate organism cultures in activated sludge systems Part III - Kinetic model. *Water SA*, **15**(2), 89–102.

Wentzel M. C., Ekama G. A., Dold P. L. and Marais G. v. R. (1990). Biological excess phosphorus removal - steady state process design. *Water SA*, **16**(1), 29–48.

Wentzel M. C., Ekama G. A. and Marais G. v. R. (1992). Processes and modelling of nitrification denitrification biological excess phosphorus removal systems - a review. *Water Science and Technology*, **25**(6), 59–82.

Witzig R., Manz W., Rosenberger S., Krüger U., Kraume M. and Szewzyk U. (2002). Microbiological aspects of a bioreactor with submerged membranes for aerobic treatment of municipal wastewater. *Water Research*, **36**, 394–402.

WRC (1984). *Theory, design and operation of nutrient removal activated sludge processes.* Ed. Wiechers HNS, TT16/84, ISBN 0 908356 13 7, Water Research Commission, Private Bag X03, Gezina 0031, South Africa, Obtainable free at http://www.wrc.org.za/Knowledge%20Hub%20Documents/Research%20Reports/TT-16-84.pdf (accessed 11 May 2018).

Yamamoto K. (2001). Membrane Bioreactor: an advanced wastewater treatment/reclamation technology and its function in excess sludge minimization. In: Advances in Water and Wastewater Treatment Technology: Molecular Technology, Nutrient Removal, Sludge Reduction and Environmental Health, T. Matsuo, K. Hanaki, S. Takizawa and H. Satoh (eds.), Elsevier, Amsterdam, pp. 229–237.

Zhang B., Yamamoto K., Ohgaki S. and Kamiko N. (1997). Floc size distribution and bacterial activities in membrane separation activated sludge processes for small scale wastewater treatment reclamation. *Water Science and Technology*, **35**(6), 37–44.

Chapter 3

MBBR and IFAS systems

H. Ødegaard

Scandinavian Environmental Technology (SET) and Norwegian University of Science and Technology (NTNU), Trondheim, Norway; SET AS, Hellemsveien 421, N 7165 Oksvoll, Norway

ABSTRACT

Even though the moving bed biofilm reactor (MBBR)- and integrated fixed film activated sludge system (IFAS)-processes are designed and used more or less as they were from the beginning (in the 1990s), there have been several advances. This chapter focuses on advances that have been made during the last 10 years in biochemical oxygen demand (BOD)- and nutrient removal, and particularly in the use of MBBR-based processes for nitrogen removal by de-ammonification, MBBRs for biological phosphate removal and MBBR-based membrane bioreactors (MBRs). Advances in MBBR biomass separation including the use of dissolved air flotation, micro-sand ballasted lamella settling (ACTIFLO®) and micro-screening (disc filtration) are presented. MBBR-based membrane bioreactors (including MBBR MBR and IFAS MBR) are discussed and a comparison is carried out in terms of volumes and footprints of a pure MBBR plant, a conventional MBR plant (based on activated sludge) and an IFAS-based MBR plant.

Keywords: Bio-P removal, de-ammonification, IFAS, MBBR, MBBR-based MBR

Editors: Giorgio Mannina, George Ekama, Hallvard Ødegaard and Gustaf Olsson
doi: 10.2166/9781780409719_0101

3.1 INTRODUCTION

The moving bed biofilm reactor (MBBR) was invented in Norway in the late 80s when focus was on nitrogen removal caused by eutrophication of the North Sea (Ødegaard *et al.*, 1994; Ødegaard, 2006). The concept was developed in close cooperation between the water treatment research group at NTNU/SINTEF in Trondheim and Kaldnes Miljøteknologi AS in Tønsberg, a company founded and based on the MBBR technology. The development was enhanced by the establishment of a research programme on nitrogen removal (Removal of nitrogen (FAN), 1988–1992 – in Norwegian) that was financed by The Norwegian State Pollution Control Authority (SFT)(Ødegaard, 1992). The research led to design criteria that are still in use (Hem *et al.*, 1994; Rusten *et al.*, 1995a,b). After a series of obstacles, the MBBR concept (reactor as well as carriers) was patent filed by Kaldnes Miljøteknologi in 1991 (with this author as the inventor) and the first plant was commissioned the same year as a small pre-denitrification plant (Lardal wastewater treatment plant [WWTP]) – the first nitrogen removal plant in Norway. Over the years the original patent was owned, and the process marketed, by a series of companies; Kaldnes Miljøteknologi AS (1991–1995), Anglian Water Ltd (1995–2002), AnoxKaldnes AB (2002–2006) and Veolia Water Systems (2006–still). Several other companies around the world are now offering MBBR systems and MBBR-based processes are used on all continents in municipal plants (including on-site) as well as industrial plants.

The MBBR is a flow-through, pure biofilm reactor, i.e. there is no recycle of biomass from the downstream separation reactor back to the MBBR. The biofilm (or attached biomass) grows on carriers that are suspended in the reactor and moving freely around with the currents set up by aeration (in aerobic reactors) and mixing (in anoxic and anaerobic reactors) – see Figure 3.1.

At steady state the net biomass growth is sloughed off the carriers due to erosion caused by the movement and mixing, and the produced biomass is separated from the water in a downstream clarification reactor. The suspended solids (SS) that are to be separated, are composed of the biomass produced as well as the incoming suspended solids that are not degraded in the reactor. And since there is no recycle, the concentration of this suspended solids is low (typically 150–250 g SS $\cdot$ m^{-3}), which allows for the use of any separation reactor alternative. Frequently compact separation alternatives are used, such as dissolved air flotation (DAF), lamella settling, floc blanket lamella settling (possibly micro-sand enhanced) as well as various kinds of filters (micro-screens, cloth filters, sand filters and membrane filters). Used in this way, MBBR plants may become extremely compact with a footprint for nitrogen removal of only 10–20% of that of a conventional activated sludge (AS) plant based on settling.

The pure MBBR is used for the removal of organic matter and nitrogen (see Figure 3.2) and is applied in municipal and industrial wastewater treatment as well as in drinking water treatment and treatment of recycled water in aquaculture.

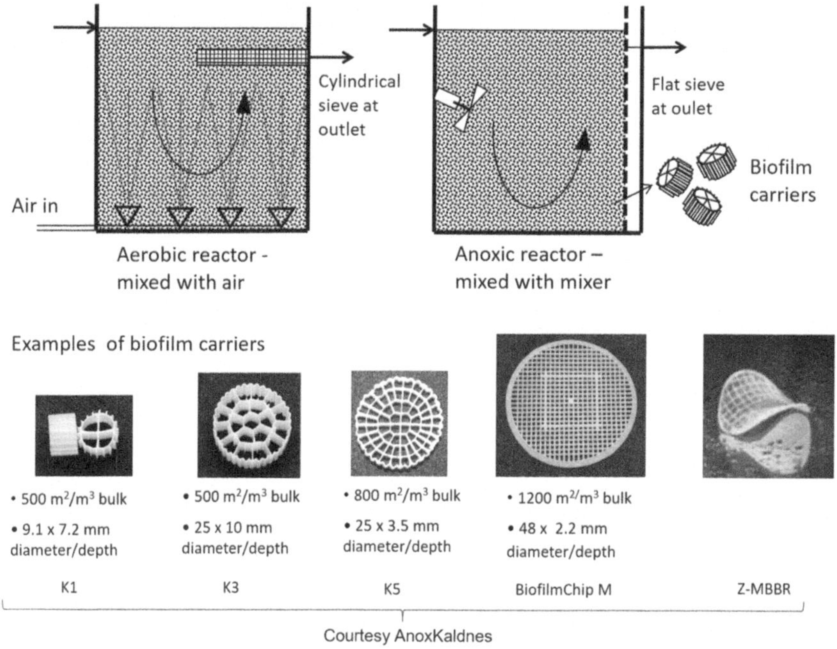

Figure 3.1 Principle of the MBBR and examples of carriers (Ødegaard, 2016).

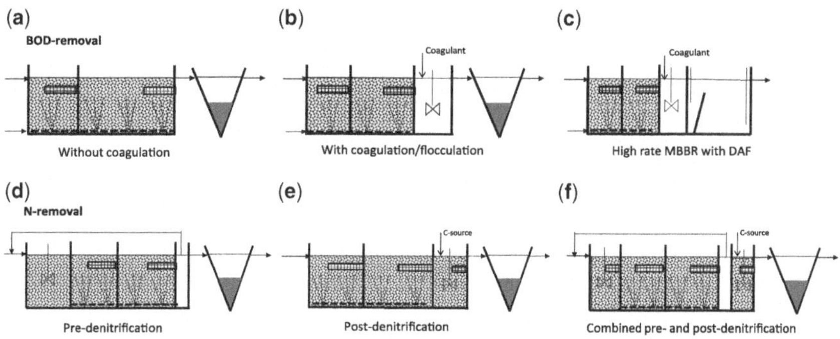

Figure 3.2 The most common principles of MBBR systems for biochemical oxygen demand (BOD)- and N-removal in pure MBBR plants.

The MBBR concept may also be used in hybrid systems, together with activated sludge, in a so-called integrated fixed film activated sludge system (IFAS for short) where MBBR carriers, with attached biomass, are present in the activated sludge tank containing suspended biomass (see Figure 3.3). In IFAS plants nitrification can be achieved at a much lower mixed liquor suspended solids sludge retention time (SRT_{MLSS}) than in conventional activated sludge plants and hence nitrification (and N-removal) is achieved at a much lower bio-reactor volume.

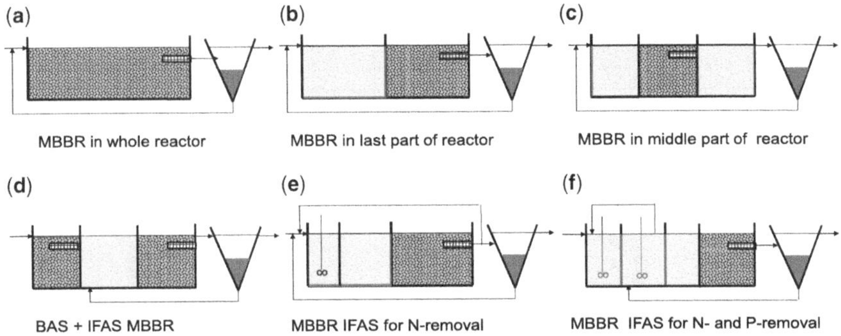

Figure 3.3 Common MBBR-based IFAS systems for different purposes (BAS – biofilm before activated sludge).

MBBR-based IFAS systems are used for a variety of reasons, including enhanced BOD removal, enhanced nitrification, enhanced N removal, enhanced biological P removal, improved settling, reduced footprint and improved operational stability (Ødegaard *et al.*, 2014).

It is fair to say that the MBBR-based processes (pure biofilm MBBR and MBBR-based IFAS) are now well established as compact, easy-to-operate wastewater treatment solutions and at this time there are least 1200 plants (about 50/50 MBBR and IFAS) of significant size around the world. The MBBR is also used in several thousand mini/on-site MBBR plants around the world.

Even though the MBBR-based processes are designed and used as they were from the beginning (in the 1990s), there have been several developments over the last 10 years. In this chapter, advances in the use of MBBR and IFAS for traditional secondary treatment (BOD- and SS-removal) and tertiary treatment (nutrient removal) will be given first, and subsequently advances in new applications of MBBR and IFAS for de-ammonification, biological P-removal and MBBR-based membrane bioreactors (MBRs) will be dealt with.

3.2 BOD-REMOVAL

The development in the area of secondary treatment (BOD- and SS-removal) has gone in the direction of increasingly more compact plants. This has been made possible partly because of the knowledge gained through research into hydrolysis of particles (Helness & Sjøvold, 2001) and partly because compact separation methods have to an increasing extent been applied. Lately the interest in de-ammonification processes for nitrogen removal (anammox) has revived the focus on high-rate BOD-removal, to be used as the C-stage in a de-ammonification process (see below).

Biodegradable, soluble organic matter is quickly degraded. Particulate organic matter is partly hydrolysed and degraded, and partly passes more or less unchanged through the reactor. The most efficient use of the MBBR for secondary treatment only is, therefore, to design for a high soluble organic loading in the first stage of the bioreactor and a lower loading in the later stages (see Figure 3.2a) – combined with subsequent coagulation/flocculation and high-rate biomass/floc separation (see Figure 3.2b). In this way one takes advantage of the fact that the biomass will operate at a higher removal rate at a high organic loading and remove the easily biodegradable organic matter first (in the first stage) before starting to hydrolyse particulate organic matter that may take place in the later stages. Subsequent coagulation/flocculation and floc separation ensures that the biomass produced by biodegradation of soluble organic matter, as well as not biodegraded, incoming particles, are removed from the water. This combines well with phosphate removal by chemical precipitation by coagulant dosing.

Design of pure MBBR systems is based on surface area loading rates, r_A (g · m^{-2} carrier surface · d) and the volume of the reactor is determined as:

$$V_{MBBR} = L_s \cdot (r_A \cdot A_{spec\cdot} \cdot f)^{-1} \cdot 100$$

V_{MBBR} = volume of MBBR (m^3)
L_s = substrate load (g · d^{-1})
r_A = design surface area loading rate (g · m^{-2} · d^{-1})
$A_{spec\cdot}$ = specific area of carrier in bulk (m^2 · m^{-3})
f = filling fraction of media (%)

Typical design values for BOD-removal are 8 g BOD_5 · m^{-2} · d at 10°C for the whole reactor (divided in at least two steps) without any post-coagulation/flocculation and 11.5 g BOD_5 · m^{-2} · d at 10°C with post-coagulation/flocculation. A typical design temperature coefficient is $\theta = 1.08$ (Norsk Vann, 2009).

Because of the compactness, the hydraulic residence time (HRT) in moving bed reactors for carbonaceous matter removal will be quite low (15–60 minutes), depending on the strength of the wastewater and the effluent requirement. It is recommended always to have a minimum of two reactor stages. Since the first stage should have a higher organic loading than the second stage, a division of the total volume of around 35% in the first reactor step and around 65% in the second is found to be optimal, based on 0.6 order kinetics in completely mixed reactors (see below).

A more sophisticated design procedure for the removal of organic matter is based on knowledge about the organic matter characteristics of the incoming water, especially with respect to particulate (suspended) organic matter versus soluble (filterable, assumed to be biodegradable) organic matter. The sizing of the reactor is made dependent on the degradation rate of the soluble BOD ($SBOD_5$) which is assumed to be 0 order at bulk $SBOD_5$ concentrations above 125 g · m^{-1} and 0.6 order at $SBOD_5$ concentrations below 125 g · m^{-1} (transformed to $SBOD_5$ from biodegradable, filterable COD (BFCOD) based on Figure 3.4). The $SBOD_5$ to be designed for, is composed of the incoming $SBOD_5$ and the $SBOD_5$ that is the result of hydrolysis of the incoming particulate BOD ($PBOD_5$). The latter is dependent on the soluble organic matter load (see Figure 3.5) and the percentage of the incoming $PBOD_5$ being hydrolysed, given by the formula in Table 3.1.

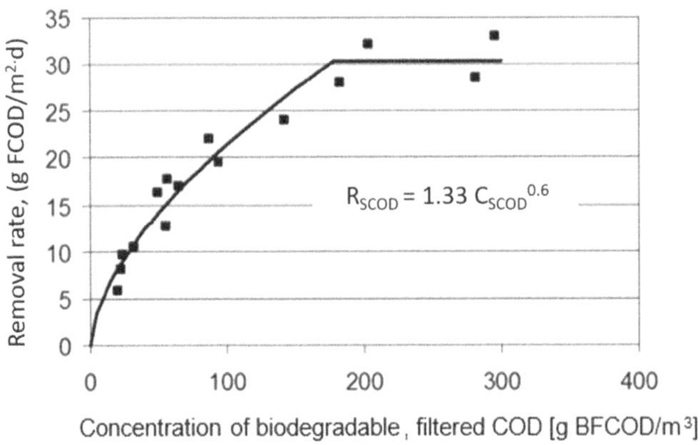

Figure 3.4 The removal rate of filtered (1 μm filtered) COD versus the concentration of filtered biodegradable COD (Helness & Ødegaard, 2005).

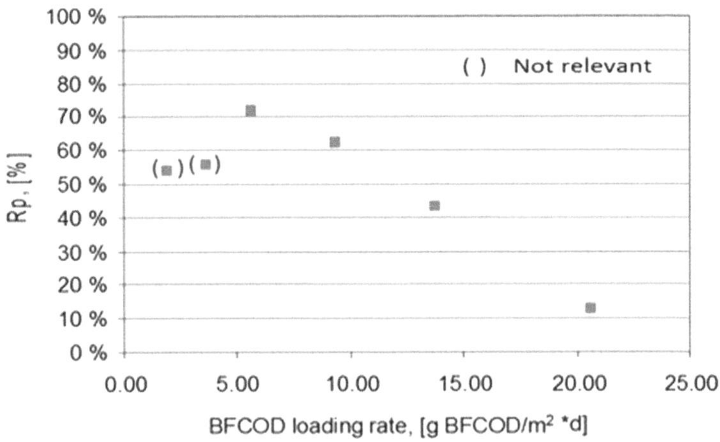

Figure 3.5 Extent of hydrolysis (R_p, %) of particulate COD being hydrolysed to soluble COD versus BFCOD load (Helness & Sjøvold, 2001).

Table 3.1 Design values for BOD-removal in MBBRs based on soluble BOD ($SBOD_5$) degradation.

Effluent Concentration, c (g $SBOD_5 \cdot m^{-3}$)	Reaction Order	Reaction Rate (15°C) g $SBOD_5 \cdot m^{-2} \cdot d^{-1}$	Biodegradation of $PBOD_5$ (%)
>125	0	$r_0 = k_0 = 20$	0
<125	0.6	$r_{0.6} = 1.15 \cdot [c]^{0.6}$	40–0.125 F^2

$F = SBOD_5$ load (g $SBOD_5 \cdot m^{-2} \cdot d^{-1}$)

The necessary area (and hence volume based on carrier characteristics and filling fraction) is determined reactor stage by reactor stage towards the effluent $SBOD_5$ targeted (for instance $SBOD_5 = 5\ g \cdot m^{-3}$ if the $BOD_{5,total}$ target is $10\ g \cdot m^{-3}$).

3.2.1 High-rate MBBR for BOD-removal

As the first stage for upgrading of existing (AS) plants (for instance in order to achieve nitrification, or as the carbon removal stage of a de-ammonification process), the so-called high-rate MBBR may be used (see Figure 3.2c) (Melin *et al.,* 2004; Helness *et al.,* 2005; Brosseau *et al.,* 2016).

The idea behind the high-rate process is that the attached biomass only degrades the soluble, most easily biodegradable organic matter, while the particulate, more slowly biodegradable organic matter passes through the reactor, more or less unchanged, and is separated from the water by coagulation/floc separation. This requires an organic loading on the MBBR that is low enough for truly soluble organic matter to be biodegraded and high enough to prevent significant hydrolysis and degradation of suspended and colloidal organic matter. In designing such a high-rate process, therefore, it is especially important to know the distribution of soluble and particulate organic matter.

In order to avoid hydrolysis of particulate organic matter, the biodegradable, filtered (i.e. soluble) organic load has to be higher than around 20–25 $gBFCOD \cdot m^{-2}_{carrier\text{-}area} \cdot d^{-1}$ (equivalent to about 15–25 g $SBOD_5 \cdot m^{-2} \cdot d^{-1}$ in a normal wastewater), as demonstrated in Figure 3.4 (Helness & Sjøvold, 2001). The reaction rate at this high load will be at its maximum (0 order) of around 30 $gBFCOD \cdot m^{-2} \cdot d^{-1}$, as demonstrated in Figure 3.4, resulting in an MBBR effluent concentration of BFCOD >150–200 $g \cdot m^{-3}$ equivalent to 100–150 $gSBOD_5 \cdot m^{-3}$. Below this, the reaction order will be close to half order, but even at 100 $gBFCOD \cdot m^{-3}$ the removal rate will be as high as 20 $gBFCOD \cdot m^{-2} \cdot d^{-1}$ corresponding to around 60 $gCOD_{total} \cdot m^{-2} \cdot d^{-1}$. At this high load only a negligible fraction of the particulate organic matter will be converted to soluble organic matter (Helness & Sjøvold, 2001) – see Figure 3.5.

Particulate matter entering the MBBR as well as biomass produced by biodegradation of easily degradable soluble organic matter in the MBBR, are separated from the water by coagulation/flocculation/biomass separation. By the use of a compact biomass separation method, such as dissolved air flotation or micro-sand ballasted lamella settling, an extremely low plant footprint will be obtained.

The separated sludge will have a high biogas potential since it consists of non-biodegraded incoming particulate organic matter and biomass resulting from a high-rate biological degradation of easily biodegradable, soluble organic matter. A combination of cationic polymer plus a small dose of iron is used in order to minimize the precipitation of metal hydroxide that does not contribute to biogas production.

A pilot plant study of the high-rate MBBR process (with DAF for biomass separation) demonstrated that secondary treatment standard (25 $gBOD_5 \cdot m^{-3}$ in effluent) could be reached at a total HRT from inlet to outlet of around 1.0 h (Melin *et al.*, 2004; Helness *et al.*, 2005). The pilot plant was operated on a quite concentrated municipal wastewater at relatively low temperatures (10–15°C). Based on this study, the following design criteria were proposed for the high-rate MBBR process:

- *Load*: 20–25 $gCOD_{filt} \cdot m^{-2} \cdot d^{-1}$ (15–20 $gBOD_{filt} \cdot m^{-2} \cdot d^{-1}$), 65–85 $gCOD_{tot} \cdot m^{-2} \cdot d^{-1}$ (45–60 $gBOD_{tot} \cdot m^{-2} \cdot d^{-1}$)
- *Sludge production*: 0.5 $gDS \cdot gCOD_{rem}^{-1}$ in MBBR step, 1.0 $gDS \cdot gSS_{rem}^{-1}$ in biomass separation step
- *Coagulant dosage*: 5 g cationic polymer $\cdot$ $kgSS^{-1}$ + 35 gFe $\cdot$ $kgSS^{-1}$

3.3 N-REMOVAL BY NITRIFICATION/DENITRIFICATION

The MBBR was developed during a time of nitrogen removal focus and several publications address this ambition (Ødegaard *et al.*, 1994; Hem *et al.*, 1994; Rusten *et al.*, 1995a,b; Ødegaard, 2014; 2016; 2017). The design criteria developed in the early 90s are, however, still valid.

3.3.1 Nitrification

After BOD-removal, nitrification design of pure MBBRs is normally based on the following model (Hem *et al.*, 1994; Rusten *et al.*, 1995a):

$$r_N = k \cdot (S_n)^n$$

r_N = nitrification rate ($gNH_4\text{-}N \cdot m^{-2} \cdot d^{-1}$)
k = reaction rate coefficient (estimated at 0.5–0.75 – depending on pretreatment)
n = reaction order constant, estimated at $n = 0.7$
S_n = rate-determining ammonium concentration, $gNH_4\text{-}N \cdot m^{-3}$ (Sn ~ (DO_{bulk} – 0.5)/3.2)

k will be dependent upon the C/N-ratio ($gBOD_5/gNH_4\text{-}N$) of the incoming water to the bioreactor and vary from 0.7 at C/N = 0.5 to 0.5 at C/N = 4.5. If the preceding

MBBR reactors are designed to remove essentially all BOD, the nitrification rate coefficient for the final nitrification reactors may be set at 0.75.

The nitrification rate is linearly dependent on the oxygen concentration up to more than 10 $gO_2 \cdot m^{-3}$ (Hem *et al.*, 1994; Æsøy *et al.*, 1998), see Figure 3.6. The dissolved oxygen (DO) strongly limits the nitrification rate even at concentrations as low as 2–3 $gO_2 \cdot m^{-3}$. Normally one is designing for DO concentrations of 4–6 $gO_2 \cdot m^{-3}$ at peak load.

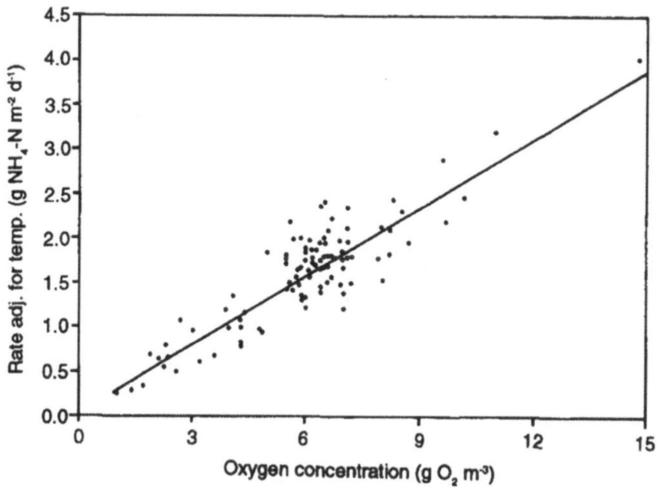

Figure 3.6 Nitrification rate (adjusted to 15°C vs. DO DO/NH_4-N ratio <2 $gO_2 \cdot gNH_4\text{-}N^{-1}$ (Hem *et al.*, 1994).

An advantage of the linear relationship between oxygen concentration and nitrification rate is that it may favourably be used for process control. If extremely low values of ammonium are required, the last reactor segment must be designed for an ammonium limited rate – in which case oxygen concentration may be low since it is no longer governing the rate of nitrification.

Oxygen transfer is enhanced by the presence of carriers as demonstrated in Figure 3.7 (Christensson, 2011). The higher the filling fraction (up to around 60%), the better the oxygen transfer. The benefit of the carriers is dependent on the design of the carrier used. It seems as if tube-like carriers are better than coin-like carriers in this respect. Carrier suppliers should, therefore, provide Standard Oxygen Transfer Rate (SOTR) data. This implies that good oxygen transfer is achieved even with a robust, medium coarse bubble aeration system, that is often used in order to achieve even air distribution and complete mixing without clogging at low head-loss (Ødegaard *et al.*, 1994).

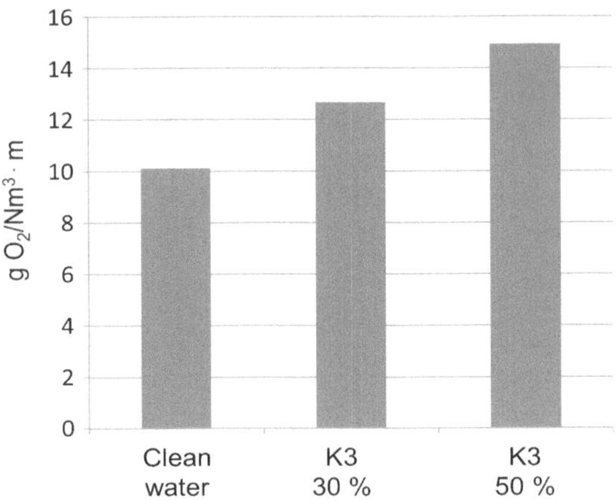

Figure 3.7 Influence of carrier filling fraction on oxygen transfer (Christensson, 2011).

3.3.2 Denitrification

Pure MBBRs are used for pre-denitrification, post-denitrification and combined pre- and post-denitrification (see Figure 3.2d–f). Too low C/N-ratio in the raw water is a common challenge when stringent effluent nitrogen concentrations are to be met in traditional pre-denitrification plants. In such situations the so-called combined pre- and post-denitrification MBBR process (see Figure 3.8) is often used. This process configuration gives a superior possibility to control towards the effluent standard aimed for – however stringent that is (Rusten & Ødegaard, 2007; Ødegaard, 2014).

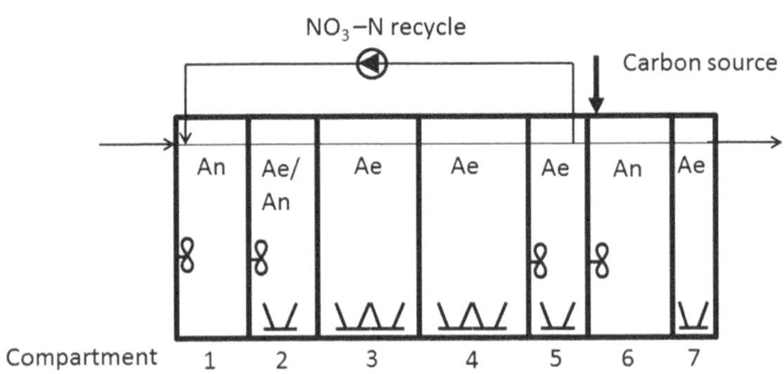

Figure 3.8 The typical build-up of a combined pre- and post-denitrification MBBR (Ødegaard, 2016).

3.3.2.1 Combined pre- and post-denitrification MBBR

The combined-denitrification (DN) MBBR plants normally have a division between aerobic and anoxic compartments of the MBBR – as indicated in Figure 3.8.

The first compartment (comp. 1) is anoxic (pre-denitrification) and receives recycled nitrate from the last aerobic compartment. The second compartment (comp. 2) may be operated both anoxic (no air) and aerobic (with air). Then follow two aerobic reactors (comp. 3 and 4), used for residual organic matter removal and nitrification. Compartment 5 also contributes to nitrification and may be operated with or without aeration. Compartment 6, to which carbon is added, is anoxic and for post denitrification. Finally there is optionally a small, aerobic compartment for oxygenation as well as for removal of any residual biodegradable organic matter, possibly caused by excess addition of external carbon source.

This bioreactor scheme may give very high nitrogen removal efficiencies and optimal flexibility of operation at minimized consumption of external carbon source (Rusten & Ødegaard, 2007). During summer operation, for instance, less nitrification volume is needed because of higher temperatures and the raw water normally has a higher DO concentration than in the winter time. In this situation compartment 2 may be used for denitrification (stirred and not aerated). More nitrate may be returned to the pre-DN tank and the need for added carbon source in the post-DN tank is lower. In summer time, therefore, most of the nitrogen can be removed by pre-denitrification, reducing the consumption of external carbon source to just that which is needed to ensure that the nitrogen removal goal is achieved. During winter operation, more nitrification capacity is needed and compartment 2 is operated with aeration. More carbon source will have to be added to the post-DN tank in order to achieve the treatment goal. The reason for the option of not aerating compartment 5, is de-oxygenation in order to minimize the return of oxygen to the pre-denitrification reactor. Because the ammonia concentration here will be low (<2 g NH_4-N · m^{-3}), the oxygen is no longer rate limiting for the nitrification rate, but rather the ammonia concentration is. Hence it does not matter that the DO is reduced which minimizes oxygen respiration in the pre-DN compartment caused by the recirculation of oxygen.

The extent of nitrification may be controlled by the oxygen level in the nitrification reactor (comp. 4) since there is a linear relationship between nitrification rate and oxygen concentration (see Figure 3.6). The extent of denitrification in the post denitrification step is governed by the availability of carbon source. Nitrate measurement in the post denitrification outlet may therefore control carbon source addition and hence the final denitrification result.

Figure 3.9 shows post-denitrification rates in a combined pre- and post-denitrification MBBR plant (Gardermoen WWTP in Norway) where three different external carbon sources were tested (Rusten *et al.*, 1996). The tests were carried out in a situation where the nitrate concentration was not rate limiting. Use of methanol and monopropylene glycol (MPG) gave similar denitrification rates. Use of ethanol doubled the denitrification rates. All three external carbon sources showed a temperature coefficient of $\theta = 1.07$. Necessary C/N-ratios were also temperature dependent, which can be explained by higher specific sludge production (lower decay) at lower temperatures. The necessary C/N-ratio increased with an average of 45% when the temperature was reduced from 15 to 5°C.

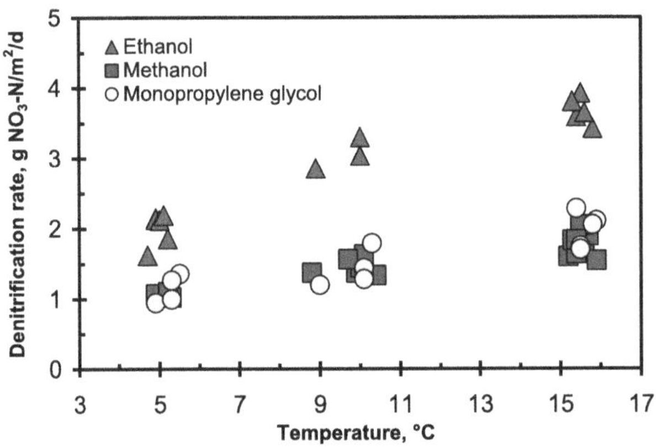

Figure 3.9 Denitrification rate versus temperature with various external carbon sources (Rusten *et al.*, 1996).

For 95% removal of nitrate, ethanol had the lowest carbon demand, with a necessary C/N-ratio of 3.9 $gCOD_{added} \cdot gNO_3\text{-}N_{equivalents}^{-1}$ at 10°C. MPG had the highest necessary C/N-ratio at 5.6 $gCOD_{added} \cdot gNO_3\text{-}N_{equivalents}^{-1}$ at 10°C (Rusten *et al.*, 1996).

3.3.2.2 Post-denitrification MBBR

Pure MBBR post-denitrification is also frequently used downstream of an existing nitrifying activated sludge plant (Aspegren *et al.*, 1998; Stinson *et al.*, 2009; Ericsson, 2011).

The post denitrification rate is totally governed by the availability of biodegradable organic matter as long as the effluent NO_3-N concentration is >~5 $gNO_3\text{-}N \cdot m^{-3}$). The $K_{s,COD}$-value is quite low, around 3 $gSCOD_{biodegradable} \cdot m^{-3}$, so as long as the $SCOD_{biodegradable}$-concentration is above ~10 g biodegradable, soluble COD (BSCOD) $\cdot m^{-3}$, the rate is not limited by organic matter (Rusten *et al.*, 1995b). If very low effluent nitrate concentrations are to be met (<3–5 $gNO_3\text{-}N \cdot m^{-3}$), the nitrate concentration will limit the denitrification rate. Reported $K_{s,\,NO3\text{-}N}$-values are in the range of 0.5–3.0 $gNO_3\text{-}N \cdot m^{-3}$ (Christensson, 2008; Rusten, 2008; Stinson *et al.*, 2009). One may conservatively design for a $K_{s,\,NO3\text{-}N}$ value of 1.5 $mgNO_3\text{-}N \cdot m^{-3}$ and a maximum design rate of 3 $gNO_x\text{-}N \cdot m^{-2}d^{-1}$ at 15°C at no substrate limitation

(BSCOD >10 g · m^{-3} and NO_3-N >5 g · m^{-3}. At a moderate nitrate limitation (effluent 3–5 gNO_3-N · m^{-3}), the rate is still quite high (>2 gNO_x-N · $m^{-2}d^{-1}$ at 15°C) while at strong limitation (NO_3-N concentrations <3 gNO_3-N · m^{-3}), the design rates should probably not be set above 1 gNO_3-N · m^{-2} · d^{-1} at 15°C. Ericsson (2011) reported maximum full-scale denitrification capacities at Rya WWTP in Sweden, of 2.2 gNO_x-N · m^{-2} · d^{-1} at around 10°C, with methanol as carbon source.

This has, above all, an impact on how the post denitrification reactor should be divided in segments. The following thumb rule is proposed:

(1) If the effluent design NO_3-N value is >3 mg · l^{-1}, use only one post-DN reactor.
(2) If the effluent design NO_3-N value is <3 mg · l^{-1}, divide the DN-step in two reactors, designed for maximum rate in the 1st reactor and half of the maximum rate in the 2nd reactor. If only one reactor is possible, design it for half of the maximum design rate.
(3) If the effluent design NO_3-N value is <1 mg · l^{-1}, one should consider three reactors in the post-DN step, with the 3rd designed for a quarter of the maximum rate.

3.3.3 N-removal in MBBR-based IFAS plants

The most frequently used objective for using MBBR-based IFAS is to enhance nitrification, with the goal of being able to fully nitrify at a lower aerobic SRT_{MLSS} than in a conventional AS system. This results in a reduced footprint for nitrifying-, N removal- and biological nutrient removal (BNR) plants. Positive side-effects are improved biomass settling and operational stability. IFAS systems are therefore, frequently used for upgrading of conventional AS (CAS) plants.

One of the benefits of the MBBR IFAS system is that one needs only add the amount of attached growth area (i.e. the amount of carriers) that is necessary. The maximum recommended filling fraction in a pure MBBR system is 67% (Ødegaard *et al.,* 1994). In MBBR IFAS systems filling fractions >60% are seldom used, and one may adapt the biofilm growth area necessary to the volume available (Ødegaard *et al.,* 2014).

As shown in Figure 3.3a–c one may add carriers to the entire reactor volume or to one or more segments of the reactor. Figure 3.10a shows a common pre-denitrification application where carriers are used in the aerobic part of the reactor train. Carriers may also be used, however, in anoxic reactors (as shown in Figure 3.10b), although this is currently less common.

Nitrification in MBBR-based IFAS systems is mainly influenced by the aerobic SRT_{MLSS}, the organic carrier area load (relative to ammonium load), the reactor DO concentration, the ammonium concentration and the temperature. In order to demonstrate this, data from pilot plants at Tau WWTP, Norway and from Broomfield WWTP, Colorado, USA are shown in Figure 3.11 (Rusten *et al.,* 2003).

The Tau pilot plant, with a filling fraction of only 18% (64 $m^2 \cdot m^{-3}$ carrier surface), achieved complete nitrification at aerobic SRT_{MLSS} as low as 2.2 d at 15°C, about 40% lower aerobic SRT_{MLSS} than without carriers (Figure 3.11a). The carrier biomass nitrification rate in the Broomfield plant (see Figure 3.11b) agreed well with the nitrification design model (see above) for pure MBBR systems, indicating that this model may also be used in MBBR IFAS systems. This pilot demonstrated that a DO concentration of 4 $g \cdot m^{-3}$ was needed to achieve an effluent of 1 $gNH_4 \cdot m^{-3}$ (Figure 3.11b) indicating the important role of the attached biomass in nitrification.

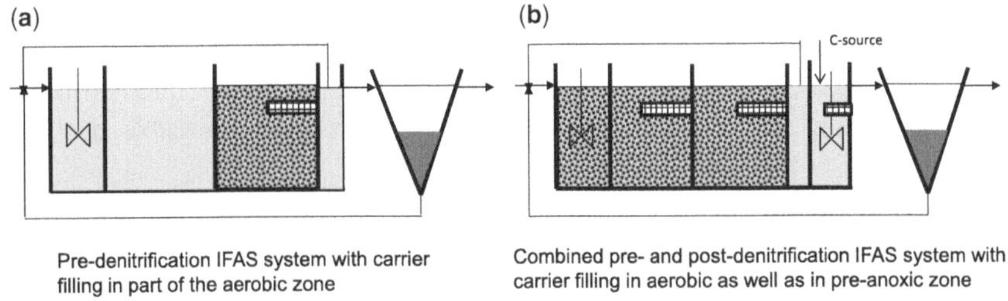

Figure 3.10 Examples of IFAS systems for nitrification/denitrification.

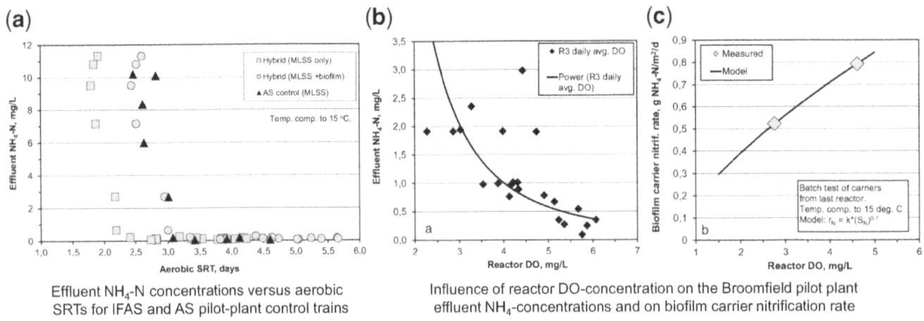

Figure 3.11 Influence of aerobic SRT_{MLSS} and DO in MBBR-based IFAS pilot plants (Rusten *et al.*, 2003).

Figure 3.12 shows the relationship between design aerobic SRT_{MLSS} and temperature for various US-plants designed for nitrification (Johnson, 2009) together with the curves for nitrifier growth rate and the Abwassertechnische Vereinigung (ATV) design curve for nitrification in AS plants. The SRTs for these full-scale MBBR IFAS systems are generally far lower than the German design curve.

Different models have been used to explain the enhancement of nitrification in IFAS systems: (a) the high attached biomass SRT, (b) an increased nitrifier biomass per unit volume, (c) an increased nitrifier activity in the attached biomass compared to the suspended biomass, and (d) the seeding of nitrifiers into the suspended biomass from the attached biomass (Ødegaard *et al.*, 2014). McQuarrie (2009) studied the nitrification activity in the suspended biomass and the attached biofilm in an MBBR IFAS system based on data from Thomas (2009). They found that the specific nitrification activity (mgNH$_4$-N · gSS^{-1} · h^{-1}) was much higher in the attached biomass than in the suspended biomass (see Figure 3.13a).

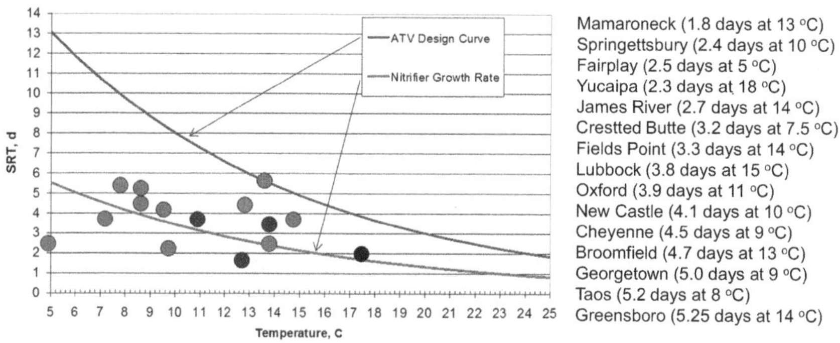

Figure 3.12 Design SRT versus temperature for full-scale IFAS systems (Johnson, 2009).

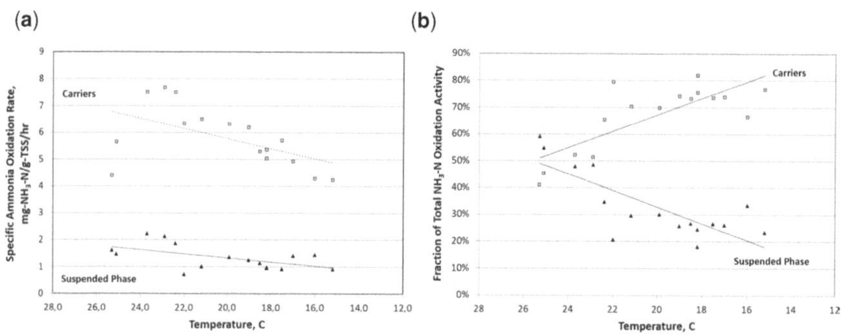

Specific nitrification activity in attached and suspended biomass versus temperature

Fraction of NH$_3$-N oxidation activity in attached and suspended biomass versus temperature

Figure 3.13 Distribution of nitrification activity in attached and suspended biomass as a function of temperature (McQuarrie, 2009; Thomas, 2009).

A probable explanation for this, is that the nitrifiers in the attached biomass are always exposed to low soluble biodegradable carbon to ammonia ratio, resulting in an attached biofilm enriched with nitrifying bacteria. McQuarrie (2009) also demonstrated that the fraction of NH$_4$-N oxidation activity taking place on the carriers, increased with decreasing temperature, while the fraction in the suspended biomass decreased with decreasing temperature. This demonstrates that the benefit of the carriers in the overall process is greater at low temperatures when it is most needed.

Denitrification (DN) in IFAS systems may be by pre-DN, post-DN and combined pre- and post-DN (Ødegaard, 2014). Pre-DN-systems, in which the anoxic reactor does not contain carriers, are most common, but there also is an interesting potential for using carriers in anoxic reactors (Choi *et al.*, 2007).

In the previously mentioned Tau pilot plant, Rusten *et al.*, (2003) found a >50% higher DN rate in the IFAS pre-anoxic reactor biomass than in the CAS pre-anoxic reactor biomass (3.15 g NO_x-N · kg $MLSS^{-1}$ · h^{-1} and 2.00 g NO_x-N · kg $MLSS^{-1}$ · h^{-1} respectively). So did Onnis-Hyden *et al.* (2011). This indicates that the low aerobic SRT of hybrid processes leads to higher DN-rates than in CAS plants. In systems with higher MLSS and SRT, much of the carbon is lost during the aerobic phase; hence, the shorter the aerobic SRT, the more carbon is available for DN.

3.4 N-REMOVAL BY DE-AMMONIFICATION IN MBBR-BASED PLANTS

Traditionally, biological nitrogen removal in wastewater is carried out through nitrification/denitrification. The challenges with this process are the high energy demand for the aerobic (autotrophic) part of the process (nitrification), and the carbon source requirement for the anoxic (heterotrophic) part of the process (denitrification). Over the last 10 years research focus has been on an alternative nitrogen removal process (de-ammonification) as a consequence of the desire to overcome the challenges of nitrification/denitrification.

De-ammonification is a biological treatment process that converts ammonia to nitrogen gas without the need for a carbon source and with much less air needed than in nitrification (Water Environment Research Foundation [WERF], 2014). De-ammonification is accomplished through two biological process steps. The first is termed nitritation, which is the aerobic oxidation of ammonia-N (NH_4-N) to nitrite-nitrogen (NO_2-N) by autotrophic, aerobic, ammonia-oxidizing bacteria (AeAOB). Nitritation is well known in wastewater treatment, as it is the initial step in biological nitrification of ammonia-N to nitrate-N. Since stoichiometrically only about half of the ammonia needs to be converted to nitrite, this is also known as partial nitritation. In the second step of de-ammonification, which is the anammox (anaerobic ammonia oxidation) reaction, NH_4-N is the electron donor and oxidized under anaerobic conditions by autotrophic, anaerobic ammonia-oxidizing bacteria (AnAOB) that can use NO_2-N as the electron acceptor. Hence the overall process is most correctly referred to as the partial nitritation/anammox process, but in this chapter the expression de-ammonification is used (for short).

There are several obvious advantages of the de-ammonification process; the energy for air is about 60% lower, no carbon source is needed, the alkalinity consumption is about 50% lower, and the sludge production is in the order of 70% lower (if organic particles are removed upstream). About 89% of the inorganic nitrogen (NH_4-N + NO_2-N) fed to the process ends up as N_2 gas and about 11% as NO_3-N (WERF, 2014). So if very low tot N-concentrations in the effluent (<~3 $gN \cdot m^{-3}$) are aimed for, a small nitrification/post-denitrification step has to be applied downstream of the de-ammonification step.

Since it is implemented most easily when the temperature is high (preferably >~25°C), the ammonium concentration is high (>500 mg NH_4-N/l) and the C/N-ratio is low, the process has been successfully implemented for sludge reject water (in side-stream), that normally represents about 25% of the nitrogen load on a typical BNR plant. It is not yet established as a proven process for nitrogen removal in the main-stream, but will represent a break-through in wastewater treatment when it is.

3.4.1 De-ammonification in the side-stream

The MBBR has demonstrated itself to be a robust and compact biofilm technology for de-ammonification in sludge reject water (side-stream) (Jardin & Hennerkes, 2012; Rosenwinkel & Cornelius, 2005; Christensson *et al.*, 2013; Malovanyy, 2017). The first full scale MBBR demonstration plant for treatment of reject water was established in Hattingen, Germany (Rosenwinkel & Cornelius, 2005) as a multi-stage MBBR process. The demonstration project here proved the MBBR de-ammonification process to be cost-effective in comparison to conventional N-removal (Jardin & Hennerkes, 2012).

In a biofilm reactor it is possible carry out both nitritation and de-nitritation by anammox in the same biofilm. Partial nitrification to nitrite and autotrophic N-removal (i.e. anammox) may occur simultaneously within the biofilm, where aerobic and anoxic zones results from oxygen mass transfer limitation under limited dissolved oxygen (DO) conditions (see Figure 3.14).

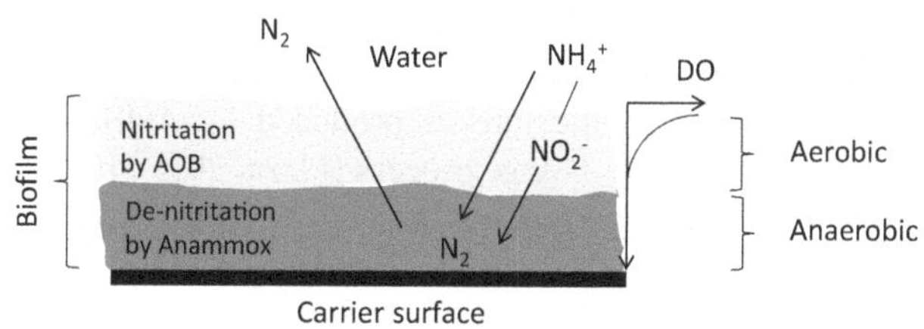

Figure 3.14 Schematic of 1-stage nitritation/anammox biological processes occurring inside a carrier's biofilm.

In an MBBR the ammonium oxidizing bacteria (AeAOB) as well as the anammox (AnAOB) bacteria are maintained in the attached biofilm on the suspended carriers retained in the reactor by the sieves, with no risk of biomass wash-out (Cema, 2009; Christensson *et al.*, 2013). AeAOB oxidize NH_4 to NO_2 in the aerobic zone of the biofilm (i.e. outer part) while AnAOB bacteria located in the anoxic zone of the biofilm (i.e. inner part) consume NO_2 produced by AeAOB together with the excess NH_4-N (see Figure 3.14). The main challenge of single-stage de-ammonification is to prevent further oxidation of nitrite to nitrate by nitrite oxidizing bacteria (NOB). One strategy is to use intermittent aeration in order to prevent nitrite oxidation by lowering the DO intermittently.

ANITA™Mox is a single-stage MBBR de-ammonification process (Christensson *et al.*, 2013; Veuillet *et al.*, 2013; 2014). In the ANITA™MOX system, a DO control system is used to prevent nitrite oxidation in the aerobic zone of the biofilm while maximizing the amount of nitrite available for the anammox bacteria. The DO set-point is automatically adjusted based on online inlet and outlet concentrations of NH_4 and NO_3 to control the NO_3 production below 11% of NH_4-N removed (i.e. stoichiometric NO_3 production by anammox) while at the same time keeping high NH_4-oxidation performance in the reactor. This real-time DO control strategy reduces the need for a mechanical mixer in the MBBR due to the continuous aeration pattern (Christensson *et al.*, 2013).

To shorten the start-up phase, new ANITA™Mox plants are seeded with a small fraction of colonized carriers, which reduce the time required for the development of a mature de-ammonification biofilm on the brand new carriers. Seeding has proven to dramatically decrease the start-up time from up to a year down to 2–3 months depending of the amount of seeding (Christensson *et al.*, 2013). A full-scale study indicated, however, that inoculum was not necessary in order to start up a full-scale de-ammonification process in the reject water from thermophilic digestion if nitrification biomass is already established. The start-up took 120 days in full-scale and 72 days in lab-scale (Kanders *et al.*, 2016).

When comparing 2-stage MBBR and 1-stage MBBR for de-ammonification on sludge reject water, Cema *et al.* (2010) found excellent nitritation in the 2-stage system, but also that it was more difficult to control the anammox stage. Nitrite is a substrate for anammox bacteria but nitrite accumulation can also inhibit their activity (Fernandez *et al.*, 2012; Lotti *et al.*, 2012). Inhibition was experienced in a 2-stage pilot-plant, while the nitrites produced by AOB were consumed immediately by anammox in the 1-stage process (Plaza, 2015). In order to prevent anammox inhibition in a 2-stage process the NO_2-N/NH_4-N ratio in the reactor effluent (influent to anammox reactor) should be kept as close as possible to the stoichiometric optimum of 1.32 (0.57 NO_2-N/0.43 NH_4-N).

In an attempt to overcome the inhibition challenge and to improve the ANITA™Mox process performances under different operating conditions, it was hypothesized that substrate transport could be enhanced by combining suspended cultures and fixed biomass into one system, i.e. an IFAS system (see Figure 3.15) (Veuillet *et al.*, 2013; 2014).

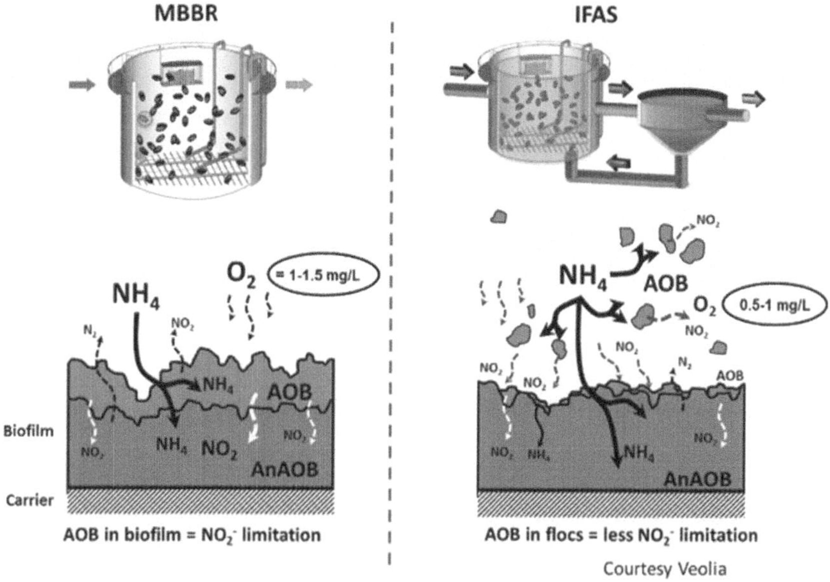

Figure 3.15 Pure MBBR versus MBBR-based IFAS for de-ammonification (Veuillet *et al.,* 2013).

In lab-scale tests on anaerobic sludge digester reject water, it was demonstrated that the nitrogen removal capacity was almost four times higher when comparing the pure 1-stage MBBR version with the IFAS 1-stage version (Christensson *et al.,* 2013; Veuillet *et al.,* 2013). With reference to Figure 3.15 the authors offered the following explanation:

"In an IFAS system the NO_2 level in the bulk liquid may be high enough to be able to fully diffuse into the deep layer of the biofilm where the AnAOB are mostly located, but not too high in order to avoid inhibition of AnAOB. Increased AOB activity in the IFAS configuration is obtained by the increased growth of AOB in the MLSS. In the IFAS mode the sludge retention time (SRT) can be increased from less than a day to several days, thereby preventing wash-out of AOBs. Substrate diffusion limitation in flocs is less apparent than in biofilms that are thicker and denser. Hence this should lead to better substrate accessibility for AOB (oxygen, NH_4) in the suspended solids in the IFAS configuration, meaning that AOB that pre-existed on the outer-layer of the biofilm gradually disappear from the biofilm due to a lack of oxygen (i.e. DO in bulk liquid is lower in IFAS than MBBR) that is now mostly consumed by AOB in the liquid phase. The biofilm is therefore almost exclusively composed of AnAOB with a very fine top layer of oxygen scavengers like AOB or normal heterotrophs as indicated in Figure 3. The larger AOB population in IFAS configuration improves the overall flux of NO_2 produced for AnAOB but also the residual concentration in the bulk improving the diffusion of the NO_2 through the basal layer of the biofilm

where AnAOB are located and therefore increasing their active fraction in the biofilm" (Veuillet *et al.*, 2014).

In the experiments, the global amount of AnAOB in the IFAS and the MBBR configuration was found to be similar. The improvement of nitrites flux and production in the IFAS mode led, however, to a N-removal rate near the maximum that was obtained during fully anoxic AnAOB activity batch tests, with non-limiting substrate levels (i.e. NH_4-N and NO_2-N) (Veuillet *et al.*, 2013).

Full-scale experiences with the ANITA™Mox process, both in pure biofilm as well as in IFAS mode, from the demonstration plant at Sjölunda WWTP in Sweden, confirm the findings in the lab- and pilot-scale tests (Veuillet *et al.*, 2014).

3.4.2 De-ammonification in the main-stream

While a number of wastewater treatment plants have implemented the side-stream de-ammonification process as a cost-effective, efficient and reliable option to treat reject water, the techniques required to sustain de-ammonification in the colder and more dilute main-stream wastewater are yet to be developed. The challenges of mastering main-stream de-ammonification have been summarized as follows (Trela *et al.*, 2014):

(1) The dominance of NOB growth at lower temperatures makes the selection of AOB over NOB challenging. Moreover, NOB suppression by free ammonia inhibition is not possible because of low ammonium concentration. If NOB out-competes AOB, nitrate accumulation results, which will significantly decrease nitrogen removal efficiency. This seems to be the biggest challenge for reaching high efficiency of nitrogen removal treating main-stream wastewater.

(2) The possible solution for out-competing NOB is to optimize operation parameters (DO, intermittent aeration phase duration, pH, inorganic carbon concentration) for stimulating AOB growth and suppression of NOB growth. Using the IFAS system instead of pure MBBR may also be a possible way to overcome NOB out-competition – as discussed above.

(3) An effective retention of the anammox biomass in a reactor is required. This is because inflow nitrogen concentration in main-stream wastewater is low (25–50 $gNH_4\text{-}N \cdot m^{-3}$) and this, together with low yield and growth rate of anammox bacteria, leads to low anammox biomass production. This is the main reason why biofilm systems (i.e. MBBR) are especially suitable.

(4) Nitrogen transformation rates are about 70–80% lower for main-stream wastewater with a yearly average temperature of 15°C, compared to supernatant treatment at 30°C based on activation energies of de-ammonification reactions. Moreover, because of lower ammonium and nitrite concentrations in the process, even lower rates can be expected.

At this time most developers and researchers work with MBBR-based IFAS systems for main-stream de-ammonification (Lemaire *et al.*, 2015; Malovanny *et al.*, 2015; Malovanny, 2017). IFAS-based ANITA™Mox is proposed for use according to two different strategies (Lemaire *et al.*, 2015), see Figure 3.16.

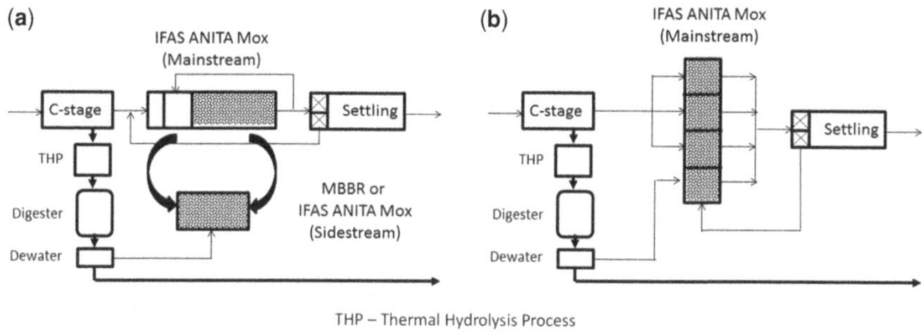

Figure 3.16 Schematic of main-stream ANITA™Mox WWTP (a) with carrier recycling concept between side-stream and main-stream ANITA™Mox reactors and (b) with alternating feed concept between side-stream and main-stream water (Ødegaard, 2016 after Lemaire *et al.*, 2015).

The main aim is to obtain bio-augmentation by moving carriers from the more robust side-stream process to the less robust main-stream process. Anammox-rich biofilm carriers can easily be transferred back and forth from the side-stream to the main-stream, for instance by the use of an air-lift pump (Figure 3.16a). This action will allow for partial regeneration of the anammox biofilm due to the more favourable conditions in the side-stream reactor (higher temperature, higher NH_4 level, lower COD level). The carrier transfer flow rate (Figure 3.16a) will have to be controlled in order to maximize the bio-augmentation effect to the main-stream IFAS ANITA™Mox reactor (Lemaire *et al.*, 2015).

In the alternative approach (Figure 3.16b) the feeding is alternated between the main-stream COD-treated effluent and the side-stream reject water to a multi-celled IFAS ANITA™Mox system. As shown in Figure 3.16b, one of the cells will receive the side-stream reject water for a period of time providing optimal conditions for anammox and is, therefore, temporarily considered as a side-stream system. The side-stream reject water is then switched to feed another IFAS cell while the main-stream COD-treated effluent is switched back to the cell which has previously been fed with the side-stream reject water. The period of time during which a given IFAS cell receives a side-stream effluent will have to be controlled.

Results from two main-stream IFAS ANITA™Mox studies were presented – one pilot-scale study in Paris treating chemically enhanced primary effluent and one full-scale prototype (reactor volume 50 m^3) study at ANITA™Mox plant in Malmö, Sweden treating effluent from a high-rate activated sludge step in the full-scale Sjölunda plant (Lemaire *et al.*, 2015). In the prototype plant the bio-augmentation strategy using alternating feed between side-stream and main-stream effluent was tested. From these studies it was concluded that the IFAS ANITA™Mox process was successfully applied as a main-stream process with removal rates obtained in the pilot-plant of 1.4 gN · m^{-2} · d^{-1} (on average) during the summer (at 23°C) and 0.5–0.8 gN · m^{-2} · d^{-1} during the start of winter (at 17°C). The bio-augmentation strategy in the prototype plant seemed to be efficient in suppressing the NOB activity while achieving N-removal rates up to 0.1–0.3 kgN · m^{-3} · d^{-1} (equal to 0.23–0.68 gN · m^{-2} · d^{-1}) at 13–18°C (Lemaire *et al.*, 2015).

Malovanyy (2017), who studied both pure MBBR and IFAS single-stage systems for main-stream de-ammonification, found an advantage of IFAS systems. He presented the mass balance for the IFAS single stage system shown in Figure 3.17 that demonstrates that heterotrophic denitrification played a significant role. The heterotrophic denitrification activity presupposes availability of organic matter that is in contradiction to the enhancement of anammox bacteria. This demonstrates the challenge in design and operation of one-stage systems.

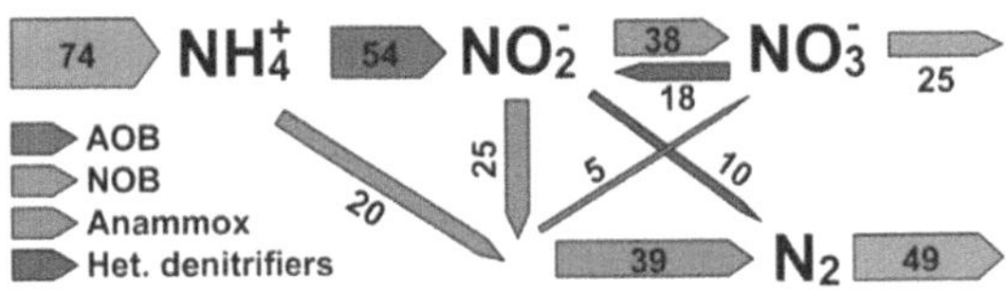

Figure 3.17 Nitrogen conversion routes in a single stage IFAS de-ammonification experiment (Malovanyy, 2017).

The development of the MBBR system over the years has demonstrated that better control and operability has been achieved when dividing the bacterial cultures in separate stages – making it possible to better optimize process conditions for each stage. The single-stage pure MBBR seems to be inferior to the single-stage MBBR-based IFAS system for de-ammonification. However, a two-stage pure MBBR-process, where nitritation and anammox is separated in two different stages, may be competitive in a main-stream process if NOB control can be secured. This may require extended BOD-removal in the C-stage, biofilm thickness control and bio-augmentation from the side- to the main-stream.

A two-stage system with one or more continuously aerated MBBRs for nitritation and an anoxic MBBR for anammox, was tested out in lab-scale by Piculell *et al.* (2016). It was hypothesized that the wash-out of NOB from the biofilm would be enhanced by having a very thin biofilm. The biofilm thickness was maintained below 200 µm by the use of a carrier especially designed for de-ammonification (AnoxKaldnes Z-200 carriers – see Figure 3.1), hence exposing a large fraction of the biomass close to bulk liquid free ammonia (FA) and/or free nitrous acid (FNA) concentrations. By doing so, it was expected that NOB establishment in the deeper biofilm layers would be prevented, and the chances of washing out NOBs from the biofilm would hence increase. The thin biofilm would also ensure high oxygen availability, which was expected to improve nitritation rates. The feed was switched periodically from low-strength, low-temperature main-stream wastewater to reject water at high temperatures and concentrations (in agreement with the strategy shown in Figure 3.16b). This sudden exposure to high substrate concentrations and temperatures was expected to inhibit NOB growth in the thin biofilm, and possibly also to boost AOB activity (Piculell *et al.*, 2016).

The lab-study demonstrated indeed that stable nitritation and anammox was achievable and the results demonstrated that periodic exposure of the biomass to high concentration reject water in order to favour AOB activity and suppress NOB growth was successful (Piculell *et al.*, 2016). In the nitritation stage, ammonia removal rates ranged between 0.25 and 0.50 $kgNH_4\text{-}N \cdot m^{-3} \cdot d^{-1}$ (equal to 1.25–2.5 $gNH_4\text{-}N \cdot m^{-2} \cdot d^{-1}$), with a nitrite accumulation ratio above 75%, and the effluent was fed to the second stage anammox reactor, with removal rates reaching 0.20 $kgNH_4\text{-}N \cdot m^{-3} \cdot d^{-1}$ (equal to 1.0 $gNH_4\text{-}N \cdot m^{-2} \cdot d^{-1}$). At an average influent concentration of 51.6 ± 5.2 $gNH_4\text{-}N \cdot m^{-3}$ effluent concentrations below 15 $gTotN \cdot m^{-3}$ could be achieved. Further studies are needed in order to optimize the scheme and evaluate the fate of the NOB, but once fully developed, this concept is expected to enable new solutions for main-stream anammox applications (Piculell *et al.*, 2016).

In a proposed flow diagram for an MBBR-based, energy neutral, advanced wastewater treatment plant for the future, Ødegaard (2016) proposed a scheme for de-ammonification as shown in Figure 3.18.

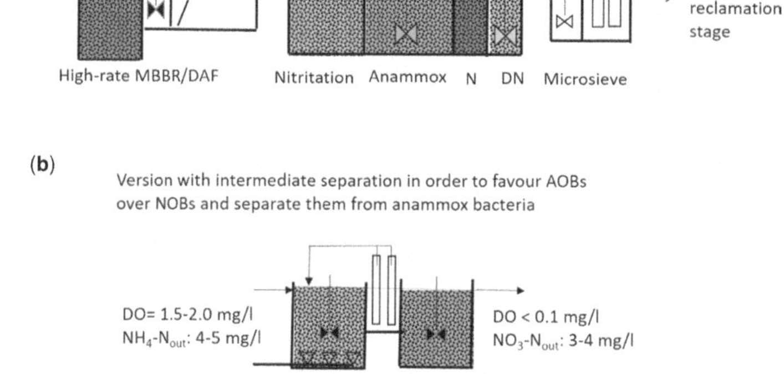

Figure 3.18 Proposed scheme for de-ammonification in an MBBR-based energy-neutral process solution for advanced wastewater treatment (based on Ødegaard, 2016).

The strategy is based on:

- Good removal (>85%) of BOD up-front of the de-ammonification reactor by a high-rate MBBR process, in order to suppress heterotrophs during de-ammonification
- De-ammonification in the main stream in two-stage MBBR (with thin biofilm carrier) in order to be able to optimize the two parallel processes:
  - Nitritation at DO = 1.5–2.0 $gO_2 \cdot m^{-3}$ and $NH_4\text{-}N_{out}$ = 4–5 $g \cdot m^{-3}$ (in order to maximize AeAOB and suppress NOB)
  - Anammox at DO <0.1 $gO_2 \cdot m^{-3}$ and $NO_3\text{-}N_{out}$ = 3–4 $g \cdot m^{-3}$ (in order to make AnAOB overrule NOB for NO_2)
- Use of bio-augmentation, as presented above by moving carriers back and forth between the de-ammonification MBBRs in side-stream and main-stream by the use of an air-lift pump (or similar) as well as returning the side-stream water to the main-stream de-ammonification MBBR
- Use of a two-step nitrification/denitrification MBBR directly after the main-stream de-ammonification MBBR in order to get down to very low Tot N-concentrations in the effluent (<3 g TotN $\cdot m^{-3}$)

Post denitrification (with carbon source addition) would have to be used in the post nitrification/denitrification step since there is very little $SCOD_{biodegradable}$ left in the recycled side-stream after anaerobic digestion and dewatering, and very little $SCOD_{biodegradable}$ left in the main-stream after the nitrification step.

Since it has been shown that the AOB activity is higher in a suspended biomass, it might be beneficial to insert a compact separation step (for instance a micro-screen, see Figure 3.18b) in between the aerobic and the anaerobic step. This may enhance AOB activity and supress NOB activity in the aerobic reactor.

3.5 P-REMOVAL

In MBBR plants P-removal is normally carried out through chemical precipitation. Also in IFAS plants chemical P-removal may be implemented (normally through simultaneous precipitation), but in IFAS plants biological P-removal may also quite easily be implemented in the same way as in conventional activated sludge (CAS) plants. Chemical P-removal in MBBR- and IFAS plants is quite established and below we shall focus on advances in biological P-removal in pure MBBR plants and only touch briefly upon chemical P-removal.

3.5.1 Chemical P-removal in MBBR and IFAS plants

In MBBR plants P-removal is normally carried out through chemical precipitation, usually by adding Al^{3+} or Fe^{3+} to the effluent directly after the MBBR and before flocculation and floc/biomass separation reactors, see Figure 3.19.

Since the SS-concentration after the MBBR is relatively low (150–250 gSS · m^{-3}) there is no need for a separate biomass separation step between the MBBR and the chemical precipitation/flocculation step. Flocculation is normally carried out in a multi-compartment flocculator with residence time (HRT) of 10–30 min depending on the upstream process. The floc/biomass separation step may be based on any particle separation method (see below), but normally it is based on settling or dissolved air flotation (DAF). If the effluent P standard is very stringent (<0.3 mg Tot P · l^{-1}) it is common to use a polishing step based on micro-straining (i.e. disc filtration), sand filtration or membrane filtration.

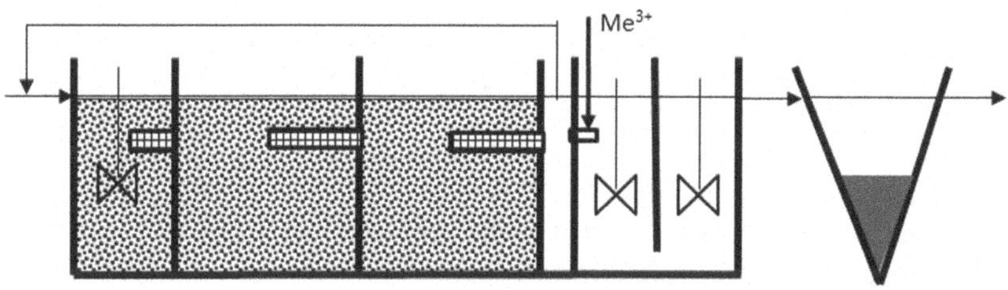

Figure 3.19 Typical principle for chemical P-removal in a pre-denitrification (as an example) MBBR plant.

In IFAS plants, chemical P-removal is commonly carried out through simultaneous precipitation by adding the precipitant directly to (or after) the IFAS-reactor, normally without any flocculation reactor in between.

3.5.2 Biological P-removal in MBBR plants

When conditions are favourable (sufficiently high concentration of easily biodegradable COD in incoming water), biological P-removal may quite easily be implemented in IFAS plants in the same way as in conventional activated sludge (CAS) plants. A University of Cape Town (UCT)-configuration is preferable. At this time it is not known whether or not any plants have been built using carriers in the anaerobic reactor.

3.5.2.1 Discontinuously operated (SBR) MBBR

Through experiments in small pilot scale, Helness demonstrated that biological phosphorus and nitrogen removal could be achieved in an MBBR operated as an SBR (Helness, 2007; Helness & Ødegaard, 2001; 2005). It was found that the SBR cycle should be tuned to achieve near complete removal of easily biodegradable soluble COD in the anaerobic phase and complete nitrification in the aerobic phase.

Pilot experiments with municipal wastewater (influent concentrations after pretreatment: 4.2 gPO_4-P · m^{-3}, 18 gNH_4-N · m^{-3}, 104 gSCOD · m^{-3} and 205 gCOD · m^{-3}) were carried out. It was demonstrated that at an SBR cycle of 6 hours (with 1 hour and 40 minutes anaerobic phase followed by a 4 hour and 20 minutes aerobic phase), less than 10 $gN_{soluble}$ · m^{-3} and 0.3 gPO_4-P · m^{-3} in the SBR effluent could be achieved. However, with this wastewater (low in biodegradable COD), removals as good as this could only be achieved, when dosing with an external carbon source (acetate). A ratio of 15 $gBSCOD/gPO_4$-P was required in order to achieve the effluent concentrations reported above. For an SBR MBBR process like this, the design criteria in Table 3.2 were recommended for a wastewater with COD:N:P ~ 100:10:2 (Helness, 2007).

Table 3.2 Proposed design values for an SBR MBBR for a wastewater with COD:N:P ~ 100:10:2 (Helness, 2007).

Parameter	Design Value
Total COD-loading rate, (gCOD · m^{-2} · d^{-1})	<5
Anaerobic BSCOD[1]-loading rate, (gBSCOD · m^{-2} · d^{-1})	<5
Aerobic ammonia loading rate, (gNH_4-N · m^{-2} · d^{-1})	<0.4
Required influent $BSCOD/PO_4$-P	20

[1]BSCOD – biodegradable soluble COD.

3.5.2.2 Continuously operated MBBR

Saltnes *et al.* (2017) presented an alternative, continuously operated, pure MBBR process for nutrient removal, based on bio-P and simultaneous nitrification/denitrification (SND). In this MBBR process the biomass experiences succeeding anaerobic and aerobic conditions by physically moving the carriers from the aerobic to the anaerobic zone by a conveyer belt – as demonstrated in Figure 3.20.

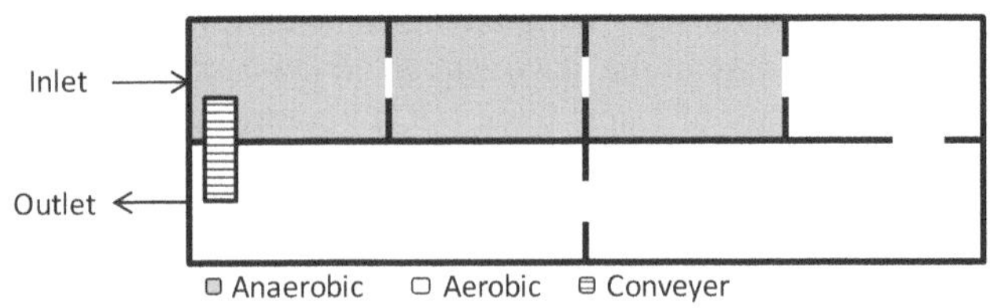

Figure 3.20 Flow scheme of the pilot used in the study of a continuous MBBR-plant for bio-P and SND-N removal (Saltnes *et al.*, 2017).

The MBBR is partitioned in several anaerobic and aerobic zones – all containing carriers. There are no sieves between the zones and the carriers are flowing, together with the water, through openings in the partition walls between the zones, first through the mechanically mixed, anaerobic zones and thereafter through the aerated zones. From the last aerobic zone the carriers (with biofilm on them) are transported mechanically by a conveyer sieve, that also separate the carriers from the wastewater, back to the first anaerobic zone, while the treated wastewater flows out of the bioreactor from the last aerobic zone.

In the experiments carried out at HIAS WWPT in Norway, the MBBR of the pilot plant had a total reactor volume of 7 m^3 and was operated with an effective carrier area of 300 $m^2 \cdot m^{-3}$ reactor volume. The anaerobic volume was varied between 30 and 40% of the total volume, and the total hydraulic residence time (HRT) was varied between 5 and 10 hours. The pilot plant was operated in two main experimental periods (referred to as pilot 1 and pilot 2). In pilot 2 the pilot plant design was improved, within the same total volume of the pilot plant, with optimized flow passages between zones and division of the aerobic volume into several volumes in order to enhance SND. The DO in the aerated zones was kept between 4 and 8 $gO_2 \cdot m^{-3}$.

The inlet wastewater to HIAS WWTP is influenced by effluent from the food processing industry in the area and has a relatively high ammonia- as well as biodegradable COD-concentration (61 $gNH_4\text{-}N \cdot m^{-3}$, 312 $gSCOD \cdot m^{-3}$, 3.6 $gPO_4\text{-}P \cdot m^{-3}$). The load and removal of PO_4-P (in $gPO_4\text{-}P \cdot m^{-2} \cdot d^{-1}$) over several months of operation in pilot 1 and for 3 weeks of operation in pilot 2, after stable enhanced biological phosphorus removal (EBPR) activity had been established in each pilot, are shown in Figure 3.21a. The results for pilot 1 are divided into groups according to the SCOD load ($gSCOD \cdot m^{-2} \cdot d^{-1}$). The removal of PO_4-P seemed to be negatively influenced by high organic loads (>5 $gSCOD \cdot m^{-2} \cdot d^{-1}$) in pilot 1, while in pilot 2 the removal was >95% even at an SCOD load of 5.2 $g \cdot m^{-2} \cdot d^{-1}$ (Saltnes *et al.*, 2017).

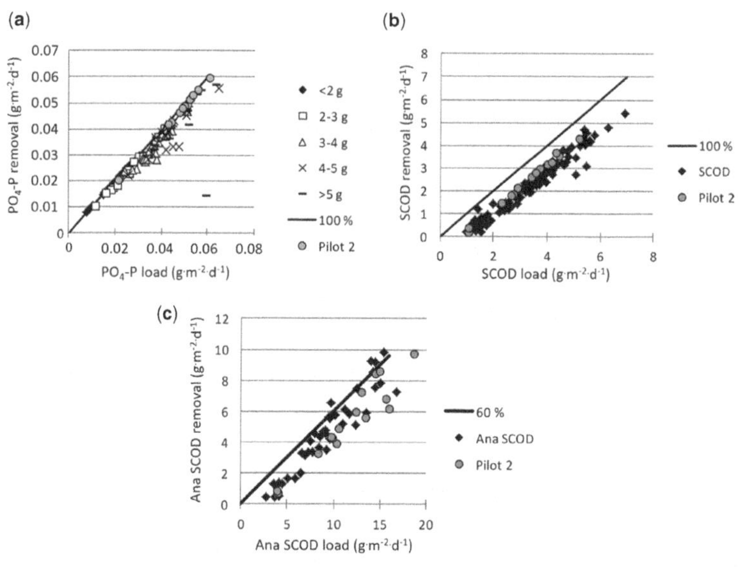

Figure 3.21 Removals of PO_4-P and SCOD in the pilot plant (Saltnes *et al.*, 2017).

Based on the experiments it was concluded that a total P concentration in the effluent of 0.4 gP · m^{-3} (the full-scale plant standard) could be reached in the full-scale plant provided that a particulate P concentration of less than 0.2 gP · m^{-3} could be maintained after biomass separation.

A steady EBPR process is dependent on phosphate accumulating organisms (PAO) to take up volatile fatty acids (VFA; produced by hydrolysis of easily biodegradable organic matter, SBCOD in the anaerobic zone), for which they use stored polyphosphate resulting in P release. The VFA taken up and stored as polyhydroxyalkanoates (PHA) are utilized for growth and P uptake in the aerobic zone and the new generation of PAO also take up P in excess resulting in P removal. If VFA or SBCOD leak to the aerobic zone, one might experience a reduction in P uptake due to reduced PAO growth. A good removal of SCOD in the anaerobic phase is important, therefore, for PAO both under anaerobic and aerobic conditions.

Figure 3.21b shows the overall SCOD load versus removal and Figure 3.21c shows the SCOD load versus removal in the anaerobic phase for some months of operation in pilot 1 and for 3 weeks of operation in pilot 2. Because of the inert SCOD-fraction (difference between the points and the 100% line), the total SCOD removal is only about 80%, but the biodegradable SCOD-removal is >90% at a removal rate >5 g · m^{-2} · d^{-1}. About 60% of the SCOD was removed in the anaerobic zone, with an anaerobic SCOD load of up to 18 g · m^{-2} · d^{-1} (Saltnes *et al.*, 2017).

During the pilot and laboratory testing of the new bio-P process (pilot 2), a significant nitrogen removal potential was discovered. Both nitrification and denitrification took place in the aerobic zone. This indicates that simultaneous nitrification/denitrification (SND) took place in different layers of the biofilm as proposed by Helness and Ødegaard (2001). The process enhances SND because most of the BOD is removed before the aerobic stage thereby improving nitrification. The BOD uptake in the anaerobic zone by denitrifying PAO secures an efficient use of wastewater carbon by anoxic uptake of PO_4-P and denitrification of NO_3-N using the same carbon. However, when PAO participate in denitrification, their P removal per COD taken up decreases. As opposed to a standard activated sludge EBPR-plant, NO_3-N from the aerobic zone is not returned to the anaerobic/anoxic zone in this process which could otherwise lead to inefficient use of carbon (Saltnes *et al.*, 2017).

Simultaneous nitrification/denitrification (SND) was studied only in the pilot 2 experimental phase and experiments are still going on. The average ammonia load, the nitrification rate and the denitrification rate in pilot 2 was found to be 0.86 gN · m^{-2} · d^{-1}, 0.54 gN · m^{-2} · d^{-1} and 0.33 gN · m^{-2} · d^{-1} respectively (Saltnes *et al.*, 2017).

3.6 ORGANIC MICRO-POLLUTANT REMOVAL

Biofilm processes, and in particular MBBR-based processes, have been found to be more efficient for the removal of organic micro-pollutants (OMP) than activated sludge processes (Falås *et al.*, 2012; 2013; 2016; Torresi *et al.*, 2017).

Through batch experiments Falås *et al.* (2012) assessed, the removal of seven active pharmaceutical substances (ibuprofen, ketoprofen, naproxen, diclofenac, clofibric acid, mefenamic acid, and gemfibrozil) by the biomass on MBBR carriers and in suspended sludge from several full-scale wastewater treatment plants (including activated sludge, MBBR-based IFAS and MBBR plants). A distinct difference between nitrifying activated sludge biomass and biomass on suspended biofilm carrier in removal of several pharmaceuticals was demonstrated. The biomass on MBBR carriers gave considerably higher removal rates per unit biomass (i.e. suspended solids for the activated sludges and attached solids for the carriers) of diclofenac, ketoprofen, gemfibrozil, clofibric acid and mefenamic acid, as compared to the suspended sludges. Even carrier biomasses with the poorest results show higher removal rates than the activated sludge biomasses with the best results (Falås *et al.*, 2012).

In a later study at the Bad Ragaz IFAS wastewater treatment plant in Switzerland, batch experiments with carriers and activated sludge from the same full-scale reactor were carried out in order to assess the micro-pollutant removal rates of the carrier biofilm under oxic conditions and the suspended sludge under oxic and anoxic conditions (Falås *et al.*, 2013). Clear differences in the micro-pollutant removal kinetics of the attached and suspended growth were demonstrated, often with considerably higher removal rates for the biofilm compared to the sludge. The results from model predictions and plant measurements showed that the removal efficiency of the process could be predicted with acceptable accuracy (~25%) for most of the modelled micro-pollutants. The model estimations indicated that the attached growth in hybrid (IFAS) processes can contribute significantly to removal of individual pharmaceutical compounds (Falås *et al.*, 2013).

Torresi *et al.* (2017) investigated different optimization strategies using MBBRs towards the removal of 23 commonly detected micro-pollutants (i.e. pharmaceuticals) in municipal wastewater. Overall it was found that biotransformation rates were significantly enhanced by the post-denitrification system as compared to the pre-denitrification and nitrification only system for the majority of micro-pollutants (~60%) suggesting the positive impact of easily biodegradable carbon sources (such as methanol, glycol or ethanol) on micro-pollutant removal. The removal of compounds such as propranolol, atenolol, citalopram, venlafaxine (under post-denitrifying conditions) and diclofenac (under aerobic conditions) was improved compared to conventional activated sludge. It was concluded that MBBR can offer a suitable technology that can be optimized for the removal of micro-pollutants in municipal wastewater under a range of operating conditions (nitrifying, pre- and post-denitrifying) (Torresi *et al.*, 2017).

3.7 SEPARATION OF BIOMASS FROM MBBR AND IFAS SYSTEMS

Separation of biomass from MBBR systems is different from that in activated sludge systems or IFAS systems. In contrast to activated sludge and IFAS systems, the biomass to be separated includes only the growth since there is no recycle. This means that the SS to be separated is in the order of 150–250 mgSS · l^{-1} in normal municipal wastewater – around 20 times less than in activated sludge and IFAS systems. This makes it possible to combine the MBBR with any of the commonly used separation technologies; settling, flotation, micro-screening, media filtration and membrane filtration. A review is given by Ødegaard *et al.* (2010).

This section focuses on advances in separation from MBBRs through compact separation methods, such as micro-sand ballasted lamella sedimentation, dissolved air flotation and micro-screening as well some comments on separation of the suspended biomass in IFAS systems. Separation by membranes is discussed in section 8.

3.7.1 Separation characteristics of MBBR biomass

The characteristics of the biomass leaving an MBBR are different from that of activated sludge with respect to morphology as well as particle size distribution.

In an activated sludge plant, the process of developing an active biomass begins with the growth of zoogleal bacterial floc which then becomes colonized by protozoa which feed on the free swimming bacteria to produce a clarified effluent. In the moving bed process, the order of colonization seems to be reversed (Mosey, 1996). High loading rates in an MBBR (around 30 gCOD · $m^{-2}d^{-1}$) produce compact bacterial biofilms, with protozoan population either absent or limited to small free-swimming protozoa and Vorticella spp. Moderate loading rates, around 10–15 gCOD · $m^{-2}d^{-1}$ promote a more "fluffy" biofilm with a rich variety of ciliated protozoa. Low loading rates (<5 gCOD · $m^{-2}d^{-1}$) promote a biofilm generally dominated by stalked ciliates (Mosey, 1996).

To a lesser extent than in activated sludge processes, therefore, one will find well defined aggregates of bacteria (flocs) and to a larger extent either small particles (single bacteria or clusters of bacteria) or rather large particles (biofilm flakes), which is a characteristic also for other biofilm processes (trickling filters, rotating biological contactors - RBCs etc).

Åhl *et al.* (2006) studied the particle size distributions in the bulk liquid of various reactor steps in a pilot plant, obtaining different curves at different MBBR loadings (i.e. different HRTs in the MBBR). In Figure 3.22 the particle size distributions presented as differential volume (large particles) and differential number (small particles) are shown.

It is demonstrated that particle aggregation takes place with a clear shift towards larger particles with increasing HRT (decreasing organic load) as shown in the differential volume % analysis. Aggregation of the smaller particles also results in a redistribution of the numbers of particles in the various size fractions where there is a relative increase in the number of smallest size fractions, particularly at HRTs higher than 2 hours (right graph in Figure 3.22).

In another study using the same pilot plant, Melin *et al.* (2005) analysed how the relative amount of COD in different filtered size fractions changed in the MBBR effluents including the soluble fraction from the membrane permeates. It was demonstrated (Figure 3.23a) that the majority of the COD was found in particles >1 μm and that there was a clear decrease in colloidal particles (0.1–1 μm fraction) when the HRT increased. The results also show (Figure 3.23b) that the majority of the particles in the MBBR effluents are above 1 μm in size and it was confirmed that there is a clear shift towards smaller particles when the HRT in the MBBR is reduced. It was observed throughout the experiments that the low-rate (long HRT) MBBR produced flocs that settled much better than the high-rate.

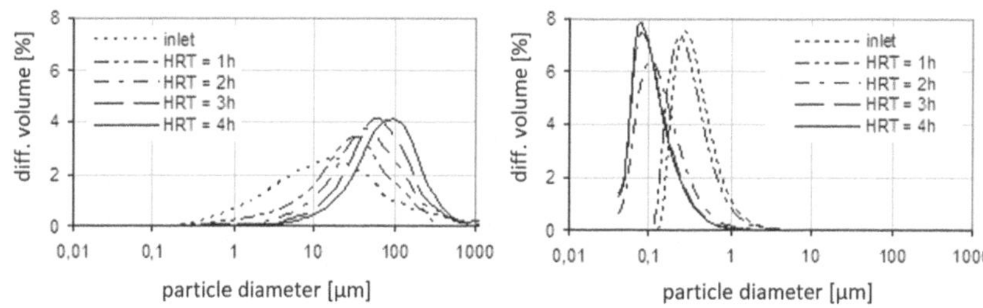

Figure 3.22 Development of particle size distributions as a function HRT (Åhl *et al.*, 2006).

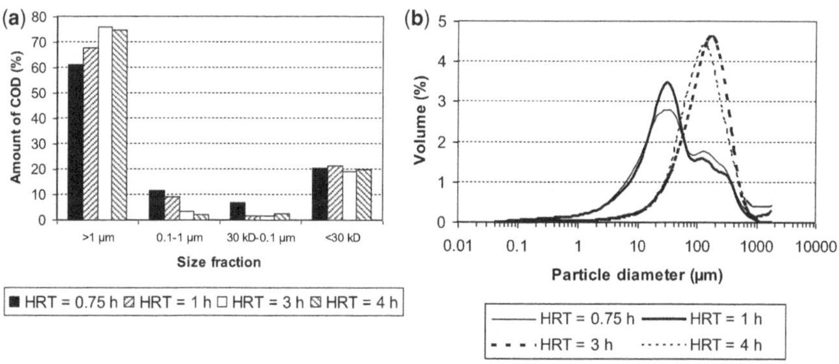

Figure 3.23 Relative amounts of COD in the different size fractions (a) and the particle size distributions (b) in the MBBR effluent (Melin *et al.*, 2005).

The above demonstrates that separation of biomass from MBBRs by conventional settling is not straight-forward and that flocculation of some sort is required to ensure optimal separation. The flocculation may be physical/chemical or biological. Quite often P-removal is required and metal salt coagulation followed by flocculation takes care of the fine particle challenge. Biological flocculation will take place in IFAS processes or in processes where the MBBR is directly followed by activated sludge (without any separation in between).

3.7.2 High-rate biomass separation after MBBRs

Settling is probably the most widely used method of biomass separation also after MBBRs. As demonstrated above, however, the characteristics of MBBR biomass are not favourable for sedimentation without pre-coagulation. The major portion of the MBBR sludge settles well, but especially in highly loaded MBBRs, the fine fractions ought to be coagulated in order to achieve a low SS in the effluent. In most countries, however, phosphate removal is required and coagulation/flocculation directly after the MBBR is implemented (see Figure 3.17).

Over the last decades some very compact, high-rate particle separation reactors have been more increasingly used in the water treatment sector, such as lamella settlers, flotation units and micro-screens. Since the SS concentration to be separated is much lower in biofilm systems than in activated sludge systems, these separation techniques have also been used for MBBR biomass separation in wastewater treatment. In particular dissolved air flotation (DAF) units, micro-sand ballasted lamella settlers (ACTIFLO™), and micro-screens (various types of drum-, disc- and belt-filters) have been used, and experiences from these applications will be focused on here.

3.7.2.1 Dissolved air flotation (DAF)

Dissolved air flotation (DAF) is favourably used for separating biomass from MBBRs. In Europe and North America alone there are more than 60 MBBR/DAF plants and additionally at least 25 such plants are used on-board cruise-ships (Wessman, 2017). Especially when combined with pre-coagulation, flotation gives a very good separation result (normally below 10 gSS · m^{-3}) in addition to a concentrated sludge phase (at least 4% DS). The combination MBBR + DAF results in a very compact overall plant.

In most municipal wastewater cases the plants consist of pretreatment, MBBR, metal salt addition for phosphate removal, flocculation (optionally with polymer addition) and flotation. In Norway alone there are around 20 such plants and typically <10 gSS · m^{-3} and <0.5 gTotP · m^{-3} is achieved on average annually.

Flotation reactors are usually designed based on surface overflow rate and dispersion water (air saturated) flow. Typically, but quite conservatively, an overflow rate of 5.0–7.5 m · h^{-1} at design flow and 10 m · h^{-1} at maximum design flow is used (Norsk Vann, 2009). The amount of air precipitated is proportional to the product of saturated water flow and pressure. The typical saturator pressure is 400–600 kPa (4–6 bar) and the typical saturated water flow (in % of maximum design flow) is 10–25% (depending on SS_{in} and air saturation). When using the lower limit for pressure, the higher limit for saturated water flow should be used – and vice versa. The dispersion equipment should have a minimum capacity equivalent to 10% of maximum design flow at a saturator pressure of 600 kPa.

Flocculation is important and normally a multi-stage (minimum two stages) flocculator with a total residence time of 15–20 min and relatively high mixing intensity (G = 70–90 s^{-1}) in all stages should be used.

In traditional DAF plants the dispersion water is made by pressurizing a return flow by the use of a compressor, a static mixer and a pressure tank (Figure 3.24a) creating bubbles in the 20–40 μm range. Nowadays it is more common to use a high pressure turbine/mixer pump together with an air saturator that can produce finer bubbles (<10 μm).

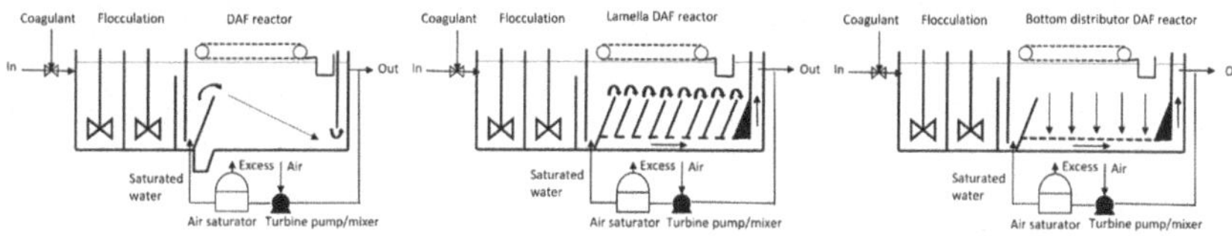

Figure 3.24 Traditional DAF and newer designs of DAF-reactors – with flow distributors.

Some newer designs of DAF-reactors are using lamellas in the clarification zone (Figure 3.24b) – mainly to ensure good hydraulic distribution. Another way of achieving the same is by introducing a reactor bottom with holes for flow distribution (Figure 3.24c).

3.7.2.2 Micro-sand ballasted lamella sedimentation

When combined with a micro-sand ballasted lamella sedimentation reactor (such as ACTIFLO™), the MBBR plant gets an extremely low footprint. Therefore, this process combination is becoming of increasing interest. Around 20 plants based on MBBR and ACTIFLO™ are in operation, especially in Norway, France and Canada, ranging in size from 4000 to 160,000 $m^3 \cdot d^{-1}$ (Wessman, 2017).

The principle of the process is shown in Figure 3.25. After metal (Al or Fe) hydroxide/phosphate precipitation, an anionic polymer is added, as well as micro-sand to the flocculation reactor. The micro-sand is enmeshed in the hydroxide/phosphate floc and kept together by the anionic polymer and flocculated (at relatively high G-values) before being separated from the water in an up-flow lamella clarifier. The main purpose of the lamellas is to ensure good hydraulic distribution. This is required because the weight of the flocs is so high, due to the enmeshed micro-sand, that it results in a very high settling velocity, in the 50–150 $m \cdot h^{-1}$ range. The micro-sand attached to the sludge is separated from the sludge in a hydro-cyclone, returned to the flocculation chamber and hence reused. However, there is a certain consumption of micro-sand, estimated by the supplier (Veolia Water Technologies) to be 2.5 $g \cdot m^{-3}$ in MBBR/ACTIFLO™ plants which means 3.75 tons per month for a 50,000 $m^3 \cdot d^{-1}$ plant. The average rise rate in the clarifier tank of the plants that are in operation is 72 $m \cdot h^{-1}$, ranging from 35 to 120 $m \cdot h^{-1}$ (Wessman, 2017). The coagulation/flocculation zone (which is especially designed for the different versions of the ACTIFLO™, is about of the same size (both in footprint and volume) as the settling zone.

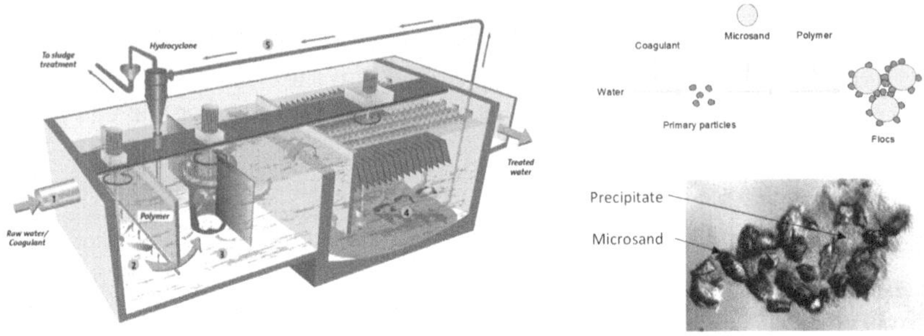

Figure 3.25 The ACTIFLO™ Turbo separator and working principle (Courtesy Veolia Water Technologies).

Most of the Norwegian MBBR/ACTIFLO™ plants are designed for secondary treatment plus P-removal and some are used in areas of extreme variation in load, for instance in touristic towns. Typically one is achieving <10 gSS · m^{-3} and <0.5 gTotP · m^{-3} at dosages of around 10–15 gFe · m^{-3} + 1–2 g anionic polymer · m^{-3}. The operational experiences with ACTIFLO™ in combination with MBBR are quite good, but it is a vulnerable process. On the one hand, if the dosage of any one of the additives (metal coagulant, anionic polymer or micro-sand) fails, the whole separation process fails. On the other hand, the process is quickly restarted after such a failure. Operation costs are high because of all the additives.

3.7.2.3 Micro-screening (disc filtration)

Micro-screening, i.e. drum and disc filtration, is mainly based on sieving of particles and thus the particle size distribution is crucial to the separation result. Nominal pore size openings normally range from 10 µm to 50 µm depending upon application. The principle of operation of most devices is relatively simple and straight-forward. There are a number of different micro-screening units (drum-, disc- and belt-filters) available. Here the experiences with the so-called Hydrotech™ disc filter are referred to, see Figure 3.26.

Disc filters are normally used as a polishing step after another separation method such as sedimentation (including ACTIFLO™) or DAF. When disc filtration follows an MBBR directly without any previous separation steps, the fine particle fraction mentioned above, is a challenge that requires pre-coagulation/flocculation in order to get low effluent SS values. This was confirmed when disc filtration without any pre-coagulation/flocculation was tested in two separate pilot tests, at Nordre Follo and at Gardermoen WWTP in Norway (Ødegaard *et al.*, 2010). In order for the capacity to be sufficiently high, considering the backwash water consumption, it was found that a mesh >20 µm ought to be used. The results in Figure 3.27 and 3.28 stem from tests with a 40 µm mesh. In these tests the applied filtration rates were in the order of 2–6 m · h^{-1} and in most cases not corresponding to maximum hydraulic capacity.

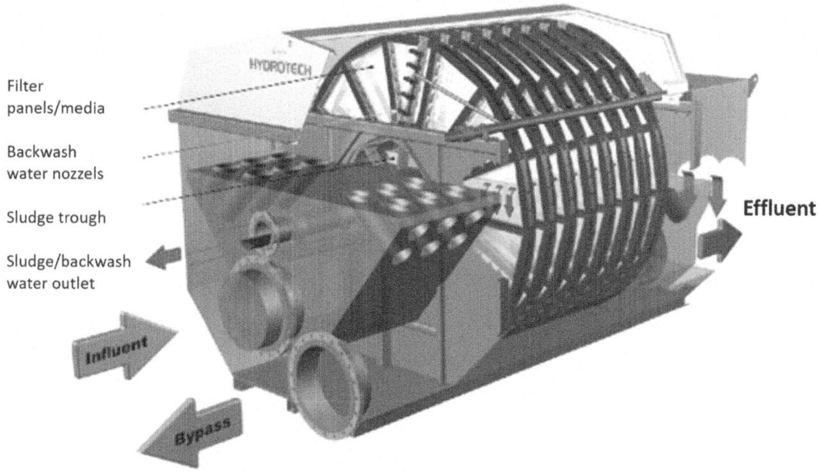

Figure 3.26 Hydrotech disc filter (Courtesy Veolia Water Technologies).

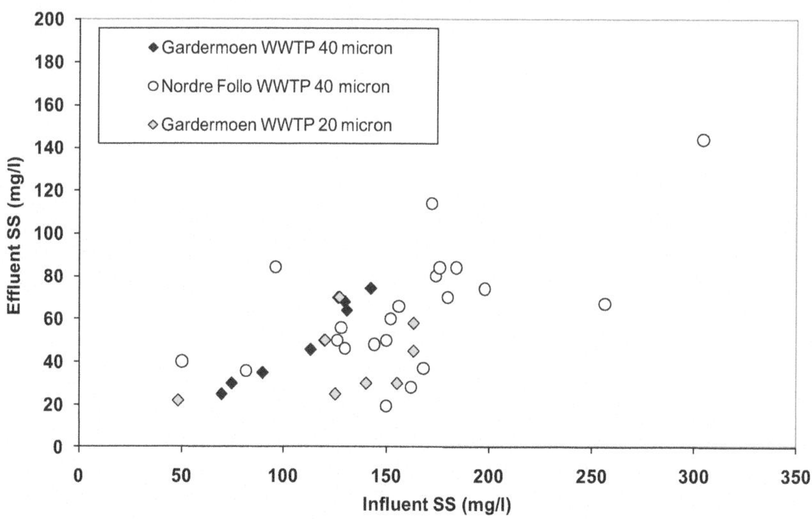

Figure 3.27 Effluent SS versus influent SS (MBBR effluent SS) without any pre-coagulation/flocculation (Ødegaard *et al.*, 2010).

Figure 3.27 shows the effluent SS versus influent SS when no coagulant was dosed, while Figure 3.28 shows the effluent SS as a function of polymer dose when a cationic polymer was dosed – as the only coagulant. When comparing the two figures, it is demonstrated that the effluent SS could be significantly reduced by the use of polymer coagulation and flocculation ahead of the disc filter (Ødegaard *et al.*, 2010).

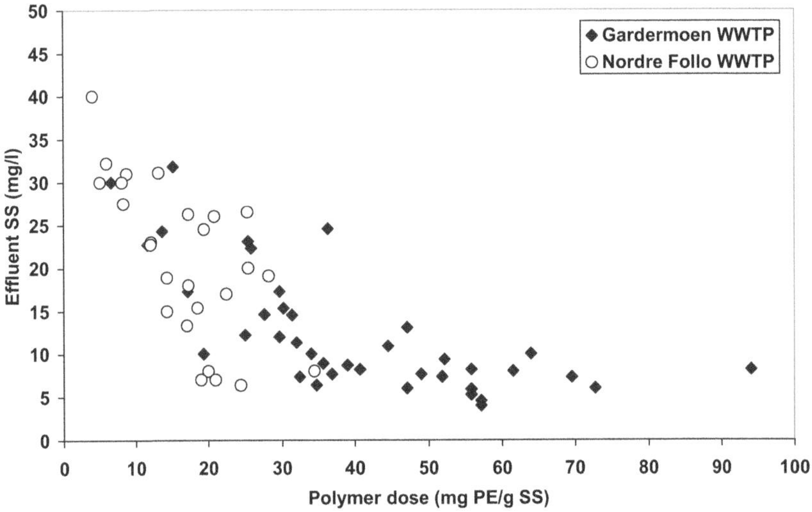

Figure 3.28 Effluent SS after disc filtration of MBBR effluent when polymer was dosed as a function of polymer dosing (Ødegaard *et al.* 2010).

Effluent-SS well below 10 mg · l^{-1} could be achieved when using cationic polymer at a relative high dose – above 35 mg polymer/g SS, which corresponds to 5–7 mg cationic polymer · l^{-1} at a typical MBBR effluent-SS of 150–200 mg SS · l^{-1}. When using even higher polymer doses, the hydraulic capacity could be further increased, but the effluent SS remained relatively constant.

Disc filtration is used directly after MBBR post-denitrification in several plants, for instance for Sjölunda WWTP as well as Rya WWTP in Sweden (Nunes *et al.*, 2013; Wilén *et al.*, 2016). At Rya WWTP Hydrotech™ disc filters in two separate lines, are used to separate suspended solids from a) an effluent from secondary settlers of the non-nitrifying activated sludge system and b) biomass from the post-denitrifying MBBRs directly. The disc filter plant consists of 16 disc filters in each line corresponding to a total filtration area (area of filter cloth) of 3584 m^2 and with 15 μm cloth pore size (Nunes *et al.*, 2013). As demonstrated in Figure 3.29, the hydraulic capacity of the disc filters varies between 5 and 10 m · h^{-1}, when the influent-SS varies between 10 and 40 gSS · m^{-3}. It is 7.5 m · h^{-1} at an influent concentration of around 20 mgSS · l^{-1} both from the post denitrifying MBBR and from activated sludge settlers – corresponding to an average particle load of 152 gSS · m^{-2} · h^{-1}.

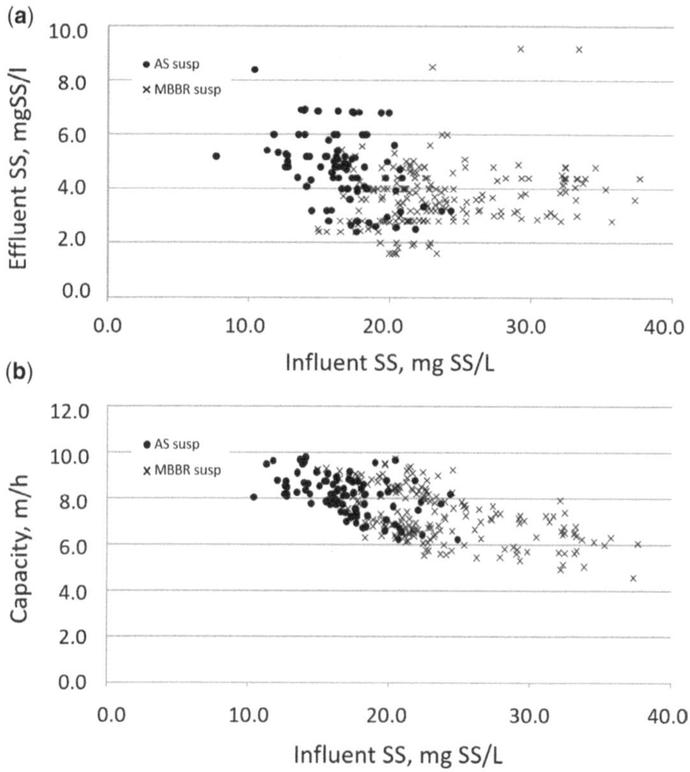

Figure 3.29 The hydraulic capacity (filtration rate) of the disc filters versus influent-SS (left) and effluent-SS versus influent-SS (right) at Rya WWTP (Nunes *et al.*, 2013).

Figure 3.29 demonstrates that the residual suspended solids concentrations were slightly lower in the effluent treating the post-denitrification MBBR biomass than that from the activated sludge settlers and well below 5 gSS · m^{-3}. The tot. P concentration was <0.2 gP · m^{-3} for approximately 60% of the days (Wilén *et al.*, 2016).

In Table 3.3 are given typical design values when using Hydrotech^{TM} disc filter for separating biomass after an MBBR (Ødegaard *et al.*, 2010). Pre-coagulation/flocculation is normally required when separating directly after an MBBR, while it is normally not needed when used as a polishing step after the earlier coagulation/flocculation/main separation step.

As the table demonstrates, the footprint is significantly lower than for methods like sand filtration and DAF. Energy consumption is low and principally related to backwashing of the filter (pressurizing backwash water to approximately 7 bar). Nunes *et al.* (2013) reported 0.013 kWh · m^{-3} at the Rya plant. The amount of backwash water is approximately 2–4% of the total flow. Regular and frequent cleaning is of utmost importance to keep backwashing at a minimum and maintain the design capacity (Wilén *et al.*, 2016).

Table 3.3 Hydraulic capacity and energy and backwash water consumption based on the Hydrotech™ disc filter (Ødegaard *et al.*, 2010).

Influent SS-concentration	Filtration Rate ($m \cdot h^{-1}$)		Footprint Requirement[1] ($m^3 \cdot m^{-2} \cdot h^{-1}$)		Energy Consumption ($Wh \cdot m^{-3}$)	
	18 μm	40 μm	18 μm	40 μm	18 μm	40 μm
<40 $gSS \cdot m^{-3}$ [2]	8		75		12	
100–200 $gSS \cdot m^{-3}$ [3]		6–8		60–75		16

[1]Surface loading based on required ground area, not including sidewalks for maintenance etc.
[2]I.e. when used directly after separate MBBR for post-denitrification or for polishing after main separation step
[3]I.e. when used directly after MBBR without previous separation and including coagulation/flocculation before disc filter with 7–9 $mgAl \cdot l^{-1}$ and 3–4 ppm cationic polymer dosage.

3.7.3 Biomass separation in IFAS systems

Several authors (Fouad & Barghava, 2005; Blank *et al.*, 2007; Briggs, 2009; Kim *et al.*, 2010; Wilson *et al.*, 2012) have all reported better AS settling properties after introducing IFAS into an existing AS plant. Figure 3.30 (Briggs, 2009) and Figure 3.31 (Wilson *et al.*, 2012) both show improvements of Sludge Volume Index (SVI) after introduction of IFAS.

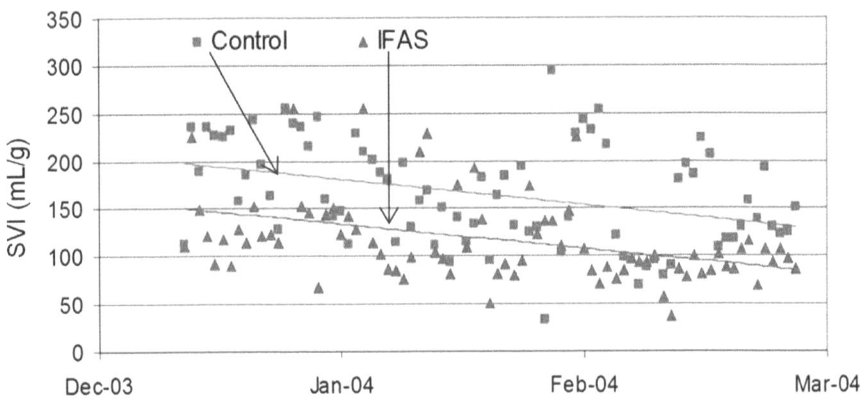

Figure 3.30 Comparison of SVI in two parallel lines, at Lakeview WWTP, Ontario, Canada (Briggs, 2009).

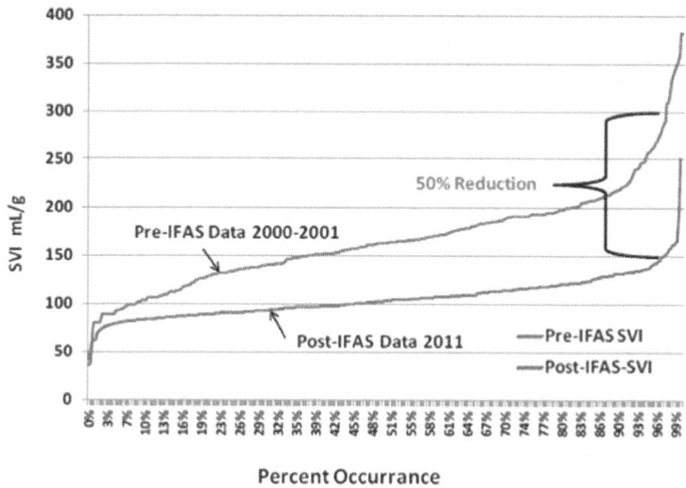

Figure 3.31 Pre- and post IFAS SVIs at Field's Point at WWTP, USA (Wilson *et al.*, 2012).

3.8 MBBR-BASED MEMBRANE BIOREACTOR (MBR) SYSTEMS

When very strict effluent standards are to me met (typically in re-use situations) membrane bioreactors (MBRs) are being favoured. Traditional MBRs are based on activated sludge with biomass separation by ultrafiltration (UF) membranes (typical pore size: 40 μm) designed for a reasonably high MLSS concentration (7–10 g MLSS · m^{-3}) and a membrane flux of around 20 l · m^{-2} · h^{-1}. This make the MBR far more compact than a traditional activated sludge plant and a competitor to MBBR with respect to compactness.

The main draw-back of the traditional MBR process is the high energy consumption caused by the extra air that is used for keeping the membranes from fouling and clogging too fast. The higher energy also results in a comparatively high operation cost. Another draw-back is the need for membrane replacement because of membrane ageing and permanent fouling.

An interesting alternative is an MBBR-based MBR system that may be placed after a pure MBBR or in an IFAS-based MBBR, see Figure 3.32. The nomenclature has been confusing in literature. MBBR-MBR is often used for an MBBR-based IFAS system, but there are clear differences between an MBR system based on a pure MBBR and one based on IFAS, see Figure 3.32.

The system in Figure 3.32a does not have recycle from the membrane reactor and hence the primary bioreactor is a pure biofilm reactor, while the primary bioreactor the system in Figure 3.32c is a hybrid activated sludge/biofilm (IFAS) bioreactor, as is also the one in Figure 3.32b since the MLSS stays in the reactor.

In this chapter, there is clear differentiation between those systems based on pure biofilm in the primary bioreactor (here called MBBR-MBR) and those based on a hybrid (IFAS) primary bioreactor (here called IFAS-MBR).

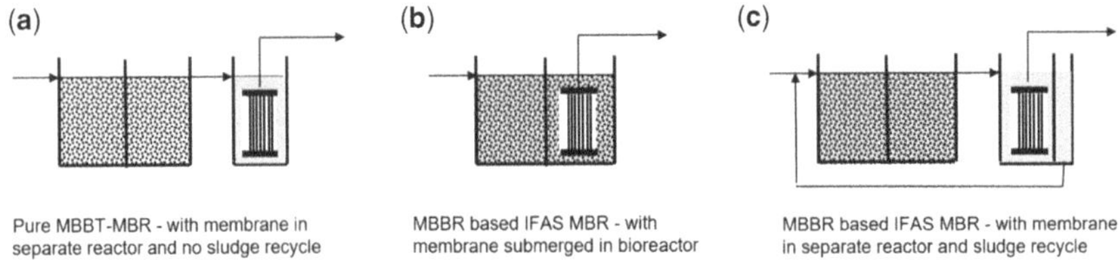

Figure 3.32 The various MBBR-based MBR systems (Ødegaard, 2017).

3.8.1 Pure MBBR + membrane (MBBR-MBR)

The combination of pure MBBR and membrane separation may be implemented in different ways. The membrane unit may be placed directly after the MBBR as an immersed membrane (Figure 3.33a and 3.33b) or as a separate, contained membrane unit (Figure 3.33c and 3.33d). All these systems have been tested.

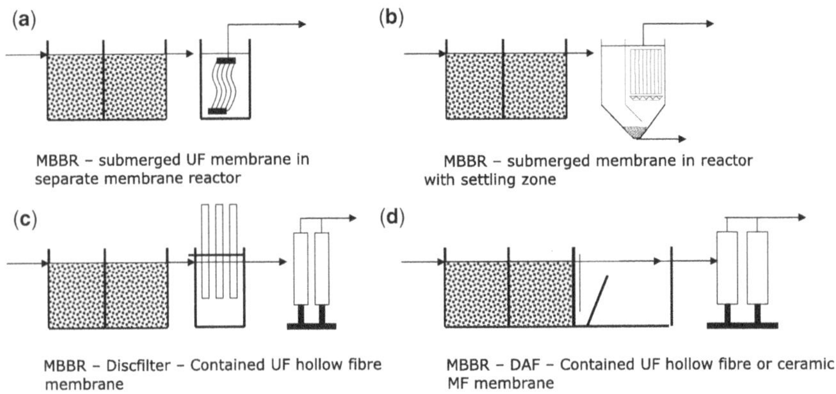

Figure 3.33 Examples of MBBR-membrane configurations.

Most of the studies on MBBR-MBR reported so far, have been based on immersed UF-membranes (Leiknes & Ødegaard, 2006; Ivanovic *et al.*, 2006; 2008; Ivanovic & Leiknes, 2012). These studies have shown that the concentration of biomass as well as the particle size distribution of the MLSS adjacent to the membrane, is influencing the fouling of the membrane and therefore several measures have been proposed in order to minimize this particle-caused fouling.

Ivanovic *et al.* (2006) demonstrated the beneficial effect of designing the membrane reactor in such a way that bio-sludge is flocculated in the reactor and continuously taken out. Melin and Ødegaard, (2007) demonstrated that coagulation of the MBBR biomass prior to separation in an immersed UF-membrane had a beneficial effect. The fouling rate in these experiments was found to be linearly proportional to the flux and it was shown that a sustainable flux (close to zero fouling rate) could be obtained at a flux of around $20\ l \cdot m^{-2} \cdot h^{-1}$ without coagulation and around $25\ l \cdot m^{-2} \cdot h^{-1}$ with coagulation (Melin & Ødegaard, 2007).

When applying membranes directly after the MBBR, contained membrane modules are not useful since the biomass concentration to be separated would be 200–300 $mgSS \cdot l^{-1}$. Ødegaard *et al.* (2012) investigated, therefore, the use of an intermediate, high-rate biomass separation step between the MBBR and the membrane unit in order to lower the MLSS concentration entering the membrane reactor. The principle of this process solution is shown in Figure 3.33 where the high-rate biomass separation step consists of a disc filter (DF) unit (Figure 3.33c) or a dissolved air flotation unit (DAF) (Figure 3.33d). It could alternatively be a micro-sand ballasted lamella settling unit (ACTIFLO™).

Ødegaard *et al.* (2012) reported on pilot-scale experiments with real municipal wastewater from the Gardermoen WWTP in Norway. The full-scale plant is a combined pre- and post-denitrification MBBR-plant with coagulation/flocculation and DAF separation. The water that was to be separated, could be taken (a) directly after the full-scale MBBR and passed through coagulation/flocculation + disc filter + UF membrane pilot module (referred to below as MBBR + DF + UF), or (b) from the effluent of the full-scale plant (with DAF separation) and passed through the UF pilot module (referred to below as MBBR + DAF + UF) – without any coagulant added directly ahead of the membrane pilot module.

Both solutions would be very compact in practice, but the use of a 40 μm disc filter (DF) for the primary biomass separation ahead of the UF would give the smallest footprint. In order for the DF alternative to be optimized, a cationic polymer had to be used, however, (either alone or in combination with a metal salt like aluminium) ahead of the disc filter (DF).

This process combination is quite risky, therefore, because of the possibility of polymer fouling of the membrane. An interesting alternative, based on DF, would be one with controlled polymer dosing ahead of the DF and aluminium dosing after the DF and before the UF. The hypothesis was that the hydroxide precipitation caused by the aluminium precipitation would sweep possible polymer residuals and hence reduce the risk of membrane fouling (Ødegaard *et al.*, 2012).

It was found, however, that the solution based on MBBR – coagulation/ flocculation – DAF – UF was the one with the lowest fouling, the highest possible operating flux, the lowest backwash water consumption and the lowest membrane cleaning chemicals consumption.

Even though a fair amount of R&D has been executed and a number of plants have been built (e.g. Darling Walk project in Sydney (Shaw, 2017), plants for pure MBBR with biomass separation (MBBR-MBR) are not yet common.

3.8.2 MBBR based hybrid MBR (IFAS MBR)

Most studies in lab- and pilot-plants on MBBR-based MBR systems have been carried out in MBBR-based hybrid MBRs (IFAS MBR). A comprehensive review is given in Leyva-Díaz *et al.* (2016).

Many research studies have been performed at lab-scale in a system such as the one in Figure 3.32b (Duan *et al.*, 2015; Leyva-Diaz, 2016; Tang *et al.*, 2016). Most new full-scale MBR plants today use, however, separate membrane tanks with sludge recycle, which is the IFAS-solution as in Figure 3.32c. The one in Figure 3.33b is restricted to small plants. One benefit that has been reported for such a system (Figure 3.33b), is reduced fouling as a consequence of reduced foulant deposition as well as abrasion of the cake layer on the membrane – caused by the carriers (Wei *et al.*, 2006; Yang *et al.*, 2009).

Most of the studies that compare the pure MBBR-MBR and the IFAS-MBR go in the favour of IFAS-MBR (Leyva-Díaz *et al.*, 2016) and most of the studies that compare the traditional MBR and the IFAS-MBR go in the favour of IFAS-MBR (Duan *et al.*, 2015). The IFAS-MBR is claimed to have the advantage of operating at higher fluxes, being more compact, having better energetic efficiencies and better membrane fouling control than the traditional MBR. Hence the IFAS-MBR should be the most interesting system, taking advantage of the best qualities of suspended biomass, attached biomass and membrane separation systems. In many studies, however, the SRT_{MLSS} and HRT in the IFAS-MBR has been about the same as in conventional MBRs (i.e. SRT_{MLSS}: 15–30 d, HRT: 10–24 h). In this author's opinion, this does not utilize the full potential of IFAS-MBR in the practical scale where size and cost matters. The benefit of IFAS is that nitrification can be obtained at less than half of the SRT_{MLSS} than what is required in traditional activated sludge plants (Ødegaard *et al.*, 2014).

Moreover, at a lower SRT_{MLSS} more carbon will be available for denitrification through hydrolysis of the MLSS and several studies (Rusten *et al.*, 2003; Onnis-Hayden *et al.*, 2011) have shown that the specific denitrification rate in IFAS systems is around twice as high as in conventional activated sludge systems. Hence there is the potential for the IFAS MBR system as compared to the conventional MBR system, to have a much smaller bioreactor since the nitrification as well as the denitrification zone may both be smaller.

3.9 A COMPARISON BETWEEN MBBR-, MBR- AND IFAS MBR SYSTEMS

In order to evaluate the practical full-scale potentials of the MBBR-based MBR processes for advanced wastewater treatment, a comparative analysis between conventional MBBR/DAF-, conventional MBR- and IFAS MBR processes was carried out (Ødegaard, 2017). The study is based on data for an existing, large wastewater treatment plant that is to be upgraded within very limited space, and with very strict effluent standard (TotN <4 g · m^{-3}, TotP <0.3 g · m^{-3}) with an average flow of 190,000 m^3 · d^{-1}, a peak flow of 15,833 m^3 · h^{-1} and the wastewater characteristics given in Table 3.4.

Three basic alternative processes were analysed (with a number of sub-alternatives (see Figure 3.34):

- MBBR with subsequent advanced particle separation based on DAF + Disc filter (DF)
- MBR (activated sludge-based MBR with UF membranes for biomass separation)
- IFAS-MBR (IFAS with carriers in aerobic reactors only and with UF membranes for biomass separation)

The MBBR alternative is based on combined pre-and post-denitrification and chemical P-removal. The MBR alternatives (MBR and IFAS-MBR) are also based on pre- and post-denitrification (with and without external carbon source), and with one sub-alternative based on biological P-removal (with external carbon source). The three basic processes were analysed with three different pretreatment alternatives: (a) no primary (fine screen only), (b) primary treatment (based on settling), and (c) chemically enhanced primary treatment (CEPT, based on coagulation/flocculation/settling). The processes are compared with respect to bioreactor volume, construction volume and construction footprint at two different climate situations/wastewater temperatures (design temperature 10°C and 18°C).

Table 3.4 Composition of the wastewater for which the process design will be made.

Parameter	Raw Water	After Primary	After CEPT	Effluent Standard 95%-ile	Average Used for Design
Biochemical Oxygen Demand (BOD_5) (mg/l)	200	130	70	<10	<5
Chemical Oxygen Demand (COD) (mg/l)	500	360	150	–	–
Total Suspended Solids (TSS) (mg/l)	250	135	30	<10	<5
Ammonia Nitrogen (NH_3-N) (mg/l)	30	28	28	<6	<3
Total Kjeldahl Nitrogen (TKN) (mg/l)	50	40	35	<8	<4
Ortho-Phosphate (OP) (mg/l)	3.0	2.7	1.0	–	–
Total Phosphorus (TP) (mg/l)	6.5	4.6	2.0	<0.5	<0.25

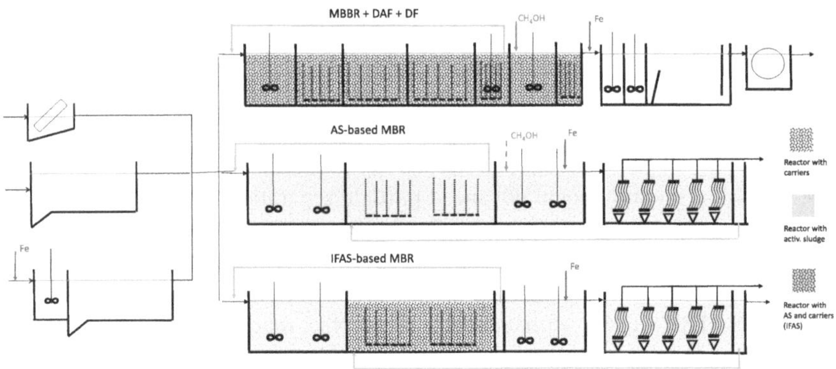

Figure 3.34 The basic wastewater treatment processes analysed and compared.

The MBBR + DAF + DF alternative is based on a pre-denitrification step in two stages, a BOD-removal step, a nitrification step in two stages plus a de-oxygenation step, a post-denitrification step in two stages and a re-aeration step. A carrier with a bulk specific surface of 800 $m^3 \cdot m^{-2}$ is used with filling fractions of 60% in aerobic and 50% in anoxic reactors respectively. The reactor volume design rates are at 10°C (18°C in brackets) – average if two stages; pre-DN: 0.6 (0.95) g NO_x-N · $m^{-2} \cdot d^{-1}$, BOD-removal: 3.6 (6.25) g $BOD_5 \cdot m^{-2} \cdot d^{-1}$, nitrification: 0.5 (1.0) g NH_4-N · $m^{-2} \cdot d^{-1}$, O_2-depletion: 0.225 (0.45) g NH_4-N · $m^{-2} \cdot d^{-1}$, post-DN: 0.6 (0.95) NH_4-N · $m^{-2} \cdot d^{-1}$, re-aeration: 3 g COD · $m^{-2} \cdot d^{-1}$. The design DO is 2 $gO_2 \cdot m^{-3}$ in the BOD-removal reactor and 5 $gO_2 \cdot m^{-3}$ and 4 $gO_2 \cdot m^{-3}$ in the two nitrification stages respectively, and 2 $gO_2 \cdot m^{-3}$ in the de-oxygenation tank. The recycle from the aerobic (deoxygenation) step to pre-denitrification is 300%. External carbon in the form of methanol is added to the post-denitrification step. Iron chloride is added to the effluent of the bioreactor before flocculation (in two stages) and biomass/floc separation in the DAF and subsequent micro-screen (disc filter).

Because of the relatively low C/N-ratio, an external carbon source has to be used in the MBBR-alternative. In the MBR- and IFAS MBR alternatives, hydrolysis will be more extensive and it may be possible to meet the required standard without an external carbon source if an extra post-anoxic volume for endogenous denitrification is included.

In the traditional MBR alternative two sub-alternatives are analysed, therefore, with and without external carbon source added. The MBR bioreactor is based on a pre-anoxic step (pre-denitrification), an aerobic step (residual BOD-removal and nitrification) and a post-anoxic step (post-denitrification). The pre-denitrification step was designed for a denitrification rate of 2.2 and 1.3 gNO_x-N · $kgMLSS^{-1} \cdot h^{-1}$ at 18°C and 10°C respectively. The post-denitrification step was designed for a high denitrification rate with external carbon source (methanol) added (6.25

and 3.6 gNO_x-N · kgMLSS^{-1} · h^{-1} at 18°C and 10°C respectively) and for a low (endogenous) denitrification rate when carbon source was not added (0.75 and 0.45 gNO_x-N · kgMLSS^{-1} · h^{-1} at 18°C and 10°C respectively). Iron chloride was added to the post-anoxic step. The nitrification step was designed based on sludge age according to accepted design rules (such as DWA A-131 and DWA – M 227) (SRT = 3.8 d at 18°C and process factor = 1.5).

When dimensioning the bioreactor in MBR systems, one may include (or not) the biomass in the membrane tank when designing for the nitrification tank volume. Membrane fouling is, however, very dependent upon the extent of organic matter degradation in the water that is to be membrane separated, which again is dependent upon the SRT_{MLSS}. Because the SRT_{MLSS} is lower in the IFAS MBR, and in order to be able to compare on equal basis, the biomass of the membrane tank is not included in the sizing of the nitrification tank. It is included, however, in the calculation of the aerobic $SRT_{MLSS.}$ The design MLSS in the aerobic part of the activated sludge reactor was set at 8 kgMLSS · m^{-3}, resulting in 6.4 kgMLSS · m^{-3} in the pre-denitrification reactor, and it was set at 10 kgMLSS · m^{-3} in the membrane reactor. The design DO in the aerobic tank of the bioreactor was set at 2 gO_2 · m^{-3}.

The IFAS-MBR alternative was designed in principle exactly in the same way as the MBR alternative, except for the fact that carriers were used in the aerobic (nitrification) stage of the bioreactor. A carrier with a bulk specific surface of 800 m^3 · m^{-2} was used, at a filling fraction 50%. No carriers were used in anoxic reactors. As for the conventional MBR system, the biomass in the membrane reactor was also not included when sizing the aerobic (IFAS) stage of the bioreactor. As mentioned above, the pre-denitrification rate in IFAS systems is higher than that in traditional CAS systems and was designed at a pre-denitrification rate of 4.0 and 2.3 gNO_x-N · kgMLSS^{-1} · h^{-1} at 18°C and 10°C respectively while the post-denitrification step was designed for the same rates as the conventional MBR system (see above).

In most of the alternatives, P-removal is carried out with iron precipitation. The use of bio-P was also evaluated and is discussed briefly below. It became clear, however, that the carbon available in the raw water was not sufficient for both denitrification and bio-P removal. So either the system has to use precipitating chemicals anyway or extra carbon source has to be dosed. Methanol is not useful for bio-P bacteria and sodium acetate was used.

The membrane step consists of a submerged membrane unit (hollow fibre) and a subsequent de-oxygenation stage on the sludge return. The volumes of the membrane tank and the flocculation/DAF tanks are very close to equal. The disc filter station (in the MBBR alternative) is placed above ground and no volume is calculated, but rather the surface space required. The MBR alternatives need a fine sieve between the bioreactor and the membrane reactor to prevent hair etc. clogging the membrane. The DF in the MBBR alternative and the fine sieve in the MBR alternatives have about the same footprint.

Below some parameters related to the size and cost are analysed. One has chosen to highlight bioreactor volume and construction volume as well as construction area (footprint). In Table 3.5 the required process volumes for the most probable process alternatives are compared for three alternatives of primary treatment at 18°C and 10°C. The volumes are given relative to the volume for the MBR without external carbon source alternative.

Table 3.5 shows that the bioreactor volume needed for the MBBR alternative (with external carbon source) is about 9% smaller than the MBR alternative without carbon source, but about 9% larger if carbon source is used also in the MBR alternative. The bio-P MBR alternative requires considerably more bioreactor volume – 24% more than the MBR alternative. The bio-P alternative is without methanol addition for denitrification, but needs acetate for bio-P. Actually this alternative cannot guarantee that the effluent tot P-standard is met.

Quite often there is a discussion regarding which alternative is the more compact one. This was investigated by analysing the construction volume as well as the construction footprint. These two parameters are the most decisive ones for the investment cost.

Table 3.5 Bioreactor volumes at different primary treatments relative to the one based on MBR without C-source and primary settling at 18°C and 10°C. Biomass in membrane tank is not included as active. In brackets: volume at 10°C relative to that at 18°C for the same process.

Treatment System	**18°C**			**10°C**		
	Primary	**No Primary**	**CEPT[6]**	**Primary**	**No Primary**	**CEPT[6]**
MBBR + DAF + DF[1]	0.91	1.14	0.82	0.83 (1.71)	1.03	0.72
MBR						
No – C-source[2]	1.00	1.19	_	1.00 (1.89)	1.30	_
C-source[3]	0.81	_	0.71	0.86 (1.99)	_	0.60
Bio-P[4]	1.24	_	_	1.23 (1.88)	_	_
IFAS-MBR[5]	0.62	0.71	0.39	0.59 (1.85)	0.68	0.37

[1]With Fe-addition for P-precipitation and with external carbon source (methanol).
[2]With Fe-addition for P-precipitation and without external carbon source.
[3]With Fe-addition for P-precipitation and with external carbon source (methanol).
[4]With bio-P removal (with acetate as external carbon source), no Fe-addition or methanol addition.
[5]With Fe-addition for P-precipitation and without external carbon source.
[6]P-precipitation carried out in CEPT. All alternatives with CEPT have external carbon source addition.

The construction volume was calculated by increasing the process volumes by 5% as well as adding 1 m from water level to top of basin (free-board). The construction area was increased by 10% of the area calculated as volume divided by depth, to allow for gangways etc. The volume of the primary was included. Traditional settling tanks with a peak flow overflow rate of 4.8 m · h^{-1} were used for primary settling and the same tanks were used in chemically enhanced primary settling – but now with lamellas in the tank securing an overflow rate based on projected area of >0.8 m · h^{-1} at peak flow. This means that the footprint surface area will be the same in the two alternatives. The water depth of the primary (as well as flocculation in CEPT) was 4 m (3 m under lamella in CEPT). The depths of the MBBR- and MBR tanks were 8 m and the DAF (including flocculation)- and membrane tanks were 3.5 m.

Table 3.6 shows the construction volumes and construction footprint at different primary treatments relative to the one based on MBR without C-source.

Table 3.6 shows that the construction volumes needed, as well as the construction footprint for the whole plant, of the MBBR alternative and the MBR alternative, are quite similar. The operation cost is estimated, however, to be about 20% lower in the MBBR alternative than in the MBR alternative even if external carbon is not used in the latter. It is interesting to note, though, that the IFAS MBR alternative is clearly the process that requires the smallest construction volume as well as the smallest footprint. To this author's knowledge, the IFAS-MBR is not used in full-scale plants yet. This analysis proves, however, that it has the potential of becoming a very competitive alternative.

Table 3.6 Construction volumes and construction footprint at different primary treatments relative to the one based on MBR without C-source (18°C). Conditions are as described in the footnotes of Table 3.5.

Treatment System	Construction Volume Required (m^3)[1]			Construction Footprint (m^2)[1]		
	Primary	No Primary	CEPT	Primary	No Primary	CEPT
MBBR + DAF + DF	0.94	0.88	0.92	0.96	1.06	0.92
MBR[1]						
No – C-source	1.00	0.92	–	1.00	1.09	–
C-source	0.87	–	0.88	0.91	–	0.92
Bio-P	1.16	–	–	1.11	–	–
IFAS-MBR[1]	0.74	0.59	0.62	0.83	0.87	0.71

[1]Biomass in membrane tank not included as contributing.

3.10 SUMMARY AND CONCLUSIONS

Since its introduction around 25 years ago, the moving bed biofilm reactor (MBBR) has established itself as a well-proven, robust and compact reactor for wastewater treatment. The efficiency of the reactor has been demonstrated in many process combinations, both for BOD removal and nutrient removal and for small as well as large plants. Even though it has been focused on municipal wastewater applications in this chapter, the reactor has been used extensively also for industrial wastewater treatment, particularly in the food-, the pulp and paper- and the aquaculture industries.

The reason for the popularity is the simplicity and the compactness of the MBBR and the solid scientific basis on which the MBBR-based process was developed. Even though the MBBR was invented and developed as a pure biofilm reactor, there is now an increasing use of the reactor principle also in hybrid systems, i.e. IFAS processes – especially for upgrading and capacity increase of existing activated sludge plants.

The primary advantage of the MBBR as compared to activated sludge reactors is its compactness and the lack of need for sludge recirculation. The advantage over other biofilm processes is its flexibility. One may use almost any reactor shape and one may choose different operating loadings in a given reactor volume, simply by choice of carrier filling fraction.

The MBBR has favourably been used for nitrification as well as for denitrification. High nitrification and denitrification rates have been demonstrated even at low temperatures. When using the MBBR process for nitrogen removal in carbon-deficient waters, a combined pre- and post-denitrification process is recommended because this process combination is superior when it comes to process control and performance.

The MBBR processes are normally combined with chemical P-removal and this process combination may result in extremely compact plants when high-rate biomass/floc separation processes are used. In this paper results and experiences are reported from the upcoming use of micro-sand ballasted lamella sedimentation (ACTIFLO™), flotation (DAF), micro-screening (disc filtration) as well as membrane (UF) filtration. The possibility of using these compact separation processes gives MBBR processes great flexibility. It is a considerable advantage over activated sludge processes, where one has to rely on sedimentation or membrane filtration (in MBRs) for biomass separation.

MBBR-based processes (MBBR and IFAS) have successfully been used in practice for de-ammonification in side-stream water (sludge reject) and lately de-ammonification has also been successfully demonstrated in main-stream. MBBR-based IFAS systems are most favourable in one-stage systems, but pure MBBR systems may be shown to be more favourable in two-stage systems.

MBBR-based processes (MBBR and IFAS) seem to be more efficient for the removal of organic micro-pollutants (OMP) than activated sludge processes.

3.11 REFERENCES

Æsøy A., Ødegaard H. and Bentzen G. (1998). The effect of sulphide and organic matter on the nitrification activity in a biofilm process. *Water Science and Technology*, **37**(1), 15–122.

Åhl R. M., Leiknes T. and Ødegaard H. (2006). Tracking particle size distributions in a moving bed biofilm membrane reactor for treatment of municipal wastewater. *Water Science and Technology*, **53**, 33–42.

Aspegren H., Nyberg U., Andersson B., Gotthardsson S. and LaCour Jansen J. (1998). Post denitrification in a moving bed biofilm reactor process. *Water Science and Technology*, **38**(1), 31–38.

Blank S. A., Hoffmann H. and Hahn H. H. (2007). Integration of suspended biofilm carriers in activated sludge plants for improvement of final settling (Translated from German. Original title: integration von schwimmenden Auf-wuchskörpern in Belebungsanlagen zur Erhöung der Sedimentationsleistung von Nachklärbecken). *KA-Abwasser Abfall*, **54**(5), 84–487 (in German).

Briggs T. (2009). Full-scale testing of IFAS at G.E. Booth (Lakeview) WWTP - lessons learned. *Presentation at Workshop on Biofilm Reactors*, WEFTEC, 11 October 2009, Orlando.

Brosseau C., Emile B., Labelle M.-A., Laflamme E., Dold P. and Comeau Y. (2016). Compact secondary treatment train combining a lab-scale moving bed biofilm reactor and enhanced flotation process. *Water Research*, **106**, 571–582.

Cema G. (2009). Comparative Study on Different Anammox Systems. PhD Thesis, KTH, Royal Institute of Technology, Stockholm, Sweden. http://rymd.lwr.kth.se/Publikationer/PDF_Files/LWR_PHD_1053.pdf.

Cema G., Trela J., Plaza, E. and Surmacz-Górska, J. (2010). Partial nitritation/Anammox process – from two-step towards one-step process. In: *Proceedings IWA World Water Congress*, 19–24 September 2010, Montreal.

Choi N., Phillips H., Johnson T. L., Butler R. and Lauro T. (2007). Pilot testing of MBBR and IFAS treatment processes for nitrification and denitrification at the Mamaroneck WWTP. *Proceedings of the 80th WEFTEC- National Conference of the Water Environment Federation*, 80, 13–17 October 2007, San Diego, CA.

Christensson M. (2008). Personal communication.

Christensson M. (2011). Moving on with MBBR. In: *Proceedings WEF/IWA Conference on Nutrient Recovery and Management*, 9–12 January 2011, Miami.

Christensson M., Ekström S., Andersson Chan A., Le Vaillant E. and Lemaire R. (2013). Experience from start-ups of the first ANITA Mox plants. *Water Science and Techonolgy*, **67**(12), 2677–2684.

Duan L., Li S., Han L., Song Y., Zhou B. and Zhang J. (2015). Comparison between moving bed-membrane bioreactor and conventional membrane bioreactor systems. Part I: membrane biofouling. *Environmental Earth Sciences*, **73**(9), 4881–4890.

Eriksson J. (2011). The influence of the nitrate load on the denitrification capacity - performed at GRYAABs MBBR plant. Eksamensarbete kemiingeniörsprogrammet. Chalmers tekniska högskola, Gothenburg.

Falås P., Baillon-Dhumez A., Andersen H. R., Ledin A. and la Cour Jansen J. (2012). Suspended biofilm carrier and activated sludge removal of acidic pharmaceuticals. *Water Research*, **46**, 1167–1175.

Falås P., Longrée P., la Cour Jansen J., Siegrist H., Hollender J. and Joss A. (2013). Micropollutant removal by attached and suspended growth in a hybrid biofilm-activated sludge process. *Water Research*, **47**, 4498–4506.

Falås P., Wick A., Castronova S., Habermacher J., Ternes T. A. and Joss A. (2016). Tracing the limits pf organic micropollutant removal in biological wastewater treatment. *Water Research*, **95**, 240–249.

Fernandez I., Dosta J., Fajardo C., Campos J. L., Mosquera-Corral A. and Mendez R. (2012). Short- and long-term effects of ammonium and nitrite on the anammox process. *Journal of Environmental Management*, **95**, S170–S174.

Fouad M. and Bhargava R. (2005). Sludge production and settleability in biofilm-activated sludge process. *Journal of Environmental Management*, **131**(3), 417–424.

Helness H. (2007). Biological Phosphorous Removal in a Moving Bed Biofilm Reactor (2007). PhD-thesis, 2007:177 Norwegian University of Science and Technology (NTNU), Trondheim, Norway.

Helness H. and Ødegaard H. (2001). Biological phosphorous and nitrogen removal in a sequencing batch moving bed biofilm reactor. *Water Science and Technology*, **41**(1), 233–240.

Helness H. and Ødegaard H. (2005). Biological phosphorus and nitrogen removal from municipal wastewater with a moving bed biofilm reactor. *Proceedings IWA Specialized Conference Nutrient Management in Wastewater Treatment Processes and Recycle Streams*, 19–21 September 2005, Krakow, pp. 435–444. ISBN 83-921140-1-9.

Helness H. and Sjøvold F. (2001). Degradation of Particulate Organic Matter in a Moving Bed Biofilm Reactor. SINTEF report STF66 F01104, 2001, SINTEF, Trondheim, Norway (restricted).

Helness H., Melin E., Ulgenes Y., Järvinen P., Rasmussen V. and Ødegaard H. (2005). High-rate wastewater treatment combining a moving bed biofilm reactor and enhanced particle separation. *Water Science and Technology*, **52**(10-11), 117–127.

Hem L., Rusten B. and Ødegaard H. (1994). Nitrification in a moving bed biofilm reactor. *Water Research*, **28**(6), 1425–1433.

Ivanovic I. and Leiknes T. (2012). The biofilm membrane bioreactor (BF-MBR)—a review. *Desalination and Water Treatment*, **37**(1-3), 288–295.

Ivanovic I., Leiknes T. and Ødegaard H. (2006). Influence of loading rates on production and characteristics of retentate from a biofilm membrane bioreactor (BF-MBR). *Desalination*, **199**, 490–492.

Ivanovic I., Leiknes T. and Ødegaard H. (2008). Fouling control by reduction of submicron particles in a BF-MBR with an integrated flocculation zone in the membrane reactor. *Separation Science and Technology*, **43**(7), 1871–1883.

Jardin N. and Hennerkes J. (2012). Full-scale experience with the deammonification process to treat high strength sludge water - a case study. *Water Science and Technology*, **65**(3), 447–455.

Johnson C. H. (2009). Personal communication.

Kanders L., Ling D. and Nehrenheim E. (2016). Rapid start-up of one-stage deammonification MBBR without addition of external inoculum. *Water Science and Technology*, **74**, 2541–2550.

Kim H., Gellner J. W, Boltz J. P, Freudenberg R. G., Gunsch C. K. and Schuler A. J. (2010). Effects of integrated fixed film activated sludge media on activated sludge settling in biological nutrient removal systems. *Water Research*, **44**(5), 1553–1561.

Leiknes T. and Ødegaard H. (2006). The development of a biofilm membrane bioreactor. *Desalination*, **202**, 135–143.

Lemaire R., Veuillet F., Zozor P., Stefansdottir D., Christensson M., Skonieczny T. and Ochoa J. (2015). Mainstream deammonification using ANITA™Mox Process (2015).

In: *Proceedings IWA conference on Nutrient Removal and Recovery*, 17–21 May 2015, Gdansk, Poland.

Leyva-Díaz J. C., Martin-Pacual J. and Poyatos J. M. (2016). Moving bed biofilm reactor to treat wastewater. *International Journal of Environmental Science and Technology*. DOI: 10.1007/s13762-016.

Lotti T., van der Star W. R. L., Kleerebezema R., Lubellob C. and van Loosdrecht M. C. M. (2012). The effect of nitrite inhibition on the anammox process. *Water Research*, **46**(8), 2559–2569.

Malovanny A. (2017). Anammox Based Systems for Nitrogen Removal From Mainstream Municipal Wastewater. PhD Thesis, KTH, Royal Institute of Technology, Stockholm, Sweden. Accepted 6 April 2017.

Malovanyy A., Yang J., Trela J. and Plaza E. (2015). Combination of UASB reactor and partial nitritation/Anammox MBBR for municipal wastewater treatment. *Bioresource Technology*, **180**, 144–153.

McQuarrie J. P. (2009). Integrated fixed film activated sludge (IFAS). *Proceedings WEFTECH 2009, Workshop W213 Biofilm Reactors*. Application to Today's Global Wastewater Challenges, Orlando, FL.

Melin E. and Ødegaard H. (2007). Pilot Tests of Kaldnes Membrane Bioreactor with Zenon ZeeWeed Membranes. Phase 2. SINTEF report No SBF IN F07303 (restricted).

Melin E., Helness H., Kenakkala T. and Ødegaard H. (2004). High-rate wastewater treatment based on moving bed biofilm reactor, polymer coagulation and flotation. In: Chemical water and wastewater treatment VIII, H. H. Hahn, E. Hoffmann and H. Ødegaard (eds), IWA Publishing, London, pp. 39–48.

Melin E., Leiknes T., Helness H., Rasmussen V. and Ødegaard H. (2005). Effect of organic loading rate on a wastewater treatment process combining moving bed biofilm and membrane reactors. *Water Science and Technology*, **51**(6-7), 421–430.

Mosey F. E. (1996). Biofilms on Kaldnes carrier particles. *Proc. from seminar organised by Anglian Water*, 27–29 November 1996, Cambridgeshire, UK.

Norsk Vann (2009). Guideline for the Design of Wastewater Treatment Plants. Report No. 168/2009. Norsk Vann, Hamar, Norway (In Norwegian).

Nunes J., Fredriksson O., Lindqvist P. and Mattsson A. (2013). Experience of full scale operation of disc filter as a final treatment step at Gryaab. *Proceedings NordIWA-Conference*, 8–10 Oktober 2013, Malmö, Sweden. (In Swedish).

Ødegaard H. (1992). Treatment of wastewater – Removal of nutrients. TAPIR Publishers, Trondheim, Norway. ISBN 82-519-1109-5 (in Norwegian).

Ødegaard H. (2006). Innovations in wastewater treatment: the moving bed biofilm process. *Water Science and Technology*, **53**(9), 17–33.

Ødegaard H. (2014). Compact wastewater treatment with MBBR. *Proceedings International DSD Conference on Sustainable Stormwater and Wastewater Management*, 12–14 November 2014, Hong Kong. http://www.dsdic2014.hk/program2.html.

Ødegaard H. (2016). A road-map for energy-neutral wastewater treatment plants of the future based on compact technologies (including MBBR). *Frontiers of Environmental Science & Engineering*, **10**(4), 02. doi: 10.1007/s11783-0835-0.

Ødegaard H. (2017). New applications for MBBR and IFAS systems. In: Frontiers in wastewater treatment and modelling, G. Mannina (ed.), Book Series: Lecture Notes in Civil Engineering, Vol. 4. Springer Int. Publ. AG, Switzerland, pp. 499–507.

Ødegaard H., Rusten B. and Westrum T. (1994). A new moving bed biofilm reactor - applications and results. *Water Science and Technology*, **29**(10-11), 157–165.

Ødegaard H., Rusten B. and Wessman F. (2004). State of the art in Europe of the moving bed biofilm reactor (MBBR) process. *Proceedings WEFTECH Conference*, 2–6 October 2004, New Orleans.

Ødegaard H., Cimbritz M., Christensson M. and Dahl P. C. (2010). Separation of biomass from moving bed biofilm reactors (MBBRs). *Proceedings WEF/IWA Biofilm Reactor Technology Conference*, 15–18 August 2010, Portland, Oregon, USA.

Ødegaard H., Mende U., Skjerping E. O., Simonsen S., Strube R. and Bundgaard E. (2012). Compact tertiary treatment based on the combination of MBBR and contained hollow fibre UF-membranes. *Desalination and Water Treatment*, **42**(1-3), 80–86. http://dx.doi.org/10.1080/19443994.2012.683145.

Ødegaard H., Christensson M. and Sørensen K. (2014). Hybrid Systems. In: Activated Sludge 100 years – and counting, Chapter 15, D. Jenkins and J. Wanner (eds), IWA publishing, London.

Onnis-Hayden A., Majed N., Schramm A. and Gu A. Z. (2011). Process optimization by decoupled control of key microbial populations: distribution of activity and abundance of polyphosphate-accumulating organisms and nitrifying populations in a full-scale IFAS-EBPR plant. *Water Research*, **45**, 3845–3854.

Piculell M., Christensson M., Jönsson K. and Welander T. (2016). Partial nitrification in MBBRs for mainstream deammonification with thin biofilms and alternating feed supply. *Water Science and Technology*, **73**(5), 1253–1260.

Plaza E. (2015). Personal communication.

Rosenwinkel K. and Cornelius A. (2005). Deammonification in the moving-bed process for the treatment of wastewater with high ammonia content. *Chemical Engineering Technology*, **28**(1), 49–52.

Rusten B. (2008). Personal communication.

Rusten B. and Ødegaard H. (2007). Design and operation of nutrient removal plants for very low effluent concentrations. *Water Practice*, **1**(5), 1–13.

Rusten B., Hem L. and Ødegaard H. (1995a). Nitrification of municipal wastewater in novel moving bed biofilm reactors. *Water Environment Research*, **67**(1), 75–86.

Rusten B., Hem L. and Ødegaard H. (1995b). Nitrogen removal from dilute wastewater in cold climate using novel moving bed biofilm reactors. *Water Environment Research*, **67**(1), 65–74.

Rusten B., Wien A. and Skjefstad J. (1996). Spent aircraft deicing fluid as external carbon source for denitrification of municipal wastewater: From waste problem to beneficial use. 51. *Purdue Ind. Waste Conf. Proc.*, Ann Arbor Press, Inc., Chelsea, Michigan.

Rusten B., Nielsen M., Welander T. and Rasmussen V. (2003). Increasing the capacity of activated sludge plants by using AS/MBBR hybrid process with Kaldnes biofilm carriers. *Proceedings 5th IWA Conference on Biofilms Systems*, Cape Town, South Africa.

Saltnes T., Sørensen G. and Eikås S. (2017). Biological nutrient removal in a continuous biofilm process. *Water Practice & Technology*, **12**(4). doi: 10.2166/wpt.2017.083.

Shaw K. (2017). Personal communication.

Stinson B., Peric M., Neupane D., Laquidara M., Locke E., Murthy S., Bailey W., Kharkar S., Passarelli N., Derminassian R., Carr J., Sultan M., Shih G., Barnard J., Daigger G., Parker D., Randall C. and Wilson T. (2009). Design and operating considerations for a post denitrification MBBR to achieve limit of technology effluent NOx < 1 mg/l and effluent TP < 0.18 mg/l. *Proceedings of the Water Environment Federation, Nutrient Removal 2009*, pp. 1225–1254(30). doi: https://doi.org/10.2175/193864709793901590.

Stinson B., Murthy S., Bott C., Wett B., Al-Omari A., Bowden G., Mokhtari Y. and De Clippeleir H. (2013). Roadmap Toward Energy Neutrality & Chemical Optimization at Enhanced Nutrient Removal Facilities. In: *Proceedings from WEF/IWA Nutrient Removal and Recovery Conference*, 28–31 July, Vancouver, Canada.

Tang B., Yu C., Bin L., Zhao Y., Feng X., Huang S., Ding J., Chen C. and Li P. (2016). Essential factors of an integrated moving bed biofilm reactor-membrane bioreactor: adhesion characteristics and microbial community of the biofilm. *Bioresource Technology*, **211**, 574–583.

Thomas W. A. (2009). Evaluation of Nitrification Kinetics for a 2.0 MGD IFAS Process Demonstration. Master of Science Thesis, Environmental Engineering Department; Virginia Polytechnic Institute and State University, Blacksburg, VA, USA.

Torresi E., Plósz B. G., Christensson M. and Smets B. F. (2017). Removal of Micropollutants in Moving Bed Biofilm Reactors (MBBRs): Microbial Diversity and Functional Relationships. PhD-dissertation, Department of Environmental Engineering, Technical University of Denmark (DTU), Kgs. Lyngby.

Trela J., Malovanyy A., Yang J., Plaza E., Trojanowicz K., Sultana R., Wilén B.-M., Persson F. and Baresel C. (2014). De-Ammonification. Synthesis Report 2014 R&D at Hammarby Sjöstadsverk. IVL-report no B 2210. http://www.ivl.se/download/18.1acdfdc8146d949da6d43f9/1413294579949/B2210 + Synthesis + report + 2014.pdf.

Veuillet F., Lacroix S., Bausseron A., Gonidec E., Ochoa J., Christensson M. and Lemaire R. (2013). IFAS ANITA™Mox process – A new perspective for advanced N-removal. Proc. of 9th IWA conf. on Biofilm Reactors, May 2013, Paris, France.

Veuillet F., Bausseron A., Gonidec E., Chastrusse S., Christensson M., Lemaire R. and Ochoa J. (2014). ANITA™Mox deammonification process: possibility to handle high COD level using the IFAS configuration. *Proceedings of IWA Water Congress & Exhibition*, 21–26 September, 2014, Lisbon, Portugal.

Wei C.-H., Huang X., Wang C.-W. and Wen X.-H. (2006). Effect of suspended carrier on membrane fouling in a submerged membrane bioreactor. *Water Science and Technology*, **53**(6), 211–220.

WERF (2014). Nutrient Challenge. IN: J. B. Neethling (ed.) Deammonification, Water Environment Research Foundation (WERF) 2014. http://www.werf.org.

Wessman F. (2017). Personal communication.

Wilén B.-M., Cimbritz M., Pettersson T. and Mattson S. A. (2016). Large scale tertiary filtration – results and experiences from the discfilter plant at the Rya WWTP in Sweden. *Water Practice & Technology*, **11**(3). doi: 10.2.166/wpt.2016.063.

Wilson M., Sandino J., Cote T., Brueckner T., Boltz J. P. and Michelsen D. (2012). Design and operational insights from the world's largest integrated fixed-film activated sludge (IFAS) process for low nitrogen limits. *Proceedings 85th WEFTEC*, New Orleans, LA, USA.

Yang S., Yang F., Fu Z. and Lei R. (2009). Comparison between a moving bed membrane bioreactor and a conventional membrane bioreactor on membrane fouling. *Bioresource Technology*, **100**(24), 6655–6657.

Chapter 4

Aerobic granular sludge: State of the art, applications, and new perspectives

S. F. Corsino[1], T. R. Devlin[2], J. A. Oleszkiewicz[2] and M. Torregrossa[1]

[1]*Department of Civil, Environmental, Aerospatial Engineering and Material, University of Palermo, Viale delle Scienze, building 8, 90128, Palermo - Italy*

[2]*Department of Civil Engineering, University of Manitoba, Winnipeg - Canada, R3T 5V6*

ABSTRACT

The aerobic granular sludge (AGS) process has attracted significant interest over the last decade and is one of the most promising wastewater treatment technologies. AGS offers several advantages over conventional activated sludge (CAS) including excellent settling and higher volumetric loading capacity. Because of porosity concentration gradients develop and stratified aerobic, anoxic, and anaerobic layers form throughout granule depth. This is the reason for simultaneous Nitrogen and phosphorus removal in a single tank.

Aerobic granulation is influenced by many parameters including wastewater characteristics and operating conditions. Among these, the anaerobic upflow feeding strategy and a properly balanced feast/famine regime have most significantly influenced granule stability and nutrient removal efficiency. Hydraulic shear forces can improve the physical characteristics of the granules, whereas selective wasting also allows undesirable biomass with low density to be washed out.

Implementation of AGS in full-scale requires some important retrofitting including adequate grit and particulate organics removal from the wastewater.

Editors: Giorgio Mannina, George Ekama, Hallvard Ødegaard and Gustaf Olsson
doi: 10.2166/9781780409719_0155

Furthermore, aerobic granular sludge cycles should be designed with flexibility to address variable influent loads. Considerations include equalization and flow balancing, dissolved oxygen control strategies, multiple draw points for selective wasting and swing-capabilities.

Several studies and full-scale implementations have demonstrated that AGS is suited for the treatment of a wide variety of industrial wastewaters, as well as municipal wastewater. Full-scale implementations of AGS treating municipal wastewater have improved process stability, attained more stringent effluent quality (i.e., total nitrogen (TN) <5 mg/L; total phosphorus (TP) <0.3 mg/L), resulting in 30% to 60% energy savings compared with CAS.

Keywords: Aerobic granular sludge, continuous flow reactor, industrial wastewater, municipal wastewater, sequencing batch reactor

4.1 INTRODUCTION

Aerobic granular sludge (AGS) is considered one of the most promising technologies in the field of biological wastewater treatment (Carucci *et al.*, 2010; Lee *et al.*, 2010). AGS offers several advantages over conventional activated sludge (CAS) including excellent settling and high biomass retention, therefore allowing for high volumetric loading. Due to the dense and compact structure of the granules, AGS offers higher resilience to drastic changes in operating conditions such as temperature or organic load, and toxic compounds. Lastly, AGS reactors have lower sludge yields due to the selection and proliferation of slow growing microorganisms (de Kreuk *et al.*, 2010; Ni & Yu, 2010).

During the first AGS workshop in 2004 aerobic granules were defined as aggregates of microbial origin which do not coagulate under reduced hydrodynamic shear and settle significantly faster than activated sludge flocs (De Kreuk *et al.*, 2005b). Aerobic granules contain active microorganisms as well as components of microbial origin (e.g., proteins, polysaccharides, humic acids, etc.). No carrier material is necessary for their formation. The minimum size, density, and surface characteristics of granules should be such that the difference between sludge volume index (SVI) after 10 min of settling and SVI after 30 min of settling is negligible. To understand the main mechanisms involving aerobic granulation, researchers worldwide have extensively studied the effects of several factors and operating conditions on aerobic granulation.

Aerobic granulation has mainly been observed in sequencing batch reactors (SBRs) (Beun *et al.*, 1999; Arrojo *et al.*, 2004; Wagner & da Costa, 2013). They differ from most activated-sludge plants, which operate in continuous flow, because SBRs combine all the treatment steps and processes (e.g., feeding, reaction, and settling) into a single basin. SBR systems offer several advantages compared to continuous-flow conventional activated sludge reactors, especially regarding aerobic granulation. The first advantage of an SBR is the possibility to select for particles with high settling velocities. This selection can be controlled through the duration of the settling phase, a selective wasting phase, or a combination of both. In a granular sludge SBR the slowly settling fraction of biomass is washed out with the effluent or with the selectively wasted sludge (Beun *et al.,* 1999), resulting in the gradual enrichment of fast settling particles in the reactor.

The second advantage is the high, pulse-load of substrate at the beginning of the cycle. During the influent feeding phase, referred to as the feast period, accumulating microorganisms convert the soluble biodegradable organic substrate into polyhydroxyalkoanates (PHA). Stored PHA will be used as a carbon source for growth in the presence of oxygen, nitrate, or nitrite when the soluble bioavailable organic substrate in solution has been consumed. This phase in the granular sludge SBR cycle, when the biodegradable organic substrate has been depleted in solution, is referred to as the famine period. The slowly degradable PHA which has been formed by accumulating microorganisms during the anaerobic influent feeding period diverts organics from rapidly growing aerobic heterotrophs. The out-selection of fast growing microorganisms leads to stable granules comprised of dense microbial aggregates with smooth surface layers (Picioreanu *et al.,* 2000). Despite multiple successful full-scale applications in SBR mode, AGS still needs to be integrated into continuous flow applications.

4.2 STRUCTURE AND COMPOSITION OF AEROBIC GRANULES

Perhaps the most defining feature of AGS, and granular sludge of any nature, is the tightly aggregated mass of microorganisms in a matrix of extracellular polymeric substances (EPS). Observing granular sludge with the naked eye demonstrates that it is completely different from CAS systems that utilize suspended biomass – Figure 4.1. In fact, AGS is more like attached growth systems, such as moving bed-biofilm reactors, that utilize inert carrier materials to promote cellular adhesion and the development of a self-maintaining biofilm. The only difference is that granular sludge does not require expensive synthetic carrier material to obtain the high-capacity benefits of other attached growth processes.

Recent interest in AGS has raised questions about what actually differentiates granules from flocs (Bruce *et al.,* 2014). Some have defined it by a strict minimum diameter while others as a function of parameters such as the ratio of SVI at different settling times (Mancell-Egala *et al.,* 2014). The latter is becoming increasingly more relevant as hybrid granule-floc systems are considered. Thus, operators will require reliable methods to define the amount of granules relative to total biomass in the system (Giesen & Thompson 2013b). For the purpose of this discussion, however, granules will refer to a dense microbial consortium typically spherical or ellipsoidal in shape with a smooth outer surface (Chen *et al.,* 2008).

4.2.1 Physical characteristics

The compact, tightly aggregated characteristics of aerobic granular sludge are responsible for most of the benefits granular sludge processes exhibit over conventional systems. For instance, AGS in SBRs can obtain comparable treatment to CAS within 70% of the footprint (Pronk *et al.*, 2015b). Furthermore, a 60% reduction in energy requirements has been demonstrated at a full-scale AGS plant in the Netherlands.

Figure 4.1 Full-scale Nereda® AGS biomass treating un-clarified municipal wastewater after one minute of settling (left) and a stereoscopic image of AGS cultivated at the University of Manitoba on municipal primary effluent (right).

The average diameter of AGS has been reported between 0.2 and 5 mm, although granules as large as 23 mm have been observed (Dangcong *et al.*, 1999; Tay *et al.*, 2001a; Tay *et al.*, 2001b; Zhu & Wilderer, 2003; Torregrossa *et al.*, 2007). The large size and high density of aerobic granules allows for excellent settling properties and low SVI from 20–80 mL/g (Su *et al.*, 2012). Values for specific gravity from 1.004 to 1.1 have been reported for AGS, further demonstrating that granules are denser than flocs which typically have specific gravities from 1.002 to 1.006 (Li *et al.*, 2009; Shi *et al.*, 2009; Su *et al.*, 2012).

Like any physical system, in situ settling velocities are affected by temperature, viscosity, and geometric properties. Large granules (i.e., average diameter = 2.3 ± 0.5 mm) from a lab-scale reactor had settling velocities of 84 m/h at 5°C and 145 m/h at 40°C while small granules (i.e., average diameter = 1.5 ± 0.3 mm) had settling velocities of 35 m/h at 5°C and 63 m/h at 40°C (Winkler *et al.*, 2012a). Granules can also be highly porous, allowing for sufficient mass transfer through the microorganism-EPS complex. Because of porosity concentration gradients develop and stratification of aerobic, anoxic, and anaerobic layers form throughout granule depth and cycle time. This in turn leads to microbial diversity across granule depth, and thus stratification of nutrient removal pathways (Winkler *et al.*, 2013a).

The mass transfer in aerobic granules can be disrupted by an overgrowth of filamentous bacteria on the outer surface layer of granules when readily available substrate is present under aerobic conditions (Devlin *et al.*, 2017). Particulate matter can also contribute to surface "fouling" by acting as a constant source of biodegradable substrate to microorganisms in the surface layer under aerobic conditions (Wagner *et al.*, 2015). Furthermore, overproduction of EPS in highly loaded granular sludge reactors can result in reduced pore space and reduced mass transfer leading to continuous breakage of granules (Corsino *et al.*, 2016c). Any detriment to granular sludge porosity will directly affect overall performance, since mass transfer to localized areas containing specialized microorganisms would be compromised.

Specific gravity and settling velocities of aerobic granular sludge are both higher than conventional activated sludge systems. Granules settle rapidly and as discrete particles, resulting in a complete overhaul of clarification principles applied to flocculent sludge used in conventional treatment. Analysis of full-scale granular biomass indicated that phosphate-accumulating organisms (PAO) had a higher density than glycogen-accumulating organisms (GAO) leading to significantly higher settling velocities for PAO dominated granules (Winkler *et al.*, 2013b).

4.2.2 Extracellular polymeric substances

EPS has been recognized as a key element that shapes and provides structural support for biofilms (Flemming & Wingender, 2003). Similarly, EPS is highly involved in AGS structure and morphology. Wang *et al.* (2006) stated that EPS acts as a buffer layer for cells in biofilms against harsh external environments and that components of EPS are also used by bacteria as carbon and energy sources during starvation periods. Seviour *et al.* (2012) found that gel-forming exopolysaccharides were major structural materials in AGS. Specifically, granular and alginate-like exopolysaccharide, both producing gel-like structures, have been proposed as the gel-forming constituents of the granules. However, other polymers should not be excluded as candidates.

Chen *et al.* (2007a) were the first to describe the distribution of EPS (i.e., as proteins, polysaccharides, and lipids) and cells (i.e., both total and dead) in aerobic granules. The authors found that in acetate-fed granules protein and β-D-glucopyranose polysaccharides formed the core whereas the cells and α-D-glucopyranose polysaccharides accumulated in the outer layers. Adav *et al.* (2008) examined the role of individual components of EPS on structural stability of phenol-fed aerobic granules. Selective enzymatic hydrolysis of proteins, lipids, and α-polysaccharides had a minimal effect upon the three-dimensional structural integrity of the granules. Conversely, selective hydrolysis of β-polysaccharides fragmented the granules. The β-polysaccharides were therefore expected to form the backbone of a network-like outer layer with embedded proteins, lipids, α- polysaccharides, and cells.

It has also been observed that alginate-like exopolysaccharides in AGS had significantly higher amounts of poly-guluronic acid blocks and lower poly-guluronic acid-mannuronic acid blocks, while alginate-like exopolysaccharides from flocculent sludge had equal amounts of these chemical blocks (Lin *et al.*, 2013). Consequently, the increased number of guluronic acid blocks were demonstrated to promote higher resistance of aerobic granules, compared to flocculent biomass, to breakage when subjected to deformation.

Xiong and Liu (2013) investigated the role of proteins and polysaccharides in maintaining the structural stability of the aerobic granules. The authors observed that hydrolysis of extracellular proteins in the EPS matrix of aerobic granules led to collapse of the EPS matrix, and subsequent disintegration of aerobic granules. These observations suggested that extracellular proteins would be essential for maintaining structural stability of the EPS matrix of aerobic granules. Similarly, a reduction in the protein to polysaccharide ratio in the EPS matrix is commonly considered a sign of weakening granule structure (Zhang *et al.*, 2011; Corsino *et al.*, 2015). Indeed, because of their simple molecular structure, polysaccharides are more sensitive to biological and chemical degradation than proteins. Therefore, granules with low protein to polysaccharide ratios have a matrix consisting of weak structural components.

Wang *et al.* (2017) conducted a comprehensive study on the correlation between EPS components and granular sludge properties in an anaerobic sulphidogenic granular sludge reactor. The authors observed that the protein to polysaccharide ratio correlated positively with the loosely-bound EPS to tightly-bound EPS ratio. As the relative amount of proteins and loosely-bound EPS increased so did the hydrophobicity. However, the mean particle size decreased with increases in the relative amount of proteins and polysaccharides. Thus, an increase in the protein content of AGS particles may lead to enhanced microbial adhesion and more compact granule particles characterized by a dense EPS structure.

Based on the studies reported in the literature, the role of individual components of the EPS matrix on AGS stability has still not been clarified. One of the reasons for this could be the extraction method. As stated by Adav and Lee (2008a), extracting EPS from aerobic granules was difficult due to their compact structure. Several extraction methods have been proposed including high-speed centrifugation, ultrasound-formamide-NaOH, or thermal, either individually or in combination. Lin *et al.* (2010) claimed that such great divergence influences the value of exopolysaccharides content obtained, since the quantity measured depends on the extraction method used. Therefore, misrepresentation may occur if the role of exopolysaccharides is assessed simply by these measured values.

4.2.3 Ion exchange and biologically induced precipitation

Initiating microbial adhesion is a critical step in attaining the compact structure of aerobic granules. Initiating microbial adhesion can be thermodynamically unfavourable in some cases since the electric repulsion of negatively charged natural surfaces such as cells and certain EPS molecules may be greater than the attractive forces (Poortinga *et al.*, 2002). For this reason, anionic and cationic polymeric compounds must interact with one another to hold microorganisms and other EPS components together. Divalent cations (e.g., Ca^{2+}, Mg^{2+}) help stabilize the microorganism-EPS complex via divalent cation binding (Lin *et al.*, 2013). On the other hand, monovalent cations (e.g., Na^{+}, K^{+}, NH_4^{+}) could exchange with divalent cations in the granular sludge complex (Sobeck & Higgins, 2002). For instance, compared to CAS and anaerobic ammonia oxidizer (AAO) granules, AGS particles had 4–9 times higher ammonium adsorption capabilities at an equilibrium ammonium concentration of 30 mg-N/L (Bassin *et al.*, 2011). In general, monovalent to divalent cation ratios above 2 can lead to poor aggregation properties.

Other chemical exchange mechanisms, such as the exchange of potassium from potassium magnesium phosphate (i.e., K-struvite) in the granular sludge complex with ammonium in bulk solution, have been observed to contribute to the overall removal of nitrogen and phosphorus (Lin *et al.*, 2012). It has been suggested that small increases in the fraction of precipitates, as low as 1% to 5% of total granule volume, would significantly increase the density of aerobic granules and thereby increase their settling velocity (Winkler *et al.*, 2013b). Phosphorus precipitates, such as biologically induced hydroxyl-apatite (i.e., $Ca_5(PO_4)_3(OH)$), have been repeatedly detected in the core of aerobic granules (Angela *et al.*, 2011; Huang *et al.*, 2015). Other studies have observed elevated levels of other calcium precipitates, such as aragonite, and thereby suggested that elevated calcium levels in the wastewater may help induce granulation (Juang *et al.*, 2010; Liu *et al.*, 2015; Li *et al.*, 2017).

Components of the EPS matrix of aerobic granules have also induced the formation of struvite, another phosphorus precipitate commonly found in aerobic granules (Lin *et al.*, 2012). Alginate-like exopolysaccharides specifically, which are characteristic of aerobic granules, were found to induce the formation of potassium/ammonium magnesium phosphates (i.e., struvite). High concentrations of calcium precipitates specifically have been reported within the cores of granular sludge. Qin *et al.* (2004a) and Liu and Tay (2004) showed that concentrations of iron, magnesium, and aluminium remained constant in the granules during start-up whereas calcium increased. Ren *et al.* (2008) and Qin *et al.* (2004b) observed that granules containing calcium precipitates had stronger structures and larger diameters, associated with a negative impact on bioactivity based on specific oxygen uptake rates.

4.2.4 Microbial community and nutrient removal capabilities

Arguably, the most interesting aspect of aerobic granular sludge is the layering and interactions of different microorganisms throughout granule depth. All granules are characterized by a concentric multi-layered structure, which must contain channels and pores for the transport of oxygen and substrates (Paul & Liu, 2012). With proper mass transfer conditions in place aerobic, anoxic, and anaerobic zones characterized by a dissolved oxygen (DO) gradient can be defined across granule depth (Bassin *et al.,* 2012b; Bassin *et al.,* 2012a). This differentiation allows for simultaneous carbon, nitrogen, and phosphorus removal by a suite of microorganisms in one granule (Wang *et al.,* 2009a).

Typically, granules consist of an outer, 100–200 μm aerobic layer over the top of anoxic/anaerobic layers. Aerobic microorganisms such as ordinary heterotrophic organisms (OHO), ammonia oxidizing bacteria (AOB), nitrite oxidizing bacteria (NOB), phosphate accumulating organisms (PAO), and glycogen accumulating microorganisms (GAO) grow in the outer layer while deeper layers contain denitrifying OHO (DOHO), denitrifying PAO (DPAO), and denitrifying GAO (DGAO) (Tay *et al.,* 2002b; Adav *et al.,* 2007; Adav *et al.,* 2008b; Adav & Lee, 2008a). Various other specialty microbes such as anaerobic ammonium oxidizers (AAO), denitrifying anaerobic methane oxidizing archaea (DAMOA), denitrifying anaerobic methane oxidizing bacteria (DAMOB), and methanogens have been observed within AGS (Bassin *et al.,* 2012b; Winkler *et al.,* 2012b). Nitrite/nitrate generated by AOB and NOB penetrates to the anoxic layer, which acts as a nitrite/nitrate sink. The oxidized inorganic nitrogen is removed by conversion to nitrogen gas through denitrification, thereby creating a nutrient gradient where nitrite/nitrate travels from the aerobic to the anoxic layer. Denitrification and phosphorus removal pathways can be coupled through DGAO and DPAO pathways. Bassin *et al.* (2012b) observed that DGAO were the main microorganisms responsible for nitrate reduction in AGS reactors operated at 20°C and 30°C. The same authors observed that a significant fraction of the nitrite was reduced to nitrogen gas via DPAO, specifically Clade II *Candidatus accumulibacter phosphatis.*

If AAO develop within the anoxic layer, nitrite can also be utilized to oxidize residual ammonium to nitrogen gas with approximately 10% residual nitrate (Cao *et al.,* 2017). The nitrate produced by AAO can be utilized by any of the facultative microorganisms to oxidize organic matter to carbon dioxide. Encouraging establishment of AAO within aerobic granules has been characterized by higher ammonium effluent concentrations and low (e.g., 0.5 g-COD/g-N, where COD = chemical oxygen demand) carbon to nitrogen ratios (Winkler *et al.,* 2012b; Isanta *et al.,* 2015). Furthermore, the ratio of NOB to AOB in AGS systems has been observed to be higher than in CAS systems, which does not benefit AAO establishment since they compete directly with NOB for nitrite (Winkler *et al.*, 2012c). In fact, activity measurements have demonstrated a threefold higher nitrite oxidizing activity than ammonium oxidizing activity in AGS. These results demonstrated that there is complex nutrient cycling within AGS particles, where some NOB are either growing mixotrophically using organics for nitrate reduction or denitrifiers are reducing nitrate to nitrite within AGS particles and thereby supplying additional nitrite for NOB without the additional ammonium for AOB.

The diverse microbial community of aerobic granular sludge, coupled with its ability to generate nutrient gradients, allows it to perform complete nutrient removal in a single tank simply by switching airflow on or off. Completely aerobic granular sludge processes, those that do not implement an anaerobic contact phase for fresh wastewater, have demonstrated excellent organic carbon and nitrogen removal capabilities. Processes that do have an anaerobic contact phase, however, have also exhibited complete biological phosphorus removal. Thus, aerobic granular sludge becomes an ideal process for removal of carbon, nitrogen, and phosphorus from wastewaters. Furthermore, the slow growth of microorganisms in stable granular sludge systems as well as the constant recycling of organic and inorganic compounds for oxidation and reduction processes throughout the granule contribute to lower sludge yield and reduced dependency on external carbon and/or metal salt addition to meet stringent nutrient limits compared to conventional activated sludge systems.

Dead cells and precipitates can also exist under anaerobic conditions near the core of aerobic granules (de Kreuk *et al.*, 2005a). Being an active microbial community, granular sludge is constantly growing and therefore increasing in diameter, unless mechanically broken apart first. As granular sludge diameter increases the extent to which oxygen, nitrite, and nitrate penetrate decreases, resulting in the development of anaerobic conditions in the interior of granular sludge particles (Bassin *et al.*, 2012b). Depending on the sludge age of the anaerobic core, it is possible that microorganisms capable of degrading EPS in the granular sludge matrix become established. It has yet to be demonstrated whether or not this is a common mechanism for self-maintenance of individual granular sludge particles or a mechanism that induces instability, ultimately leading to the breakdown of larger granular sludge particles.

Since the aerobic layer does not usually exceed more than 200 μm, the diversity of the microbial community should increase as average granule diameter surpasses 400 μm. Furthermore, as bulk DO concentrations decrease, the aerobic layer is expected to decrease in size, thereby leading to more diverse microbial communities in smaller granules. Microbial diversity as well as community structure would directly influence the performance of aerobic granules. For instance, analysis of full-scale granular biomass indicated that PAO had a higher density than GAO leading to significantly higher settling velocities for PAO dominated granules (Winkler *et al.*, 2013b). Thus, out-selection of GAO could improve AGS performance. Furthermore, the kinetic rates and stoichiometry of individual metabolic pathways would govern granule stability – Table 4.1.

Table 4.1 Microbial kinetic rates and stoichiometry – developed from Yagci *et al.* (2004), Chen *et al.* (2014), and Envirosim (2017).

Microbial Designation	Yield (g-COD/ g-substrate)	Max Net Growth Rate (1/d)	Conditions, Substrate Type
AOB	0.15	0.73	Aerobic, ammonium
NOB	0.09	0.53	Aerobic, nitrite
AAO	0.11	0.19	Anoxic, ammonium/nitrite
OHO	0.67	2.6	Aerobic, COD
	0.54	3.0	Anoxic, COD
	0.10	0.67	Anaerobic, COD
PAO	0.57	0.85	Aerobic, COD_{VFA}
	0.46	0.85	Anoxic, COD_{VFA}
GAO	0.29	0.90	Aerobic, COD_{VFA}
	0.23	0.90	Anoxic, COD_{VFA}
Acetogens	0.10	0.20	Anaerobic, COD
Methanogens	0.10	0.17	Anaerobic, $COD_{acetoclastic}$
	0.10	1.3	Anaerobic, $COD_{hydrogen}$
DAMOA	0.071	0.036	Anoxic, $COD_{methane}$
DAMOB	0.055	0.043	Anoxic, $COD_{methane}$

The higher the microbial community yield, the more quickly the control volume (i.e., the granule volume) would be "replaced" by new biomass. Higher microbial community yields would therefore make granules more susceptible to mass transfer limitations due to excessive growth of microorganisms and their by-products, such as EPS (Devlin *et al.*, 2017; di Biase *et al.*, 2017). Thus, continuously aerated AGS systems at high bulk DO concentrations would be less stable than AGS systems with anaerobic upflow influent distribution operated at low bulk DO concentrations. In particular, the elimination of aerobic/anoxic substrate utilization by OHO would be desired, while anoxic utilization of substrate by PAO and GAO would be preferred.

Barr *et al.* (2016) demonstrated on a molecular level that AGS communities are much more efficient than flocculent communities at utilizing nutrients. Based on the number of significantly enriched proteins assigned to Clusters of Orthologous Groups, it was shown that the AGS metagenome had significantly higher transport and metabolism related proteins (i.e., for carbohydrates, amino-acids, coenzymes, lipids, and inorganic ions) while flocculent communities had more proteins related to translation, ribosomal structure, biogenesis, energy production, and conversion. In general, this means that the flocculent community was actively growing without expending too much energy on substrate utilization while the AGS community had to invest energy into adenosine triphosphate consuming transport mechanisms for substrates with low energy return (Hosie & Poole, 2001).

Contrary to common depictions, granular sludge is not always made up of uniform spherical particles with smooth surfaces. In practice the structure of granular sludge, including the microorganisms in the outer surface layer, is dependent on both wastewater characteristics and operational strategies. For instance, uneven granule surfaces can develop when aerobic granular sludge processes are used to treat wastewater with high content of slowly biodegradable particulate organic loads (Wagner *et al.*, 2015). This is because slowly biodegradable particulates sorb to the surface of granular sludge aggregates and gradually release bioavailable organic matter via hydrolysis. The hydrolysed organic matter is then used directly within the vicinity of the original particulate organic matter that had sorbed to the granule, resulting in a microcosm of aerobic heterotrophic activity. At quasi-steady-state an equilibrium would be obtained between new slowly biodegradable particulate matter attaching to the surface of granular sludge and older particulate matter being utilized by the aerobic heterotrophs surrounding it.

When the load of slowly biodegradable particulate matter is low, there would be no significant effect on granule surface structure and the outer layer may appear smooth and compact, as long as other operating parameters are within bounds. As the quantity of slowly biodegradable particulate matter increases, however, the rate of degradation by aerobic heterotrophs will be lower than the rate of attachment, resulting in the retention of particulate organic matter to the outer surface layer and the appearance of an uneven granular surface structure (de Kreuk *et al.*, 2010). The excessive substrate sorbed to granule surfaces will also encourage the development of filamentous microorganisms feeding off the aerobically hydrolysed substrate. Furthermore, the presence of septic wastewater could also encourage the growth of filamentous microorganisms at the granule surface (Strom & Jenkins, 1984; Devlin *et al.*, 2017). The presence of filamentous microorganisms will increase the drag force of granular sludge and subsequently decrease the settling velocity of granules resulting in poor-quality effluent or even washout of biomass.

4.3 FACTORS AFFECTING GRANULE FORMATION AND STABILITY

Several studies have analysed the main mechanisms contributing to the granulation process. According to these studies aerobic granulation is a gradual process evolving from flocculent inoculum to compact granules without the use of carrier material. Tay *et al.* (2002a) recognized that aerobic granulation was a process of microbial self-immobilization without carrier support, including four subsequent steps:

(1) microbial contact to form small aggregates by hydrodynamic, diffusive, gravity and/or thermodynamic forces;
(2) attraction of such small aggregates to form bigger aggregates by physical, chemical and biochemical forces;
(3) EPS secretion that promotes the thickening of such structures; and
(4) hydrodynamic shear force to stabilize the three dimensional structure of the granule.

Granulation is influenced by many parameters including wastewater characteristics and operating conditions such as seed sludge, substrate composition, organic loading rate (OLR), feeding strategy, reactor design and hydrodynamics, settling time, selective wasting, volume exchange ratio (VER), and aeration intensity. The current bottleneck of AGS technology is the stability of the granules during long-term reactor operation. As reported by several authors aerobic granules would disintegrate after prolonged operation (Tay *et al.*, 2002a; Adav *et al.*, 2008; Verawaty *et al.*, 2013). Loss of granule stability in the long-term is due to several factors, some of which simultaneously contribute to granule instability and breakage. A brief discussion, based on the literature, about the effects of the main operating conditions on aerobic granulation is reported.

4.3.1 Alternating "feast" and "famine" conditions

Biomass with high internal substrate storage response, such as accumulation of PHA by PAO and GAO, are thought to benefit from alternating conditions of excess substrate (i.e., feast phase) and lack of substrate (i.e., famine phase) – Figure 4.2.

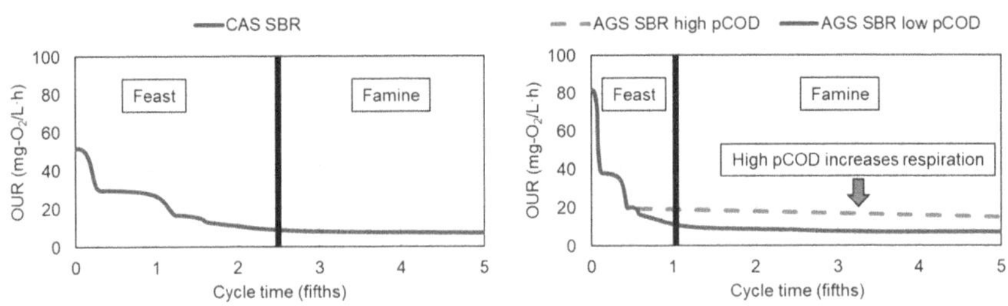

Figure 4.2 Depiction of feast and famine phases by oxygen uptake rate (OUR) profiles in a continuously aerated CAS SBR system (left) and continuously aerated AGS SBR system (right) with and without extended feast conditions due to high particulate COD in the influent. The endogenous respiration phase in OUR profiles can be used to monitor the development and duration of the famine phase.

It has been suggested that this is because rapidly growing OHO experience higher decay under extended aerobic conditions, or the famine phase. Thus, growth of microorganisms capable of PHA storage during the feast phase would be favoured, especially since PHA storing microorganisms usually have lower decay rates. It is unlikely that PHA storing microorganisms would be selected for in a continuously aerated AGS SBR. Simulations using BioWin^{TM} software (Envirosim, 2017) of a completely aerobic plug-flow system (i.e., five equally sized bioreactors in series) demonstrated that the ratio of PAO to OHO would be $\ll 1\%$. Rather, selection of PHA storing microorganisms would depend on maximizing the availability of volatile fatty acids (VFA) to PAO and GAO under anaerobic conditions.

It should be noted that wastewaters deficient in nitrogen, phosphorus, or other elements for cellular growth could encourage PHA storage (Ding *et al.,* 2015). In municipal wastewater, there is typically not enough easily available chemical oxygen demand (COD) to fully remove total nitrogen (TN) or total phosphorus (TP). By applying AGS to wastewaters that have excess COD and low TN and TP, such as brewery wastewaters, PHA may be encouraged under completely aerobic conditions (di Biase *et al.,* 2017). If the ratio of COD to TN to TP is above 100:5:1 there should be no nitrogen or phosphorus deficiency (Envirosim, 2017).

The development of a famine phase depends on wastewater characteristics and OLR (de Kreuk *et al.,* 2005a). For instance, the presence of slowly hydrolysable biodegradable matter may extend the duration of the feast phase by supplying a constant source of readily biodegradable organics. Furthermore, treatment plants that are nearing design capacity may also lack the required quantity of biomass and contact time required to rapidly reduce the concentration of bioavailable substrate, leading to shorter starvation periods and potentially higher process instability. Corsino *et al.* (2017b) noted that an imbalance in the feast and famine duration, with prolonged feast phase at the expense of a short famine phase, led to granule disintegration in an AGS SBR with upflow anaerobic influent distribution. The same authors suggested that the length of feast period should not exceed 25% of cycle length.

4.3.2 Hydrodynamic shear forces

Hydrodynamic shear forces may be defined as forces generated due to the friction between the bulk liquid and granules surface. The hydrodynamic shear force has typically been quantified by the upflow superficial air velocity in lab-scale systems. This parameter is calculated as the airflow divided by the cross-sectional area of the aerated zone (Devlin *et al.,* 2017). In aerobic heterotrophic biofilm systems, higher hydrodynamic shear forces have resulted in stronger biofilms, whereas biofilm tends to develop a weaker structure when the shear force is too low (Kwok *et al.,* 1998). Liu and Tay (2002) did not observe granulation when upflow superficial air velocities were less than 1.2 cm/s. Similarly, Chen *et al.* (2007b) found that unstable granules developed when upflow superficial air velocities were less than 2.4 cm/s. In a subsequent study, McSwain *et al.* (2004) observed that AGS was not stable at superficial upflow air velocities below 1 cm/s. The authors stated that low aeration rates create oxygen limitation during the reaction, increasing the duration of the feast period and shortening the famine period. However, most of the studies linking hydrodynamic shear force to granulation potential were completed in continuously aerated AGS systems.

More recently, Devlin *et al.* (2017) demonstrated that high hydrodynamic shear forces are not always a necessity for granulation although, they may benefit granular sludge formation and stability depending on wastewater characteristics and reactor configuration. Besides, Devlin and co-authors stated that shear might prevent the overgrowth of fast growing OHO, as well as the accumulation of particulates originating from the influent or EPS, from fouling granule surfaces. At higher applied OLR and food to microorganism ratio (F/M) the potential for fast growing OHO to become established in the bulk solution or on the surface of granules is larger.

Likewise, the potential for particle organic matter and EPS to accumulate on granule surfaces is also higher since hydrolysis and degradation rates have a maximum ceiling. In these cases, higher hydrodynamic shear forces would act to displace detrimental matter from granule surfaces and prevent the formation of a loose outer structure that would lead to poor settling properties and therefore instability. Thus, it can be concluded that high hydrodynamic shear force is not a requirement for granulation. In continuously aerated AGS systems operated with high OLR and F/M ratio, the importance of hydrodynamic shear force is more significant. However, many small-scale studies have successfully generated stable AGS at upflow superficial air velocities below 0.5 cm/s (Devlin *et al.*, 2017).

4.3.3 Influent distribution

In general, AGS SBR systems may be categorized into one of two types of influent distribution: (1) continuously aerated; and (2) upflow anaerobic influent distribution. The method of influent distribution will influence the morphology and microbial community of aerobic granules. De Kreuk and van Loosdrecht (2004) observed that a long feeding time, up to 50 min, through a settled bed of granules was favourable over a short feeding time during mixing. The two most important reasons for this were that the application of long influent distribution periods at large-scale treatment plants simplifies SBR scheduling, as well as the selection of PAO and GAO, due to the alternating aerobic and anaerobic phases – Figure 4.3. Thwaites *et al.* (2017) did granulate CAS under a 20 min anaerobic-40 min aerobic split feed strategy, although the granules exhibited poor phosphorus removal. According to Pronk *et al.* (2015a), the basic principle of stable aerobic granulation is the selection of slower growing bacteria and distributing the substrates throughout the granule during the anaerobic influent distribution period.

In the case of wastewater rich in easily biodegradable soluble substrates, when influent is distributed anaerobically with a plug-flow regime, such substrates are taken up by PAO or GAO and converted to internal storage polymers, such as PHA, that are used for growth at a relatively slow rate in subsequent aerobic/anoxic phases (de Kreuk & van Loosdrecht, 2004). At the beginning of the cycle accumulating microorganisms such as PAO and GAO near the granule surface take up easily biodegradable substrates. As anaerobic feeding continues, the PAO and GAO near the surface become saturated with internally stored polymer and the easily biodegradable substrate penetrates further into the granule. Thus, substrate reaches PAO and GAO communities residing deeper within the granule. This leads to uniform biomass production throughout the granule ensuring granule stability. While no oxygen will be present in deeper regions of the granules, even during vigorous aeration, the storage polymers can be oxidized by other electron acceptors such as nitrite and nitrate, ensuring optimal phosphate and nitrogen removal (de Kreuk *et al.*, 2005b).

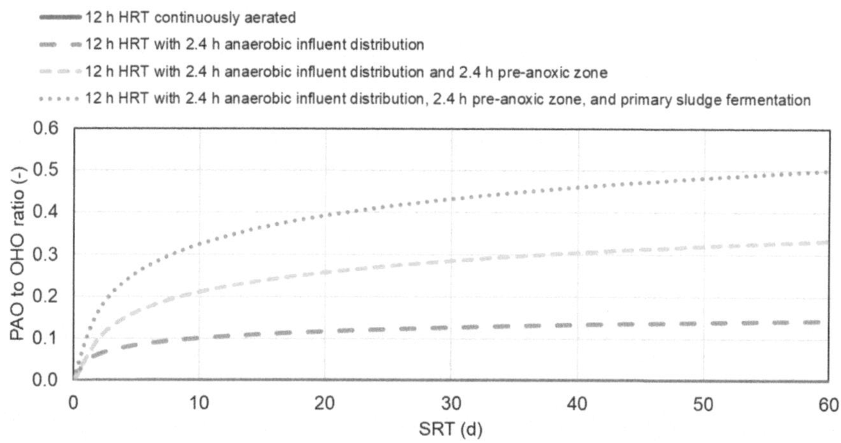

Figure 4.3 Estimated PAO to OHO ratio for a modelled biological treatment facility with increasing solids retention time (SRT) and various configurations at a 12 h hydraulic retention time (HRT) (Envirosim, 2017). All models had typical primary influent characteristics of 500 mg-COD/L, 40 mg-TN/L, and 10 mg-TP/L as input. Fermentation was modelled at a 3 d SRT receiving primary sludge from a clarifier removing 70% of influent total suspended solids (TSS).

Conversely, an aerated influent distribution strategy could lead to diffusion limitations, commonly characterized by a loose granule structure and surface layer of filamentous microorganisms. During the feast phase, when bulk liquid substrate concentration is high, the substrate would be rapidly utilized for simultaneous growth and formation of storage polymers limited to the outer aerobic layer of the granule (Beun *et al.*, 2002). The rapid, aerobic consumption of septic substrates in the presence of oxygen encourages the formation of filamentous microorganisms, which could foul the surface of granules (Strom & Jenkins, 1984). Filamentous growth will have a detrimental effect on the settling properties of the sludge and thus on the effluent quality (Martins *et al.*, 2011; Pronk *et al.*, 2015a). In these cases, increased hydrodynamic shear is required to ensure smooth and stable granulation. Devlin *et al.* (2017) noted significant surface fouling due to the proliferation of filamentous microorganisms when organic loads exceed capacity for anaerobic uptake, even with pure acetate as substrate. The same authors observed that higher hydraulic shear forces, by increasing upflow air velocity, could prevent surface fouling and limit filamentous overgrowth. Overall, granules formed under an aerobic fast-feeding strategy are more prone to degranulation (Beun *et al.*, 2002).

Particulate substrates, often present in high quantities in industrial wastewaters, also impact granule sludge formation and stability. Particulate substrates are mainly hydrolysed at the surface of the granules during steady state (de Kreuk *et al.*, 2010). The porous structure of granular sludge acts as a filter, retaining particles larger than the pore at the granule surface. Hydrolysis products during the anaerobic influent feeding period will be converted into storage polymers, thereby selecting for PAO and GAO like organisms. However, depending on the anaerobic hydrolysis rate as well as the quantity of particulate matter, aerobic hydrolysis will also occur. Aerobic hydrolysis will supply a steady stream of biodegradable matter and extend the duration of the feast phase. The aerobic hydrolysis product will be directly used for growth by heterotrophic organisms at the surface of the granules, generating substrate diffusion limitation gradients (Mosquera-Corral *et al.*, 2005). This will induce filamentous outgrowth, less stable granule formation, and higher suspended solids in the liquid phase. In this respect, Corsino *et al.* (2016b, 2017a) noted that an extended famine regime, by increasing the cycle length, could enhance aerobic granule stability. Indeed, a proper balance of feast/famine conditions inhibits the proliferation of filamentous microorganisms and favours the growth of self-aggregating bacteria capable of internal substrate storage.

Lastly, if easily biodegradable substrates are fed slowly in a mixed aerobic environment, this will lead to severe substrate diffusion limitation gradients, providing very good conditions for the proliferation of filamentous organisms. AGS fed under these conditions will therefore quickly deteriorate. Breakage of the granules will occur as the inside will not receive any substrate and die. Filamentous growth will have a detrimental effect on the settling properties of the sludge and thus on the effluent quality (Martins *et al.*, 2011; Pronk *et al.*, 2015a). It can be concluded that aerobic substrate gradients within the granule structure are favourable to the growth of fast growing heterotrophic organisms that lead to granules with irregular and weak structures that are more prone to breakage. Maximizing anaerobic utilization of biodegradable substrate prevents the proliferation of fast growing aerobic heterotrophs and encourages uniform granules with smooth and compact surfaces.

4.3.4 Selective wasting

In most granular sludge related studies, when the authors described the selection pressure, they usually referred to the settling time. The settling time in AGS reactors is considered a crucial parameter for granulation when an independent selective wasting regime is not applied (Qin *et al.*, 2004b). Several studies have demonstrated that shorter settling times enhance aerobic granulation (Jiang *et al.*, 2002; Hu *et al.*, 2005). Among these, Adav *et al.* (2009) compared granulation from three identical column-type reactors with settling times of 10, 7, and 5 min. These corresponded to 4.2 m/h, 3.0 m/h and 2.1 m/h in terms of critical settling velocities, respectively. The authors noted that very short settling times led to premature washout of certain non-flocculated strains, thereby yielding low biodiversity for the granules that were cultivated. It was also observed that too rapid a decrease in settling time leads to washout of desired bacterial guilds such as PAO, AOB, and NOB (Weissbrodt *et al.*, 2013). A higher initial washout of biomass induces the production of excess quantities of EPS to be secreted by functional strains, without a strong challenge for substrate by competing strains that have been washed out. Lochmatter and Holliger (2014) suggested that an option to avoid excessive biomass loss during start-up would be to link the settling time directly to the observed washout and to recirculate sludge from a secondary clarifier in case of a massive washout. Moreover, the same authors did not observe an enhanced washout below a settling time of 10 min.

Indeed, a short settling time causes the washout of poorly settling suspended biomass and retains only well settling granules (Qin *et al.*, 2004b). However, short settling times discharge the selectively wasted light fraction of biomass with the effluent. This can result in high effluent solids concentrations greater than discharge limits. A more conservative method of granulation is to employ a discrete selective wasting regime apart from the discharge of effluent (Winkler *et al.*, 2011; Bassin *et al.*, 2012c; Henriet *et al.*, 2016). This also allows for an additional stage of microbial manipulation. For instance, Winkler *et al.* (2011) demonstrated via fluorescence in situ hybridization (FISH) analysis that more dense granules located at the bottom of the settled sludge bed contained considerably more PAO compared to the lighter granules on top which were dominated by GAO. By selectively wasting biomass exclusively from the top of the settled sludge bed the same authors could increase TP removal efficiency from 71% to near 100%. By separating selective wasting from the effluent discharge, treated wastewater solids can be less than 5 mg/L and higher removal efficiencies can be obtained (Niermans *et al.*, 2014).

4.3.5 Organic loading rate

OLR is defined as the ratio between organic substrate load, calculated as the product of flow and substrate concentration, and volume of the reactor receiving the substrate. The F/M takes the volumetric representation of OLR and converts it to the ratio of organic substrate load and "active" biomass, usually as volatile suspended solids (VSS), in the reactor. In continuously aerated AGS systems, where the entire cycle including the influent distribution is under aerated conditions, high OLR and F/M seemed to play a key role in aerobic granulation. Specifically, Kim *et al.* (2008) found that the optimum value of OLR for granule formation was around 2.5 kg-COD/m^3·d. Li *et al.* (2011) studied the formation of aerobic granules in four SBRs fed with glucose-based substrate at different COD concentrations. This resulted in different F/M from 0.3 to 1.1 g-COD/g-VSS·d. The authors observed that a higher F/M promoted faster formation of granules with large average diameter. The lower range of F/M studied generated smaller granules at a slower rate of formation. Similarly, Li *et al.* (2008) analysed the granulation process at OLRs between 1.5 and 4.5 kg-COD/m^3·d. The authors observed that higher OLR resulted in faster formation of larger and looser granules, while a lower loading rate resulted in slower formation of smaller and denser granules. Granules formed at different OLR had different morphology, structural properties, and even bacterial species distribution. Liu and Tay (2002) observed that granulation could not be achieved at an OLR lower than 2 kg-COD/m^3·d in continuously aerated AGS systems. In contrast, Wang *et al.* (2009a) successfully cultivated granules at an OLR of 1 to 1.7 kg-COD/m^3·d in a completely aerobic system, however almost one year was needed to obtain mature granules.

The role of OLR and F/M appears to diminish with the introduction of an anaerobic upflow influent distribution period. Under this operating configuration, it was demonstrated that stable aerobic granules have been generated by applying lower OLRs compared with completely aerobic systems. Pronk *et al.* (2015b) operated a full-scale reactor with OLR ranging from 0.65–0.85 kg-COD/m^3·d. Similarly, Devlin *et al.* (2017) generated stable granules on proteinaceous substrate within a month at an OLR of 1.4 kg-COD/m^3·d. Corsino *et al.* (2017b) also generated stable AGS within three weeks at an OLR of 1.4 kg-COD/m^3·d while treating effluent from an anaerobic moving bed biofilm reactor (MBBR) treating brewery wastewater. Based on the reported studies, it can be concluded that OLR and F/M are not crucial parameters for aerobic granulation. However, the impact of high OLRs and F/M is more significant in continuously aerated AGS systems. These results clearly indicate that there are more important factors influencing aerobic granulation than the OLR.

4.3.6 Other environmental factors

The main environmental factors affecting biological processes, including AGS, are DO, oxidation-reduction potential (ORP), pH, and temperature. So far, the literature has suggested that the DO was not a dominating factor for granulation despite claims that DO, instead of shear stress, governed the granulation process (Dulekgurgen *et al.*, 2008). Instead, DO is suggested to have an impact on the layering and stratification of microbial communities. Since an oxygen gradient develops across the depth of granular sludge particles, anaerobic, anoxic as well as aerobic layers could coexist within a single granule at any given time.

Stratification clearly impacts nutrient removal processes due to the simultaneous presence of varying ORP conditions and therefore the capacity for simultaneous nitrification-denitrification (SND) with DOHO or DPAO. The aerobic layer of granules was demonstrated to be strictly related to bulk DO concentrations. Li *et al.* (2008) demonstrated that for granules of equal diameter oxygen penetrated further into the granule when bulk DO concentrations were higher. Thus, if DO in the bulk liquid was too high the development of anoxic or anaerobic layers could be inhibited, especially for smaller granules. de Kreuk *et al.* (2005b) observed that reduction of DO to 40% saturation during the aeration phase significantly improved total nitrogen removal efficiency however it led to decreasing granule density and ultimately to disintegration of granules.

Concerning pH, it was observed that fungi dominated aerobic granules at a pH of 4 and that the granule size was as large as 7 mm. Fungi resulted in faster granulation by providing a filamentous matrix for the microbial community. At pH higher than 8 granules were dominated by bacteria and granule size only reached 4.8 mm (Yang *et al.*, 2008). Some authors found that acidic and neutral solutions in AGS reactors had a significant impact on granulation (Rezasoltani *et al.*, 2015). However, if the fungal growth is not controlled, their irregular patterns can lead to system instability. On the other hand, a neutral medium of the bioreactor resulted in bacterial domination of granules that had a more compact structure. Lashkarizadeh *et al.* (2016) observed that operating the reactors under alkaline pH induced granule disintegration and resulted in an increased solids concentration in the effluent, as well as a significant decrease in the size of the particles. The same authors observed that acidic pH did not have significant impacts on granule stability. Changes in chemical structure and composition of the EPS matrix were suggested as the main factors inducing granule instability under high pH.

Temperature, in addition to affecting bacterial growth kinetics, also influences the settling velocity because of changes in the viscosity of the bulk solution. de Kreuk *et al.* (2005b) noted that, in an SBR operated at 8°C, granules with irregular shape and excessive filamentous bacteria developed which caused unstable operation by severe biomass washout. The authors concluded that aerobic granulation at low temperatures was not practically feasible. All microbial processes run slower at lower temperatures which could have limited the granulation at a lower temperature (Lettinga *et al.*, 2001). Winkler *et al.* (2012a) showed that not only are microbial factors hindered at lower temperatures, but also granule settling velocity is significantly reduced at lower temperatures. On the other hand, operating at high temperature (e.g., >30°C) could also inhibit granulation by preventing agglomeration of EPS and increasing the solubility of chemical species (e.g., calcium, iron, magnesium, phosphorus) that have been detected as precipitates in the core of AGS particles (Ab Halim *et al.*, 2016).

4.3.7 Design considerations and control strategies

Aerobic granulation can occur in a wide range of scenarios with different operating parameters. However, certain variables do seem to be more significant for aerobic granulation and long-term granule stability. Based on the literature, several operating parameters and design considerations were drawn – Table 4.2. Among these, upflow anaerobic influent feeding regimes are almost unanimously considered essential for the formation of stable aerobic granules (de Kreuk & van Loosdrecht, 2004; Pronk *et al.,* 2015a). Duration of the upflow anaerobic influent distribution period will depend on the OLR and wastewater characteristics. Influent distribution should be designed to maximize the conversion of organic matter in the influent into intracellular polymeric substances, such as PHA. Fifty minutes to one hour is typically sufficient for the upflow anaerobic influent distribution phase when treating municipal wastewater.

Adav *et al.* (2008b, 2010), observed that the high OLRs encouraged proliferation of filamentous bacteria that initially favours a rapid increase in the size of the granules. The filamentous bacteria, however, also resulted in granules with loose structures and therefore lower settling velocities. In this respect, de Kreuk and van Loosdrecht (2004) stated that selection of slow growing organisms (i.e., phosphate or glycogen accumulating bacteria) could enhance granule stability. The same authors suggested that a long feeding time through a settled sludge bed appeared feasible and was favourable over a short feeding time.

Verawaty *et al.* (2013) hypothesized that a certain "critical" granule size will be achieved at steady state under certain operational conditions. Due to the consequences of particle collision and structural weaknesses, granules larger than the critical size are expected to break apart until reduced to the critical size or smaller. Conversely, the growth of small granules in steady-state AGS reactors equals the rate at which large granules are being reduced by breakage or attrition. According to the authors, this cycle of granulation and degranulation becomes a dynamic process that is constantly evolving. Similarly, Zhu *et al.* (2013) demonstrated that a solids retention time (SRT) of granular sludge longer than 14 days leads easily to the deterioration of aerobic granules. The same authors and Zhang *et al.* (2015) suggested that a selective discharge of the aged granules from the bottom of the reactors (5% of the settled bed daily), favours the stability of AGS in the long term.

The reason for larger granules weakening is partly due to mass transfer limitations leading to a weakened inner core of the granule. Zheng *et al.* (2006) previously suggested that the mass transfer limitation of large granules could produce an anaerobic core and thereby stimulate the activities of anaerobic strains, hence breaking apart large granules. Lemaire *et al.* (2008) claimed that the breaking of mature granules was attributable to the clogging of pores and channels in the granules, hence hindering the nutrient intake by the microbes. In this context, Corsino *et al.* (2017a) demonstrated that excessive EPS production contributed to the clogging of pore space in the granules, causing their breakage in the long run. An extended famine period encouraged greater EPS consumption by bacteria, thus limiting the clogging of pores, and allowed for the long-term operation of stable aerobic granules. Adav *et al.* (2010) noted granule disintegration at high OLR (i.e., >21.3 kg-COD/m^3·d), because bacteria lost their capacity for producing proteins or polysaccharides at the applied OLR.

Table 4.2 Summary of the main operating conditions affecting aerobic granulation and their qualitative impact on the process.

Operating Parameter and Other Factors	Impact on the Process	Comments
Organic loading rate	✓	Higher OLRs accelerate start-up; effects on physical characteristics.
Substrate composition	×	Aerobic granules successfully developed with several organic substrates.
Hydraulic shear forces	✓✓✓	Improvement of physical characteristics; prevent filamentous overgrowth.
Selective wasting	✓✓✓✓	Allow selection of bigger fast settling particles, while washing-out the smaller ones.
Anaerobic upflow feeding	✓✓✓	Improvement of granule stability and nutrient removal efficiency.
Feast/famine regime	✓✓✓	Allow selection of bacteria with storage capability.
SRT	✓	Prevent excessive ageing of granules that are more prone to breakage.
DO	✓✓	Not too low to prevent the core weakening; Not too high to favour a proper layered structure.
pH	×	–
Temperature	✓	Low temperature delays the granulation process.
Un-ionized ammonia	✓	Ammonia Loading Rate higher than 0.80 kg/m^3·d resulted in granule instability.

Legend: ✓✓✓✓ = fundamental; ✓✓✓ = very important; ✓✓ = important; ✓ = not crucial; × = not relevant.

Particulate organic matter and other slowly biodegradable organic substrates can offset the benefits of anaerobic influent feeding. Due to slow anaerobic hydrolysis rates, slowly biodegradable substrates may not be fully converted and subsequently utilized for PHA storage during the upflow anaerobic influent distribution phase. Therefore, the slowly biodegradable substrates would contribute to a continuous supply of bioavailable substrate via aerobic hydrolysis during aeration (de Kreuk *et al.*, 2010). The growth of aerobic heterotrophs would thereby be promoted in a confined location since particulate and colloidal substrates would be concentrated in the bulk solution or near the granule surface. Overgrowth at the granule surface could lead to mass transfer limitations and the loss of nutrient removal (Devlin *et al.*, 2017).

Two approaches have been suggested to prevent the proliferation of microorganisms on the surface of granules that feed off aerobically hydrolysed substrate. The first approach is to apply an extended famine phase by increasing the duration of the aerobic period. The feast phase, the period of time in which hydrolysis is occurring and substrate is available near the granule surface, should not be more than 20% of the aerobic period (Valentino *et al.*, 2013; Corsino *et al.*, 2017b). The second approach would be to increase the applied hydrodynamic shear force in order to detach filamentous overgrowth at the granule surface (Devlin *et al.*, 2017). From a design point of view, the first approach implies an increase in capital cost due to larger reactor volumes while the second implies an increase in operating costs for electricity to power blowers or mixers.

The last crucial parameter for aerobic granulation is the selective wasting. It is considered a fundamental parameter because it allows for the out-selection of slowly settling flocculent biomass and retention of rapidly settling granular sludge. Taking into account settling time, design characteristics of reactors, as well as upflow fluid velocities used in experimental studies, the average settling velocity applied for selective wasting is approximately 10–14 m/h. This means that only particles with higher settling velocities will be retained within the reactor.

4.4 APPLICATION OF AEROBIC GRANULAR SLUDGE TO MUNICIPAL WASTEWATER

Full-scale implementations of AGS have improved process stability, attained stringent effluent quality (i.e., TN <5 mg/L; TP <0.3 mg/L), and have resulted in 30% to 60% energy savings (Niermans *et al.*, 2014). Energy requirements less than 14 kWh/(PE·y) were recorded for a full-scale AGS facility which is approximately 60% lower than CAS facilities in the Netherlands. The AGS technology has been described as a paradigm shift for municipal wastewater treatment because of the capital and operational savings it provides. In this section, a summary of municipal wastewater treatment applications and design considerations for AGS are presented.

4.4.1 Municipal wastewater characteristics

Municipal wastewater solids are made up of both soluble and particulate organic and inorganic matter. Proteins, carbohydrates, lipids, and fibres contribute to the particulate organic fraction, while readily available soluble organic matter is comprised of volatile fatty acids, alcohols, and other simple carbohydrates (Metcalf & Eddy, 2014). Depending on the configuration of sewer systems, municipal wastewater can contain low or high quantities of bioavailable substrates. Larger sewersheds located on relatively flat terrain in warmer climates will generally yield septic wastewater with lower biodegradable fractions, having already been utilized within the sewer system (Oleszkiewicz *et al.,* 2015). On the other hand, municipalities in mountainous regions with steep terrain and lower temperatures will provide treatment facilities with fresh, unaltered wastewater.

The fraction of soluble biodegradable substrate in the wastewater is particularly important for AGS processes. The higher the fraction of soluble biodegradable substrate, the more likely the AGS process is to be stable. This is because fresh wastewater can be introduced into an anaerobic contact phase where readily bioavailable substrate is taken up and converted to internal storage products by accumulating bacteria. Thus, there will be less residual organic matter in bulk solution passing through to aerobic conditions if more of the substrate is soluble biodegradable. By diverting a majority of the initial organic matter to internal storage products, it is less likely that a fast growing aerobic heterotrophic culture will develop on granule surfaces, resulting in mass transfer limitations and process instability (de Kreuk & van Loosdrecht, 2004; Devlin *et al.,* 2017).

Furthermore, municipal wastewater can be classified as low-, medium-, or high-strength depending on how water conservative municipalities are, as well as how much inflow and infiltration is experienced. Municipalities that use less water and have separate sewers tend to have stronger wastewater that contains higher concentrations of organic and inorganic matter. It has been speculated that aerobic granular sludge, which derives benefits from the development of nutrient gradients across granule depth, performs better on higher strength wastewaters than low-strength, diluted wastewaters (Zhang *et al.,* 2015). Higher concentrations facilitate diffusion and therefore promote more rapid uptake and utilization of nutrients by aerobic granular sludge.

4.4.2 Optional and required pretreatment of municipal wastewaters

The extent of municipal wastewater pretreatment is still largely debated in literature and practice. Personal communication with leading environmental consultants has yielded contradictory responses. Some swear by removing as much solid content as possible up front through primary treatment while others have argued that AGS only works on primary influent. All of these claims are based on individual experiences with demonstration facilities likely confounded by site-specific variables that have not been accounted for in design and operation of the demonstration facilities. Full-scale Nereda® installations have worked successfully for many years on primary influent, demonstrating that the AGS SBR process is a robust technology capable of handling the additional particulate loads and variability associated with primary influent (Giesen & Thompson 2013b; Pronk *et al.*, 2015b).

It is expected that the only necessary pretreatment before any AGS process would be screening and grit removal. Due to the nature of AGS processes, which select dense particles, the breakthrough of grit from pretreatment could result in significant accumulation of undesirable inert material. Thus, it is very important that good grit removal with sufficient redundancy be designed for AGS processes. Optional pretreatment for AGS processes includes the removal of particulate organics from the wastewater. A summary of the pretreatment options and their applications is provided in Table 4.3.

Table 4.3 Summary of AGS SBR pretreatment configurations based on municipal wastewater characteristics.

Pretreatment	Process Diagram	Comments
Screening + Grit removal	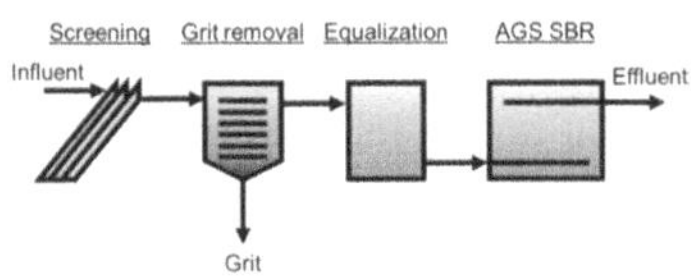	For municipal wastewater with TSS and VSS capable of adsorption and degradation by AGS without disrupting process stability.
Screening + Grit removal + Sedimentation	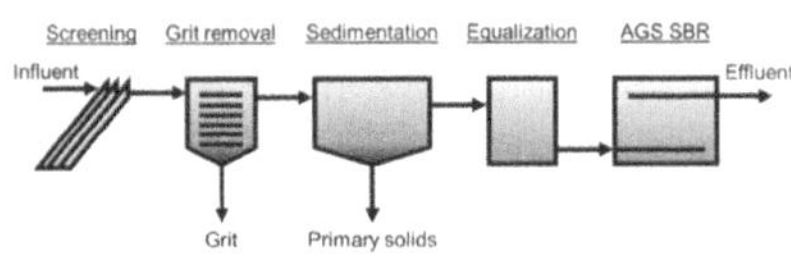	For municipal wastewater with TSS and VSS exceeding the capacity of AGS, having adequate residual COD to TN ratio for nutrient removal.
Screening + Grit removal + Sedimentation + Fermentation	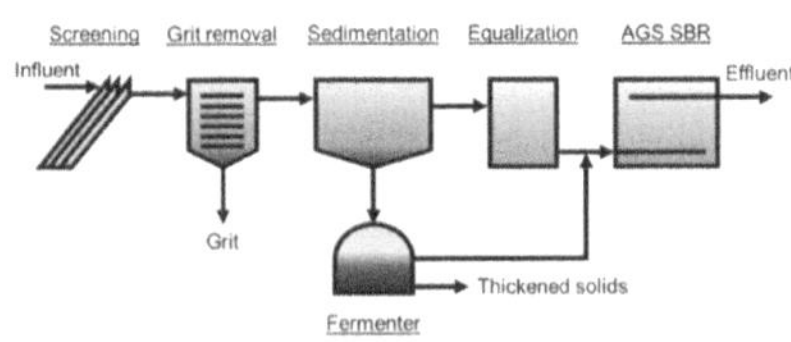	For municipal wastewater with TSS and VSS exceeding the capacity of AGS, as well as COD deficiency for nutrient removal.

The removal of additional influent solids, besides grit, before aerobic granular sludge reactors has many plausible arguments both for and against it. Reasons for good solids removal up front include the separation of slowly biodegradable matter, which can lead to uneven surface structure, and the removal of dense inert particles that could accumulate in granular sludge reactors. Diversion of settleable organics away from mainstream processes has been practised for increased mainstream capacity and beneficial reuse in the side-stream (Oleszkiewicz *et al.*, 2015). On the other hand, it has been argued that higher wastewater solids content can help initiate, or is even necessary for, granulation. High density, unbiodegradable inert particles found in primary influent, which mostly (i.e., 50–70% removal) settle out during primary treatment, may contribute to more rapidly settling granular sludge particles if the inert particles become associated with the granular sludge matrix. It has also been suggested that allowing more slowly biodegradable particulate organic matter through to the granular sludge process can alleviate problems associated with low-strength wastewaters. Nevertheless, with properly designed cycles that account for the peak organic load in the design year, AGS SBRs can adequately treat primary influent with particulate organic substrates to exceptional effluent quality (López-Palau *et al.*, 2012; Corsino *et al.*, 2017b). Thus, the application of properly balanced feast and famine phases allows for additional organics in the primary influent to be beneficially used for nutrient removal without the particulate load negatively influencing the AGS process.

Generally, particulate substrates are hydrolysed at the surface of granules during steady state in properly designed AGS SBRs. Municipal wastewater with abnormally high particulate substrate can negatively impact AGS stability (de Kreuk *et al.*, 2010). The porous structure of granular sludge acts as a filter, retaining particles larger than the pore size at the granule surface. Hydrolysis products during the anaerobic influent feeding period would be converted into storage polymers, thereby selecting for PAO and GAO. However, depending on anaerobic hydrolysis rates as well as the quantity of particulate matter, aerobic hydrolysis will also occur. Aerobic hydrolysis would supply a steady stream of biodegradable matter that would be used by heterotrophic organisms at the surface of granules (Mosquera-Corral *et al.*, 2005; Devlin *et al.*, 2017). This has induced filamentous outgrowth, resulted in less stable granules, and generated higher suspended solids in the liquid phase. If wastewater characterization determined that influent wastewater was high in particulate matter, primary treatment should be considered. Primary treatment could include sedimentation, micro-screening, or chemically enhanced primary treatment (CEPT). The removal of organic matter may result in low carbon to nitrogen ratios, and therefore poor nutrient removal even with SND. If the wastewater was determined to be high in particulate matter as well as carbon deficient after primary treatment, fermentation of primary solids could be used to increase the amount of readily available substrate sent to the AGS process (Leong *et al.*, 2016). Fermentation products, such as acetate and propionate, are easily utilized by AGS and contribute to their stability (Liu *et al.*, 2011; Milia *et al.*, 2016).

4.4.3 Operational considerations for municipal wastewater treatment

An efficient, stable AGS SBR process treating municipal wastewater must consider a few design factors. The following discussion will include operational considerations for municipal wastewater treatment with AGS as it pertains to: (1) cycle configuration; (2) OLR; (3) selective wasting; and (4) upflow anaerobic influent distribution.

Cycle timing must be designed to provide adequate oxidation of soluble and particulate organic compounds within diurnal flow and load variation. If higher loads of organics exceed the anaerobic uptake capacity of the biomass during a single cycle the residual organics in the bulk solution would pass on to the aerated react phase and/or subsequent cycles. Under these conditions residual organics could promote the growth of fast growing aerobic heterotrophs, ultimately destabilizing the process if they are allowed to proliferate further (Pronk *et al.*, 2015a). Thus, aerobic granular sludge cycles should be designed with flexibility in order to address variable influent loads. Depending on the degree of flow and load variation observed, different design considerations can be made. These could include equalization and flow balancing, DO control strategies to ensure adequate "famine" conditions develop at higher observed loads, multiple draw points for selective wasting by a pump with variable frequency drive (i.e., remove more or less biomass considered to be part of the light fraction and therefore not desirable), and swing-capabilities allowing portions of the upflow anaerobic influent distribution period to be aerated (López-Palau *et al.*, 2012; Thwaites *et al.*, 2017).

The literature varies in terms of recommended OLRs. For instance, the reported range on municipal wastewater has been from 0.6–1.4 kg-COD/m^3·d (Pronk *et al.*, 2015b; Devlin *et al.*, 2017). Most of the reported data from full-scale Nereda® facilities may not represent the design OLR, since most of their installations are relatively recent (i.e., first installation in 2012). Rather than a set OLR, AGS processes for municipal wastewater treatment should be designed and operated based on F/M. The flow and load of wastewater will continue to increase for municipalities experiencing population growth. Thus, the OLR will always be increasing. For this reason, operators of AGS facilities must understand the optimal biomass concentrations, relative to the incoming load, that will provide adequate nutrient removal and energy efficiency. In general, F/M ratios from 0.3 to 1.1 g-COD/g-VSS·d have been adequate for AGS operation (Li *et al.*, 2011). However, operating above 1 g-COD/g-VSS·d may not be suitable as the risk for hydrous bulking increases.

Another important consideration includes selective wasting and the decoupling of granular solids retention time from free-living microorganisms. Ideally, the SRT of free-living microorganisms would be designed to be less than their minimum SRT. By doing so, the proliferation of fast growing heterotrophs would ideally never occur. Given the inherent variability and characteristics of municipal wastewater, some substrate will always be available under aerobic conditions, thereby promoting the growth of free-living microorganisms and the development of a persistent light fraction of biomass. For this reason, selective wasting is important in order to manage, rather than completely out-select, the light fraction of biomass (Derlon *et al.*, 2016). Designers and operators should consider that the more aggressive the selective wasting strategy (i.e., lower applied SRT), the higher the yield of the light fraction of biomass. Thus, selective wasting should be optimized to aid in the diversion of substrate to anaerobic utilization by microorganisms within AGS particles. The only way to decrease the yield of biomass, and therefore improve the performance of AGS, is to divert substrate away from direct aerobic utilization and towards anaerobic uptake (Pronk *et al.*, 2015a; Devlin *et al.*, 2017).

In order to effectively limit the growth of fast growing aerobic heterotrophs, AGS SBRs should be designed to divert as much carbon as possible to internal storage products during the anaerobic influent introduction phase. In general, an anaerobic contact time of 50 min has been suggested as sufficient for anaerobic substrate conversion while treating municipal wastewaters (de Kreuk & van Loosdrecht, 2004). However, more dilute wastewaters in areas with higher degrees of inflow, infiltration, and wet weather may require less anaerobic contact time (Thwaites *et al.*, 2017). Arguably, one of the most important factors is the configuration of influent distributors in AGS SBRs with upflow anaerobic influent distribution. Without proper hydraulic design, short-circuiting could occur and thereby disrupt optimal conditions for anaerobic uptake of substrate in the settled sludge bed. This is detrimental since anaerobic diversion of organic carbon to internal storage products is desired (Liu *et al.*, 2004; Devlin *et al.*, 2017). It is well known that the breakthrough of organic substrate in bulk solution to aerobic conditions negatively affects AGS stability in SBRs. This could be due to many reasons, including the overgrowth of filamentous organisms on the surface of granules or an accumulation of particulate matter. Both cases lead to mass transfer limitations and therefore increase the potential for granule disintegration. For municipal wastewater treatment, where stringent discharge limits must be achieved (e.g., TSS <5 mg/L, biochemical oxygen demand (BOD_5) <5 mg/L, TN <5 mg/L, and TP <0.5 mg/L), an AGS SBR configuration with upflow anaerobic influent distribution would be preferred – Table 4.4.

4.4.4 Case study: Nereda® technology

Research on AGS began at Delft University of Technology in 1993. The technology matured as an outcome of Dutch concerted research and development (Morgenroth *et al.*, 1997). The final product was Royal Haskoning DHV's product Nereda®, the first and only commercially available AGS technology. The Nereda® aerobic granular sludge process is a constant level sequencing batch reactor system. Following a settling period, raw wastewater is introduced to the bottom of the reactor under anaerobic conditions. Addition of raw wastewater simultaneously

Table 4.4 Comparison of continuously aerated AGS SBRs to AGS SBRs with upflow anaerobic influent distribution.

Configuration	Process Diagram	Comments
Continuously aerated	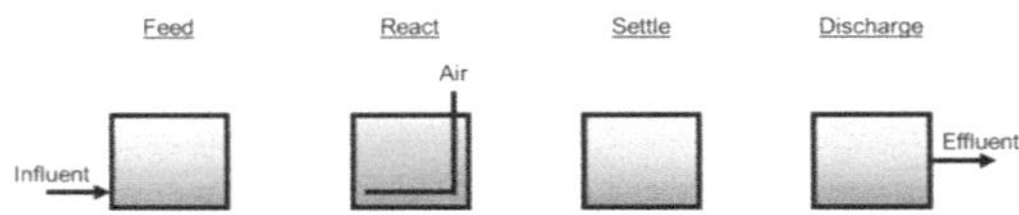	Influent is supplied over a very short period (i.e., <5 min) during aeration. The settling time is also short (i.e., <10 min) and selectively wasted biomass is discharged with the effluent. Feed and discharge occur separately.
Upflow anaerobic influent distribution	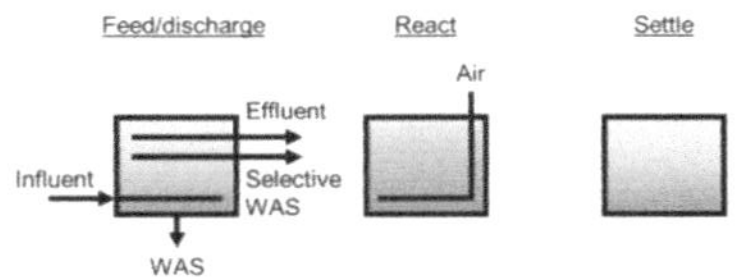 *WAS: waste activated sludge	Influent is supplied gradually over an extending period of time (i.e., 50 min) under anaerobic conditions. Discharge occurs with feed. During the feed, selectively wasted biomass is also removed, separating the light solids from the effluent.

displaces treated effluent in the reactor which is discharged from the top through overflow weirs. Influent addition and discharge of effluent is stopped after the appropriate amount of raw wastewater has been introduced to the reactor and the aeration system is turned on resulting in the simultaneous removal of carbon, nitrogen, and phosphorus. Installations of Nereda® have demonstrated significant improvements with regard to process stability, effluent quality (i.e., TN <5 mg/L; TP <0.3 mg/L), and energy savings (i.e., 30% to 60%) compared to traditional activated sludge processes (Niermans *et al.*, 2014). Other advantages of the technology included lower capital and operational expenses, and a small physical footprint (de Bruin *et al.*, 2004).

In general, the Nereda® process can be configured in one of four ways. Two of the configurations are recommended for greenfield applications, while the other two are for brownfield (i.e., retrofit or optimization) installations. The first greenfield configuration consists of at least three SBRs in parallel, such that at least one SBR is supplied influent at all times. This configuration is the only truly continuous mode of operation for the Nereda® process since influent distribution and effluent discharge occur simultaneously. The second greenfield application includes a buffer tank followed by two SBRs in parallel. Buffering is required to store the influent when influent distribution to the SBRs is not available. The first brownfield application is the retrofit of existing SBR or CAS basins into Nereda® SBRs. This is achieved by implementing influent distribution grids at the bottom of the reactor and effluent weirs equally distributed across the reactor surface. The last brownfield configuration is installing a parallel Nereda® SBR to the existing CAS process. The parallel Nereda® SBR treats a portion of the flow, and excess granular sludge is returned to the existing CAS process. This bioaugmentation significantly reduces the SVI of CAS processes, as well as their nutrient removal capabilities.

The first full-scale application of Nereda® began in 2005 with a retrofitted milk storage tank treating 250 m^3/d of cheese factory waste in Ede, Netherlands (Giesen *et al.*, 2013a). The technology was scaled up for municipal wastewater treatment in 2008 with a 5,000 m^3/d demonstration facility at the Gansbaai sewage treatment plant in South Africa (Giesen & Thompson, 2013b). Although only required to attain TN less than 15 mg/L and TP less than 10 mg/L, the Gansbaai installation reliably achieved effluent TN of 10 mg/L and TP of 3.2 mg/L (Niermans *et al.*, 2014). Success at the Ede and Gansbaai facilities proved that AGS technology was feasible for full-scale applications of both industrial and municipal wastes. The Gansbaai demonstration in particular allowed applications of Nereda® to expand, starting with the design and construction of Epe WWTP, Netherlands in 2010–2011 (Giesen & Thompson 2013b).

Operational since September 2011, the Epe WWTP consists of inlet works with screens and grit removal, followed by three Nereda® reactors and effluent polishing via gravity sand filters (Giesen & Thompson, 2013b). At an average flow of 8,000 m^3/d and peaking factor of 4.5 the facility has been able to achieve effluent TN of 4 mg/L and TP of 0.3 mg/L with only 60% of the original energy requirements (i.e., 3,500 kWh/d to 2,000–2,500 kWh/d after installation; Niermans *et al.*, 2014). Other Nereda® installations have been installed on municipal and industrial discharges, successfully treating highly variable wastewater. For instance, the Nereda® facility in Ryki, Poland can receive as much as 60% of the total organic load from hauled in sources and the campaign vegetable processing industry, achieving effluent below TN 10 mg/L and TP below 1 mg/L. Furthermore, the Ryki WWTP seasonally receives wastewater below 8°C, demonstrating the capability for cold climate wastewater treatment.

The largest Nereda® installation operating to date was commissioned in 2013 at the Garmerwolde WWTP, Netherlands. The additional capacity expected from two 9,500 m^3 Nereda® reactors was 30,000 m^3/day (i.e., 15 h HRT) with a peaking factor of approximately 3.4. The tank sizes of the AGS reactors are similar to the world's largest SBR tanks. The Garmerwolde WWTP has passed the one-year performance test to validate that combined effluent meets the target of less than 7 mg/L of TN and less than 1 mg/L of TP. The Nereda® installation at the Garmerwolde WWTP now treats 50% more wastewater than originally planned (i.e., 10 h HRT) and has resulted in significant energy savings.

A retrofit underway at the Ringsend WWTP in Dublin, Ireland will soon exceed the capacity of Garmerwolde and become the world's largest Nereda® installation with a design flow of 600,000 m^3/d and peaking factor of 2. The existing CAS SBRs at Ringsend are overloaded at approximately 0.7 kg-COD/m^3·d, and the WWTP struggles to achieve discharge requirements associated with COD and TSS. The Nereda® retrofit will incorporate complete nutrient removal up to a design load of approximately 0.9 kg-COD/m^3·d within the existing footprint. There are 24 existing SBRs with a volume of 13,500 m^3/d, resulting in an HRT of 13 h at the average design flow of 600,000 m^3/d.

The Nereda® technology has also shown potential for retrofitting existing continuous sludge systems after a successful demonstration at the 70,000 m^3/d Frielas WWTP in Lisbon, Portugal (Niermans *et al.*, 2014). The demonstration SBR was setup in one of the six existing CAS bioreactors and operated in parallel with the remaining CAS bioreactors (Giesen *et al.*, 2013c). Data from 2012 has indicated 30% energy savings for aeration alone when compared with the existing system (Inocêncio *et al.*, 2013). Combining this with the energy savings that granules bring by not using settling tanks, sludge recirculation pumps, and post-filtration units, the overall electricity demand was expected to be 50% of the existing CAS process (Niermans *et al.*, 2014).

A unique application of Nereda® for municipal wastewater treatment is at the Vroomshoop WWTP in The Netherlands. The Vroomshoop WWTP receives an average flow of 3000 m^3/d, which is then split 50/50 between an old, racetrack-type continuous flow activated sludge system and the new Nereda® installation. The peak factor of the entire facility is approximately 3.2. Excess sludge from the Nereda® process was returned to the old activated sludge process, effectively bioaugmenting the continuous system with rapidly settling granular sludge aggregates. Within a year and 6 months, the SVI of the old activated sludge process dropped from 100 to 60 mL/g, while the SVI of the Nereda® process operates below 40 mL/g.

There are a variety of other Nereda® applications for municipal treatment directly resulting from the Epe WWTP success. With more than 20 WWTPs in the international pipeline, the technology seems to be on the path to satisfy future demand, and may prove to be a new standard for municipal wastewater treatment. However, Nereda® technology does not offer an immediate solution for existing plug-flow systems that are commonly found across the world.

4.5 APPLICATION OF AEROBIC GRANULAR SLUDGE TO INDUSTRIAL WASTEWATERS

The ability of granules to tolerate and degrade toxic organic compounds such as phenol and antibiotics adds to the attractiveness of the process (Liu *et al.*, 2011; Amorim *et al.*, 2013). It has even been suggested that harsh environmental conditions could enhance granulation since bacteria are stimulated to produce EPS that then act as a protective shield for microorganisms within the granules (Zhang & Tay, 2016). Nevertheless, most toxic and recalcitrant compounds have adverse effects on biological performance, as well as on granule stability (Gao *et al.*, 2017). Thus, properly designed AGS systems should be well-suited for the treatment of industrial wastewaters which can contain high concentrations of organic substrate, as well as recalcitrant and xenobiotic substances within the bounds that does not affect long-term granule stability. In the following sections a summary of industrial wastewater treatment by AGS is reported.

4.5.1 Agro-food wastewater

The agro-food industries generate large volumes of wastewater from both the production line and cleaning practices. Some of these industries generate high salinity wastewater (i.e., fish-canning), wastewater with high suspended solids, as well as wastewater with high organic substrate concentrations. High-throughput treatment systems are therefore required to comply with discharge limits (e.g., into the municipal sewer or direct discharge to surface water) or to obtain an effluent that is suitable for reuse within the industry or other nearby sites.

4.5.1.1 Brewery wastewater

Wastewater produced by breweries is characterized by high concentrations of biodegradable COD derived from organic components (e.g., yeast, sugars, soluble starch, etc.). Nitrogen and phosphorus concentrations are typically dependent on the amount of yeast in the effluent (Fillaudeau *et al.*, 2006). Thus, brewery wastewater can be nitrogen and phosphorus deficient. Wang *et al.* (2007) successfully cultivated AGS on brewery wastewater. The authors obtained stable granules after nine weeks of operation. After granulation, high and stable removal efficiencies (i.e., >88%) for organic carbon and nitrogen were achieved at the volumetric exchange ratio of 50% and cycle duration of 6 h.

Operating under similar process conditions, Corsino *et al.* (2017b) noted that, although aerobic granules successfully developed in less than 4 weeks, the granular structure deteriorated and became irregular because the feast and famine periods were not properly balanced. A 16 h HRT was necessary to provide an adequate starvation phase that benefitted granule stability. The same authors noted that granules developed at lower OLR were structurally more compact than granules developed at higher OLR. Nevertheless, further investigations are necessary to focus on decreasing the volumetric exchange ratio or extending the total duration of the SBR cycle in the treatment of high-strength wastewater.

4.5.1.2 Dairy wastewater

Dairy wastewater is characterized by high concentrations of TSS and particulate substrates. Arrojo *et al.* (2004) evaluated the feasibility of treating dairy wastewater with AGS. High removal efficiencies in terms of COD and nitrogen were obtained at steady state when full granulation was achieved. However, the authors noted that in certain operating conditions, the amount of TSS in the effluent significantly increased. The presence of TSS in the effluent was strongly affected by either the length of the withdrawal period or by the particulate COD (pCOD) to biomass ratio applied to the systems. The authors also observed that by decreasing the withdrawal time and limiting the pCOD/VSS applied to the systems the TSS in the effluent improved.

Val del Río *et al.* (2012) observed that the maximum OLR treated with stable granular biomass was 4 kg-COD/m^3·d. Schwarzenbeck *et al.* (2004) found that process efficiency, in terms of volume exchange ratio (VER) and cycle duration, was limited to the development of adequate feast and famine conditions. The presence of slowly biodegradable organic carbon led to extended feast phases since endogenous respiration rates were not observed during a single SBR cycle. Thus, granules always showed structural deficiencies at high OLR due to filamentous outgrowths on aerobically hydrolysed organic matter. Decreasing the VER, extending the total duration of the SBR cycle, as well as incorporating an anaerobic hydrolysis period prior the AGS reactor could therefore enhance process efficiency of high TSS wastewaters.

4.5.1.3 Fish-canning wastewater

Wastewater generated by fish-canning industries is characterized by high salinity (i.e., up to 150 g-NaCl/L), high COD, and TN loads, as well as by a large amount of TSS. Biological treatment of fish-canning wastewater must therefore resist osmotic pressures, generated by high salinity, on cellular membranes that could cause plasmolysis. Capodici *et al.* (2018) stated that shortcut nitrification-denitrification could be the most effective solution for nitrogen removal in saline wastewater. The operational approaches that are generally implemented to inhibit the NOB activity, including high temperature, low DO, decoupled SRT, and high free ammonia concentration are not necessary since salt concentration higher than 20 g-NaCl/L is sufficient for NOB inhibition.

Figueroa *et al.* (2008) treated fish-canning wastewater with up to 30 g-NaCl/L. The AGS withstood the saline conditions of the saline wastewater and achieved complete organic matter removal with the presence of nitritation-denitritation activities. The ammonia was mainly oxidized to nitrite and the efficiency of the denitritation process was limited to values between 20% and 55%. Corsino *et al.* (2016a) successfully achieved halotolerant aerobic granular sludge by increasing the salinity in the influent wastewater stepwise. The authors observed that high salt concentrations (i.e., >20 g-NaCl/L) led to the accumulation of nitrite due to the inhibition of NOB. The same authors reported that simultaneous nitritation-denitritation was successfully sustained at salinities up to 50 g-NaCl/L. The total nitrogen concentration in the effluent was less than 10 mg/L at salinities up to 50 g-NaCl/L, revealing a removal efficiency over 95%. However, nitritation efficiency collapsed above 50 g-NaCl/L, likely because oxygen diffusion within the liquid bulk was limited by salt concentration. In contrast, organic matter removal was not affected by salinity, but rather by the organic loading rate. Both COD and BOD removal efficiencies were over 90%, and specific analyses carried out on the treated wastewater revealed that aerobic granules were able to remove more than 95% of the particulate organic matter. The same authors noted that salinity significantly influenced the physical properties of the aerobic granules (Corsino *et al.*, 2017). Particularly, with increasing salinity the EPS structure was gradually modified. The fraction of EPS that was not bound increased with salinity, while the amount of proteinaceous EPS decreased. The modification of the physical structure was more noticeable at salinity above 50 g-NaCl/L, when the granules irreparably lost their stability.

4.5.1.4 Palm oil wastewater

Palm oil production has increased rapidly because of its multiple uses in both food and other industries, especially in the South-East Asia. Direct discharge of palm oil mill effluent into the environment is forbidden due to the high COD concentrations of up to 50,000 mg/L, typically with 50% as BOD. Abdullah *et al.* (2013) successfully cultivated AGS at different OLRs ranging between 1.5–3.5 kg-COD/m^3·d. Different volumetric loading rates led to differences in morphology and structural features of the AGS, in which higher loading rates promoted the formation of larger and looser granules. An interesting observation was that low bacterial speciation was observed in the reactor with high OLR. The bacterial population in the aerobic granular sludge moved away from its initial population representing a permanent change. The authors stated that the shift in microbial population was strictly related to operating conditions. Indeed, it is likely that at high organic loads, the greatest demand for electron acceptors is such that fast growing heterotrophic organisms have many competitive advantages over slower growing autotrophic aerobes.

4.5.1.5 Winery wastewater

López-Palau *et al.* (2012) treated winery wastewater in an AGS SBR. Winery wastewater is characterized by a high variability of its organic content. The authors, in order to take into account the fluctuations in organic load, developed a system to automatically control the HRT based on the anticipated organic load. The relationship between the slope of ORP and DO profiles was used to apply a proper ratio of feast to famine conditions by changing the HRT. At higher loads, a longer HRT was applied in order to develop an adequate famine phase for the longer feast phase applied. The authors considered that optimal conditions were achieved when the famine phase was twice as long as the feast phase (i.e., famine phase of 67% of the aerated period). Under such operating conditions, the AGS SBR reached 95% removal of organic matter under a wide range of organic matter concentrations (i.e., 1 to 8 g-COD/L). Furthermore, the AGS SBR showed high stability and a rapid response to sudden changes in the COD influent concentration.

4.5.2 Petrochemical and oily wastewater

Petrochemical and oily wastewaters are characterized by poor biodegradability due to the presence of hydrocarbons, metal salts, sulphides, phenol, and other substances. Zhang *et al.* (2011) gradually adapted aerobic granules, previously cultivated with synthetic wastewater, to petrochemical wastewater by gradually increasing the proportion of petrochemical wastewater in the influent. Following the addition of petrochemical wastewater, granule settleability became poor and the EPS matrix began to weaken. When petrochemical wastewater was 100% granule settleability and removal of pollutants deteriorated sharply. The authors observed that adding additional co-metabolites (e.g., propionate) to stimulate the degradation pathways was an effective way to ensure effective removal of nutrients and maintain a compact structure of aerobic granules. Corsino *et al.* (2015) observed that biomass fed with saline oily wastewater generated from the washing of oil tankers (i.e., slop) granulated faster compared with conventional synthetic wastewater. However, the granules appeared slightly unstable and more susceptible to breaking. The combined effect of salinity and hydrocarbons caused the inhibition of the autotrophic biomass, with the consequence that nitrification was absent. Hydrocarbons were removed by a physical-biological mechanism, involving an initial adsorption stage followed by the biological degradation of adsorbed hydrocarbons. The adsorption of hydrocarbons within the pores of granules provided longer contact time between microorganisms and pollutant, thereby ensuring the degradation of recalcitrant compounds. Total petroleum hydrocarbons were successfully removed at steady state with removal of over 90%.

Milia *et al.* (2016) evaluated the feasibility of treating petrochemical wastewater characterized by high TOC and ammonium nitrogen concentrations (i.e., up to 490 and 630 mg/L, respectively). The authors observed that aerobic granules cultivated from non-acclimated biomass irreversibly deteriorated when the percentage of petrochemical wastewater in the influent was raised to 30%. A faster and more stable granulation was observed when activated sludge already acclimated to toxic substances was used as inoculum. The same authors stated that due to higher superficial air velocities and to the different sludge used as inoculum, the aerobic granules exhibited a better physical structure, as well as a higher removal efficiency in terms of nitrogen and total organic carbon. In a recent study, Caluwé *et al.* (2017) analysed two different reactor configurations to treat petrochemical wastewater by means of AGS. Particularly, a completely aerated SBR with a feast/famine regime and an SBR operated with an anaerobic feast/aerobic famine strategy. The authors observed that cultivation of aerobic granules with petrochemical wastewater was possible, confirming what was previously observed by Corsino *et al.* (2015). Furthermore, the feeding regime did not affect the granule formation.

4.5.3 Landfill leachate

Landfill leachate is one of the most significant contamination problems at landfill sites. Leachate contains salts, heavy metals, organic matters, and xenobiotic organic compounds (Renou *et al.,* 2008). In addition to xenobiotic organic compounds, landfill leachate is characterized by high ammonium nitrogen loads (Varank *et al.,* 2011). An AGS SBR was studied for landfill leachate treatment by Wei *et al.* (2012). Their results demonstrated that COD removal rates decreased as the influent ammonium concentration increased. Concerning nitrogen, it was found that the removal mechanism depended on the influent concentration. When the total ammonium nitrogen (TAN) concentration in the landfill leachate was 366 mg-NH_4-N/L, the dominant nitrogen removal process in the AGS SBR was SND. Under ammonium concentrations of 788 mg-NH_4-N/L, nitrite accumulation occurred and the accumulated nitrite was reduced to nitrogen gas by the shortcut denitrification process. When the influent TAN increased to a higher level of 1105 mg-NH_4-N/L, accumulation of nitrite and nitrate persisted throughout the cycle, and the removal efficiencies of ammonium decreased to 40%. Similar results were also obtained by Di Bella and Torregrossa (2014), which pointed out the decisive role of ammonium concentration. In particular, the organic and nitrogen removal efficiencies decreased when ammonium concentration increased. For this reason, a specific ammonium pretreatment must be applied before AGS application. Furthermore, the authors emphasized the role of the start-up phase, suggesting that the initial cultivation of granules should be carried out with small increments of leachate, in order to optimize the bacteria acclimation and specification. Yuan *et al.* (2016) demonstrated that AGS was able to treat a mixture of wastewater with landfill leachate at high ammonium concentration of 900 mg-NH_4-N/L. The COD removal was highly dependent on the ammonium concentration in the influent and on the biodegradability of the organic matter.

4.5.4 Wastewater contaminated by emerging micropollutants

Many chemicals used in everyday pharmaceutical and personal care products (PPCPs) are among the most commonly detected compounds in surface waters throughout the world. Such compounds may have significant biological effects, such as enzyme inhibitors, modifiers of cell to cell communication, disrupters of membrane transport, and processes for energy generation (Duque *et al.*, 2011). Continued exposure to antibiotics can enhance the selection of resistant bacterial strains in the environment (Kümmerer, 2009). AGS was successfully used for the treatment of aqueous streams containing antibiotics and several PPCPs. Duque *et al.* (2011) demonstrated that conventional AGS was not able to remove 2-fluorophenols. Applying bioaugmentation with a specialized bacterial strain able to degrade 2-fluorophenols resulted in full degradation of the compound by the AGS. Overall, the AGS SBR was proven to be robust, exhibiting a high performance after bioaugmentation with the 2-fluorophenols degrading strain. Amorim *et al.* (2013) studied the performances of an AGS reactor treating synthetic wastewater containing fluoroquinolones, a class of antibiotics. No evidence of fluoroquinolones biodegradation was observed, but the compound did adsorb to the AGS and was gradually released into the medium in successive cycles after fluoroquinolones were removed from the influent. Overall, neither COD nor ammonium removal were affected during the shock loadings.

Moreira *et al.* (2015) investigated the effects of fluoxetine on an AGS reactor and noted that fluoxetine adsorption/desorption to granules continuously occurred. Fluoxetine shock loads did not show a significant effect on the removal of COD or ammonium, whereas a significant decrease in denitrification removal efficiency was observed. Moreover, a shift in microbial community occurred probably due to fluoxetine exposure, which induced adaptation/restructuring of the microbial population. Similarly, Zhao *et al.* (2015) noted changes in the microbial profiles with time in response to the presence of PPCPs. Initially, PPCP addition negatively affected granular quality, but after microbial adaptation, the system recovered and effectively removed certain PPCPs.

4.6 AEROBIC GRANULAR SLUDGE IN CONTINUOUS FLOW REACTORS

One of the more recent areas of AGS research focuses on applications in continuous flow configurations. Although the Nereda® technology solves the issue of continuous discharge and fixed liquid level by having at least three SBRs in parallel with simultaneous influent distribution and effluent discharge, it does not provide a solution to existing plug-flow configurations; unless significant concrete work or greenfield development were to be done. In this section the literature and prospects for granulation in continuous plug-flow systems is discussed.

4.6.1 Operation under continuous flow

Operational considerations for continuous plug-flow aerobic granular sludge are essentially the same as considerations for operation in SBRs. There should be sufficient anaerobic contact of fresh wastewater with biomass, promoting as much anaerobic uptake of soluble organics as possible. For the same reason, pretreatment methods to remove particulate matter should be considered such that the characteristics of wastewater supplied to the aerobic granular sludge process are able to be converted to readily bioavailable substrate and converted to internal storage products by accumulating microorganisms.

Granulation of biomass in a continuous flow would require representation of what is known from granulation in SBRs, as well as from the full-scale Nereda® process. It has widely been suggested that SBRs, which cycle from high to low substrate concentrations, benefit the formation of AGS (Pronk *et al.*, 2015b). Due to the nature of plug-flow, where a mass of mixed liquor travels along the length of the reactor, the changing substrate gradients with reactor length would be representative of SBR operation. However, conventional plug-flow BNR processes have numerous recycle lines (i.e., at least one internal recycle and return activated sludge) which dilute substrate concentrations and therefore repress the substrate gradient experienced by the biomass. It would therefore be important to minimize the number of recycle lines as well as the quantity of flow recycled (i.e., no internal recycle, and return activated sludge no more than 70% of the influent flow).

It is well known that the breakthrough of organic substrate in bulk solution to aerobic conditions negatively impacts AGS stability in SBRs. This could be due to many reasons, including the overgrowth of filamentous organisms on the surface of granules or an accumulation of particulate matter. Both cases lead to mass transfer limitations and therefore increase the potential for granule disintegration (de Kreuk *et al.*, 2010; Wagner *et al.*, 2015). Thus, it would be important to consider designing the plug-flow configuration such that maximum anaerobic utilization of substrate occurred. This could be done by increasing the volume of the anaerobic zone to account for up to one third of the reactor volume. Under such conditions the hydrolysis of complex biodegradable influent substrate, as well any residual particulate matter carried over from the return activated sludge line, would be enhanced leading to more anaerobic substrate utilization. Overall, anaerobic substrate utilization promotes uniform growth across the granule and contributes to lower sludge yields.

Allowing for transition of anaerobic contact to an aerobic zone with high oxygen uptake rate and then a second aerobic zone with low oxygen uptake rate, under all influent variability, should promote granulation, assuming that a bulk of the organic carbon is converted to internal storage products in the anaerobic zone. The purpose would be to prevent conditions where fast growing aerobic heterotrophs would be allowed to proliferate in the bulk solution, which could lead to operational problems. As well, the transition from a high-rate aerobic zone to a low-rate aerobic zone allows the microbial community to cycle through a phase of EPS production followed by EPS consumption. Therefore, continuous stirred tank reactor conditions should be avoided in continuous design. Rather, plug-flow or a series of at least two aerobic tanks would be desired.

To move from a hybrid-AGS system (i.e., flocculent biomass and granules) to a completely granular system there must be some degree of selective wasting. Essentially, the SRT of the light biomass fractions (i.e., free-living cells, flocculent biomass) must be decoupled from the SRT of granules and controlled at completely different ends of the spectrum. At 20°C the minimum SRT for aerobic heterotrophic organisms approaches 9 h (Envirosim, 2017). Thus, to effectively out-select all free-living heterotrophic organisms in suspension, the aerobic SRT of the light biomass fraction would have to be 9 h at 20°C. In practice, this may be too aggressive an SRT and it is more likely that the range of SRT for the selectively wasted light fraction of biomass would be 1 to 4 d. On the other hand, previous lab-scale results as well as modelling estimations have demonstrated that the optimum SRT of granules should be controlled at 20 to 50 d. At long granule SRT desirable metabolic pathways with low yields are selected for, as well as a higher fraction of PAO to OHO (Pronk *et al.*, 2015a; Devlin *et al.*, 2017).

The selective wasting device should be designed with flexibility to allow for diurnal variation in flow, as well as changing flows as population in the sewer catchment increases. For this reason, hydrocyclones are not the most desirable device since the selection pressure in a single hydrocyclone would change with flow. Likewise, physical screening (i.e., 0.2–0.5 mm) is unlikely to provide a robust solution, since granules may be subjected to breakage and not necessarily complete disintegration. A much better selective wasting device would decouple the selective wasting criteria from flowrate such that additional operational parameters could be controlled to add system flexibility.

The final consideration for AGS in continuous plug-flow would be the degree of variability allowed. It is well known that wastewater varies with season as well as time of day. In fact, in smaller communities the peak hour flow can be three times the average daily flow. With variable flow also comes variable load. Since AGS is sensitive to the breakthrough of organic substrate to the aerobic zone, design should consider minimizing the risk of extended "feast" phases (Corsino *et al.*, 2016c).

4.6.2 Current designs and outlook for the future

Demonstration facilities for aerobic granular sludge have included innovative designs, such as airlift membrane bioreactors, as well as simple modifications to existing continuous flow processes in order to selectively waste free-living microorganisms that make up the lower density fraction of biomass (Corsino *et al.*, 2016a). For instance, both screens and hydrocyclones have been applied to the biomass wasting stream of continuous flow processes. Screens act to physically retain large particles from waste biomass while hydrocyclones remove biomass in the overflow and retain dense biomass in the underflow (Winckel *et al.*, 2016). Screens are problematic for a number of reasons. For instance, screens would not be effective during the granulation process since there would be no large granules to selectively retain. Furthermore, selection via a single size may pose a problem if process perturbations were to occur, resulting in degranulation and subsequently a washout of biomass via the waste stream.

Hydrocyclones have also proven to be inefficient in some cases, particularly where process influent has contained excessive grit or other dense, inorganic particles (Ford *et al.*, 2016). Without proper removal of dense inorganic matter up front, hydrocyclones essentially work to retain material like grit. Under these conditions the fraction of grit and other inorganic material can quickly become a significant fraction of solids in the reactor. Furthermore, there is a general consideration to be made for both screens and hydrocyclones on the waste stream. In order to effectively prevent the development of a fast growing aerobic heterotrophic population the sludge wasting rate must exceed the minimum solids retention time of aerobic heterotrophs. Otherwise, the wasting system will act to select for a high-rate biomass instead of granular sludge alone (Isanta *et al.*, 2015).

In general, there is a need to develop more flexible technology for selective wasting of the light fraction of biomass. New selective wasting technologies will need to be developed in order to address flow and load variability on a daily, seasonal, as well as design life basis. The implementation of properly designed selective wasting devices will bolster the anaerobic uptake capabilities of the AGS process in continuous plug-flow by minimizing the quantity of the light fraction of biomass. However, plug-flow configurations must also be rearranged in order to minimize recycle lines and prevent the dilution effects observed in conventional plug-flow BNR systems. Determining proper design parameters, such as return activated sludge rates, anaerobic tank sizes, and aeration strategies to optimize anaerobic utilization of substrate are necessary for energy efficient AGS development in continuous flow. Designers and operators should keep in mind that most of the advantages of AGS come from carbon efficient metabolic pathways, and that operation to minimize aerobic utilization of substrate should benefit both AGS, as well as hybrid floc-AGS systems.

4.7 CONCLUSION

Aerobic granular sludge (AGS) technology offers several advantages over conventional activated sludge (CAS) systems. AGS attains comparable or better treatment than CAS within 70% of the footprint, while achieving a 30% to 60% reduction in energy requirements. Substrate gradients that develop within the granules promote the development of a rich and varied microbial community, which in turn allows for simultaneous nutrient removal in a single tank.

Full-scale implementations of AGS have already demonstrated its effectiveness and feasibility. Retrofits of existing CAS systems with AGS technology have attained stringent effluent quality, especially concerning nitrogen and phosphorus removal.

Interest in AGS is also generated by the encouraging results obtained in the treatment of industrial wastewaters. Since the granules can tolerate and degrade toxic organic compounds, AGS is suited for the treatment of a wide variety of industrial wastewaters. Therefore, properly designed AGS systems could be well-suited for full-scale implementations at industrial facilities containing compounds of concern.

Long-term stability of granules remains the current bottleneck of AGS technology. Different operating parameters seem to be more significant for long-term granule stability. Well defined design and operating parameters, in the light of available information from the literature, would enable stable granulation during long-term operation and instil confidence in design engineers who are responsible for the future of wastewater treatment.

4.8 REFERENCES

Abdullah N., Yuzir A., Curtis T. P., Yahya A. and Ujang Z. (2013). Characterization of aerobic granular sludge treating high strength agro-based wastewater at different volumetric loadings. *Bioresource Technology*, **127**, 181–187. doi: 10.1016/j.biortech.2012.09.047

Ab Halim M. H., Nor Anuar A., Abdul Jamal N. S., Azmi S. I., Ujang Z. and Bob M. M. (2016). Influence of high temperature on the performance of aerobic granular sludge in biological treatment of wastewater. *Journal of Environmental Management*, **184**, 271–280. doi: 10.1016/j.jenvman.2016.09.079

Adav S. S. and Lee D.-J. (2008a). Extraction of extracellular polymeric substances from aerobic granule with compact interior structure. *Journal of Hazardous Materials*, **154**, 1120–1126.

Adav S. S. and Lee D.-J. (2008b). Extraction of extracellular polymeric substances from aerobic granule with compact interior structure. *Journal of Hazardous Materials*, **154**, 1120–1126. doi: 10.1016/j.jhazmat.2007.11.058

Adav S. S., Lee D.-J. and Lai J. Y. (2007). Effects of aeration intensity on formation of phenol-fed aerobic granules and extracellular polymeric substances. *Applied Microbiology and Biotechnology*, **77**, 175–182.

Adav S. S., Lee D.-J. and Tay J.-H. (2008). Extracellular polymeric substances and structural stability of aerobic granule. *Water Research*, **42**, 1644–1650. doi: 10.1016/j.watres.2007.10.013

Adav S. S., Lee D.-J. and Lai J.-Y. (2009). Aerobic granulation in sequencing batch reactors at different settling times. *Bioresource Technology*, **100**, 5359–5361. doi: 10.1016/j.biortech.2009.05.058

Adav S. S., Lee D.-J. and Lai J.-Y. (2010). Potential cause of aerobic granular sludge breakdown at high organic loading rates. *Applied Microbiology and Biotechnology*, **85**, 1601–1610. doi: 10.1007/s00253-009-2317-9

Amorim C. L., Maia A. S., Mesquita R. B. R., Rangel O. S. S., Van Loosdrecht M. C. M., Elizabeth M. and Castro P. M. L. (2013). Sciencedirect performance of aerobic granular sludge in a sequencing batch bioreactor exposed to ofloxacin, norfloxacin and ciprofloxacin. *Water Research*, **50**, 101–113. doi: 10.1016/j.watres.2013.10.043

Angela M., Béatrice B. and Mathieu S. (2011). Biologically induced phosphorus precipitation in aerobic granular sludge process. *Water Research*, **45**, 3776–3786. doi: 10.1016/j.watres.2011.04.031

Arrojo B., Mosquera-Corral A., Garrido J. M. and Méndez R. (2004). Aerobic granulation with industrial wastewater in sequencing batch reactors. *Water Research*, **38**, 3389–3399. doi: 10.1016/j.watres.2004.05.002

Barr J. J., Dutilh B. E., Skennerton C. T., Fukushima T., Hastie M. L., Gorman J. J., Tyson G. W. and Bond P. L. (2016). Metagenomic and metaproteomic analyses of accumulibacter phosphatis-enriched floccular and granular biofilm. *Environmental Microbiology*, **18**, 273–287. doi: 10.1111/1462-2920.13019

Bassin J. P., Pronk M., Kraan R., Kleerebezem R. and Van Loosdrecht M. C. M. (2011). Ammonium adsorption in aerobic granular sludge, activated sludge and anammox granules. *Water Research*, **45**, 5257–5265. doi: 10.1016/j.watres.2011.07.034

Bassin J. P., Kleerebezem R., Dezotti M. and van Loosdrecht M. C. M. (2012a). Measuring biomass specific ammonium, nitrite and phosphate uptake rates in aerobic granular sludge. *Chemosphere*, **89**, 1161–1168. doi: 10.1016/j.chemosphere.2012.07.050

Bassin J. P., Kleerebezem R., Dezotti M. and van Loosdrecht M. C. M. (2012b). Simultaneous nitrogen and phosphate removal in aerobic granular sludge reactors operated at different temperatures. *Water Research*, **46**, 3805–3816. doi: 10.1016/j.watres.2012.04.015

Bassin J. P., Winkler M. -K. H., Kleerebezem R., Dezotti M. and van Loosdrecht M. C. M. (2012c). Improved phosphate removal by selective sludge discharge in aerobic granular sludge reactors. *Biotechnology and Bioengineering*, **109**, 1919–1928.

Beun J. J., Hendriks A., Van Loosdrecht M. C. M., Morgenroth E., Wilderer P. A. and Heijnen J. J. (1999). Aerobic granulation in a sequencing batch reactor. *Water Research*, **33**, 2283–2290. doi: 10.1016/S0043-1354(98)00463-1

Beun J. J., Van Loosdrecht M. C. M. and Heijnen J. J. (2002). Aerobic granulation in a sequencing batch airlift reactor. *Water Research*, **36**, 702–712. doi: 10.1016/S0043-1354(01)00250-0

Bruce S. C. R., Downing L., Young M. and Nerenberg R. (2014). Floc or granule? evidence of granulation in a continuous flow system. *Water Environment Federation*. Proceedings of the Water Environment Federation, WEFTEC 2014: Session 202 through Session 209, 2891–2897(7). doi: 10.2175/193864714815942503.

Caluwé M., Dobbeleers T., D'aes J., Miele S., Akkermans V., Daens D., Geuens L., Kiekens F., Blust R. and Dries J. (2017). Formation of aerobic granular sludge during the treatment of petrochemical wastewater. *Bioresource Technology*, **238**, 559–567. doi: 10.1016/j.biortech.2017.04.068

Cao Y., van Loosdrecht M. C. M. and Daigger G. T. (2017). Mainstream partial nitritation–anammox in municipal wastewater treatment: status, bottlenecks, and further studies. *Applied Microbiology and Biotechnology*, **101**, 1365–1383. doi: 10.1007/s00253-016-8058-7

Capodici M., Corsino S. F., Torregrossa M. and Viviani G. (2018). Shortcut nitrification-denitrification by means of autochthonous halophilic biomass in an SBR treating fish-canning wastewater. *Journal of Environmental Management*, **208**, 142–148. doi: 10.1016/j.jenvman.2017.11.055

Carucci A., Milia S., Cappai G. and Muntoni A. (2010). A direct comparison amongst different technologies (aerobic granular sludge, SBR and MBR) for the treatment of wastewater contaminated by 4-chlorophenol. *Journal of Hazardous Materials*, **177**, 1119–1125. doi: 10.1016/j.jhazmat.2010.01.037

Chen M. Y., Lee D.-J. and Tay J.-H. (2007a). Distribution of extracellular polymeric substances in aerobic granules. *Applied Microbiology and Biotechnology*, **73**, 1463–1469. doi: 10.1007/s00253-006-0617-x

Chen Y., Jiang W., Liang D. T. and Tay J.-H. (2007b). Structure and stability of aerobic granules cultivated under different shear force in sequencing batch reactors. *Applied Microbiology and Biotechnology*, **76**, 1199–1208. doi: 10.1007/s00253-007-1085-7

Chen Y., Jiang W., Liang D. T. and Tay J.-H. (2008). Aerobic granulation under the combined hydraulic and loading selection pressures. *Bioresource Technology*, **99**, 7444–7449.

Chen X., Guo J., Shi Y., Hu S., Yuan Z. and Ni B. (2014). Modeling of simultaneous anaerobic methane and ammonium oxidation in a membrane bio film reactor. *Environmental Science & Technology*, **48**(16), 9540–9547. doi: 10.1021/es502608s

Corsino S. F., Campo R., Di Bella G. and Torregrossa M. (2015). International biodeterioration & biodegradation cultivation of granular sludge with hypersaline oily wastewater. *International Biodeterioration & Biodegradation*, **105**, 192–202. doi: 10.1016/j.ibiod.2015.09.009

Corsino S. F., Campo R., Di Bella G., Torregrossa M. and Viviani G. (2016a). Study of aerobic granular sludge stability in a continuous-flow membrane bioreactor. *Bioresource Technology*, **200**, 1055–1059. doi: 10.1016/j.biortech.2015.10.065

Corsino S. F., Capodici M., Morici C., Torregrossa M. and Viviani G. (2016b). Simultaneous nitritation–denitritation for the treatment of high-strength nitrogen in hypersaline wastewater by aerobic granular sludge. *Water Research*, **88**, 329–336. doi: 10.1016/j.watres.2015.10.041

Corsino S. F., Capodici M., Torregrossa M. and Viviani G. (2016c). Fate of aerobic granular sludge in the long-term: the role of EPSs on the clogging of granular sludge porosity. *Journal of Environmental Management*, **183**, 541–550. doi: 10.1016/j.jenvman.2016.09.004

Corsino S. F., Capodici M., Torregrossa M. and Viviani G. (2017a). Physical properties and extracellular polymeric substances pattern of aerobic granular sludge treating hypersaline wastewater. *Bioresourse Technology*, **229**, 152–159. doi: http://doi.org/10.1016/j.biortech.2017.01.024

Corsino S. F., di Biase A., Devlin T. R., Munz G., Torregrossa M. and Oleszkiewicz J. A. (2017b). Effect of extended famine conditions on aerobic granular sludge stability in the treatment of brewery wastewater. *Bioresource Technology*, **226**, 150–157. doi: 10.1016/j.biortech.2016.12.026

Dangcong P., Bernet N., Delgenes J. P. and Moletta R. (1999). Aerobic granular sludge-a case report. *Water Research*, **33**, 890–893.

de Bruin L. M. M., de Kreuk M. K., van der Roest H. F. R., Uijterlinde C. and van Loosdrecht M. C. M. (2004). Aerobic granular sludge technology: an alternative to activated sludge? *Water Science and Technology*, **49**, 1–7.

de Kreuk M. K. and van Loosdrecht M. C. M. (2004). Selection of slow growing organisms as a means for improving aerobic granular sludge stability. *Water Science and Technology*, **49**, 9–17.

de Kreuk M. K., Pronk M. and van Loosdrecht M. C. M. (2005). Formation of aerobic granules and conversion processes in an aerobic granular sludge reactor at moderate and low temperatures. *Water Research*, **39**, 4476–4484. doi: 10.1016/j.watres.2005.08.031

de Kreuk M. K., Heijnen J. J. and Van Loosdrecht M. C. M. (2005a). Simultaneous COD, nitrogen, and phosphate removal by aerobic granular sludge. *Biotechnology and Bioengineering*, **90**, 761–769. doi: 10.1002/bit.20470

de Kreuk M. K., McSwain B. S., Bathe S., Tay S. T. L. and Schwarzenbech N. (2005b). *Aerobic Granular Sludge*. Selected proceedings of the 1st IWA-workshop Aerobic Granular Sludge organised by the Institute of Water Quality Control and Waste Management of the Technical University of Munich (TUM). Water and Environmental Management Series. IWA Publishing. ISBN: 1-84339-509-6, ISSN: 1476–1785

de Kreuk M. K., Kishida N., Tsuneda S. and van Loosdrecht M. C. M. (2010). Behavior of polymeric substrates in an aerobic granular sludge system. *Water Research*, **44**, 5929–5938. doi: 10.1016/j.watres.2010.07.033

Derlon N., Wagner J., da Costa R. H. R. and Morgenroth E. (2016). Formation of aerobic granules for the treatment of real and low-strength municipal wastewater using a sequencing batch reactor operated at constant volume. *Water Research*, **105**, 341–350. doi: 10.1016/j.watres.2016.09.007

Devlin T. R., di Biase A., Kowalski M. and Oleszkiewicz J. A. (2017). Granulation of activated sludge under low hydrodynamic shear and different wastewater characteristics. *Bioresource Technology*, **224**, 1–7. doi: 10.1016/j.biortech.2016.11.005

Di Bella G. and Torregrossa M. (2014). Aerobic granular sludge for leachate treatment. *Chemical Engineering Transactions*, **38**, 493–498. doi: 10.3303/CET1438083

di Biase A., Devlin T. R., Kowalski M. S. and Oleszkiewicz J. A. (2017). Performance and design considerations for an anaerobic moving bed biofilm reactor treating brewery wastewater: impact of surface area loading rate and temperature. *Journal of Environmental Management*, **216**, 392–398. doi: 10.1016/j.jenvman.2017.05.093

Ding Z., Bourven I., Guibaud G., van Hullebusch E. D., Panico A., Pirozzi F. and Esposito G. (2015). Role of extracellular polymeric substances (EPS) production in bioaggregation: application to wastewater treatment. *Applied Microbiology and Biotechnology*, **99**, 9883–9905. doi: 10.1007/s00253-015-6964-8

Dulekgurgen E., Artan N., Orhon D. and Wilderer P. A. (2008). How does shear affect aggregation in granular sludge sequencing batch reactors? relations between shear, hydrophobicity, and extracellular polymeric substances. *Water Science and Technology*, **58**, 267–276. doi: 10.2166/wst.2008.382

Duque A. F., Bessa V. S., Carvalho M. F., de Kreuk M. K., van Loosdrecht M. C. M. and Castro P. M. L. (2011). 2-Fluorophenol degradation by aerobic granular sludge in a sequencing batch reactor. *Water Research*, **45**, 6745–6752. doi: 10.1016/j.watres.2011.10.033

Envirosim (2017). BioWin 5.0.

Figueroa M., Mosquera-Corral A., Campos J. L. and Méndez R. (2008). Treatment of saline wastewater in SBR aerobic granular reactors. *Water Science and Technology*, **58**, 479. doi: 10.2166/wst.2008.406

Fillaudeau L., Blanpain-Avet P. and Daufin G. (2006). Water, wastewater and waste management in brewing industries. *Journal of Cleaner Production*, **14**, 463–471. doi: 10.1016/j.jclepro.2005.01.002

Flemming H. C. and Wingender J. (2003). Extracellular polymeric substances (EPS): structural, ecological and technical aspects. *Book Chapter in: Encyclopedia of Environmental Microbiology*. doi: 10.1002/0471263397.env292

Ford A., Rutherford B., Wett B. and Bott C. (2016). Implementing Hydrocyclones in Mainstream Process for Enhancing Biological Phosphorus Removal and Increasing Settleability through Aerobic Granulation. *Proceedings of WEFTEC 2016*, Water Environment Federation, New Orleans, Louisiana, USA, 2016, pp. 2809–2822.

Gao M., Diao M.-H., Yuan S., Wang Y.-K., Xu H. and Wang X.-H. (2017). Effects of phenol on physicochemical properties and treatment performances of partial nitrifying granules in sequencing batch reactors. *Biotechnol Reports*, **13**, 13–18. doi: 10.1016/j.btre.2016.12.002

Giesen A. and Thompson A. (2013b). Aerobic granular biomass for cost-effective, energy efficient and sustainable wastewater treatment. *Proceedings in: 7th European Waste Water Management Conference*, Manchester, England, 14th–15th October, 2013.

Giesen A., de Bruin L. M. M., Niermans R. P. and ven der Roest H. F. (2013a). Advancements in the application of aerobic granular biomass technology for sustainable treatment of wastewater. *Water Practice and Technology*, **8**, 47–54.

Giesen A., van Loosdrecht M., de Bruin B., van der Roest H. and Pronk M. (2013c). Full-scale Experiences with Aerobic Granular Biomass Technology for Treatment of Urban and Industrial Wastewater. *Proceedings in: International Water Week*, Singapore, 2nd–4th June, 2013.

Henriet O., Meunier C., Henry P. and Mahillon J. (2016). Improving phosphorus removal in aerobic granular sludge processes through selective microbial management. *Bioresource Technology*, **211**, 298–306. doi: 10.1016/j.biortech.2016.03.099

Hosie A. H. F. and Poole P. S. (2001). Bacterial ABC transporters of amino acids. *Research in Microbiology*, **152**, 259–270. doi: 10.1016/S0923-2508(01)01197-4

Hu Z., Ferraina R. A., Ericson J. F., MacKay A. A. and Smets B. F. (2005). Biomass characteristics in three sequencing batch reactors treating a wastewater containing synthetic organic chemicals. *Water Research*, **39**, 710–720. doi: 10.1016/j.watres.2004.11.018

Huang W., Huang W., Li H., Lei Z., Zhang Z., Tay J. H. and Lee D. J. (2015). Species and distribution of inorganic and organic phosphorus in enhanced phosphorus removal aerobic granular sludge. *Bioresource Technology*, **193**, 549–552. doi: 10.1016/j.biortech.2015.06.120

Inocêncio P., Coelho F., van Loosdrecht M. C. M. and Giesen A. (2013). The future of sewage treatment: nereda technology exceeds high expectations. *Water*, **21**, 28–29.

Isanta E., Reino C., Carrera J. and Perez J. (2015). Stable partial nitritation for low-strength wastewater at low temperature in an aerobic granular reactor. *Water Research*, **80**, 149–158. doi: 10.1016/j.watres.2015.04.028

Jiang H. L., Tay J. H. and Tay S. T. L. (2002). Aggregation of immobilized activated sludge cells into aerobically grown microbial granules for the aerobic biodegradation of phenol. *Letters in Applied Microbiology*, **35**, 439–445. doi: 10.1046/j.1472-765X.2002.01217.x

Juang Y. C., Adav S. S., Lee D. J. and Tay J. H. (2010). Stable aerobic granules for continuous-flow reactors: precipitating calcium and iron salts in granular interiors. *Bioresource Technology*, **101**, 8051–8057. doi: 10.1016/j.biortech.2010.05.078

Kim I. S., Kim S. M. and Jang A. (2008). Characterization of aerobic granules by microbial density at different COD loading rates. *Bioresource Technology*, **99**, 18–25. doi: 10.1016/j.biortech.2006.11.058

Kümmerer K. (2009). Antibiotics in the aquatic environment – A review – part I. *Chemosphere*, **75**, 417–434. doi: 10.1016/j.chemosphere.2008.11.086

Kwok W. K., Picioreanu C., Ong S. L., van Loosdrecht M. C. M., Ng W. J. and Heijnen J. J. (1998). Influence of biomass production and detachment forces on biofilm structures in a biofilm airlift suspension reactor. *Biotechnology and Bioengineering*, **58**(4), 400–407.

Lashkarizadeh M., Munz G. and Oleszkiewicz J. A. (2016). Impacts of variable pH on stability and nutrient removal efficiency of aerobic granular sludge. *Water Science and Technology*, **73**, 60–68. doi: 10.2166/wst.2015.460

Lee D.-J., Chen Y.-Y., Show K.-Y., Whiteley C. G. and Tay J.-H. (2010). Advances in aerobic granule formation and granule stability in the course of storage and reactor operation. *Biotechnology Advances*, **28**, 919–934. doi: 10.1016/j.biotechadv.2010.08.007

Lemaire R., Webb R. I. and Yuan Z. (2008). Micro-scale observations of the structure of aerobic microbial granules used for the treatment of nutrient-rich industrial wastewater. *ISME Journal*, **2**, 528–541. doi: 10.1038/ismej.2008.12

Leong J., Rezania B. and Mavinic D. S. (2016). Aerobic granulation utilizing fermented municipal wastewater under low pH and alkalinity conditions in a sequencing batch reactor. *Environmental Technology*, **37**, 55–63. doi: 10.1080/09593330.2015.1063704

Lettinga G., Rebac S. and Zeeman G. (2001). Challenge of psychrophilic anaerobic wastewater treatment. *Trends in Biotechnology*, **19**, 363–370.

Li A., Yang S., Li X. and Gu J. (2008). Microbial population dynamics during aerobic sludge granulation at different organic loading rates. *Water Research*, **42**, 3552–3560. doi: 10.1016/j.watres.2008.05.005

Li X. M., Liu Q. Q., Yang Q., Guo L., Zeng G. M., Hu J. M. and Zheng W. (2009). Enhanced aerobic sludge granulation in sequencing batch reactor by Mg2+ augmentation. *Bioresource Technology*, **100**, 64–67.

Li A. J., Li X. Y. and Yu H. Q. (2011). Effect of the food-to-microorganism (F/M) ratio on the formation and size of aerobic sludge granules. *Process Biochemistry*, **46**, 2269–2276. doi: 10.1016/j.procbio.2011.09.007

Li X., Luo J., Guo G., Mackey H. R., Hao T. and Chen G. (2017). Seawater-based wastewater accelerates development of aerobic granular sludge: a laboratory proof-of-concept. *Water Research*, **115**, 210–219. doi: 10.1016/j.watres.2017.03.002

Lin Y., de Kreuk M., van Loosdrecht M. C. M. and Adin A. (2010). Characterization of alginate-like exopolysaccharides isolated from aerobic granular sludge in pilot-plant. *Water Research*, **44**, 3355–3364. doi: 10.1016/j.watres.2010.03.019

Lin Y. M., Bassin J. P. and Van Loosdrecht M. C. M. (2012). The contribution of exopolysaccharides induced struvites accumulation to ammonium adsorption in aerobic granular sludge. *Water Research*, **46**, 986–992. doi: 10.1016/j.watres.2011.11.072

Lin Y. M., Sharma P. K. and van Loosdrecht M. C. M. (2013). The chemical and mechanical differences between alginate-like exopolysaccharides isolated from aerobic flocculent sludge and aerobic granular sludge. *Water Research*, **47**, 57–65. doi: 10.1016/j.watres.2012.09.017

Liu Y. and Tay J. H. (2002). The essential role of hydrodynamic shear force in the formation of biofilm and granular sludge. *Water Research*, **36**, 1653–1665.

Liu Y. and Tay J. H. (2004). State of the art of biogranulation technology for wastewater treatment. *Biotechnology Advances*, **22**, 533–563.

Liu Y., Yang S.-F. and Tay J.-H. (2004). Improved stability of aerobic granules by selecting slow-growing nitrifying bacteria. *Journal of Biotechnology*, **108**, 161–169.

Liu L., Sheng G. P., Li W. W., Tong Z. H., Zeng R. J., Liu J. X., Xie J., Peng S. C. and Yu H. Q. (2011). Cultivation of aerobic granular sludge with a mixed wastewater rich in toxic organics. *Biochemical Engineering Journal*, **57**, 7–12. doi: 10.1016/j.bej.2011.07.005

Liu Y. Q., Lan G. H. and Zeng P. (2015). Excessive precipitation of $CaCO_3$ as aragonite in a continuous aerobic granular sludge reactor. *Applied Microbiology and Biotechnology*, **99**, 8225–8234. doi: 10.1007/s00253-015-6727-6

Lochmatter S. and Holliger C. (2014). Sciencedirect optimization of operation conditions for the startup of aerobic granular sludge reactors biologically removing carbon, nitrogen, and phosphorous. *Water Research*, **59**, 58–70. doi: 10.1016/j.watres.2014.04.011

López-Palau S., Pinto A., Basset N., Dosta J. and Mata-Álvarez J. (2012). ORP slope and feast-famine strategy as the basis of the control of a granular sequencing batch reactor treating winery wastewater. *Biochemical Engineering Journal*, **68**, 190–198. doi: 10.1016/j.bej.2012.08.002

Mancell-Egala A., Haydée D. C., Murthy S. and Novak J. (2014). Metrics for Settling of Flocculent and Granular Solids. In: *WEFTEC 2014*. pp. 839–846.

Martins A. M. P., Karahan Ö. and van Loosdrecht M. C. M. (2011). Effect of polymeric substrate on sludge settleability. *Water Research*, **45**, 263–273. doi: 10.1016/j.watres.2010.07.055

McSwain B. S., Irvine R. L. and Wilderer P. A. (2004). The influence of settling time on the formation of aerobic granules. *Water Science and Technology*, **50**, 195–202.

Metcalf & Eddy (2014). Wastewater Engineering: Treatment and Resource Recovery, 5th edn. McGraw-Hill, United States.

Milia S., Malloci E. and Carucci A. (2016). Aerobic granulation with petrochemical wastewater in a sequencing batch reactor under different operating conditions. *Desalin Water Treat*, **3994**, 1–10. doi: 10.1080/19443994.2016.1191778

Moreira I. S., Amorim C. L., Ribeiro A. R., Mesquita R. B. R., Rangel A. O. S. S., Van Loosdrecht M. C. M., Tiritan M. E. and Castro P. M. L. (2015). Removal of fluoxetine and its effects in the performance of an aerobic granular sludge sequential batch reactor. *Journal of Hazardous Materials*, **287**, 93–101. doi: 10.1016/j.jhazmat.2015.01.020

Morgenroth E., Sherden T., van Loosedrecht M. C. M., Heijnen J. and Wilderer P. (1997). Aerobic granular sludge in a sequencing batch reactor. *Water Research*, **31**(12), 3191–3194.

Mosquera-Corral A., de Kreuk M. K., Heijnen J. J. and van Loosdrecht M. C. M. (2005). Effects of oxygen concentration on N-removal in an aerobic granular sludge reactor. *Water Research*, **39**, 2676–2686. doi: 10.1016/j.watres.2005.04.065

Ni B.-J. and Yu H.-Q. (2010). Mathematical modeling of aerobic granular sludge: a review. *Biotechnology Advances*, **28**, 895–909. doi: 10.1016/j.biotechadv.2010.08.004

Niermans R., Giesen A., Van Loosdrecht M. C. M. and De Buin B. (2014). Full-scale Experiences with Aerobic Granular Biomass Technology for Treatment of Urban and Industrial Wastewater. In: *WEFTEC 2014*. pp. 2347–2357.

Oleszkiewicz J. A., Kruk D. and Devlin T. (2015). *Options for Improved Nutrient Removal and Recovery From Municipal Wastewater in the Canadian Context*. Technical Report in Canadian Water Network, September 2015.

Paul E. and Liu Y. (eds) (2012). *Biological Sludge Minimization and Biomaterials/Bioenergy Recovery Technologies*. Wiley, Hoboken, NJ. ISBN: 978-0-470-76882-2 July 2012, 536 pages.

Picioreanu C., Van Loosdrecht M. C. M. and Heijnen J. J. (2000). Effect of diffusive and convective substrate transport on biofilm structure formation: a two-dimensional modeling study. *Biotechnology and Bioengineering*, **69**, 504–515. doi: 10.1002/1097-0290(20000905)69:5 < 504::AID-BIT5 > 3.0.CO;2-S

Poortinga A. T., Bos R., Norde W. and Busscher H. J. (2002). Electric double layer interactions in bacterial adhesion to surfaces. *Surface Science Reports*, **47**, 1–32.

Pronk M., Abbas B., Al-zuhairy S. H. K., Kraan R., Kleerebezem R. and van Loosdrecht M. C. M. (2015a). Effect and behaviour of different substrates in relation to the formation of aerobic granular sludge. *Applied Microbiology and Biotechnology*, **99**, 5257–5268. doi: 10.1007/s00253-014-6358-3

Pronk M., de Kreuk M. K., de Bruin B., Kamminga P., Kleerebezem R. and van Loosdrecht M. C. M. (2015b). Full scale performance of the aerobic granular sludge process for sewage treatment. *Water Research*, **84**, 207–217. doi: 10.1016/j.watres.2015.07.011

Qin L., Liu Y. and Tay J. H. (2004a). Effect of settling time on aerobic granulation in sequencing batch reactor. *Biochemical Engineering Journal*, **21**, 57–52.

Qin L., Liu Y. and Tay J. H. (2004b). Effect of settling time on aerobic granulation in sequencing batch reactor. *Biochemical Engineering Journal*, **21**, 47–52. doi: 10.1016/j.bej.2004.03.005

Ren T. T., Liu L., Sheng G. P., Liu X. W., Yu H. Q., Zhang M. C. and Zhu J. R. (2008). Calcium spatial distribution in aerobic granules and its effects on granule structure, strength and bioactivity. *Water Research*, **42**, 3343–3352.

Renou S., Givaudan G., Poulain S., Dirassouyan F. and Moulin P. (2008). Landfill leachate treatment: review and opportunity. *Journal of Hazardous Materials*, **150**, 468–493. doi: 10.1016/j.jhazmat.2007.09.077

Rezasoltani S., Shayegan J. and Jalali S. (2015). Effect of pH on aerobic granulation and treatment performance in sequencing batch reactors. *Chemical Engineering & Technology*, **38**, 851–858. doi: 10.1002/ceat.201400400

Schwarzenbeck N., Erley R. and Wilderer P. A. (2004). Aerobic granular sludge in an SBR-system treating wastewater rich in particulate matter. *Water Science and Technology*, **49**, 41–46.

Seviour T., Yuan Z., van Loosdrecht M. C. M. and Lin Y. (2012). Aerobic sludge granulation: a tale of two polysaccharides? *Water Research*, **46**, 4803–4813. doi: 10.1016/j.watres.2012.06.018.

Shi X. Y., Yu H. Q., Sun Y. J. and Huang X. (2009). Characteristics of aerobic granules rich in autotrophic ammonium-oxidizing bacteria in a sequencing batch reactor. *Chemical Engineering Journal*, **147**, 102–109.

Sobeck D. C. and Higgins M. J. (2002). Examination of three theories for mechanisms of cation-induced bioflocculation. *Water Research*, **36**, 527–538. doi: 10.1016/S0043-1354(01)00254-8

Strom P. F. and Jenkins D. (1984). Identification and significance of filamentous microorganisms in activated sludge. *Water Pollution Control Federation*, **56**, 449–459.

Su B., Cui X. and Zhu J. (2012). Optimal cultivation and characteristics of aerobic granules with typical domestic sewage in an alternating anaerobic/aerobic sequencing batch reactor. *Bioresource Technology*, **10**, 125–129.

Tay J. H., Liu Q. S. and Liu Y. (2001a). Microscopic observation of aerobic granulation in sequential aerobic sludge blanket reactor. *Journal of Applied Microbiology*, **91**, 168–175.

Tay J. H., Liu Q. S. and Liu Y. (2001b). The effects of shear force on the formation, structure and metabolism of aerobic granules. *Applied Microbiology and Biotechnology*, **57**, 227–233.

Tay J. H., Liu Q. S. and Liu Y. (2002a). Aerobic granulation in sequential sludge blanket reactor. *Water Science and Technology*, **46**(4–5), 13–18.

Tay S. T., Ivanov V., Yi S. and Tay J. H. (2002b). Presence of anaerobic bacterioudes in aerobically grown microbial granules. *Microbial Ecology*, **44**, 278–285.

Thwaites B. J., Reeve P., Dinesh N., Short M. D. and van den Akker B. (2017). Comparison of an anaerobic feed and split anaerobic-aerobic feed on granular sludge development, performance and ecology. *Chemosphere*, **172**, 408–417. doi: 10.1016/j.chemosphere.2016.12.133

Torregrossa M., Di Bella G., Viviani G. and Gnoffo A. (2007). Performances of a granular sequencing batch reactor (GSBR). *Water Science and Technology*, **55**, 125–133.

Val del Río A., Figueroa M., Arrojo B., Mosquera-Corral A., Campos J. L., García-Torriello G. and Méndez R. (2012). Aerobic granular SBR systems applied to the treatment of industrial effluents. *Journal of Environmental Management*, **95**: S88–S92. doi: 10.1016/j.jenvman.2011.03.019

Valentino F., Brusca A. A., Beccari M., Nuzzo A., Zanaroli G. and Majone M. (2013). Start up of biological sequencing batch reactor (SBR) and short-term biomass acclimation for polyhydroxyalkanoates production. *Journal of Chemical Technology and Biotechnology*, **88**, 261–270. doi: 10.1002/jctb.3824

Varank G., Demir A., Top S., Sekman E., Akkaya E., Yetilmezsoy K. and Bilgili M. S. (2011). Migration behavior of landfill leachate contaminants through alternative composite liners. *Science of the Total Environment*, **409**, 3183–3196. doi: 10.1016/j.scitotenv.2011.04.044

Verawaty M., Tait S., Pijuan M., Yuan Z. and Bond P. L. (2013). Breakage and growth towards a stable aerobic granule size during the treatment of wastewater. *Water Research*, **47**, 5338–5349. doi: 10.1016/j.watres.2013.06.012

Wagner J. and da Costa R. H. R. (2013). Aerobic granulation in a sequencing batch reactor using real domestic wastewater. *Journal of Environmental Engineering*, **139**, 1391–1396. doi: 10.1061/(ASCE)EE.1943-7870.0000760

Wagner J., Weissbrodt D. G., Manguin V., Ribeiro da Costa R. H., Morgenroth E. and Derlon N. (2015). Effect of particulate organic substrate on aerobic granulation and operating conditions of sequencing batch reactors. *Water Research*, **85**, 158–166. doi: 10.1016/j.watres.2015.08.030

Wang Z. W., Li Y., Zhou J. Q. and Liu Y. (2006). The influence of short-term starvation on aerobic granules. *Process Biochemistry*, **41**, 2373–2378. doi: 10.1016/j.procbio.2006.06.009

Wang S. G., Liu X. W., Gong W. X., Gao B. Y., Zhang D. H. and Yu H. Q. (2007). Aerobic granulation with brewery wastewater in a sequencing batch reactor. *Bioresource Technology*, **98**, 2142–2147. doi: 10.1016/j.biortech.2006.08.018

Wang F., Lu S., Wei Y. and Ji M. (2009a). Characteristics of aerobic granule and nitrogen and phosphorus removal in a SBR. *Journal of Hazardous Materials*, **164**, 77–86.

Wang S.-G., Gai L.-H., Zhao L.-J., Fan M.-H., Gong W.-X., Gao B.-Y. and Ma Y. (2009b). Aerobic granules for low-strength wastewater treatment: formation, structure, and microbial community. *Journal of Chemical Technology and Biotechnology*, **84**, 1015–1020. doi: 10.1002/jctb.2127

Wang B., Wu D., Ekama G. A., Huang H., Lu H. and Chen G. H. (2017). Optimizing mixing mode and intensity to prevent sludge flotation in sulfidogenic anaerobic sludge bed reactors. *Water Research*, **122**, 481–491. doi: 10.1016/j.watres.2017.06.018

Wei Y., Ji M., Li R. and Qin F. (2012). Organic and nitrogen removal from landfill leachate in aerobic granular sludge sequencing batch reactors. *Waste Management*, **32**, 448–455. doi: 10.1016/j.wasman.2011.10.008

Weissbrodt D. G., Schneiter G. S., Fürbringer J. M. and Holliger C. (2013). Identification of trigger factors selecting for polyphosphate- and glycogen-accumulating organisms in aerobic granular sludge sequencing batch reactors. *Water Research*, **47**, 7006–7018. doi: 10.1016/j.watres.2013.08.043

Winckel T. Van, Mancellegala A., Rahman A., Bott C., Sturm B., Vlaeminck S. E., Al-omari A., Authority S., Tech V., Roads H. and District S. (2016). Balancing flocs and granules by external selectors to increase capacity in high-rate activated sludge systems. *Proceedings of the Water Environment Federation, WEFTEC 2016: Session 400 Through Session 409*, New Orleans, September 24, 2016–September 28, 2016, pp. 5617–5622.

Winkler M.-K. H., Bassin J. P., Kleerebezem R., de Bruin L. M. M., van den Brand T. P. H. and van Loosdrecht M. C. M. (2011). Selective sludge removal in a segregated aerobic granular biomass system as a strategy to control PAO–GAO competition at high temperatures. *Water Research*, **45**, 3291–3299. doi: 10.1016/j.watres.2011.03.024

Winkler M.-K. H., Bassin J. P., Kleerebezem R., van der Lans R. G. J. M. and van Loosdrecht M. C. M. (2012a). Temperature and salt effects on settling velocity in granular sludge technology. *Water Research*, **46**, 3897–3902. doi: 10.1016/j.watres.2012.04.034

Winkler M.-K. H., Kleerebezem R. and van Loosdrecht M. C. M. (2012b). Integration of anammox into the aerobic granular sludge process for main stream wastewater treatment at ambient temperatures. *Water Research*, **46**, 136–144. doi: 10.1016/j.watres.2011.10.034

Winkler M. K. H., Bassin J. P., Kleerebezem R., Sorokin D. Y. and Van Loosdrecht M. C. M. (2012c). Unravelling the reasons for disproportion in the ratio of AOB and NOB in aerobic granular sludge. *Applied Microbiology and Biotechnology*, **94**, 1657–1666. doi: 10.1007/s00253-012-4126-9

Winkler M.-K. H., Kleerebezem R., De Bruin L. M. M., Verheijen P. J. T., Abbas B., Habermacher J. and Van Loosdrecht M. C. M. (2013a). Microbial diversity differences within aerobic granular sludge and activated sludge flocs. *Applied Microbiology and Biotechnology*, **97**, 7447–7458. doi: 10.1007/s00253-012-4472-7

Winkler M.-K. H., Kleerebezem R., Strous M., Chandran K. and Van Loosdrecht M. C. M. (2013b). Factors influencing the density of aerobic granular sludge. *Applied Microbiology and Biotechnology*, **97**, 7459–7468. doi: 10.1007/s00253-012-4459-4

Xiong Y. and Liu Y. (2013). Importance of extracellular proteins in maintaining structural integrity of aerobic granules. *Colloids Surfaces B Biointerfaces*, **112**, 435–440. doi: 10.1016/j.colsurfb.2013.07.060

Yagci N., Insel G., Artan N. and Orhon D. (2004). Modelling and calibration of phosphate and glycogen accumulating organism competition for acetate uptake in a sequencing batch reactor. *Water Science and Technology*, **50**, 241–250.

Yang, S. F., Li, X. Y. and Yu, H. Q. (2008). Formation and characterisation of fungal and bacterial granules under different feeding alkalinity and pH conditions. *Process Biochemistry*, **43**(1), 8–14.

Yuan Q., Jia H. and Poveda M. (2016). Study on the effect of landfill leachate on nutrient removal from municipal wastewater. *Journal of Environmental Science*, **43**, 153–158. doi: 10.1016/j.jes.2015.10.023

Zhang Y. and Tay J. H. (2016). Alternated phenol and trichloroethylene biodegradation in an aerobic granular sludge reactor. *Biochemical Engineering Journal*, **106**, 1–10. doi: 10.1016/j.bej.2015.10.026

Zhang H., He Y., Jiang T. and Yang F. (2011). Research on characteristics of aerobic granules treating petrochemical wastewater by acclimation and co-metabolism methods. *Desalination*, **279**, 69–74. doi: 10.1016/j.desal.2011.05.060

Zhang C., Zhang H. and Yang F. (2015). Diameter control and stability maintenance of aerobic granular sludge in an A/O/A SBR. *Separation and Purification Technology*, **149**, 362–369. doi: 10.1016/j.seppur.2015.06.010

Zhao X., Chen Z., Wang X., Li J., Shen J. and Xu H. (2015). Bioresource technology remediation of pharmaceuticals and personal care products using an aerobic granular sludge sequencing bioreactor and microbial community profiling using solexa sequencing technology analysis. *Bioresource Technology*, **179**, 104–112. doi: 10.1016/j.biortech.2014.12.002

Zheng Y. M., Yu H. Q., Liu S. J. and Liu X. Z. (2006). Formation and instability of aerobic granules under high organic loading conditions. *Chemosphere*, **63**, 1791–1800. doi: 10.1016/j.chemosphere.2005.08.055

Zhu J. and Wilderer P. A. (2003). Effect of extended idle conditions on structure and activity of granular activated sludge. *Water Research*, **37**, 2013–2018.

Zhu L., Yu Y., Dai X., Xu X. and Qi H. (2013). Optimization of selective sludge discharge mode for enhancing the stability of aerobic granular sludge process. *Chemical Engineering Journal*, **217**, 442–446. doi: 10.1016/j.cej.2012.11.132

Chapter 5

Membrane-based processes

B. Jefferson, E. McAdam and M. Pidou
Cranfield University, Bedfordshire, UK

ABSTRACT

The membrane in membrane bioreactors (MBRs) rejects all solids including microorganisms and a high proportion of macromolecules and colloids. The result is an intensified biological process, requiring a smaller footprint whilst generating a superior effluent quality. However, the use of membranes generates a cost and energy penalty that is only worth paying if the challenge is sufficient. This includes meeting very tight discharge standards (beyond the capability of activated sludge systems), water reuse or capacity upgrade requirements without available land. All are becoming more commonplace and, as such, MBR technology continues to experience substantial market growth. Developments are further reducing the cost and energy penalties through innovative use of new materials, low energy shear generation and improvements in optimisation and control. Looking a little further ahead, the membrane also facilitates a switch to anaerobic biotreatment even in temperate climates. This reverses the energy position with the sewage works becoming a net producer as well as enabling a switch toward resource factory ideas.

Keywords: anMBR energy, MBR

 Advances in Wastewater Treatment
Editors: Giorgio Mannina, George Ekama, Hallvard Ødegaard and Gustaf Olsson
doi: 10.2166/9781780409719_0205

5.1 INTRODUCTION

In the context of wastewater treatment, a membrane is a material that removes physical and chemical components whilst water passes through its structure (a filter). The level of separation is defined by the size of the pores within the material. This ranges from the rejection of suspended solids with porous microfiltration membranes up to charged monovalent ions in the case of reverse osmosis. The latter is applied as a polish, post wastewater treatment in cases where water recovery is required for potable, non-potable or industrial reuse and is outside the scope of this chapter. Whereas, porous membranes can be used in conjunction with or after the main biological process. The most common, and likely to drive advancements in wastewater treatment is when the membrane is directly integrated with the main bioreactor to generate a hybrid membrane process called membrane bioreactors (MBRs). In most cases the bioreactor is operated under aerobic conditions akin to the activated sludge process with emerging alternatives based on algae or anaerobic redox conditions. The membrane replaces the sedimentation tanks to facilitate complete retention of the particulate matter as well as a significant proportion of the colloids and macromolecules. Importantly, this decouples the hydraulic and solids retention times enabling extended sludge retention times and an intensification of the mixed liquor concentration beyond that achievable in the activated sludge process. However, the membrane also influences the composition of the microbial community within the bioreactor as it alters the selection pressures exerted on the community. The impact has been shown at microbial community level, higher organism level and with regard to the heterotrophic bio-kinetics (Ghyoot & Verstraete, 2000; Smith *et al.*, 2003; Hu *et al.*, 2012).

For instance, in a pilot-scale trial assessing the impact of sludge age on both MBRs and activated sludge, the MBR was shown to operate with a constant maximum specific growth rate (μ_{max}) whereas the equivalent values for the activated sludge system decreased in line with established theory (Figure 5.1). Differences are also seen with regard to the other bio-kinetic constants (k_d, Y) with an overall trend towards the two processes converging at higher sludge ages. Higher organism concentrations reported for MBRs tend to be lower than those reported for activated sludge. However, one study where both systems were operated at the same sludge retention time (SRT) actually showed an increase in flagellates and ciliates consistent with the slightly higher k_d values reported by Smith *et al.* (2003). This aligns to the increased importance assigned to predation in relation to the steady state biomass concentration that develops (van Loosdrecht & Henze, 1999). However, in the case of the half saturation constant (K_s), the values for the MBR increased from 1.9 mg L^{-1} at a sludge age of 8 days up to 73 mg L^{-1} at a sludge age of 30 days. The equivalent data for the activated sludge systems decreased from 125 mg L^{-1} to 11 mg L^{-1} over the same sludge age range (Smith *et al.*, 2003).

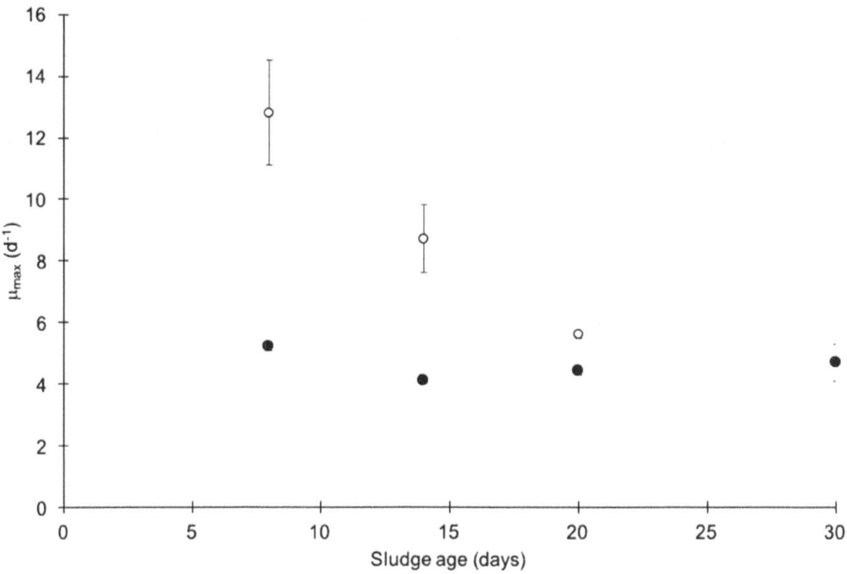

Figure 5.1 Comparison of the maximum specific growth (μ_{max}) as a function of sludge age for activated sludge (white circles) and membrane bioreactors (black circles).

The greater biomass substrate affinity for the MBR reflects differences in not only microbial diversity and community structure but also mass transfer. The operation of the MBR plants requires higher levels of aeration, commonly a combination of fine and coarse bubble systems, such that the aggregates equilibrate to a higher shear environment resulting in much smaller flocs. To illustrate, during the comparative trail (Smith *et al.,* 2003), the median particle size (measured by laser diffraction) was 97 μm for the MBR compared to 428 μm for the activated sludge system. The consequence is a 230% increase in the specific surface area of the MBR flocs at 0.098 m^2 g^{-1} compared to 0.041 m^2 g^{-1}. At the same time the respective population diversity indices for the two systems as measured by denaturing gradient gel electrophoresis (DGGE) were 6.6 and 5.8 for the MBR and activated sludge, respectively, indicating the potential significance of understanding both microbial and physical impacts.

Unfortunately, application of traditional methods for measuring the size, shape and strength (Jarvis *et al.,* 2005) of biomass flocs is significantly hampered by the concentration and active nature of the components. Most particle sizing instruments require significant dilution and are laboratory bound providing significant time for the aggregates to alter during transportation and storage. The consequence is that there is a paucity of robust data on the nature of biomass flocs in all bioreactors including MBRs which tends to minimise consideration of mass transfer impacts when analysing systems. Previous work at Cranfield comparing systems revealed the MBR flocs grew when exposed to shear environments within jar testers below 150 rpm with the same observed with the activated sludge biomass only at much lower shear rates (<50 rpm). The difference reflects the shear environments present in the two systems. Further, the flocs decreased in size as the shear rate was increased with a corresponding reduction in the fractal dimension of the MBR flocs from 2.34 to 2.13 compared to the flocs in the activated sludge systems which remained between 2.05 and 2.08 irrespective of size (Figure 5.2).

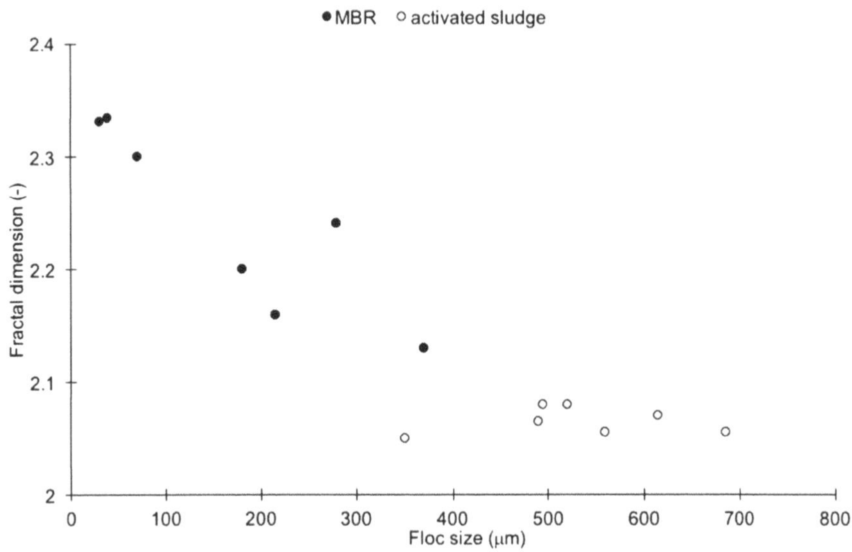

Figure 5.2 Comparison of the size and fractal dimension of flocs from an MBR and an activated sludge system run at the same SRT exposed to different shear rates.

5.1.1 MBR advantage over activated sludge?

The overall advantages of MBRs over activated sludge are commonly described with regard to superior organics removal, greater resilience and robustness of nutrient removal, lower sludge production, effluent disinfection, suitability for coupling to reverse osmosis (RO) for water reuse and smaller footprint requirements. In practice, it is difficult to validate such claims as the two processes are not commonly operated over similar conditions. However, removal of biochemical oxygen demand (BOD) and total chemical oxygen demand (COD) should always be superior due to the differences in the solid removal between the two processes. A number of studies have also shown that this extends to a 5–20% improvement in soluble organics removal, attributed in part to a rejection of macromolecules by the membrane. The majority of such studies are limited to pilot- and demonstration-scale operation with the exception of a recent study of a 250,000 PE plant split roughly evenly between an MBR and an activated sludge system (Bertanza *et al.*, 2017). The loading rates of the plants were low enabling both to be compared under non-stressed conditions. In this case the difference in performance reduced with effluent COD of 22 and 12.5 mg L^{-1} for the CAS and the MBR, respectively, accounted for by the difference in suspended solids at 12.9 and 0.7 mg L^{-1} (Bertanza *et al.*, 2017).

Overall, the main benefits of the technology are the ability to deliver excellent and stable water quality, including a permeate free of pathogens and other microbial organisms, and an intensification of the overall systems. Accordingly, the majority of municipal applications relate to either tight discharge consents, reuse or capacity upgrade drivers. The exception to this relates to hazardous chemicals, especially organic micropollutant where there is conflicting evidence as to whether MBRs out-perform activated sludge systems. This reflects the wide range of possible compounds whose removal is associated with a combination of sorption, biodegradation and volatilisation across the bioreactor. Overall, no difference in performance between MBRs and activated sludge systems can be particularly expected for easily degradable or recalcitrant compounds (Besha *et al.*, 2017). However, for compounds that exhibit an intermediate level of recalcitrance, MBRs offer potential due to their ability to operate at long sludge ages, with large specific surface areas of biomass and the ability to be relatively easily integrated with additional steps such as powdered activated carbon or dense membranes (nanofiltration - NF, RO) (Table 5.1).

Table 5.1 Example removal of micropollutants in activated sludge and MBRs (adapted from Besha *et al.*, 2017).

Compound	Removal (%)		
	Activated Sludge	**MBR**	**MBR + PAC***
Carbamazepine	0–9.5	−42 to 51	80–99
Diclofenac	2–51	15–87	>98
EE2	ND-87	ND-93.5	86.7
Metoprolol	6.5–65	29.5–58.7	>99

*PAC = powdered activated carbon.

The higher sludge age that MBRs are operated at, has been stated as an advantage for micropollutant removal with evidence of improved removal at elevated SRTs for acidic pharmaceuticals such as ketoprofen and clofibric acid (Kimura *et al.*, 2007). This has also been demonstrated for gemfibrozil and 17a-ethynylestradiol (EE2) but not for compounds such as phenacetine and caffeine (Maeng *et al.*, 2013). The improvement is often related to increased microbial diversity increasing the chance for specialist degraders. However, diversity is known to increase with reactor volume congruent to the ecological theories of island biogeography (van der Gast *et al.*, 2006) with no evidence that volume impacts removal, possibly reflecting the very high levels of inherent microbial diversity found in bioreactors treating municipal sewage. Indeed, recent evidence is starting to refute such a hypothesis placing more importance on the overall stability of the reactor which favours the inherent robustness of MBRs. Such suggestions align to recent findings across all technologies trialled during a UK-wide programme investigating delivery of the water framework directive requirements associated with micropollutants and phosphorus (Craig, 2017; Sunner, 2017). Looking forward, the challenge is a combination of the likelihood of a compound entering the sewage works and its contribution to exceeding acceptable levels in the receiving bodies. In the UK, particular concern remains for a limited number of compounds including perfluorooctanesulfonic acid (PFOS), perfluorooctanoic acid (PFOA), hexabromocyclododecane (HBCCD), polybrominateddiphenylethers (BDE) and Cypermethrin (Craig, 2017). All are very stable in water in relation to both sorption and biodegradation and, as such, MBRs are unlikely to offer any particular advantage against these systems beyond their inherent robustness/resilience and preparing the effluent water for advanced downstream treatment.

In balance to the advantages described, selection of MBRs over activated sludge systems can incur a cost penalty. In the case of low strength municipal sewage, MBRs are around 50% higher in terms of annualised investment costs (Krzeminski *et al.*, 2017). To illustrate, in a direct comparison at full scale it was estimated that the MBR treatment train costs 12–28 € PE^{-1} y^{-1} more than the activated sludge system (Bertanza *et al.*, 2017).

However, for cases where tight discharge consents are set, the difference is significantly diminished or even reversed. This was reported by Young *et al.* (2014) through the impact of changing phosphorus standard with/without total nitrogen where the MBR became cost even at a TP consent of 0.3 mg L^{-1}. This extended to a 13% lower total lifecycle cost when meeting a 0.1 mg L^{-1} TP with no TN requirements.

Overall, MBRs appear economically favourable for enhanced nutrient removal or water reuse applications where the reduced capital costs outweigh an increase in operating costs. A key sensitivity in the case of MBRs is the peak loading factor as this determines the overall membrane sizing. In such cases reduction of the peak flow through flow equalisation offers potential to reduce total costs.

An alternative to the suspended sludge based MBR systems, is the moving bed biofilm (MBBR) based MBR systems. One may differentiate between the MBBR-MBR where the membrane module is placed directly after the MBBR reactor and where there is no sludge recycle, and an IFAS-based MBBR where an integrated fixed film activated sludge (IFAS) reactor replaces the traditional activated sludge reactor in a standard MBR. These systems are described and discussed in more detail in Chapter 3 of this book, to where the reader is referred.

5.2 AEROBIC MEMBRANE BIOREACTORS (ACTIVATED SLUDGE BASED)

Aerobic MBRs can now be considered an established, mature technology for both municipal and industrial wastewater treatment. For instance, the total amount of wastewater treatment by MBRs is expected to exceed 5 million $m^3\ d^{-1}$ by 2019 with the largest plant treating a capacity of 864,000 $m^3\ d^{-1}$ once commissioned in 2018 (Krzeminski *et al.*, 2017). Comparing to the largest plant in 2004 which treated 45,000 $m^3\ d^{-1}$, this demonstrates that improvements in design and reduction in core cost components has extended the economic viability of the technology to larger plant sizes. Nowadays there are increasing numbers of sites treating in excess of 100,000 $m^3\ d^{-1}$ including 34 since 2012 of which 50% have been associated with upgrading treatment capacity of the existing plant (Krzeminski *et al.*, 2017). The impact of such expansion in treatment capacity is a significant year by year market growth with estimation of the total market value exceeding \$3b (10^9) by 2019 compared to around \$1.2b (10^9) in 2010 (Judd, 2016). Overall the compound annual growth rate between 2014 and 2019 is 12.8% which is significantly lower than forecast at 22.4% but still represents one of the fastest growing technologies within the sector (Judd, 2016). Analysis by region indicates the fastest growing sectors are in Asia-Pacific and North America both demonstrating double digit growth.

The growth in the market is driven by one of three main issues: (1) Stricter legislation related to either discharge or reuse, (2) Space limitation or (3) Retrofitting/upgrading conventional WWTPs. For instance, water reuse legislation in Mediterranean countries includes a suspended solids limit of 10 mg L^{-1}. Such levels cannot be guaranteed through the use of activated sludge systems without substantial post treatment. In comparison, the membrane in MBRs provides an absolute barrier to solids including microorganisms. This includes applications to urban reuse as the disinfection stage is reported to be limited by the presence of particles (Winward *et al.*, 2008). Equivalent challenges are reported when meeting phosphorus consents of 0.1 mg L^{-1} where it is predicted the corresponding suspended solids levels will need to be around 2–5 mg L^{-1} (Sunner, 2017).

In addition to matching well for water reuse applications, growth in the uptake of the technology is also driven by the cost and/or availability of land. Examples include Nordkanal and perhaps most impressively, Henriksdal WWTP. In the latter case, the WWTP is located inside a rock and the more compact nature of MBRs prevents the need for further excavation of the rock structure (Judd, 2018). Application related to space limitations are likely to become even more common as the development footprint of towns and cities around the world starts to engulf the WWTPs restricting the ability to expand. This can include utilisation of the existing assets, as retrofitting an external membrane tank to existing activated sludge bioreactors offers the potential to increase treatment capacity by up to three times (Krzeminski *et al.*, 2017).

Overall, MBRs still remain a niche technology at present with, for instance, only around 0.5% of Europeans being serviced by MBRs for municipal wastewater treatment. Whilst the growth areas described above will clearly show this expanding, the expectation is that it will remain below 1% in the coming years (Krzeminski *et al.*, 2017). Cost plays a significant role in this with MBRs representing a higher cost option unless very strict discharge consents are required. The additional costs relate to the installation and maintenance of the membranes, more expensive pretreatment requirements, higher levels of automation and a high energy requirement. In the case of energy, MBRs can be expected to operate at around 0.4–1.6 kWh m^{-3} compared to 0.3–0.6 kWh m^{-3} for activated sludge. The lower end of the MBR range represents the optimal operation with most plants significantly exceeding it. This is most exemplified with small, decentralised systems where the downscaling results in an extra energy penalty that can exceed 3 kWh m^{-3} (Lesjean *et al.*, 2011). Many plants in practice are not loaded to full flow capacity while aeration flow remains high because of fouling control, resulting in high kWh m^{-3} values (Lesjean *et al.*, 2011).

In addition, historical operating problems associated with stressing the systems, especially operating flux, remain barriers through poor receptivity. Whilst developments have now overcome such issues, the uptake of the technologies where people have had such experiences will likely remain negligible until the discharge requirements become sufficiently challenging.

5.2.1 The membrane in aerobic MBR systems

Membranes generally incorporate a porous support backing to provide mechanical stability onto which a thin surface layer provides the required selectivity. Ideally the membranes deliver both a narrow pore size distribution and a high overall porosity. In addition, the material has to have a reasonable degree of both chemical and thermal stability. Whilst a wide array of materials could be used, the range of materials used and pore sizes offered remains fairly consistent (Table 5.2). For instance, in a recent MBR survey, around half of all the available membranes were based on polyvinylidene fluoride (PVDF) with the rest mainly made up of polypropylene, polyethylene (PE) or polyethersulphone (PES). Similarly, most membranes have a pore size range between 0.03 and 0.4 μm with a decreased effective pore size in relation to rejection once the membrane is in operation due to the formation of the fouling layer. The selected materials are configured into one of three main membrane geometries used in MBRs: flat sheets, hollow fibres and multitubes. The individual membrane elements are then packed within a housing to generate a module through which the water flows (Figure 5.3). Detailed information on the full range is provided by Judd through either his recent book (Judd & Judd, 2011) or the very useful online resource "The MBR site – www.mbrsite.com" (Judd, 2018).

Table 5.2 Example membranes used in MBRs (Judd, 2018).

Company	Configuration	Material	Pore Size (μm)	Number of Sites Treating >100 MLD
Berghof Membrane Technology GmbH & co.	Multitube	PVDF, PES	0.03–5 kDa	–
Kubota	Flat sheet	Chlorinated PE	0.4 max, 0.2 ave	1
Origin Water	Hollow fibre	PVDF	0.1, 0.3	24
Suez Water Technologies and Solutions	Hollow fibre	PVDF, Braided	0.02–0.04	15
Koch Membrane Systems	Hollow fibre	PVDF	0.03	1
Huber SE	Flat sheet	PES	0.038	–

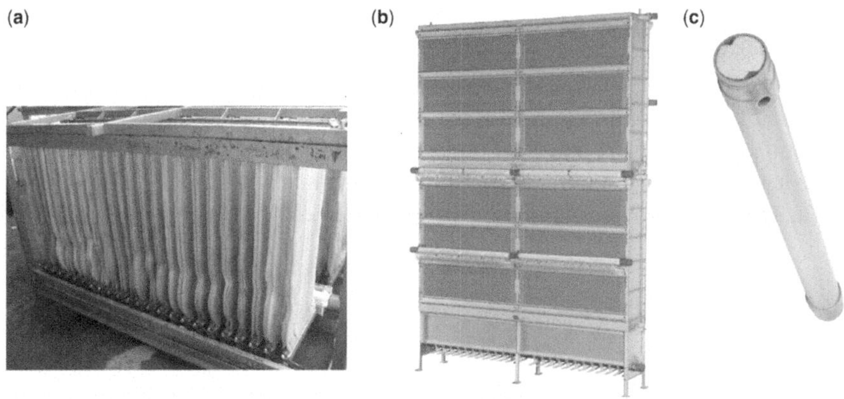

Figure 5.3 Example membrane modules: (a) a memstar hollow fibre, (b) a Kubota flat plate and (c) a Pentair multitube. (Images from Judd & Jefferson, 2003.)

Whilst the material and pore sizes used remain fairly consistent, the market growth levels experienced by MBRs has generated an expansion in the number of available commercial products with around 70 available to date (Judd, 2016). The lower production cost of hollow fibre membranes has resulted in that configuration dominating the larger-scale plants, accounting for the vast majority of the sites treating 100,000 m^3 d^{-1} or greater. At more moderate scales, the market is more spread with greater application of flat-sheet systems especially. For instance, in 2009, Kubota treated an installed capacity of 900,000 m^3 d^{-1} in 3300 separate plants including a number of large systems in the Middle East (Judd & Judd, 2011).

5.2.2 Fouling and its management

The application of membranes for enhanced particulate and colloidal rejection in MBRs results in an accumulation of material at or near the membrane surface. This leads to the formation of a polarised layer that restricts flow through additional resistance measured as an increase in the transmembrane pressure and reported as a fouling rate (dTMP dt^{-1}) or permeability decline. The mass rate at which materials approach the membrane is governed by the operating flux, normally defined in L m^{-2} h^{-1} (normally written as LMH) which then in turn impacts on the operational strategy and cleaning approaches taken. Fouling remains the preferred topic for research and a number of reviews have comprehensively summarised developments as they have progressed (Chang *et al.*, 2002; LeClech *et al.*, 2006; Drews, 2010; Meng *et al.*, 2017). Traditional approaches to assessing appropriate flux rates are based on a combination of critical and sustainable flux measures (Figure 5.4). The formal definitions and measurement methods vary although at a practical level, they represent the flux at which rapid fouling does not occur (critical flux) and the flux at which a low and stable fouling rate is observed for a prolonged time period (sustainable flux). Both are determined by monitoring the TMP during either short time periods with incremental changes in the flux (Figure 5.4) or prolonged time periods at a fixed flux. In the case of the latter, the TMP increases gradually up to a certain point and then "jumps" (Pollice *et al.*, 2005). The increase in TMP prior to the jump is referred to as the sub-critical fouling rate and together with the time to reach the jump provides insight into the requirements for fouling mitigation.

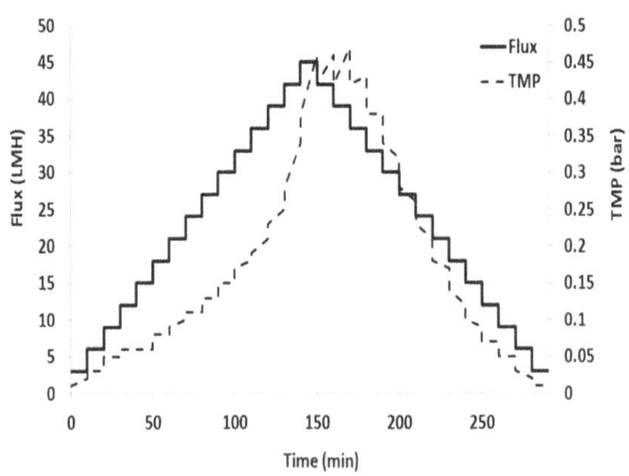

Figure 5.4 Illustration of critical flux.

The critical fouling components in MBRs are found in the soluble fraction and are composed of a range of carbohydrates, proteins and humic acids that can interact to form clusters and gels through cross linking. Consensus now exists on the importance of polysaccharides in the formation of such gels during early stage fouling (pre the TMP jump), thought in part to be due to their greater rejection and lower biodegradation than protein and humic substances within the bioreactor (Meng *et al.,* 2011). Gelling is enhanced in the presence of metal cations with equivalent work related to algal organic matter showing that the level of network bonding is related to the acidity of the carbohydrate (Rao *et al.,* 2018). The majority of the work to date has focused on carbohydrate-carbohydrate interactions with much less investigation into protein-carbohydrate interactions. The bonding in such cases is non-covalent and is governed by electrostatics, hydrophobic interactions and hydrogen bonding and has been linked to irreversible fouling (Neemann *et al.,* 2013; Pivokonsky *et al.,* 2016). Overall, the system is based on a mixed foulant array with likely synergistic effects. The consequence is that attempts to isolate a key foulant or universal surrogate biomass characteristic to identify fouling potential, are likely to remain only the hopes of research scientists and that operational focus needs to remain on minimisation of fouling impacts and reduction in the cost of its mitigation.

The fouling layer compresses as it continues to grow and as such the operating flux is important in terms of the mass flux rate and the overall pressure exerted on the layer. Typical flux rates have reduced in recent years to ensure stable operation of the plants and are now between 15–20 LMH based on average flow and 20–30 LMH based on peak flow for municipal sewage treatment. Management of the foulant layer is then accomplished through a combination of membrane scouring, relaxation, backwashing, chemically enhance backwashing and chemical cleaning (Wang *et al.,* 2014). The specific combination used depends on the membrane material and geometry with, for instance, most flat sheet membranes not able to backwash. Both backwashing and relaxation are generally incorporated into normal operation during which air scouring is continuously occurring in a cycle that lasts a few minutes. Relaxation and backwashing alter the force balance on the foulant layer and enable easy withdrawal by the scouring air. The key consideration is then related to the energy used to maintain the membrane operation through the generation of shear. In immersed systems, the shear is generated through gas sparging and so the specific aeration demand with respect to the membrane area (SADm) becomes important. Typical values for hollow fibre systems are between 0.2–0.6 $Nm^3\ m^{-2}\ h^{-1}$ whereas with flat sheet systems the values range from over 0.7 $Nm^3\ m^{-2}\ h^{-1}$ for single stack systems to between 0.2 and 0.8 for double deck configurations (Judd, 2016). Reduction in operational SADm has contributed to the significant development of the technology through improved module design and the development of cyclic and sequential sparging. Recent reviews on the overall energy demand of MBR sites reveal a wide range of reported values between 0.4–3.0 $kWh\ m^{-3}$ based on a yearly average specific energy consumption (Krzeminski *et al.*, 2017). In recent times, effort has been made to demonstrate the ability to operate MBRs at between 0.4–0.5 kWh m^{-3} through a variety of reduction and optimisation measures such as increasing flux, lowering aeration rates, use of gravity flow where possible, reduction in mixed liquor concentrations, inclusion of intermittent operation and sparging cycles.

Other areas of development involve altering the foulant-membrane interactions by either adjusting the nature of the foulants or by coating the membrane. The latter is a relatively recent development using either chemical modification or inclusion of nanoparticles such as TiO_2, ZnO and carbon nanotubes (Meng *et al.*, 2017). The modifications provide control pathways through photocatalytic or antimicrobial activity as well as hydrophilicity. The alternative is to dose chemicals into the bioreactors such as coagulants, zeolites, activated carbon, and cationic polymers as well as biological control systems based on quorum quenching or enzymatic degradation of the fouling layer (Meng *et al.*, 2017). Promising evidence has been published on significant fouling reduction although the effects do not appear universal, congruent with the complex array of potential foulants that may initiate fouling. Importantly, evidence is required as to whether the generated fouling reduction justifies the cost of the chemicals. Some of the additions (carbon, plastic beads, biofilm carriers) provide additional mechanically assisted scour and improved operation (Table 5.3). The approach works best for fixed geometries (tube and flat plate) although the long-term impact on the membrane is yet to be understood but represents an area of significant promise.

Table 5.3 Examples of mechanically assisted scour.

Particle	Size	Load	Impact	Reference
GAC*	0.3–0.6 mm	0.5–2.0 g L^{-1}	60% decrease in filtration resistance	Johir *et al.* (2011)
Plastic beads	2–5 mm	4 kg m^{-3}	20% higher flux	Siembida *et al.* (2010)
Biofilm carriers	12 × 10 × 7 mm	20% fill ratio	Extend filtration run by 8 times	Fu *et al.* (2016)

*GAC = Granular activated carbon.

Analysis of recent surveys from "the MBR site" (Judd, 2016) reveals that chief concerns amongst the MBR community relate to (1) screening/pretreatment, (2) clogging, (3) overloading/under design, and (4) membrane surface fouling. The first three have seen a decreasing importance placed on them since the first survey in 2010 with increasing prominence placed regarding energy demand, operator knowledge, sludge handling and uneven aeration. These remain areas that are under research with preference still being assigned to fouling.

5.2.3 Future outlook

The future outlook identifies a number of relatively novel configurations that are being considered and/or commercialised (Krzeminski *et al.*, 2017). These include the use of shear enhanced filtration through rotational MBR, reciprocating MBR, helical membrane modules, inclusion of intermittent electric fields, osmotic MBR, and bioelectrical chemical reactors. Rotating MBRs have been tested with flat plate, tubular and hollow fibre membranes with established commercial products such as the Huber membrane filtration VRM® and the Grundfos Biobooster which utilise ceramic discs. Trials with rotating MBRs suggest lower fouling rates and/or the ability to run at much lower SADm offering the potential for reduced energy consumption. Similar benefits have been observed with vibrating and reciprocating membranes with additional suggestions about negating the need for air scour in horizontal reciprocating systems (Wang *et al.*, 2014). These systems offer much promise but the trials to date have been limited to small scale with modest mixed liquor suspended solids (MLSS) concentrations.

Whilst ceramic materials currently remain a niche application, the number of products has expanded recently with, for instance, four new flat sheet ceramic products available since 2010 (Judd, 2016). The increased resistance to mechanical and chemical attack and an inherently lower fouling propensity will surely see rapid growth. One obvious synergy is with mechanical scouring particles where the risk to the membrane integrity will no longer be so critical. Whilst the future success of the novel configurations remains unclear they do demonstrate a continued level of innovation and development which provides optimism for the future outlook of MBRs.

5.3 ANAEROBIC MEMBRANE BIOREACTORS

Optimisation of conventional aerobic sewage treatment can only reduce energy demand and carbon emissions by approximately 5–20%. It is believed that this will be insufficient to meet long-term aspirations set by governments regarding renewable energy, energy demand and carbon footprint whilst also meeting future effluent consents. Accordingly, there is increasing belief that alternative approaches are required with the use of anaerobic sewage treatment seen as one of the most promising options. Anaerobic biotreatment utilises a range of bacteria and archaea to convert organic material into methane, carbon dioxide and water. The initial breakdown of the large organic polymers (hydrolysis) is the rate determining step in many cases and known to be strongly influenced by temperature. Accordingly, it is common to operate anaerobic biotechnology at elevated temperatures compared to aerobic systems, typically 35°C or above for extended retention times. The approach is common for processing of sludge and industrial wastewaters where the elevated carbon content generates large biogas quantities and the reactors are economic to heat to maintain the required temperature. Extending this to the main flow of wastewater treatment obviates the need for aeration with only 10–20% of the processed organic matter assimilated into biomass. This reduces energy demand and sludge production compared to activated sludge processes, freeing up space in anaerobic sludge digesters, obviating the need for advanced pretreatment of waste activated sludge as well as generating additional methane through the main treatment stage. The switch to anaerobic sewage treatment offers the potential for energy neutral sewage treatment beyond sites where sludge imports provide additional feedstock material for sludge digesters. The approach is not new, with wide spread implementation of anaerobic sewage treatment (non-membrane based) in countries such as India and Brazil (Chernicharo, 2006) where the temperature is warmer, the feed generally stronger and the discharge consents less onerous. The driver in such situations relates to negating the need for the associated aeration infrastructure and a reliable source of electricity as well as reduced sludge processing rather than electricity generation such that in all but large-scale operations the biogas is commonly flared.

Delivery of anaerobic sewage treatment in temperate climates challenges previously accepted wisdom that anaerobic processes cannot operate effectively under low temperatures and on weak substrates. However, there is growing consensus that when configured as an anaerobic membrane bioreactor, the membrane provides the necessary reactor stability, generates the required sludge retention times and meets the more challenging discharge standards (Martinez-Sosa *et al.*, 2011; Robles *et al.*, 2013). Whilst more common for industrial applications, recent research has demonstrated the potential of the technology to work in temperate climates treating settled sewage (Fawehinmi *et al.*, 2007; Martin Garcia *et al.*, 2013; Shin & Bae, 2018). The membrane retains all particles and a large proportion of the macromolecules, generating an effluent stream with no solids or microorganisms yet remaining rich in nutrients such as ammonia and phosphorus. Accordingly, the use of anMBRs is also being seen as a pivotal technology in the transformation of wastewater treatment facilities into resource factories consistent with the concepts of the circular economy (Jefferson, 2012).

5.3.1 AnMBR treatment performance and options

A number of flow sheet configurations are possible showing variation with regard to the management of the upfront and post-treatment options as well as the temperature of operation, bioreactor configuration and the role of the membrane (Figure 5.5). Influenced by past perceptions that municipal sewage could not be treated anaerobically, as the organic concentration was too low to support growth, the majority of trials have focused on treating crude or even fortified influent (Eusebi *et al.*, 2013; Shin & Bae, 2018). Whilst this maximises the organic loading rate it does so by increasing the solids concentration in the feed, which can exceed the solids treatment capacity due to the slow hydrolysis rates that are possible at ambient temperatures in temperate climates. The consequence is poor conversion rates and the chance of untreated solids interacting with the membrane, greatly increasing the fouling propensity of the system. As such, a number of authors advocate the treatment of settled sewage with the solids fraction treated in sludge digesters where conditions are optimised for processing of solid phase material (Kim *et al.*, 2011; Martin Garcia *et al.*, 2013; Gouveia *et al.*, 2015). This is particularly relevant in cases where digestion assets already exist, as a lower total cost can be generated by ensuring the majority of the solids enter the digesters.

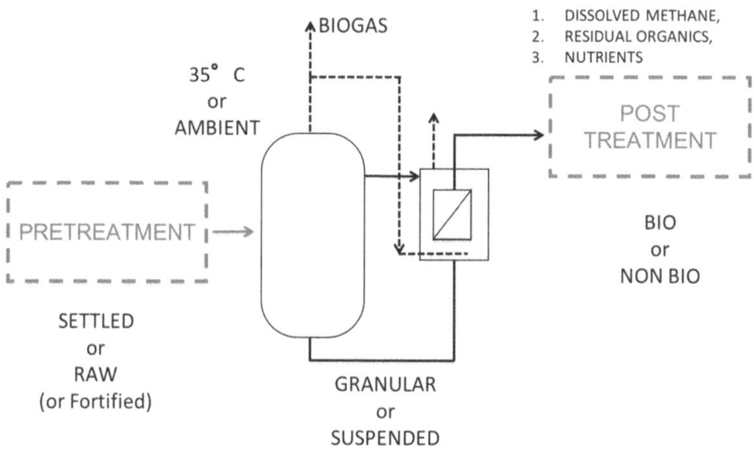

Figure 5.5 Schematic of main variants considered with anMBRs.

Unlike sludge digesters, the main anaerobic reactor needs to operate under ambient conditions due to the vast energy demand required to heat the main flow. Previous analysis has indicated that heating the flow can only generate a net benefit once the COD concentration exceeds 4–5 g L^{-1} in temperate climates thus limiting this to industrial treatment applications (Martin Garcia *et al.*, 2011). Successful pilot trials have been reported with operating temperatures between 7 and 33°C and average influent CODs between 221 and 978 (Table 5.4). A negative consequence of low temperature (sub 15°C) operation is an increase in effluent COD linked to production or decreased biodegradation of soluble microbial products. This does not translate to BOD which has been shown to remain below 15 mg L^{-1} across a temperature range of 2 to 21°C, indicating the increase is unlikely to be due to volatile fatty acid (VFA) accumulation (Martin Garcia *et al.*, 2103). In fact, multiple pilot trials in the UK have shown successful treatment and biogas generation with minimum working temperatures of around 6°C. In one study, temperature was controlled to 35°C, 22°C and 12°C, treating municipal sewage with a COD from 250–305 mg L^{-1} down to consistently below 50 mg L^{-1} even at 12°C, reaching a minimum of 25 mg L^{-1} (Fawehinmi *et al.*, 2007). The other major

Table 5.4 Example anMBRs treating real wastewater in pilot plants.

Ref	Source	Screen mm	Mem. Type, Pore Size μm	T °C	A_{mem} m^2	Flux LMH	SGDm $m^3\ m^{-2}\ h^{-1}$	Sparge Cycle %	HRT h	dTMP/dt mbar d^{-1}	COD_{IN} mg L^{-1}	COD_{out} mg L^{-1}	E_{con} kWh·m^{-3}
Flocculent													
Martinez Soas *et al.* (2011)	Crude	–	FS, 0.038, Microdyn-Nadir	20	3.5	7	1.22	100	19.2	2.61	630	80	1.66
Robles *et al.* (2013, 2012)	Crude	0.5	HF, 0.05 Puron, Koch	33	60	13	0.23	100	21–7	<960	445	77	0.24
Gouveia *et al.* (2015)	Crude	1.0	HF, 0.04 Suez GE	18	0.9	12	0.5–1.0	100	10–14	1–3	978	136	0.25
Dong *et al.* (2015)	Crude	3.0	HF, 0.04 Suez, GE	23	5.4	17	0.146	100	8.5	1.4	383	24	0.1
Shin *et al.* (2014)	Settled	–	HF, 0.03 Cheil ind.	9–30	39.5	4–7.5	75*	n/a	4.6–6.8	0.8	233	20	0.14
Granular													
Fawehinmi *et al.* (2007)	Settled	mac	HF, 0.1 Mitsubishi rayon	12	0.9	9	0.002	100	6	5.7–10	300	36	–
Martin Garcia *et al.* (2013)	Settled	mac	HF, 0.04, Suez GE	8–22	0.9	6	1.2	10	16	1.44	338	47	0.16
Cranfield A^	Settled	3.0	HF, 0.04, Suez GE	7.5–21	0.9	15	8.6	10	12	3	265	72	0.07**
Cranfield B^	Settled	6.0	HF, 0.04, Suez GE	10	0.9	8.4	1.1	10	8	3.6	221	41	0.133

* upflow velocity; ** estimated for a 10,000PE plant, ^ unpublished data, mac = macerated, FS = flat sheet, HF = hollow fibre.

impact of temperature is on the water viscosity and its impact on flux rates. For instance, operating at a flux of 7 LMH at 7°C has a normalised flux (to 20°C) (J_{20}) equivalence of 9.5 LMH and so when assessing achievable flux between studies it is important to frame the achievable rates with respect to temperatures and always consider the lowest temperature that the plant will need to operate at.

The bioreactor can be operated with flocculent or granular biomass without impact on the treatment efficacy (Martin Garcia *et al.*, 2013). In fact, three main configurations have been reported in the literature: flocculent continuous stirred tank reactor (CSTR), flocculent upflow anaerobic sludge blanket reactor (UASB) and granular UASB. Flocculent CSTRs (Figure 5.6a) are the most reported systems, generally seeded with digester sludge from a local sewage works and operated at mixed liquor concentrations up to 30 g L^{-1} (Robles *et al.*, 2013); much higher than typically utilised in aerobic MBRs. In such cases, the mass flux onto the membrane is very high, requiring continuous gas sparging to maintain permeability. The alternative is to use a UASB reactor (Figure 5.6b) with either flocculent (Gouveia *et al.*, 2015), granular (Fawehinmi *et al.*, 2007; Martin Garcia *et al.*, 2013) or fluidized GAC as a biomass carrier (Kim *et al.*, 2011). In the granular reactors, a separate flocculent layer forms above the granule bed which acts as a sludge blanket layer. The material is predominantly unhydrolysed influent solids such that the reactor works best with reduced solids in the feed (i.e. settled sewage) and requires periodic desludging to prevent carry over to the membrane tank. Analysis of wasted material revealed its biomethane potential to remain intact such that it can be readily mixed with primary sludge for digestion (Jefferson, 2012).

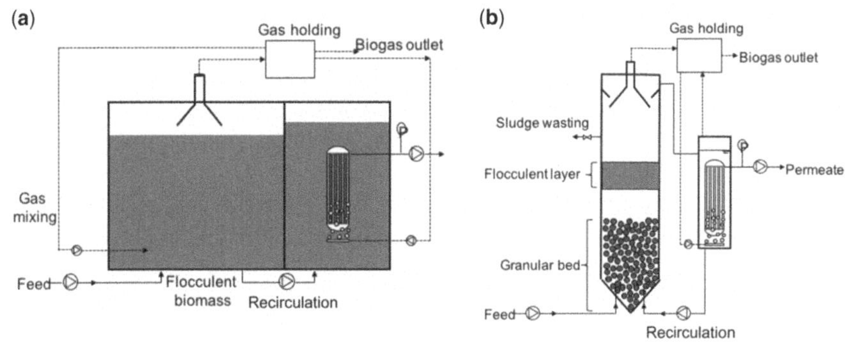

Figure 5.6 anMBR configurations: (a) CSTR or (b) UASB.

Irrespective of reactor configuration, pilots with anMBRs for sewage treatment have revealed consistent performance with average effluent COD ranging between 20 and 136 mg L^{-1} coupled with solid-free effluents by virtue of the membrane (Table 5.4). Operational hydraulic retention times (HRTs) range between 4 and 21 hours with SRTs up as high as 70–100 days. One of the other benefits of anaerobic systems are the low solids yields generated with reported levels from effectively negligible to 0.46 gVSS $gCOD_{rem}^{-1}$ when treating crude wastewater. In the case of settled wastewater the yields are more consistently low between 0.003 and 0.051 gVSS $gCOD_{rem}^{-1}$ (Fawehinmi *et al.*, 2007; Shin & Bae, 2018). In the case of the low temperature, settled sewage trials with granular USAB, the granule bed maintained its integrity and volume over prolonged operation periods demonstrating both resilience and a lack of substantial growth.

Reported biogas production rates vary between 0.04 and 0.27 $L_{methane}$ $gCOD_{rem}^{-1}$; all significantly below the theoretical value of 0.389 $L_{methane}$ $gCOD_{rem}^{-1}$ (Fawehinmi *et al.*, 2007; Shin & Bae, 2018). Two factors need to be considered in such cases. The first is that due to the operating temperature and potentially large water volumes processed, large proportions (45–80%) of the produced methane can remain in the liquid phase and can exist at levels above the saturation concentration (Cookney *et al.*, 2012, 2016). The dissolved methane needs to be removed due to requirements concerning energy recovery, process emission, and health and safety. A number of technologies exist including spray aeration, tray aerators, packed columns and membrane contactors to achieve this (Heile *et al.*, 2017).

The generation of a solids-free effluent from the anMBR aligns itself to the more intensive processes such as membrane contactors where recovery is practical at concentrations that remain directly usable with contact times of around 300 s (McLeod *et al.*, 2016). The second factor relates to the impact of sulphate reducing bacteria (SRB) that compete for organics and hence decrease methane yields. Sulphate reduction becomes stoichiometrically met once the COD:SO_4-S mass ratio exceeds 2 mgCOD $mgSO_4\text{-}S^{-1}$ (Gimenez *et al.*, 2011). To illustrate the impact, in the case of a flocculent CSTR with an influent sulphate concentration of 99 mg L^{-1} and a COD of 445 mg L^{-1}, the measured biogas yield was 0.069 $L_{methane}$ $gCOD_{rem}^{-1}$ with complete reduction of the incoming sulphate. However, if the yield was adjusted to exclude COD removed by the SRBs then the yield becomes 0.294 $L_{methane}$ $gCOD_{rem}^{-1}$ (Gimenez *et al.*, 2011).

Overall, the application of anMBRs enables wastewater treatment to become a net energy producer (McCarty *et al.*, 2011; Pretel *et al.*, 2015). The overall balance needs to take into account the downstream treatment of the residual nutrients which can be achieved in traditional technologies such as activated sludge and trickling filters coupled with coagulant dosing (Chernicharo, 2006). Total nitrogen removal is however difficult due to a lack of available carbon (Eusebi *et al.*, 2013) although some groups have utilised the dissolved methane as a carbon source which can equate to around 60–70 mg L^{-1} COD depending on temperature (Gimenez *et al.*, 2011). The alternative is to utilise the water quality generated by the anMBR to enable low energy, low cost recovery through processes such as ion exchange (Figure 5.7).

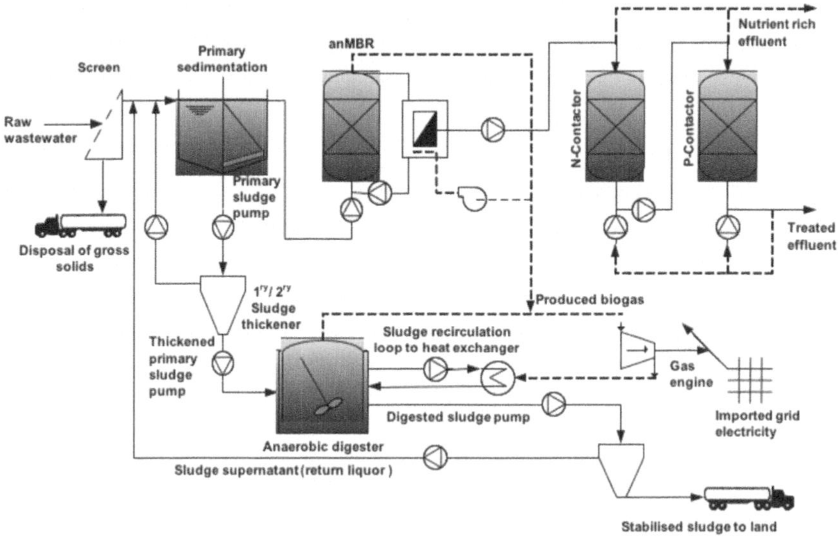

Figure 5.7 Resource factor flowsheet based on anMBR and ion exchange contactors (Jefferson, 2012).

5.3.2 The membrane in anaerobic membrane bioreactor systems

Anaerobic biomass is commonly characterised as exhibiting a higher viscosity, a higher colloidal fraction and consequently a higher fouling propensity compared to aerobic MBR sludge (Hu *et al.*, 2012). The impact is lower critical fluxes and higher sub-critical fouling rates which have been observed in both hollow fibre and tubular systems (Martin Garcia *et al.*, 2011). Reported pilot anMBR studies indicate relatively high MLSS in the CSTR reactors (Table 5.4) and this is known to increase fouling rates or sparging requirements. For instance, in a pilot trial using full-scale hollow fibre (HF) modules, critical flux was measured at 10 and 13 LMH at MLSS of 28 and 23 g L^{-1} respectively when operating at a specific gas demand ($SGD_{m)}$ of 0.23 m^3 m^{-2} h^{-1} (Robles *et al.*, 2012). This currently represents the most sustainable CSRT anMBR reported to date with an energy demand of 0.24 kWh m^{-3}. In comparison, small-scale module studies indicate critical fluxes of 4 LMH for an HF membrane operated at an SGDm of 0.19 m^3 m^{-2} h^{-1} with an MLSS of 7.7 g L^{-1} (Martin Garcia *et al.*, 2011). This mirrors established experience in aerobic systems and the need to be careful when comparing systems at difference scales with significant improvements expected when using full-scale modules. Other small scale trials have indicated critical fluxes of 8

LMH for tubular (cross flow velocity of 0.7 m s^{-1}) and 10 LMH for flat plates (SGD_m of 6 m^3 m^{-2} h^{-1}) indicating that hollow fibre systems are likely to operate at the lowest overall energy demand (Martin Garcia *et al.*, 2011; Mahmoud & Liao, 2017).

Operation of granular UASB configurations offer additional opportunities for energy reduction as the solids flux rates onto the membrane are minimised. This builds on the pioneering work of Odegaard and co-workers on the use of biofilm MBR for use in aerobic systems (Leiknes & Odegaard, 2007) and enables higher sustainable fluxes to be achieved whilst using periodic sparging (McAdam *et al.*, 2011; Cerón-Vivas *et al.*, 2012; Martin Garcia *et al.*, 2013). For instance, previous work at Cranfield on a pilot plant of the presented flowsheet (Figure 5.6) has revealed a sustainable flux of 15 LMH (7.5–21°C) with a sparging frequency of only 10% although at a high SGD_m of 8.6 m^3 m^{-2} h^{-1} (Cranfield A). However, recent work has shown that the intermittent gas sparge works best when the membrane has been put into a relaxation cycle enabling the use of low sparge frequency coupled with low net SGDm (Cranfield B). The corresponding energy demands are between 0.07–0.16 kWh m^{-3} which has been achieved with plants operating as low as 7°C. Low energy demands have also been reported in anMBRs through the use of coagulant as a fouling inhibitor (Dong *et al.*, 2015), scouring particles (Shin *et al.*, 2014) or rotating membrane units (Ruigomez *et al.*, 2016) as seen more consistently with aerobic MBRs.

Average energy production from anMBRs is approximately 0.34 kWh m^{-3} (max is 0.8 kWh m^{-3}) which is around the level of energy demand of aerobic MBRs. Accordingly, the anMBR is capable of being a net energy producer even when treating settled sewage in temperate climates. Accordingly, the outlook is bright even at this relatively early stage in the development of anMBRs. However, there is room for improvement as operating fluxes remain 33–50% lower whilst utilising SGD_m levels that are 50–300% higher than used in typical aerobic MBRs. Looking forward, it is important for more commercial-scale trials to establish a more robust baseline for performance and assessment of anMBRs. This includes understanding more operational issues such as the screening requirements and the chemical usages. For instance, a couple of the long-term pilot trials operated without the need for chemical cleaning. If this can be realised at full scale it will add a significant benefit to the technology.

5.3.3 Economics and future challenges

The lack of full-scale plants and the immaturity of the development of the technology mean that the cost of deployment of anMBRs is likely to fluctuate significantly during the initial schemes whilst established practice becomes evident. A key feature is the influent sulphate concentrations with economic modelling of a CSTR anMBR system revealing that it is better to divert the solids through primary sedimentation to an anaerobic digester when the sulphate levels are high (Pretel *et al.*, 2015). In contrast, if the sulphate levels are low, then treatment of the full load is the more economic option. In both cases the optimum represented a net energy production of 0.09–0.1 kWh m^{-3} and an annualised total cost of €0.05 m^{-3}.

Comparative economic assessment of the flowsheet depicted in Figure 5.6 to that of an activated sludge based system sized for a 10,000 PE works revealed the anMBR based flowsheets to be economically competitive. The anMBR flowsheet presented a higher CAPEX but a lower OPEX and an overall net energy production of 0.03 kWh m^{-3}. This incorporates the energy generated through anaerobic digestion of the solids fraction, evidencing that independent energy neutrality is possible at a sewage works through the application of anMBRs. The membranes were identified as the main cost component and analysis suggested a number of targets that would ensure confidence in the economics of the flowsheet:

(1) Ability to operate the USAB at an HRT of around 7 hours during average daily flow.
(2) The design flux of the membranes needs to be 12 LMH at the average daily flow.
(3) The membrane flux needs to be able to be turned up to meet flow variation in the plant.

The key challenge relates to the ability to turn up/down the membrane flux in response to changes in the flowrate though the plant. This compares to standard practice where the membrane systems are normally design for full flow treatment. Applying that approach to anMBRs makes the total MBR costs prohibitive as it doubles or trebles the required membrane area (depending on the peak load factor). This also has a knock on effect on the operating costs associated with gas sparging due to the very large membrane areas used. However, the nature of the fouling that occurs when operating the granular UASB configuration appears to present a potential solution. As the fouling is predominately particulate in nature, the formed cake can be eroded by applying elevated shear as long as a critical volume (or mass) has not been exceeded (McAdam *et al.*, 2007; Lewis *et al.*, 2012). Practically, this means either increasing the SGDm or the sparging frequency which increases the specific critical flux of the system and erodes and deposited cake. For instance during the trials of Cranfield A, increasing the sparge frequency from 10% to 50% enabled stable operation at a flux of 25 LMH for 100 hours, substantially beyond any peak flow duration (Jefferson, 2012).

A similar result can be achieved by increasing the SGDm and can be applied before, during or after the start of the increased flow. Optimum performance appears to occur when the sparging is adjusted during the flux increase and maintained for 2 hours after the flux returns to its original value. The approach has been shown to work in other systems as long as a critical volume is not exceeded, beyond which the cake compresses limiting the impact of elevated shear rates. Interestingly, during the trial of the presented flowsheet, operation beyond the critical volume occurred. Permeability was able to be restored through prolonged elevated shear and relaxation without the need for chemicals. Accordingly, the operating fluxes presented in Table 5.4 can potentially be based on average flow, manifesting a considerable reduction in capital and operating costs. However, the membrane remains a major cost item which will inevitably restrict widespread application of the technology as it has in aerobic MBRs. As such there is a need for further development that delivers lower cost, higher throughput options. One promising example is the use of low cost porous materials with pore sizes between 1 and 200 μm where dynamic membranes develop from the particle load. Early testing is promising with elevated flux rates and COD rejections of up to 99% once the cake layer has formed (Alibardi *et al.*, 2014).

Further development is clearly still required before the technology becomes widely adopted as reflected in the small market share that anMBRs occupy [around 1% of the total MBR market (Krzeminski *et al.*, 2017)]. However, recent company acquisitions indicate MBR suppliers are preparing for future market acceptance and substantial growth. This is also reflected in the number of demonstration schemes under consideration or operational indicating the appetite is there from the end user perspective if the technology can deliver on its promises.

5.4 CONCLUSIONS

The incorporation of membranes into bioreactors form an advanced wastewater treatment process, membrane bioreactors, that remain unparalleled in terms of the effluent quality achievable from a single treatment process. Use of aerobic MBRs is now established for applications where stringent discharge requirements must be met (including water reuse applications) or for plant upgrade (capacity or quality) where expansion of the existing footprint is impossible or undesirable. The success of the technology is seen in its continued market growth, although wider spread application is limited until one of the above drivers are identified due to issues around costs, energy and complexity.

Looking forward the outlook is even more favourable as the use of anaerobic MBRs (anMBRs) begins to take shape. This new advancement brings about the ability to run sewage treatment without external energy input or comprising on effluent quality. Further, anMBRs represent the pivotal technology in the delivery of the transformation from wastewater treatment to resource factory that will undoubtedly see a golden age of the use of membranes for wastewater treatment in the coming decades.

5.5 REFERENCES

Alibardi L., Cossu R., Saleem M. and Spagni A. (2014). Development and permeability of a dynamic membrane for anaerobic wastewater treatment. *Bioresource Technology*, **161**, 236–244.

Bertanza G., Canato M., Laera G., Vaccari M., Svanstrom M. and Heimersson S. (2017). A comparison between two full-scale MBR and CASE municipal wastewater treatment plants: techno-economic-environmental assessment. *Environmental Science and Pollution Research*, **24**, 17383–17393.

Besha A. T., Gebreyohannes A. Y., Tufa R. A., Bekele D. N., Curcio E. and Giorno L. (2017). Removal of emerging micropollutants by activated sludge process and membrane bioreactors and the effects of micropollutants on fouling: A review. *Journal of Environmental Chemical Engineering*, **5**, 2395–2414.

Cerón-Vivas A., Morgan-Sagastume J. M. and Noyola A. (2012). Intermittent filtration and gas bubbling for fouling reduction in anaerobic membrane bioreactors. *Journal Membrane Science*, **423–424**, 136–142.

Chang I-S., Le Clech P., Jefferson B. and Judd S. (2002). Membrane fouling in membrane bioreactors for wastewater treatment. *Journal of Environmental Management*, **128**, 1018–1029.

Chernicharo C. A. L. (2006). Post-treatment options for the anaerobic treatment of domestic wastewater. *Reviews in Environmental Science and Bio/Technology*, **5**, 73–92.

Cookney J., Cartmell E., Jefferosn B. and McAdam E. J. (2012). Recovery of methane from anaerobic process effluent using polydimethylsiloxane membrane contactors. *Water Science and Technology*, **65**, 604–610.

Cookney J., Mcleod A., Mathioudakis V., Ncube P., Soares A., Jefferson B. and McAdam E. J. (2016). Dissolved methane recovery from anaerobic effluents using hollow fibre membrane contactors. *Journal of Membrane Science*, **502**, 141–150.

Craig M. (2017). Chemical Investigation Programme 2. Available at http://events.wwtonline.co.uk/wastewater/wp-content/uploads/sites/145/2017/02/Mark-Craig.pdf (accessed 20 February 2018)

Dong Q., Parker W. and Dagnew M. (2015). Impact of $FeCL_3$ dosing on AnMBR treatment of municipal wastewater. *Water Research*, **80**, 281–293.

Drews A. (2010). Membrane fouling in membrane bioreactors – characterisation, contradictions, cause and cures. *Journal of Membrane Science*, **363**, 1–28.

Eusebi A. L., Martin-Garcia I., McAdam E. J., Jefferson B., Lester J. N. and Cartmell E. (2013). Nitrogen removal from temperate anaerobic-aerobic two stage biological systems: impact of reactor type and wastewater strength. *Journal of Chemical Technology and Biotechnology*, **88**, 2107–2114.

Fawehinmi F., Jefferson B., Chan T. and Rogalla F. (2007). Submerged Anaerobic Membrane Bioreactors (SAnMBR): ready for the big ball? In: *Proceedings of Water Environment Federation, WEFTEC 2007*, **9**, pp. 6393–6401.

Fu C., Bi X. J. and Ng H. Y. (2016). Effects of bio-carriers on membrane fouling mitigation in moving bed membrane bioreactor. *Journal of Membrane Science*, **499**, 134–142.

Ghyoot W. R. and Verstraete W. H. (2000). Reduced sludge production in a two stage membrane assisted bioreactor. *Water Research*, **34**, 2015–2215.

Gimenez J. B., Robles A., Carretero I., Duran F., Ruano M. V., Gatti M. N., Ribes J., Ferrer J. and Seco A. (2011). Experimental study of the anaerobic urban wastewater treatment in a submerged hollow-fibre membrane bioreactor at pilot scale. *Bioresource Technology*, **102**, 8799–8806.

Gouveia J., Plaza F., Garralon G., Fdz-Polanco F. and Pena M. (2015). Lon term operation of a pilot scale anaerobic membrane bioreactor (AnMBR) for the treatment of municipal wastewater under psychrophilic conditions. *Bioresource Technology*, **185**, 225–233.

Heile S., Chernicharo C. A. L., Brandt E. M. F. and McAdam E. J. (2017). Dissolved gas separation for engineered anaerobic wastewater systems. *Separation and Purification Technology*, **189**, 405–418.

Hu M., Wang X., Wen X. and Xia Y. (2012). Microbial community structures in different wastewater treatment plants as revealed by 454-pyrosequencing analysis. *Bioresource Technology*, **117**, 72–79.

Jarvis P., Jefferson B., Gregory J. and Parsons S. A. (2005). A review of floc strength and breakage. *Water Research*, **39**, 3121–3137.

Jefferson B. (2012). Self Sufficient wastewater treatment: an achievable aspiration within 5 years?. In: 9th Leading edge conference on water and wastewater technologies, 2–6th June, Brisbane, Australia.

Johir M. A. H., Aryal R., Vigneswaran S., Kandasamy J. and Grasmick A. (2011). Influence of supporting media in suspension on membrane fouling reduction in submerged membrane bioreactor (SMBR) with and without the addition of the different particle size of GAC as suspended medium. *Journal of Membrane Science*, **374**, 121–128.

Judd S. J. (2016). The status of industrial and municipal effluent treatment with membrane bioreactor technology. *Chemical Engineering Journal*, **305**, 37–45.

Judd S. J. (2018). The MBR site. Available at http://www.thembrsite.com (accessed 20 February 2018).

Judd S. J. and Jefferson B. (2003). Membranes for Industrial Wastewater Recovery and Re-use. Elsevier, Oxford.

Judd S. J. and Judd C. (2011). The MBR Book, 2nd edn. Elsevier, Oxford.

Kim J., Kim K., Ye H., Lee E., Shin C., McCarty P. L. and Bae J. (2011). Anaerobic fluidized bed membrane bioreactor for wastewater treatment. *Environmental Science & Technology*, **45**, 576–581.

Kimura K., Hara H. and Watanabe Y. (2007). Elimination of selected acidic pharmaceuticals from municipal wastewater by an activated sludge system and membrane bioreactors. *Environmental Science & Technology*, **41**, 3708–3714.

Krzeminski P., Leverette L., Malamis S. and Katsou E. (2017). Membrane bioreactors – A review on recent developments in energy reduction, fouling control, novel configurations, LCA and market penetration. *Journal of Membrane Science*, **527**, 207–227.

LeClech P., Chen V. and Fane T. A. G. (2006). Fouling in membrane bioreactors used in wastewater treatment. *Journal of Membrane Science*, **284**, 17–53.

Leiknes T. and Odegaard H. (2007). The development of a biofilm membrane bioreactor. *Desalination*, **202**, 135–143.

Lesjean B., Tazi-Pain A., Thaure D., Moesland H. and Buisson H. (2011). Ten persistent myths and the realities of membrane bioreactor technology for municipal applications. *Water Science Technology*, **63**, 32–39.

Lewis W. J. T., Chance R. M. J., Wilcox M. C., Chew Y. M. J. and Brid M. R. (2012). Investigations of cake fouling during the cross-flow microfiltration of a model suspension: Influence of buoyancy on deposition and shear-induced removal. *Procedia Engineering*, **44**, 603–606.

Maeng S. K., Choi B. G., Lee K. T. and Song K. G. (2013). Influences of solid retention time, nitrification and microbial activity on the attenuation of pharmaceuticals and estrogens in membrane bioreactors. *Water Research*, **47**, 3151–3162.

Mahmoud I. and Liao B. (2017). Effects of sludge concentration and biogas sparging rate on critical flux in a submerged anaerobic membrane bioreactor. *Journal of Water Process Engineering*, **20**, 51–60.

Martinez-Sosa D., Helmreich B., Netter T., Paris S. and Bischof F. (2011). Anaerobic submerged membrane bioreactor (AnSMBR) for municipal wastewater treatment under mesophilic and psychrophilic temperature conditions. *Bioresource Technology*, **102**, 10377–10385.

Martin Garcia I., Monsalvo V., Pidou M., LeCLech P., Judd S. J., McAdam E. J. and Jefferson B. (2011). Impact of membrane configuration on fouling in anaerobic membrane bioreactors. *Journal of Membrane Science*, **382**, 41–49.

Martin Garcia I., Mokosch M., Soares A., Pidou M. and Jefferson B. (2013). Impact on reactor configuration on the performance of anaerobic MBRs: Treatment of settled sewage in temperate climates. *Water Research*, **47**, 4853–4860.

McAdam E. J., Judd S. J., Cartmell E. and Jefferson B. (2007). Influence of substrate on fouling in anoxic immersed membrane bioreactors. *Water Research*, **41**, 3859–3867.

McAdam E. J., Lu D., Eusebi A. L., Lester J. N., Jefferson B. and Cartmell E. (2011). Integrating anaerobic processes into wastewater treatment. *Water Science and Technology*, **63**, 1459–1466.

McCarty P. L., Bae J. and Kim J. (2011). Domestic wastewater treatment as a net energy producer – can this be achieved? *Environmental Science & Technology*, **45**, 7100–7106.

McLeod A., Jefferson B and McAdam E. J. (2016). Toward gas-phase controlled mass transfer in micro-porous membrane contactors for recovery and concentration of dissolved methane in the gas phase. *Journal of Membrane Science*, **510**, 466–471.

Meng F., Zhou Z., Ni B-J., Zheng X., Huang G., Jia X., Li S., Xiong Y. and Kraume M. (2011). Characterisation of the size-fractionated biomacromolecules: tracking their role and fate in a membrane bioreactor. *Water Research*, **45**, 4661–4671.

Meng F., Zhang S., Oh Y., Zhou Z., Shin H-S. and Chae S-R. (2017). Fouling in membrane bioreactors: An updated review. *Water Research*, **114**, 151–180.

Neemann F., Rosenberger S., Jefferson B. and McAdam E. J. (2013). Non-covalent protein-polysaccharide interactions and their influence on membrane fouling. *Journal of Membrane Science*, **446**, 310–337.

Pivokonsky M., Naceradska J., Kopecka I., Baresova M., Jefferson B., Li X. and Henderson R. K. (2016). The impact of algogenic organic matter on water treatment plant operation and water quality: a review. *Critical Reviews in Environmental Science & Technology*, **46**, 291–335.

Pollice A., Brookes A., Jefferson B. and Judd S. (2005). Sub-critical flux fouling in membrane bioreactors – a review of recent literature. *Desalination*, **174**, 221–230.

Pretel R., Duran F., Robles A., Ruano M. V., Ribes J., Serralta J. and Ferrer J. (2015). Designing an AnMBR-based WWTP for energy recovery from urban wastewater: The role of primary settling and anaerobic digestion. *Separation and Purification Technology*, **156**, 132–139.

Rao N. R. H., Yap R., Whittaker M., Stuetz R. M., Jefferson B., Peirson W. L., Granville A. M. and Henderson R. K. (2018). The role of algal organic matter in the separation of algae and cyanobacteria using a novel "posi" – dissolved air flotation process. *Water Research*, **130**, 20–30.

Robles A., Ruano M. V., Ribes J. and Ferrer J. (2012). Sub-critical long term operation of industrial scale hollow fibre membranes in a submerged anaerobic MBR (HF-SAnMBR) system. *Separation and Purification Technology*, **100**, 88–96.

Robles A., Ruano M. V., Ribes J. and Ferrer J. (2013). Factors that affect the permeability of commercial hollow fibre membranes in a submerged anaerobic MBR (HF-SAnMBR) system. *Water Research*, **47**, 1277–1288.

Ruigomez I., Vera L., Gonzalez E., Gonzalez G. and Rodriguez-Sevilla J. (2016). A novel rotating HF membrane to control fouling on anaerobic membrane bioreactors treating wastewater. *Journal of Membrane Science*, **501**, 45–52.

Shin C. and Bae J. (2018). Current status of the pilot scale anaerobic membrane bioreactor treatments of domestic wastewater: A critical review. *Bioresource Technology*, **247**, 1038–1046.

Shin C., McCarty P. L., Kim J. and Bae J. (2014). Pilot scale temperate climate treatment of domestic wastewater with a staged anaerobic fluidized membrane bioreactor (SAF-MBR). *Bioresource Technology*, **159**, 95–103.

Siembida B., Cornel P., Krause S. and Zimmermann B. (2010). Effect of mechanical cleaning with granular material on the permeability of submerged membranes in the MBR process. *Water Research*, **44**, 4037–4046.

Smith S., Judd S., Stephenson T. and Jefferson B. (2003). Membrane bioreactors – hybrid activated sludge or a new process? *Membrane Technology*, December, 5–8.

Sunner N. (2017). Is the water industry ready for the challenge of operating wastewater treatment works for low phosphorus consents?. *The Big phosphorous conference*, 4–5th July 2017, Manchester, UK.

Van der Gast C. J., Jefferson B., Resid E., Robinson T., Bailey M. J., Judd S. J. and Thompson I. P. (2006). Bacterial diversity is determined by volume in membrane bioreactors. *Environmental Microbiology*, **8**, 1048–1055.

Van Loosdrecht M. C. M. and Henze M. (1999). Maintenance, endogenous respiration, lysis, decay and predation. *Water Science and Technology*, **39**(1), 107–117.

Wang Z., Ma J., Tang C. Y., Kimura K., Wang Q. and Han X. (2014). Membrane cleaning in membrane bioreactors: a review. *Journal of Membrane Science*, **468**, 276–307.

Winward G. P., Avery L. M., Stephenson T. and Jefferson B. (2008). Chlorine disinfection of grey water for reuse: Effect of organics and particles. *Water Research*, **42**, 483–491.

Young T., Smoot S., Peeters J. and Cote P. (2014). Cost effectiveness of membrane bioreactors treatment systems for low level phosphorus reduction from municipal wastewater. *Water Practice Technology*, **9**, 316–323.

Chapter 6

Organic micropollutant control

Hansruedi Siegrist[1], Adriano Joss[1], Marc Boehler[1], Christa S. McArdell[1] and Thomas Ternes[2]

[1]Eawag, Swiss Federal Institute of Aquatic Science and Technology, CH-8600 Dübendorf, Switzerland

[2]BfG, Federal Institute of Hydrology, D-56068 Koblenz, Germany

ABSTRACT

About 100,000 chemical substances have economic uses, of which 4,600 substances are produced annually in quantities exceeding 1,000 tons in the OECD member states. During production and disposal, as well as after regular use, a significant portion of these substances will unavoidably enter the aquatic environment, where they can have ecotoxicological effects, such as fish feminization. Organic compounds are increasingly being detected in water bodies and in sewage sludge in low concentration ranges (µg/L and ng/L), being referred to as organic micropollutants. The removal of micropollutants by conventional municipal wastewater treatment is incomplete and a multitude of compounds are only slightly abated. To significantly reduce eco-toxicological effects of micropollutants in receiving waters and water resources, additional treatment steps (e.g. ozonation or activated carbon adsorption) are required in combination with measures taken at the source. Ozonation requires complete nitrification and it might not be suitable for wastewater with high industrial wastewater fractions and high bromide concentrations due to the production of bromate and increased oxidation by-products. Powdered activated carbon (PAC) addition is already broadly applied. PAC can be recycled to the activated sludge treatment to improve micropollutant removal and therefore reduce PAC consumption. Recently granular activated carbon

Editors: Giorgio Mannina, George Ekama, Hallvard Ødegaard and Gustaf Olsson
doi: 10.2166/9781780409719_0231

(GAC) filters are increasingly investigated and built in full scale for micropollutant and dissolved organic carbon (DOC) removal. Additional measures in wastewater treatment should be based on careful micropollutant mass flux studies to identify critical catchment areas so that relevant hot spots can be addressed.

Keywords: activated carbon adsorption, GAC, indicator compounds, micropollutant, ozonation, PAC, process control, transformation products

6.1 INTRODUCTION

The scientific literature contains descriptions of over 80 million chemical substances, of which a good 100,000 have economic uses. The member states of the Organization for Economic Co-operation and Development (OECD) produce approximately 4,600 substances annually in quantities exceeding 1,000 tons. Of these 4,600 chemicals only about 1,000 had been tested and assessed by the end of 2010. Data are also lacking for the estimated 30,000 substances, whose market volume in Europe exceeds one ton (Federal office for the environment, 2015).

During production and disposal, as well as after regular use, a significant portion of these substances will unavoidably enter the aquatic environment. Owing to constantly improved methods of chemical analysis, organic compounds are increasingly being detected in water bodies and in sewage sludge in low concentration ranges (µg/L and ng/L and lower), being referred to as organic micropollutants.

In the last 25 years the focus of environmental chemistry research has been extended from the more "classic", mostly apolar environmental micropollutants such as polychlorinated biphenyls (PCBs), dichlorodiphenyltrichloroethane (DDT), dioxins and pesticides to polar and ionizable chemicals, so called "pollutants of emerging concern", although these have been released for much longer. Prominent among these pollutants are pharmaceuticals, hormones, and ingredients of personal care products (PPCPs) as well as biocides that enter the environment mainly through regular domestic use and in municipal wastewater (Ternes *et al.*, 2006). Additional compounds leaching out from consumer products (flame retardants), or used to create inert surfaces (perfluorinated compounds) also give reason for concern. Hospital wastewaters contribute significantly but not to a major extent, to the total pharmaceutical and biocide loads (Kovalova *et al.*, 2013; Verlicchi, 2017). Compounds not readily degraded in municipal wastewater treatment plants (WWTPs) are discharged to the receiving waters either in the dissolved states or sorbed onto suspended solids (TSS). Therefore, the removal efficiency at WWTPs is crucial and is significantly determining the contamination of the receiving water bodies. In surface water and during underground passage, further elimination is observed by photochemical, sorption and biological processes, but micropollutants can still enter groundwater where they can act as tracers for anthropogenic pollution arising from wastewater. Some polar pollutants (e.g. lipid regulators, anti-epileptics such as gabapentin and carbamazepine, and iodinated X-ray contrast media) might even pass through common drinking water treatment processes.

An early German study (Ternes, 1998) investigated the occurrence of 55 pharmaceutically active ingredients and 9 metabolites in the discharge of 49 WWTPs and in the receiving water bodies. In total, 36 active agents and 5 metabolites were detected in the WWTP effluent in concentrations of up to several µg/L. Even in surface waters, some pollutant (e.g. beta blockers and anti-epileptic drugs) peak concentrations were measured in excess of 1 µg/L. In 2010 an EU-wide monitoring survey on 156 emerging polar organic contaminants in 90 WWTP effluents showed the presence of 125 substances (80% of the target compounds) in European wastewater effluents, in concentrations ranging from low ng/L to mg/L (Loos *et al.,* 2013).

There has been much discussion in recent years of new environmental effects, such as fish feminization (Jobling & Tyler, 2003). This is partly a result of the chronic introduction of endocrine disrupting chemicals. These include, along with the body's natural hormones, pharmaceuticals used for their hormonal effect, such as contraceptives and anti-diabetic drugs. A hormonal side effect is also attributed to certain pharmaceuticals such as β-sitosterol (control of cholesterol level, also called phytoestrogen as it is widely distributed in plants) and clenbuterol (asthma relief agent). Endocrine effects are reported also for nonylphenol (surfactant metabolite) and industrial chemicals such as bisphenol A and phthalates.

Diclofenac and four pesticides (naproxen, clothianidin, imidacloprid, flonicamid) drive most of the acute risk for macroinvertebrates during dry weather conditions downstream of Swiss municipal WWTPs (Munz *et al.,* 2017). All these effects are significantly reduced by measures at the source and additional technical treatment (ozonation or activated carbon adsorption) at the WWTP. In Switzerland, a new water protection act entered into force as of January 2016 with the objective to improve surface water quality by reducing the load of micropollutants from WWTPs. A selection of WWTPs will need to be upgraded in the future with an additional treatment step for micropollutant abatement (Eggen *et al.,* 2014).

6.2 FATE OF MICROPOLLUTANTS IN MUNICIPAL WWTPS

The removal of micropollutants in WWTPs depends on the level of the biological treatment and the efficiency of the TSS separation methods. In the last 40 years, biological wastewater purification has been adapted step by step to more stringent effluent conditions for chemical oxygen demand (COD), TSS and nutrient removal. The most important micropollutant elimination processes are (Siegrist *et al.,* 2014):

- Sorption to TSS in mechanical and biological treatment, and removal by sedimentation and filtration;
- Stripping by aeration;
- Biological degradation and transformation in the activated sludge or biofilm systems of the WWTP; this may achieve biological mineralization but is mostly limited to compound transformation (Wick *et al.,* 2011a).

For **sorption** of organic micropollutants, two processes can be differentiated:

- Hydrophobic interactions of the aliphatic and aromatic groups of a compound with the lipophilic cell membrane of the microorganisms and the lipid fractions of the sludge (compounds with high octanol–water partition coefficients (K_{OW}), e.g. tonalide in Figure 6.1);
- Electrostatic interactions of positively charged groups of chemicals with the negatively charged surfaces of bacteria (e.g. norfloxacin in Figure 6.1).

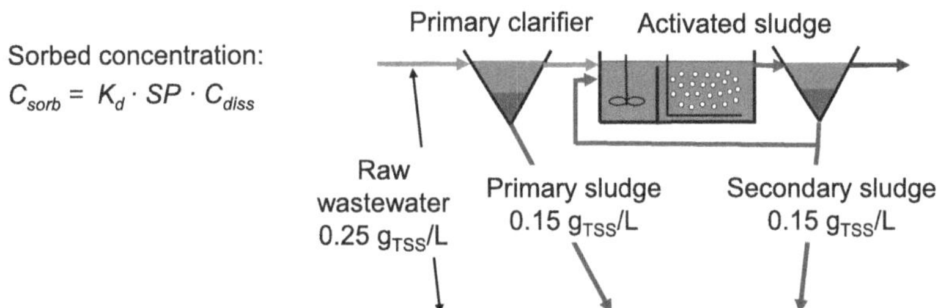

Compound	K_d (L/g_{TSS})	sorbed fraction = 100%·K_d·SP/(1+ K_d·SP)		
Diclofenac	0.5 / 0.02	11	7	<1
Ethinylestradiol	0.3	7	4	4
Tonalide	5 / 2	56	43	23
Norfloxacin	2.5 / 35	38	27	84

Figure 6.1 Sorption of some compounds in raw wastewater, primary and secondary sludge based on the K_d values given in Joss *et al.*, (2006a) and Golet *et al.*, (2003) assuming no biological degradation. The sorbed fraction is related to the inlet of the primary clarifier and the inlet to biological treatment, respectively. K_d values are given for primary and secondary sludge. (Reproduced from Siegrist *et al.*, 2014, with permission from the copyright holders, IWA publishing.)

The sorbed concentration of a substance C_{sorb} (µg/L) can be expressed by a simplified linear model (Joss *et al.*, 2006b; Wick *et al.*, 2011b). It depends on the sorption constant K_d (L/g_{TSS}), the TSS concentration or the amount of sludge produced SP (g_{TSS}/L) to which the substance can adhere, and the dissolved substance concentration C_{diss} (µg/L):

$$C_{sorb} = K_d \cdot SP \cdot C_{diss} \tag{6.1}$$

Simplified, for a fully mixed system the removal efficiency η_{sorb} (–) is:

$$\eta_{sorb} = \frac{C_{sorb}}{C_{diss} + C_{sorb}} = \frac{K_d \cdot SP}{1 + K_d \cdot SP} \tag{6.2}$$

K_d can be roughly estimated from the octanol–water distribution coefficient for non-polar compounds. For polar and charged compounds with electrostatic interactions, it must be determined by means of sorption tests. A substance with a low K_{OW} that sorbs relatively well to TSS is the antibiotic norfloxacin. The sorption is based to a large extent on electrostatic interactions between the positively charged amino group of norfloxacin and the negatively charged surfaces of the microorganisms. In a study carried out in the Zurich WWTP, Golet *et al.*, (2003) confirmed that more than 80% of the norfloxacin in the primary effluent was sorbed to the secondary sludge (Figure 6.1), while removal in primary treatment was significantly lower.

Primary and secondary sludge production are similar, but the primary sludge contains fewer microorganisms and has a larger lipid fraction. Therefore, microorganisms in the secondary sludge represent the greater proportion of the TSS compared to primary sludge, resulting in a relatively high sorption constant ($K_d \approx 35$ L/g$_{TSS}$) for norfloxacin in secondary sludge, while for the primary sludge the sorption constant is much lower ($K_d \approx 2.5$ L/g$_{TSS}$). Thus, only 27% of norfloxacin is sorbed to the primary sludge. In contrary, the hydrophobic musk fragrance tonalide sorbs more strongly to the lipophilic primary sludge than to the secondary sludge (Figure 6.1). For other substances, such as the anti-inflammatory diclofenac and estrogens, the sorbed proportion is significantly smaller (Figure 6.1). K_d values for other pharmaceuticals, personal care products and biocides are given in Joss *et al.*, (2006a) and Wick *et al.*, (2011b).

For membrane bioreactors (MBRs), the effluent is virtually free of TSS and higher airflows per volume of treated wastewater are required for membrane fouling control than in conventional WWTPs. Therefore, the removals by sorption and stripping (see below) are slightly higher.

Stripping due to aeration in biological treatment and grit, removal is mostly rather low because of the low volatility of the large, polar and rather hydrophilic micropollutants under consideration. For substances with a dimensionless Henry's law constant $H < 1$, describing the equilibrium between the gas ($C_{\text{air,off}}$) and water phase (C_{diss}), the rising gas bubble is in equilibrium with the dissolved concentration and:

$$C_{\text{air,off}} = H \cdot C_{\text{diss}} \tag{6.3}$$

The stripping efficiency is therefore only dependent on the Henry's law constant and the airflow. For the relatively volatile musk fragrance tonalide, having a constant of $H_{\text{Tonalide}} \sim 0.005$ and an air to water flow ratio $q_{\text{air}} = Q_{\text{air}}/Q_{\text{water}} = 8\text{–}10$ for conventional plants and 20–30 for MBRs, the stripping efficiency is in the order of 4–15%:

$$\begin{aligned}\eta_{\text{stripp}} &= \frac{H \cdot C_{\text{diss}} \cdot q_{\text{air}}}{C_{\text{diss}} + H \cdot C_{\text{diss}} \cdot q_{air}} = \frac{H \cdot q_{\text{air}}}{1 + H \cdot q_{\text{air}}} \\ &\cong H_{\text{Tonalide}} \cdot q_{\text{air}} = 0.04 - 0.15\end{aligned} \tag{6.4}$$

Biological degradation and transformation of a substance can take place under aerobic, anoxic and/or anaerobic conditions. Due to their low concentrations (ng/L to µg/L), single substrate growth on micropollutants is not assumed possible and we have to distinguish between (Siegrist *et al.*, 2003):

- Co-metabolism, when the bacteria only transform but do not mineralize the trace substance and do not use it as a carbon and energy source; and
- Mixed substrate growth, in which the bacteria use the trace substance as a carbon and energy source, and hence can totally mineralize it.

Both removals are thought to arise through the by random affinity of a trace substance with the bacterial enzymes in biological treatment. To some extent, the transformation or mineralization increases with the age of the sludge SRT or the biofilm, probably because slow growing bacteria establish themselves, thereby increasing the metabolic diversity (Figure 6.2 and Falås *et al.*, 2013).

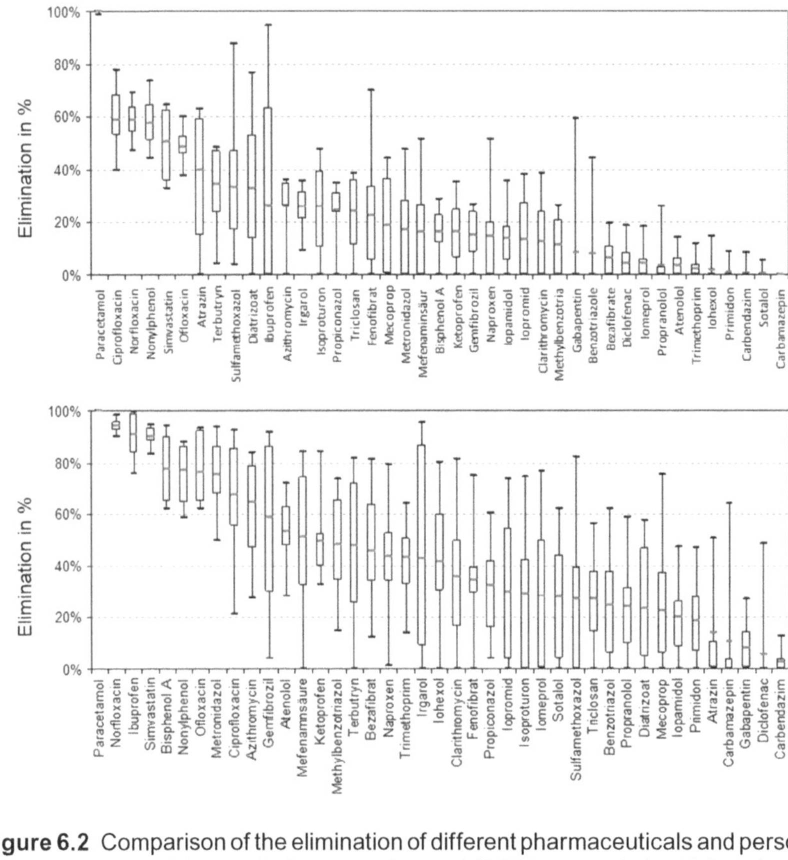

Figure 6.2 Comparison of the elimination of different pharmaceuticals and personal care products in a biochemical oxygen demand (BOD) removal plant (above) and a nitrifying plant (below) fed with the same municipal wastewater (Margot *et al.*, 2011; Margot *et al.*, 2013).

Degradation also depends on the redox state of the process. This is demonstrated by the contraceptive 17α-ethinylestradiol (Figure 6.3), which requires nitrifying conditions for degradation and is hardly degraded under anoxic and anaerobic conditions (Andersen *et al.*, 2003).

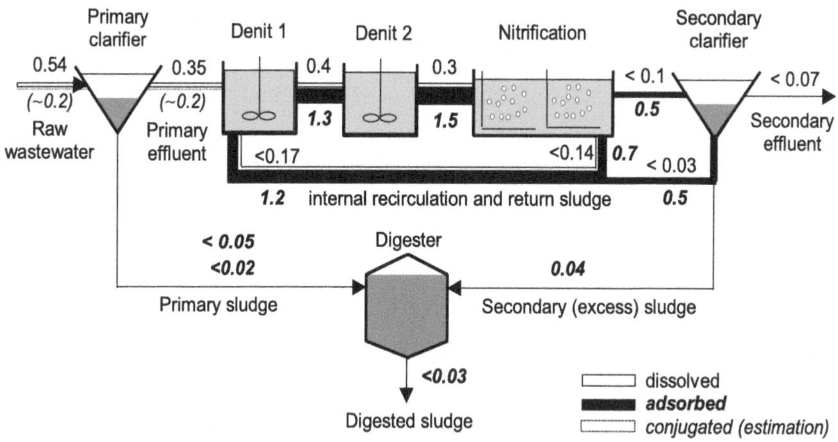

Figure 6.3 Substance flow and breakdown of the contraceptive 17α-ethinylestradiol in the Wiesbaden WWTP, Germany. The data are in g/d. The value in the inflow includes both dissolved 17α-ethinylestradiol and its conjugated form. (Reprinted with permission from Andersen *et al.*, 2003. Copyright 2018, American Chemical Society.)

While MBRs sometimes have significantly higher sludge ages than conventional plants, their micropollutant removal is only slightly increased above sludge ages of 15 d (Joss *et al.*, 2006a).

Due to the low concentrations of trace substances, the elimination occurs mostly at a reaction rate r_{biol} (μg/L, d), which can be described as the product of the first order reaction rate constant k_{biol} (L/g_{TSS}, d), the concentration of activated sludge X_{SS} (g_{TSS}/L) and the dissolved concentration C_{diss} (μg/L) (Wick *et al.*, 2009; Joss *et al.*, 2006b):

$$r_{biol} = k_{biol} \cdot X_{TSS} \cdot C_{diss} \tag{6.5}$$

Because of the first order reaction, a compartmentalized aeration basin results in lower discharge concentrations compared to a single completely mixed basin at the same total hydraulic retention time.

Joss *et al.* (2006b) presented a classification of degradable pharmaceuticals and personal care products based on first order removal rate coefficients in typical nutrient removal WWTPs (Figure 6.4).

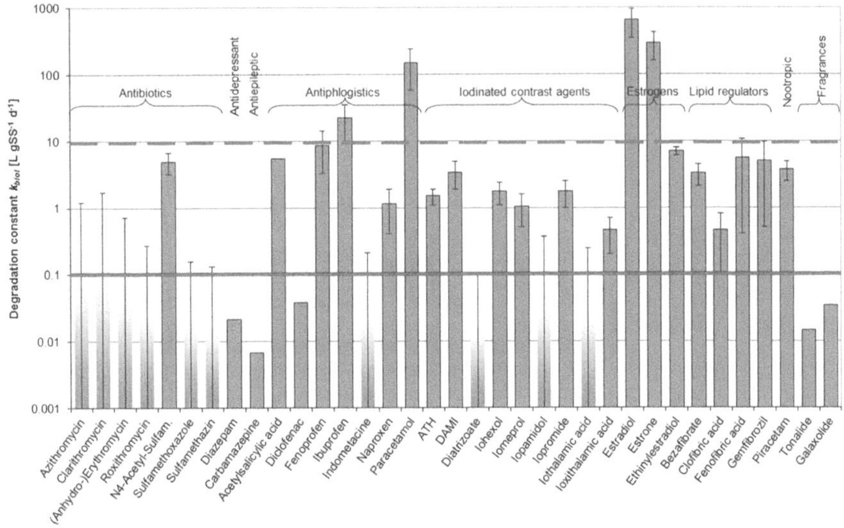

Figure 6.4 Degradation rate constants of 35 pharmaceuticals, hormones and PPCPs observed in activated sludge treatment of municipal nutrient-removal WWTPs. The error bars indicate the 95% confidence interval. The lines at k_{biol} 0.1 and 10 indicate the limits of <20% and >90% removal for a nitrifying activated sludge system with pre-denitrification (sludge age 11–15 d). The fading grey bars indicate high uncertainty because the measured concentrations are close to the detection limit. (Reprinted from Joss *et al.*, 2006b with permission from Elsevier.)

Falås *et al.* 2013 investigated the removal of organic micropollutants in a hybrid biofilm-activated sludge process with batch experiments, modeling, and full-scale measurements. Batch experiments were performed with biomass from a full-scale reactor on one hand with carriers under oxic conditions and on the other hand with sludge alone under oxic and anoxic conditions. Clear differences in the micropollutant removal kinetics of the attached and suspended biota growth were demonstrated, often with considerably higher removal rates for the biofilm compared to the sludge. For several micropollutants, the removal rates were also affected by the redox conditions, i.e. oxic and anoxic. Removal rates obtained from the batch experiments were used to model the micropollutant removal in the full-scale process. The results from the model and plant measurements showed that the removal efficiency of the process can be predicted with acceptable accuracy (±25%) for most of the modeled micropollutants. Furthermore, the model estimations indicate that the attached growth in hybrid biofilm activated sludge processes can contribute significantly to the removal of individual compounds, such as diclofenac.

On the basis of a large dataset obtained from 15 diverse biological reactor configurations over more than 10 years of total reactor operation time, and the associated analysis of a broad range of organic micropollutants in several hundred wastewater samples, the following key conclusions were drawn (Falås *et al.*, 2016):

- Micropollutant degradation is not strongly affected by sludge-age increases from 25 to 80 days and long-term exposure to organic micropollutants at typical municipal wastewater concentrations is generally not a necessary trigger for the micropollutant degradation in biological wastewater treatment.
- The presence of the main substrates for microbial growth is generally neither a main trigger nor a strong inhibitor of micropollutant degradation in biological wastewater systems, although exceptions exist.
- Many micropollutants, such as acyclovir, bezafibrate and atenolol, are almost ubiquitously degraded in aerobic wastewater treatment processes, whereas certain compounds such as trimethoprim, diuron and diclofenac seem to require quite specific aerobic treatment conditions for their degradation or transformation.
- Demethylation and deiodination of some target micropollutants with high aerobic persistence can be achieved under anaerobic conditions.
- The spectrum of organic micropollutants susceptible to biological degradation at WWTPs can be broadened by combining different aerobic and anaerobic treatment conditions. This improved removal is, however, restricted to a limited number of compounds, and many micropollutants must be considered stable in biological processes for municipal WWTPs.

6.3 BIOLOGICAL TRANSFORMATION PRODUCTS

Even though transformation products (TPs) have gained increasing interest as water contaminants, only a few studies have investigated the formation and fate of biological transformation products of pharmaceuticals in contact with biologically active matrices such as activated sludge or sediments (DWA, 2014). One reason for this is the challenge of the structural elucidation of TPs present at low concentrations in natural matrices. Sophisticated analytical techniques are needed such as hybrid high-resolution mass spectrometry and nuclear magnetic resonance (NMR) (Kormos *et al.*, 2009). High resolution mass spectrometry (MS) (e.g. LC-LTQ-Orbitrap MS) can be used to elucidate TP of simple transformation reactions (e.g. hydrolysis of amides and esters). However, confirmation of the chemical structures of TPs is difficult, even though the parent compound is known, as reference standards of TPs are mostly not available. The TP structure suggestions based on exact masses and mass fragments have to be confirmed by alternative analytical methods or by specific chemical reactions that take place with the new functional moieties formed. Possibilities of analytical methods include a range of NMR spectroscopic methods and to a much lesser extent infrared spectroscopy. A drawback of both types of technique is the elevated quantity and the high purity needed for isolated standards. In cases where no authentic standard is available and only the MS spectra of the TPs have been obtained, we might better define the suggestions of the TP chemical structures as "tentative identification". Schymanski *et al.* (2014) proposed a level system to ease the communication of identification confidence based on different confirmation criteria. A comprehensive overview of the literature up to 2008 on the detection and identification of pharmaceutical TPs has been provided by Celiz *et al.* (2009).

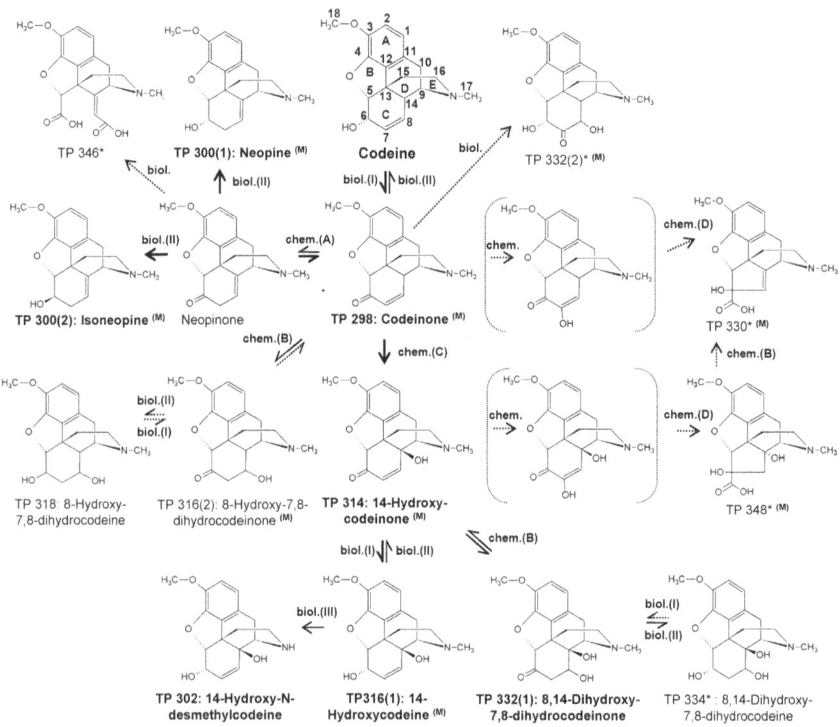

Figure 6.5 Proposed transformation pathways of codeine determined from laboratory batch experiments with codeine and codeine TPs in diluted activated sludge as well as in autoclaved controls under aerobic conditions. Major TPs are printed in bold. The dashed arrows indicate predicted reactions which could not be directly confirmed in batch experiments. Tentative TP chemical structures are marked with an asterisk. The index (M) indicates that the corresponding TPs were also detected for morphine (O-CH_3 at C3 replaced by OH). Proposed intermediates in the formation of TPs are shown in parentheses. The roman numerals I-III and the letters A-D indicate biotic and abiotic reactions, respectively: (I) oxidation, (II) reduction, (III) N-demethylation; (A) double bond shift, (B) nucleophilic addition, (C) hydroxylation, (D) benzylic acid rearrangement. (Reprinted with permission from Wick *et al.*, 2011a. Copyright 2018, American Chemical Society.)

Several recent studies have indicated that the majority of pharmaceutical TPs formed under aerobic conditions have a slightly modified molecular structure with increased polarity due to the introduction of hydroxyl-, carboxyl- or keto-moieties (Neilson & Allard, 2007; Helbling *et al.*, 2010). Based on the similarity of their molecular structure to the parent compound, some TPs are expected to possess comparable biological activities to their chemical precursors (van Zelm *et al.*, 2010; Achermann *et al.*, 2018). However, the enhanced polarity should improve the permeability of these compounds during several water treatment processes such as adsorptive filtration (e.g. activated carbon), underground soil passage or bank filtration. As a consequence, the likelihood increases that TPs are contaminating ground- and drinking water (Kormos *et al.*, 2010).

Several enzyme-catalyzed reactions seem to be quite common in the transformation of pharmaceuticals. These include: mono- and dihydroxylation, alcohol and aldehyde oxidation, ester and amide hydrolysis, *N*-dealkylation, *N*-deacetylation and decarboxylation. As an example the results of the complex codeine transformation study by Wick *et al.* (2011a) are described in Figure 6.5. This example indicates that during biological treatment a combination of biotic, enzymatic controlled, and abiotic reactions lead to the formation of a wide range of transformation products.

The pathway prediction system enviPath tries to provide different relative reasoning models to predict likely bio-transformation pathways and products (https://envipath.org/). However, the prediction is limited, especially if abiotic reactions are involved.

Although biological transformation products are increasingly found in ground and drinking water due to their higher polarity, their human health and eco-toxicological effects are in most cases considered to be less problematic due to their lower hydrophobicity and therefore lower bioaccumulation potential and the low concentration (Boxall *et al.*, 2004).

6.4 ADDITIONAL TREATMENT TO CONTROL MICROPOLLUTANT REMOVAL

Many micropollutants and their intermediates are polar and are often poorly biologically degraded; furthermore, their sorption is also limited (Wick *et al.*, 2011a). Therefore, they are only partly removed in the WWTPs and are discharged to the receiving water in the plant effluent. Using a mass flux model, which includes chemical consumption, transformation and excretion in the human body, removal in the WWTP, as well as the hydraulic conditions in the receiving waters during dry weather conditions, the concentration ranges in Swiss river systems were modeled and validated with measurements (Ort *et al.*, 2009). This study in rivers downstream of more than 100 WWTPs showed that the concentration of several eco-toxicologically critical compounds exceeded the proposed environmental quality standards because of the low dilution of the treated wastewater. This situation is likely to become even more critical with climate change, since drier and warmer summers are expected.

To reduce the load of micropollutants in the aquatic environment, **measures at the source** are feasible, e.g. eco-labeling (Klaschka *et al.*, 2004; Agerstrand *et al.*, 2009), ban restricted use of critical compounds, and separate treatment of concentrated waste streams (e.g. hospital wastewater, Kovalova *et al.*, 2013). However, for a more effective reduction of loads and to protect specific sensitive receiving waters and water resources, **technical measures at the WWTPs** using advanced treatment methods are required (Abegglen & Siegrist, 2012). The following treatments have been investigated and tested in pilot and full scale:

- Ozonation with biological post-treatment, e.g. sand filtration,
- Powdered activated carbon (PAC) addition to a contact tank followed by a clarifier for PAC elimination, or direct addition to activated sludge treatment followed by a polishing filter, or direct addition to sand filter,
- Granular activated carbon filtration (GAC),
- Process combinations: Ozonation followed by GAC filtration or ozonation followed by PAC dosage to a sand filter.

6.4.1 Ozonation followed by biological filters to remove oxidation by-products

The dosage of 3-10 g/m^3 ozone to biologically treated wastewater with a background dissolved organic carbon (DOC) of 5–15 mg/L, corresponding to 0.4–0.6 gO_3/gDOC, is a cost-effective method for more than 80% abatement and inactivation by partial oxidation of pharmaceuticals and pesticides (Figure 6.6). However, oxidation by-products are formed mainly by reactions with background DOC (Abegglen *et al.*, 2009; Benner & Ternes, 2009; Krauss *et al.*, 2010; Lee & von Gunten, 2016). To eliminate possible negative eco-toxicological effects caused by biodegradable oxidation by-products, a biofilm system is needed, e.g. a sand filter following the ozonation step (Huber *et al.*, 2005; Hollender *et al.*, 2009; Stalter *et al.*, 2010, 2013; Zimmermann *et al.*, 2011). The effectiveness of ozone depends on the background load of dissolved organic carbon compounds in the wastewater and the chemical properties of the micropollutants (Huber *et al.*, 2002). Ozone reacts specifically with electron-rich moieties, e.g., tertiary amines or activated aromatics (von Sonntag & von Gunten, 2012). During ozonation, hydroxyl radicals are produced from interaction with the background DOC, which also degrade chemicals that are not attacked by ozone (Hollender *et al.*, 2009; Zimmermann *et al.*, 2011).

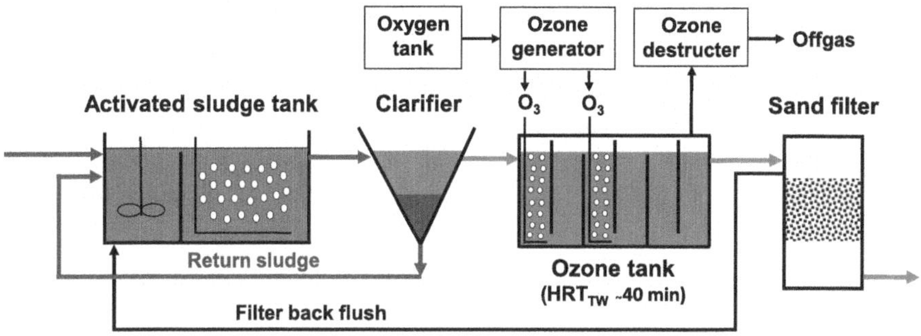

Figure 6.6 Ozonation of secondary effluent. Hydraulic retention time (HRT) in the cascaded ozone tank preventing short circuits should be about 40 minutes during maximum dry weather flow to prevent ozone loss to sand filter during storm water, when HRT is only 10–15 minutes. As post-treatment, a biological system, e.g. sand filter, is required to abate degradable reaction products. For high ozone dosages a two-step dosage is advantageous to reduce ozone consumption. A fine bubble aeration system and a reactor depth of >5 m is required to prevent too much ozone loss via off-gas.

For municipal wastewater with a high fraction of industrial wastewater, ozonation might not be suitable (Wunderlin *et al.,* 2015). For example, high bromide concentration (>100 μg/L) can result from off-gas treatment of municipal waste incineration (Soltermann *et al.,* 2016). Schindler-Wildhaber *et al.,* (2015) suggest a relatively cheap and straightforward modular laboratory decision tool to test the feasibility for ozonation. These modules consist of investigations to assess (i) the ozone stability, (ii) the efficiency of micropollutant removal, (iii) the oxidation by-product formation, as well as (iv) bioassays to measure specific and unspecific toxicity of the treated wastewater.

Ozonation has a strong disinfecting effect and achieves the EU bathing water standards for *E. coli* (<500 counts per 100 mL), but does not achieve a complete hygienization for typical doses (DWA KA 8.1, 2008; Abegglen *et al.,* 2009). However, ozonation requires a **well-functioning nitrification** because nitrite reacts rapidly with ozone, consuming 3.4 $gO_3/g\ NO_2^-N$, thus significantly increasing the ozone dosage.

At the Neugut WWTP, the first Swiss plant with an ozonation step went into operation in spring 2014. Due to the well-functioning activated sludge treatment for biological N and P removal, reaching nitrite effluent concentrations of less than 0.05 gN/m^3, a dosage of 0.4 $gO_3/gDOC$ was sufficient to reach an average 80% elimination of all 12 indicator compounds selected by the Swiss Federal Office for the Environment (Bourgin *et al.,* 2018; Table 6.1 and 6.2).

Table 6.1 Elimination of all 12 indicator compounds (see Table 6.2) in the Neugut WWTP upgraded with ozonation in relation to the ozone dosage or specific ozone dosage (Böhler *et al.*, 2017; Bourgin *et al.*, 2018).

Elimination of 12 Indicator Compounds	DOC_{inlet} g/m^{-3}	UV_{inlet} E m^{-1}	Ozone Dosage gO_3 m^{-3}	Specific Ozone Dosage gO_3/gDOC	Specific Ozone Dosage gO_3m^{-3}/(E m^{-1})
average > 80%	3.5 – 6	8 – 13	1.5 – 2.5	0.40	0.20
all > 80%	3.5 – 6	8 – 13	2.2 – 3.9	0.65	0.31

Table 6.2 Indicator compounds suggested by the Swiss Federal Office for the Environment (FOEN, 2016). An overall elimination of 80% for at least 6 selected or all 12 compounds has to be reached over the whole WWTP.

Indicator Compound	Elimination with Ozone, PAC, GAC	Type of Chemical
Amisulpride	very good	Psychotropic drug
Carbamazepine	very good	Antiepileptic drug
Citalopram	very good	Antidepressant drug
Clarithromycin	very good	Antibiotic drug
Diclofenac	very good	Analgetic drug
Hydrochlorothiazide	very good	Diuretic drug
Metoprolol	very good	Beta blocker drug
Venlafaxine	very good	Antidepressant drug
Benzotriazole	good	Corrosion inhibitor
Methylbenzotriazole	good	Corrosion inhibitor
Candesartan	good	Antihypertensive drug
Irbesartan	good	Antihypertensive drug

Two sampling campaigns (48-h flow proportional samples) were carried out at Neugut WWTP to investigate the abatement of an exhaustive list of 550 micropollutants over the whole treatment chain at a specific ozone dose of 0.55 gO_3/gDOC, corresponding to 2.0–3.3 gO_3 m^{-3} (Figure 6.7). In the primary effluent (WWTP-INF), 175 compounds were found (>Limit of quantification, LOQ) with a total concentration of 371 µg/L. Pharmaceuticals (133 µg/L; mainly metformin and paracetamol), caffeine (130 µg/L) and food additives (85 µg/L, mainly cyclamate) were found in the highest concentration. The total concentration of target compounds was decreased by 85% to 54 µg/L in biological treatment. The four compounds caffeine, cyclamate, metformin and paracetamol which were most abundant in the primary effluent, were eliminated by 99%, dominating the overall degradation. Without the contribution of these four compounds, the abatement of the remaining micropollutant load drops from 85% to 54% during the biological treatment. During ozonation, the load of micropollutants was reduced by 58%, leading to a total concentration of 23 μg/L. No significant further elimination was observed in sand filtration. Highest concentrations in the final effluent of >0.9 μg/L were found for acesulfame, sucralose, metformin, iopromide and tris(2-chloropropyl)phosphate, all compounds that showed poor abatement in ozonation. Over the whole treatment chain, a 94% reduction of micropollutant load was observed (Figure 6.7).

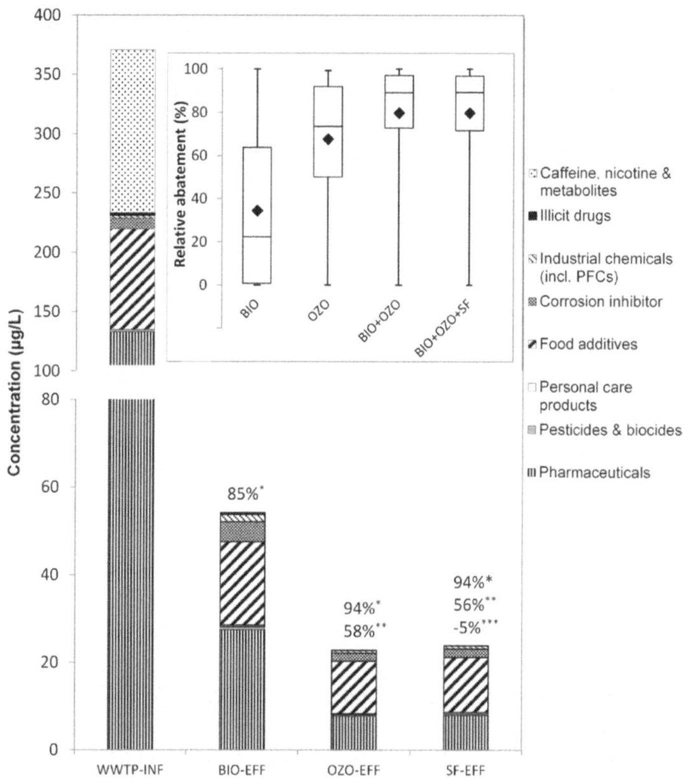

Figure 6.7 Sum of concentrations of micropollutants at various treatment steps at Neugut WWTP at a specific ozone dose of 0.55 $gO_3/gDOC$ (2.0–3.3 gO_3m^{-3}): *: abatement of total concentration relative to WWTP-INF (effluent of primary clarifier); **: abatement relative to BIO-EFF (effluent of conventional biological treatment); ***: abatement relative to OZO-EFF (effluent of ozonation reactor), SF-EFF (sand filter). Box-whisker plot depicts the distribution of calculated abatements for all substances (minimum, 25th percentile, median, 75th percentile, maximum and (♦) average) of micropollutants analyzed during all treatment steps. Because nearly half of the compounds were not detected anymore after ozonation, the values given for OZO, BIO + OZO and BIO + OZO + SF represent minimal abatements relative to the LOQ. Negative abatements were taken into account as zero abatements. (PFC = perfluorinated compound) (Reprinted from Bourgin *et al.*, 2018 with permission from Elsevier.)

Alternatively, an average abatement of all substances can be calculated from the abatement of each single substance. With this calculation, the overall average abatement over the whole WWTP was calculated to be >79% (Figure 6.7, inset). Since many substances (64 out of 147) were not detected anymore after ozonation, only a minimal abatement relative to the LOQ could be calculated. Ozonation contributes with the highest portion to the overall abatement by >67%, while only 34% of abatement was observed in biological treatment.

The ozonation by-products bromate (formed from bromide) and nitrosamines, e.g., N-nitrosodimethylamine NDMA (formed of precursors such as hydrazines and sulfamides), are known to be (possible) human carcinogens (Tricker & Preussmann, 1991; Soltermann *et al.,* 2016). Therefore, the formation of these by-products needs to be measured to attest the feasibility of ozonation. At Neugut WWTP, NDMA and bromate concentration were clearly below (potential) drinking water standards (10 μg/L for bromate and 10 ng/L for NDMA) in the discharged receiving water.

Beside sand filtration, other biological post-treatments such as a moving bed or a fixed bed can be used after ozonation. In a study comparing all three systems (Bourgin *et al.,* 2018), sand filtration was found to perform best in terms of removal of dissolved organic carbon (DOC), assimilable organic carbon (AOC) and total suspended solids (TSS). As an alternative, a granular activated carbon filter could be used as post-treatment. Even at very high run-times, e.g. after 50,000 bed volumes, corresponding to more than two years of operation, certain micropollutants are still found to be reduced (Bitterwolf *et al.,* 2017; Bourgin *et al.,* 2018). Beside sorption, a biological degradation may take place (Reungoat *et al.,* 2012), rendering the GAC filter as a promising alternative to sand filtration (see Chapter 6.4.3).

Ozone transforms organic compounds that absorb light within the UV and visible spectra to reaction products with lower UV absorbance, leading to a decrease of absorbance in ozonation. This decrease can be used as an indicator of micropollutant removal efficiency in wastewater during ozonation. As seen from Figure 6.8, specific ozone dosage (O_3/UV_{in}) and delta UV (UV_{in}-UV_{out}) correlates well with the elimination of micropollutants. Monitoring UV and delta UV allows therefore an economic control of ozone dosage for treatment plants with daily and weekly variations of nitrite (nitrite sorbs no UV but uses ozone) and DOC (Wittmer *et al.,* 2015).

For WWTPs with an effluent DOC of 10 g/m^3, the operation and investment costs (incl. sand filter) are in the range of about 10 US-cents/m^3 of wastewater for a plant size of 100,000 PE (Abegglen & Siegrist, 2012; Bourgin *et al.,* 2018). The energy consumption for ozonation, including pure oxygen production (18 kWh/kgO_3) is 0.1 kWh/m^3 for 5 gO_3/m$_3$ and is therefore significant in comparison with the total energy consumption of a WWTP.

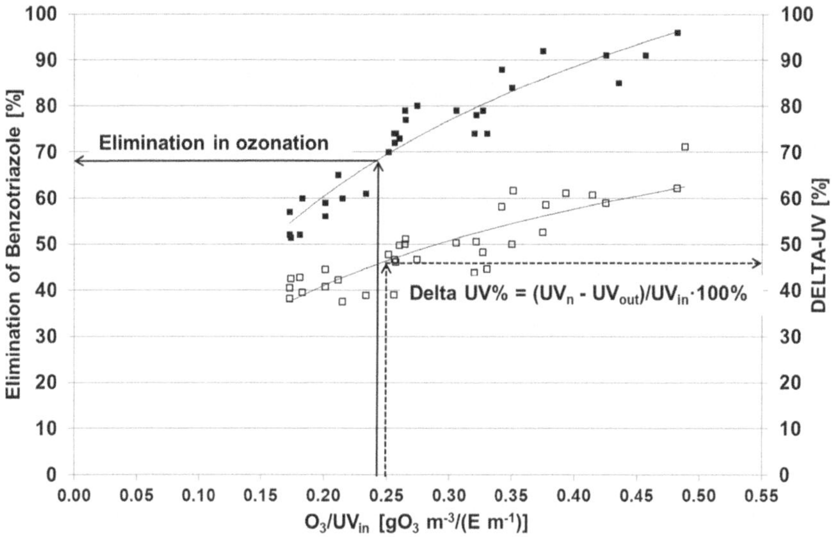

Figure 6.8 Correlation between the elimination of benzotriazole (full symbols) versus O_3-dosage/UV_{in} and delta UV absorption at 254 nm (empty symbols; Boehler *et al.* 2017). An elimination of 68% was reached at a dosage of 0.25 gO_3/UV_{in}, achieving an overall elimination over the whole plant of 90% due to additional biological elimination in activated sludge treatment.

6.4.2 Powdered activated carbon (PAC) addition

Micropollutants were found to sorb efficiently to activated carbon by hydrophobic and electrostatic interactions. A sufficient contact time of 1–2 days is required to achieve an equilibrium between dissolved and sorbed micropollutants. Such high PAC retention times can economically only be obtained by retaining PAC in a clarifier or filter system and returning it to the contact tank (Figure 6.9a). This way, an economic size of the contact tank with hydraulic retention time (HRT) in the order of a half an hour is sufficient. With pilot- and full-scale plants consisting of a contact tank (HRT_{min} = 0.5 h) and a sedimentation tank (HRT_{min} = 2 h) PAC retention times of a few days were reached. By direct addition of PAC to a two-layer sand filter (Figure 6.9b), the PAC also remains about 1–2 days in the filter because the filter is only back-flushed in intervals of 1–3 days. Less capital cost is needed if PAC is directly added to the activated sludge treatment (Figure 6.9c), however, this application requires about 50% more PAC. For all flow schemes a post-filtration (sand- or textile-filter) is required to reduce PAC loss to receiving waters (Boehler *et al.*, 2012; Obrecht *et al.*, 2015; Metzger, 2010; Löwenberg *et al.*, 2016).

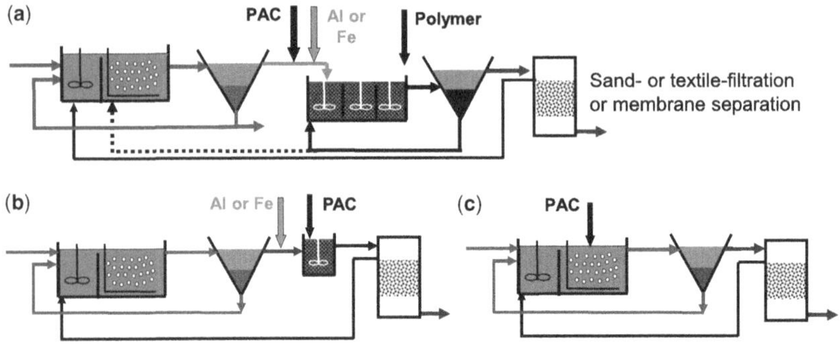

Figure 6.9 Different flow schemes for PAC applications: (a) PAC addition to contact tank, followed by a clarifier to retain and return PAC; (b) direct addition to sand filtration; and (c) activated sludge treatment. Flocculant addition is necessary for a good settling and an efficient filtration of the PAC. Post-filtration (sand/textile filter) is required to reduce PAC loss to receiving waters.

The addition of 10–15 g PAC/m^3 (5–10 gDOC/m^3) to biologically treated wastewater removed up to 70% of most of the investigated pharmaceuticals and X-ray contrast agents (Metzger, 2010; Boehler *et al.,* 2012). Returning the used PAC from the PAC addition step to the activated sludge treatment, achieving a two-step countercurrent flow of the PAC, increases micropollutant removal to 90% (Figure 6.10) and provides a 40–50% reduction of the background DOC (Boehler *et al.,* 2012). If an effluent fee has to be paid for DOC, such a two-step treatment can partly compensate the PAC treatment costs. PAC cannot be regenerated and needs to be disposed of. When returning into activated sludge treatment is applied, PAC can be incinerated together with the excess sludge in the case where sludge is not re-used in agriculture. PAC processes with separate contact and sedimentation tanks have already been in full-scale operation for several years for the de-colorization of municipal wastewater with a substantial contribution from the textile industry. Today, several WWTPs in Southern Germany have such an additional PAC step in operation for removal of micropollutants (Metzger, 2010). For efficient coagulation and sedimentation of the PAC, about 0.4 gFe/gPAC of coagulant followed by a polymer has to be added, yielding a sludge production of 15–20 gTSS/m^3, equal to about 1.5 times the PAC addition.

Boehler *et al.* (2012) successfully tested in full scale a direct dosage of 15 gPAC/m^3 to a two-layer sand filter with a backflush interval of one day (Figure 6.9b). A lower flocculent addition (0.1 gFe/gPAC) compared to sedimentation and no polymer were required. After recycling the filter backwash to the activated sludge system for counter current use of PAC, similar removal efficiencies were achieved as with the contact tank and clarifier. Löwenberg *et al.,* (2016) tested the direct addition of PAC to a two-layer sand filter in full scale for about two years. They achieved stable operation with a significant reduction of suspended solids and phosphate in the effluent compared to the filters without PAC addition operated in parallel. A small contact tank with a minimum HRT_{min} of 10–15 minutes was sufficient for flocculation. During storm water flow, PAC addition was reduced to prevent short backflush intervals. The resulting lower micropollutant elimination was compensated with the PAC recycling to biological treatment.

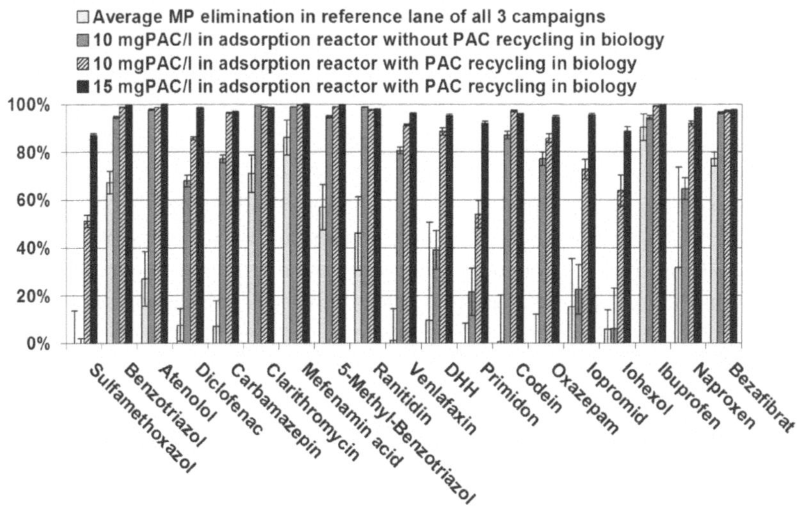

Figure 6.10 Overall elimination of selected micropollutants in activated sludge and PAC treatment with and without PAC recycling to biological treatment. (Reproduced from Boehler *et al.*, 2012, with permission from the copyright holders, IWA publishing.)

Direct addition of PAC to the nitrification part of the activated sludge tank (Figure 6.9c) was tested in pilot scale (Boehler *et al.*, 2012) and full scale (Obrecht *et al.*, 2015). PAC/DOC ratio was 30–50% higher than for two stage treatment (Figure 6.9a) but no additional flocculant was required, as precipitant was added already before for P-removal.

A basic elimination of micropollutants is maintained with direct addition or recycling of PAC to activated sludge treatment even if PAC addition is temporarily interrupted or reduced during storm water conditions due to the long retention time (similar to sludge age of the activated sludge) of PAC in biological treatment (Obrecht *et al.*, 2015). Biofilm systems (moving bed and fixed bed reactors) have not been tested yet, but are probably not suitable for PAC addition.

PAC dosage increases with the background DOC. If the sludge is recycled to the activated sludge treatment, a dosage of about 1.5 gPAC/gDOC is usually sufficient to reach an overall 80% removal of the 12 Swiss indicator compounds (Table 6.2). For direct application to activated sludge treatment a ratio of 2–3 gPAC/gDOC is required (Metzger, 2010; Boehler *et al.*, 2012; Obrecht *et al.*, 2015).

The recycling of PAC increases the secondary sludge production by about 10%, leading to a lower activated sludge age. However, this can partly be compensated by higher sludge concentrations due to better settling of the activated sludge. Corrosion problems resulting from PAC application can be reduced by using high density polyethylene or stainless steel pipes and pumps.

The sorption to activated carbon depends on the properties of the micropollutants as well as of the activated carbon. Simple physical chemical properties of the micropollutants (like the speciation corrected octanol-water distribution coefficient D_{ow}) only partly determines their sorption behavior. Compounds with log D_{ow} >2 are usually removed very well. However, the sorption of charged compounds with lower D_{ow} cannot easily be predicted (Kovalova *et al.*, 2013).

The conventional PAC parameter (e.g. iodine and methylene-blue number) are not suitable to explain the varying adsorption capacity for micropollutants for the different PAC batches. A short-term laboratory test should therefore be developed to better control PAC quality after each batch delivery (Zietzschmann *et al.*, 2015; Alt *et al.*, 2016). Also the origin of PAC has to be controlled to prevent off-gas and health problems during production.

Capital and operating costs are in the range of 15 US-cents/m^3 for a plant size of 100,000 PE (Abegglen & Siegrist, 2012). The primary energy of PAC application (30–35 kWh/kgPAC) results in 0.3–0.7 kWh/m^3 wastewater for 10–20g PAC/m^3, producing a greenhouse gas emission of 15–30 gCO_2/m^3 wastewater (12–15 kgCO_2/kgPAC). The production of PAC from renewable resources would therefore be advantageous (Alt *et al.*, 2016).

6.4.3 Granular activated carbon (GAC) filters

Recently, GAC filters (Altmann *et al.*, 2016a; Benstoem *et al.*, 2017) and fluidized beds operated with smaller GAC grains ($d_{GAC} = 0.5 \pm 0.2$ mm) than for filters (Casazza, 2016) are increasingly being investigated for micropollutant removal due to the high GAC regeneration potential (80–90%), the low regeneration energy, the lower greenhouse gas emission compared to PAC applications and the easy operation.

Originally GAC filters were built to reduce the effluent DOC of WWTPs with high fractions of industrial wastewater. However, results for micropollutant removal were also very promising (Alt *et al.*, 2016). For biologically well treated wastewater and an empty bed contact time (EBCT) of >25 minutes, filter runs of up to 20–25,000 bed volumes (BV) are possible (Figure 6.11) ensuring a sufficient micropollutant removal of >80% (Benstoem *et al.*, 2017; Bitterwolf *et al.*, 2017; Wunderlin *et al.*, 2017).

Low suspended solids concentrations (5–10 gTSS/m^3) and sufficient oxygen (provided by a slight pre-aeration or an overfall) will reduce backflush intervals and GAC loss, and improve the biological regeneration of the GAC filters.

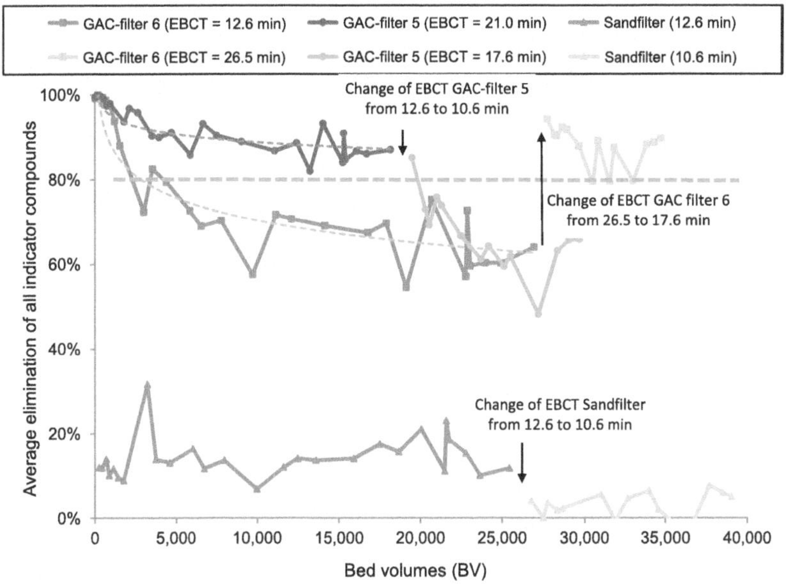

Figure 6.11 Sorption capacity of two full-scale GAC filters operated with different EBCT at Bülach WWTP, Switzerland (DOC_{outlet} = 4–5 g/m^3) in comparison with a sand filter. The average elimination of 12 selected indicator compounds (Table 6.2) depends strongly on EBCT. An EBCT of >25 minutes allows a filter run of 25–30,000 BV until elimination falls below the 80% average elimination of micropollutants. GAC filter 6 operated with a low EBCT recovered after EBCT increase (Bitterwolf *et al.*, 2017; Wunderlin *et al.*, 2017).

For an average hydraulic load of 3–8 m/h during dry weather conditions (up to 15 m/h during storm water) and an average EBCT of 25 minutes a filter depth of 2.5–3 m is required, which is substantially higher than for a conventional sand filter for suspended solids removal.

When operating in full-scale on WWTPs, multiple GAC filter units are usually run in parallel. The GAC reactivation is done sequentially, always renewing the unit with the longest operating time to extend the overall operating time while still meeting threshold values. This parallel operation of filters allows significantly higher filter run times of up to 30,000 BV, which corresponds to a GAC consumption of 10–15 g/m^3 of wastewater, similar to the one for PAC applications. Accordingly, a GAC exchange interval of more than one year is sufficient. This has to be taken into account to estimate GAC costs, primary energy and CO_2 emission (Benstoem *et al.*, 2017).

Regenerated GAC has a similar sorption capacity as new GAC but requires significantly less resources and reduces greenhouse gas emission by a factor of 3–4, which is a crucial advantage compared to PAC addition, that does not allow regeneration of the activated carbon. GAC production from renewable resources (e.g. wood or coconut) would also strongly lower primary energy consumption and greenhouse gas emission, but it is not clear yet if they are suitable for GAC or only for PAC production.

The DOC sorption capacity of GAC filters decreases significantly after a few thousand bed volumes, but remains at a level of 15–20% thereafter, similar to sand filtration (Bourgin *et al.*, 2018). DOC elimination is therefore significantly lower than for PAC applications (see Chapter 6.4.2).

The filters units should allow an easy GAC exchange, e.g. outlet pipes at the bottom of the filter, also reducing exchange time to less than one day (Alt, 2015). GAC filters are sensitive to high hydraulic loads. Storm water conditions will lower the EBCT and therefore temporarily reduce micropollutant elimination. This effect can be avoided by spaciously designed filters. The water protection agencies supervising the effluent conditions of the WWTP could also allow the compensation of a lower elimination during storm water events, when receiving waters are less affected by micropollutants due to the higher dilution, with a higher elimination during dry weather conditions.

The application of GAC (with a smaller grain size than for filters: $d_{GAC} = 0.5 \pm 0.2$ mm) in a fluidized bed requires two storage tanks for the used and the new GAC, respectively, because GAC has to be added daily (about 1% of GAC exchange per day) to avoid a reduced elimination efficiency. The size of the fluidized bed is similar to the GAC filters. The dosage can be adapted to the organic and hydraulic load variation (e.g. storm water conditions). GAC consumption in fluidized beds is similar to that for GAC-filter applications and GAC can also be regenerated (Cassaza, 2016), but suspended solids are hardly removed.

6.4.4 Process combinations

Recently, process combinations have been increasingly investigated for municipal WWTPs with high fractions of industrial wastewater, where full ozonation has the potential to produce increased amounts of problematic or toxic by-products e.g., high bromate concentrations. Bromate formation from bromide starts to increase strongly above 0.4 $gO_3/gDOC$ (Soltermann *et al.*, 2016). Therefore, low ozone doses should be applied if high bromide concentrations are present.

At the ProRheno WWTP in Basel (Switzerland) a combination of ozonation followed by PAC addition and sand filtration was successfully investigated at pilot scale and will be built in full scale for about 300,000 pe. Such a combination allows a higher flexibility at varying high fractions of wastewater from chemical industry.

A promising combination is ozonation followed by GAC filtration. The sand filtration that needs to be built after ozonation in any case to reduce transformation products of ozonation (Chapter 6.4.1) can be replaced by a GAC filter. However, a GAC filter requires a deeper filter bed than a normal sand filter (see Chapter 6.4.3), which makes overall operation and investment costs higher than for the single treatment process, although GAC consumption is lower and less ozone can be applied than for full ozonation. However, the combined process is more adaptable for wastewater conditions with a variable industrial fraction.

6.4.5 Control of operation

In Switzerland 12 indicator substances (Table 6.2) have been selected by the Swiss Federal Office for the Environment (FOEN) for monitoring the removal efficiency of micropollutants in the WWTPs. These compounds were selected because they are regularly consumed, they have wastewater concentration levels clearly above the detection limits, and are not degraded or show little degradation in conventional biological treatment. Therefore, they are suitable for monitoring and controlling the efficiency of ozone and/or activated carbon treatment. The elimination efficiency of at least six of these indicator compounds (4 from group I with very good elimination and 2 from group II with good elimination) have to be selected by the cantonal protection agency and their elimination efficiency measured 4–12 times per year depending on the size of the WWTP. The selection of the indicator substances is regularly adapted depending on their consumption and occurrence in municipal wastewater.

Different research groups are investigating UV absorption as a suitable online signal to control ozone and PAC dosage because the decrease of the UV signal at 254 nm correlates with the required ozone and PAC dosage (see Figure 6.8, Wittmer *et al.*, 2015; Altmann *et al.*, 2016b). Recently it was observed that delta UV absorption difference also correlates with the average relative elimination of the indicator compounds in GAC filtration processes (Figure 6.12).

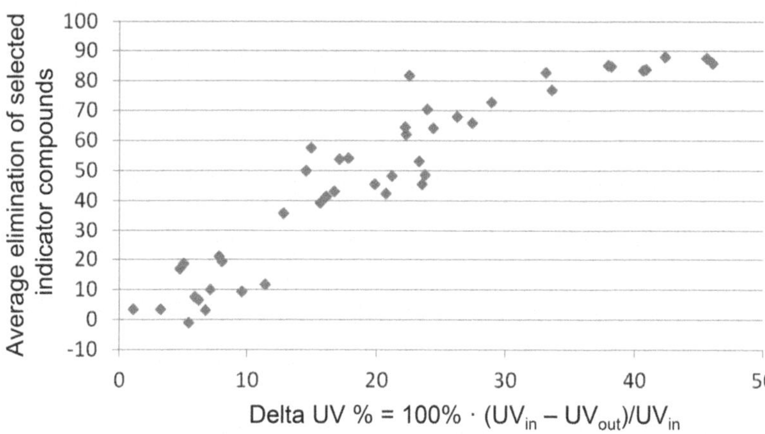

Figure 6.12 Correlation between the average relative elimination of the 12 selected indicator compounds and delta UV at 254 nm in GAC filtration at different run times (bed volumes) at Bülach WWTP (Rensch *et al.*, 2016).

UV absorption might therefore also be suitable to control the micropollutant removal efficiency of activated carbon adsorption processes and to optimize GAC exchange and PAC dosage as well as the operation of combined processes.

6.5 CONCLUSIONS AND OUTLOOK

The removal of micropollutants by conventional municipal wastewater treatment due to sorption, stripping and biological degradation is incomplete and a multitude of compounds are only slightly abated. To significantly reduce eco-toxicological effects of micropollutants and their transformation products in receiving waters and water resources, additional treatment steps (e.g. ozonation or activated carbon adsorption) are required in combination with measures taken at the source (e.g. at chemical production and manufacturing by separation and separate treatment of concentrated industrial wastewater streams).

Ozonation requires a prior complete nitrification with low effluent nitrite concentrations due to the high ozone consumption of nitrite. It might not be suitable for wastewater with high industrial wastewater fractions and high bromide concentrations due to the production of bromate and increased oxidation by-products.

PAC addition is already broadly applied. If PAC is recycled to the activated sludge treatment to improve micropollutant removal or to reduce PAC consumption it may reduce the activated sludge age if the additional excess sludge cannot be compensated by a better sludge settling and therefore may result in a higher activated sludge concentration. PAC recycling to biological treatment is not suitable if the excess sludge is used in agriculture due to the high micropollutant concentration in the PAC enriched sludge.

Recently also GAC filters have been increasingly investigated and built in full scale for micropollutant and DOC removal. The combination of partial ozonation with activated carbon adsorption (PAC or GAC) is investigated for municipal wastewater with varying high fractions of industrial wastewater, where full ozonation may produce toxic by-products.

To optimize micropollutant removal processes, monitoring and control with sensors for UV absorption is advantageous and will reduce the costs for ozone and activated carbon application.

Additional measures in wastewater treatment should be based on careful studies of the mass flux of micropollutant and their environmental effects to identify critical catchment areas so that relevant hot spots can be addressed.

6.6 REFERENCES

Abegglen C. and Siegrist H. (2012). Mikroverunreinigungen aus kommunalem Abwasser - Verfahren zur weitergehenden Elimination auf Kläranlagen (Micropollutants from municipal wastewater, additional processes for an improved elimination in wastewater treatment plants). Federal Office for the Environment, Bern, Umweltwissen No. 1214, http://www.bafu.admin.ch/publikationen/publikation/01661/index.html (accessed March 2018).

Abegglen C., Escher B., Hollender J., Koepke S., Ort C., Peter A., Siegrist H., von Gunten U. and Zimmermann S. (2009). Ozonung von gereinigtem Abwasser Schlussbericht Pilotversuch Regensdorf (Ozonation of biologically treated wastewater, final report of pilot experiments at WWTP Regensdorf), Federal Office for the Environment, Bern, https://www.micropoll.ch/fileadmin/user_upload/Redaktion/Dokumente/01_Berichte/02_Technische_Verfahren/02_Ozonung/2009_Schlussbericht_Regensdorf.pdf (accessed March 2018).

Achermann S., Bianco V., Mansfeldt C. B., Vogler B., Kolvenbach B. A., Corvini P. F. X. and Fenner K. (2018). Biotransformation of Sulfonamide Antibiotics in Activated Sludge: Formation of Pterin-Conjugates Leads to Sustained Risk. *Environmental Science and Technology*, **52**, 6265–6274.

Agerstrand M., Wester M. and Ruden C. (2009). The Swedish Environmental Classification and Information System for Pharmaceuticals – An empirical investigation of the motivations, intentions and expectations underlying its development and implementation. *Environment International*, **35**(5), 778–786.

Alt K. (2015). Einsatz von granulierter Aktivkohle zur Elimination von Spurenstoffen (Use of granulated activated carbon on WWTPs for the elimination of micropollutants). Seminar, Project meeting, WWTP Bülach, 23 September.

Alt K., Benstöm F., Biebersdorf N., Böhler M., Bornemann C., Hiller C., Jedele K., Jekel M., Lyko S., Metzger S., Nahrstedt A., Remy C. and Wintgens T. (2016). Aktiv-kohleeinsatz auf kommunalen Kläranlagen zur Spurenstoffentfernung (Use of activated carbon on municipal WWTPs for the elimination of micropollutants). Report of the DWA-working group KA-8.6. *KA-Korresp. Abwasser, Abfall*, **63**(12), 1062–1067.

Altmann J., Rehfeld D., Träder K., Sperlich A. and Jekel M. (2016a) Combination of granular activated carbon adsorption and deep-bed filtration as a single advanced wastewater treatment step for organic micropollutant and phosphorus removal. *Water Research*, **92**, 131–139.

Altmann J., Massa L., Sperlich A., Gnirss R. and Jekel M. (2016b) UV_{254} absorbance as real-time monitoring and control parameter for micropollutant removal in advanced wastewater treatment with powdered activated carbon. *Water Research*, **94**, 240–245.

Andersen H., Siegrist H., Halling-Sørensen B. and Ternes T. (2003). Fate of estrogens in a municipal sewage treatment plant. *Environmental Science and Technology*, **37**, 4021–4026.

Benner J. and Ternes T. A. (2009). Ozonation of propranolol: formation of oxidation products. *Environmental Science and Technology*, **43**(13), 5086–5093.

Benstoem F., Nahrstedt A., Boehler M., Knopp G., Montag D., Siegrist H. and Pinnekamp J. (2017). Performance of granular activated carbon to remove micropollutants from municipal wastewater – A meta-analysis of pilot- and large-scale studies. *Chemosphere*, **185**, 105–118.

Bitterwolf S., Boehler M., Siegrist H. and Joss A. (2017). Elimination von Spurenstoffen durch granulierte Aktivkohle (GAK) Filtration (Elimination of micropollutants with granulated activated carbon filtration): Full-scale investigation on WWTP Furt-Bülach, Extended interim report, January.

Boehler M., Zwickenpflug B., Hollender J., Ternes T., Joss A. and Siegrist H. (2012). Removal of micropollutants in municipal wastewater treatment plants by powder-activated carbon. *Water Science and Technology*, **66**(10), 2115–2121.

Boehler M., Blunschi M., Czekalski N., Fleiner J., Kienle C., Langer M., McArdell C. S., Teichler R. and Siegrist H. (2017). Biologische Nachbehandlung von kommunalem Abwasser nach Ozonung (Biological post-treatment of municipal wastewater after ozonation). Final report for the Federal office for the environment, in the frame of the technology promotion project ReTREAT, Eawag.

Bourgin M., Beck B., Boehler M., Borowska E., Fleiner J., Salhi E., Teichler R., von Gunten U., Siegrist H. and McArdell C. (2018). Evaluation of a full-scale wastewater treatment plant upgraded with ozonation and biological post-treatments: Abatement of micropollutants, formation of transformation products and oxidation by-products. *Water Research*, **129**, 486–498.

Boxall A. B. A., Sinclair C. J., Fenner K., Kolpin D. and Maud S. J. (2004). When synthetic chemicals degrade in the environment. *Environmental Science and Technology*, **38**, 368A–375A.

Casazza R. (2016). Pilot-project Micro-GAC WWTP Penthaz (Switzerland). VSA-Fortbildungskurs Mikroverunreinigung, Emmetten, Switzerland, 2–4 November.

Celiz M., Tso J. and Aga D. (2009). Pharmaceutical metabolites in the environment: analytical challenges and ecological risks. *Environmental Toxicology and Chemistry*, **28**(12), 2473–2484.

DWA Arbeitsgruppe KA-8.1 (2014). Bedeutung von Transformationsprodukten für den Wasserkreislauf (Importance of transformation products in the water cycle), DWA, Hennef, ISBN 978-3-944328-76-8.

DWA Working group KA-8.1 (2008). Anthropogene Spurenstoffe im Wasserkreislauf (Anthropogenic micropollutants in the water cycle), DWA (Deutsche Vereinigung für Wasserwirtschaft), Hennef 2008, ISBN 978-3-940173-74-4.

Eggen R., Hollender J., Joss A., Scharer M. and Stamm C. (2014). Reducing the discharge of micropollutants in the aquatic environment: the benefits of upgrading wastewater treatment plants. *Environmental Science and Technology*, **48**(14), 7683–7689.

Falås P., Longrée P., la Cour Jansen J., Siegrist H., Hollender J. and Joss A. (2013). Micropollutant removal by attached and suspended growth in a hybrid biofilm-activated sludge process. *Water Research*, **47**, 4498–4506.

Falås P., Wick A., Castronovo S., Habermacher J., Ternes T. and Joss A. (2016). Tracing the limits of organic micropollutant removal in biological wastewater treatment. *Water Research*, **95**, 240–249.

Federal Office for the Environment (2015). Chemicals: in brief, https://www.bafu.admin.ch/bafu/en/home/topics/chemicals/in-brief.html (accessed May 2018).

FOEN (2016) Verordnung zur Überprüfung des Reinigungseffekts von Massnahmen zur Elimination von organischen Spurenstoffen bei Abwasserreinigungsanlagen (Regulation to control the treatment efficiency of technical processes for the elimination of organic micropollutants), https://www.admin.ch/opc/de/classified-compilation/20160123/index.html (accessed May 2018).

Golet E., Xifra I., Siegrist H., Alder A. and Giger W. (2003). Environmental exposure assessment of fluoroquinolone antibacterial agents from sewage to soil. *Environmental Science and Technology*, **37**, 3243–3249.

Helbling D. E., Hollender J., Kohler H. P. and Fenner K. (2010). Structure-based interpretation of biotransformation pathways of amide-containing compounds in sludge-seeded bioreactors. *Environmental Science and Technology*, **44**, 6628–6635.

Hollender J., Zimmermann S., Koepke S., Kraus M., McArdell C., Ort C., Singer H., von Gunten U. and Siegrist, H. (2009). Elimination of organic micropollutants in a municipal WWTP upgraded with a full-scale post-ozonation followed by sand filtration. *Environmental Science and Technology*, **43**, 7862–7869.

Huber M., Canonica S., Park G. and von Gunten U. (2002). Oxidation of pharmaceuticals during ozonation and advanced oxidation processes. *Environmental Science and Technology*, **37**, 1016–1024.

Huber M., Göbel A., Joss A., Hermann N., Löffler D., McArdell C., Ried A., Siegrist H., Ternes T. and von Gunten U. (2005). Oxidation of pharmaceuticals during ozonation of municipal wastewater effluents: a pilot study. *Environmental Science and Technology*, **39**, 4290–4299.

Jobling S. M. and Tyler C. (2003). Endocrine disruption in wild freshwater fish. *Pure and Applied Chemistry*, **75**, 2219.

Joss A., Carballa M., Kreuzinger N., Siegrist H. and Zabczynski S. (2006a). Chapter 6: Wastewater Treatment. In: Human Pharmaceuticals, Hormones and Fragrances - The challenge of micropollutants in urban water management, T. Ternes & A. Joss (eds). IWA Publishing, London, ISBN10 1843390930, pp. 243–292.

Joss A., Zabczynski S., Göbel A., Hoffmann B., Löffler D., McArdell C., Ternes T., Thompsen A. and Siegrist H. (2006b). Biological degradation of pharmaceuticals in municipal wastewater treatment: proposing a classification scheme. *Water Research*, **40**, 1686–1696.

Klaschka U., Liebig M., Moltmann J. F. and Knacker T. (2004). Potential environmental risks by cleaning hair and skin. Eco-label - a possibility to reduce exposure to personal care products. In: Pharmaceuticals in the Environment. Sources, Fate, Effects and Risks, K. Kümmerer (ed.), 2nd edn, Springer Verlag, Berlin, pp. 411–430.

Kormos J. L., Schulz M., Wagner M. and Ternes T. (2009). Multistep approach for the structural identification of biotransformation products of iodinated X-ray contrast media by liquid chromatography/hybrid triple quadrupole linear ion trap mass spectrometry and (1)H and (13)C nuclear magnetic resonance. *Analytical Chemistry*, **81**(22), 9216–9224.

Kormos J. L., Schulz M., Kohler H. P. and Ternes T. (2010). Biotransformation of selected iodinated X-ray contrast media and characterization of microbial transformation pathways. *Environmental Science and Technology*, **44**(13), 4998–5007.

Kovalova L., Siegrist H., von Gunten U., Eugster J., Hagenbuch M., Wittmer A., Moser R. and McArdell C. S. (2013). Elimination of micropollutants during posttreatment of hospital wastewater with powdered activated carbon, ozone, and UV. *Environmental Science and Technology*, **47**(14), 7899–7908.

Krauss M., Longree P., Van Houtte E., Cauwenberghs J. and Hollender J. (2010). Assessing the fate of nitrosamine precursors in wastewater treatment by physicochemical fractionation. *Environmental Science and Technology*, **44**(20), 7871–7877.

Lee, Y. and von Gunten, U. (2016). Advances in predicting organic contaminant abatement during ozonation of municipal wastewater effluent: reaction kinetics, transformation products, and changes of biological effects. *Environmental Science: Water Research & Technology*, **2**(3), 421–442.

Loos R., Carvalho R., António D., Comero S., Locoro G., Tavazzi S., Paracchini B., Ghiani M., Lettieri T., Blaha L., Jarosova B, Voorspoels S., Servaes K., Haglund P., Fick J., Lindberg R., Schwesig D. and Gawlik B. (2013). EU-wide monitoring survey on emerging polar organic contaminants in wastewater treatment plant effluents. *Water Research*, **47**, 6475–6487.

Löwenberg J., Zenker A., Krahnstöver T., Boehler M., Baggenstos M., Koch G. and Wintgens T. (2016). Upgrade of deep bed filtration with activated carbon dosage for compact micropollutant removal from wastewater in technical scale. *Water Research*, **94**, 246–256.

Margot J., Magnet A., Thonney D., Chevre N., de Alencastro F. and Rossi L. (2011). Traitement de micropolluant dans les eaux usées, rapport final sur les essais pilotes a la step de Vidy, VSA-Plattform "Verfahrenstechnik Mikroverunreinigungen": (Elimination of micropollutants in wastewater treatment, final report of the pilot experiments at the WWTP Vidy, Lausanne), https://www.micropoll.ch/fileadmin/user_upload/Redaktion/Dokumente/01_Berichte/02_Technische_Verfahren/02_Ozonung/2011_Rapport_final_STEP_Vidy.pdf (accessed May 2018).

Margot J., Kienle C., Magnet A., Weil M., Rossi L., de Alencastro L., Abegglen C., Thonney D., Chevre N., Schaerer M. and Barry D. (2013). Treatment of micropollutants in municipal wastewater: ozone or powdered activated carbon? *The Science of the Total Environment*, **461–462**, 480–498.

Metzger S. (2010) *Einsatz von Pulveraktivkohle zur weitergehenden Reinigung von kommunalem Abwasser (Application of powder activated carbon for an advanced treatment of municipal wastewater)*. Dissertation at TU Berlin, Oldenbourg Industrieverlag München, ISBN 978-3-8356-3231-8.

Munz N., Burdon F., de Zwart D., Junghans M., Melo L., Reyes M., Schönenberger U., Singer H., Spycher B., Hollender J. and Stamm Ch. (2017) Pesticides drive risk of micropollutants in wastewater-impacted streams during low flow conditions. *Water Research*, **110**, 366–377.

Neilson A. H. and Allard A. S. (2007) Environmental Degradation and Transformation of Organic Chemicals. CRC Press, Boca Raton, FL, USA.

Obrecht J., Thomann M., Stoll J., Frank K., Sobaszkiewicz M., Boller M. and Freisler P. (2015). PAK Dosierung ins Belebungsbecken – Alternative zur nachgeschalteten Pulveraktivkohleadsorption (PAC dosage into the activated sludge tank – alternative to post treatment applications). *Aqua & Gas*, **2**, 20–32.

Ort C., Hollender J., Schaerer M. and Siegrist H. (2009). Model-based evaluation of reduction strategies for micropollutants from wastewater treatment plants in complex river networks. *Environmental Science and Technology*, **43**, 3214–3220.

Rensch D., Margot J., Schachtler M. and Wunderlin P. (2016) Betriebsüberwachung von Anlagen zur Elimination von Mikroverunreinigungen (Operation control of treatment steps for the elimination of micropollutants). VSA-Fortbildungskurs Mikroverunreinigung, Emmetten, 2–4 November.

Schindler-Wildhaber Y., Mestankova H., Schärer M., Schirmer K., Salhi E. and von Gunten U. (2015). Novel test procedure to evaluate the treatability of wastewater with ozone. *Water Research*, **75**, 324–35.

Schymanski E. L., Jeon J., Gulde R., Fenner K., Ruff M., Singer H. P. and Hollender J. (2014). Identifying small molecules via high resolution mass spectrometry: communicating confidence. *Environmental Science and Technology*, **48**(4), 2097–2098.

Siegrist H., Joss A., Alder A., McArdell C., Göbel A. and Ternes T. (2003). Micropollutants - New challenge in wastewater disposal. *EAWAG News*, **57**, 7–10.

Siegrist H., Joss A. and Ternes T. (2014). Chapter 7 Micropollutants removal. In: Activated sludge – 100 years and counting, D. Jenkins & J. Wanner (eds). IWA Publishing, London, ISBN 9781780404936, pp. 117–130.

Soltermann F., Abegglen C., Goetz C. and von Gunten U. (2016). Bromide sources and loads in Swiss surface waters and their relevance for bromate formation during wastewater ozonation. *Environmental Science and Technology*, **50**(18), 9825–9834.

Stalter D., Magdeburg A. and Oehlmann J. (2010). Comparative toxicity assessment of ozone and activated carbon treated sewage effluents using an in vivo test battery. *Water Research*, **44**(8), 2610–2620.

Stalter D., Magdeburg A., Quednow K., Botzat A. and Oehlmann J. (2013). Do contaminants originating from state-of-the-art treated wastewater impact the ecological quality of surface waters? *Plos One*, **8**(4), 1–10.

Ternes T. (1998). Occurrence of drugs in German sewage treatment plants and rivers. *Water Research*, **32**, 3245–3260.

Ternes T., Giger W. and Joss A. (2006) Chapter 1 Introduction. In: Human Pharmaceuticals, Hormones and Fragrances - The challenge of micropollutants in urban water management, T. Ternes & A. Joss (eds). IWA Publishing, London, ISBN10 1843390930, pp. 1–13.

Tricker A. R. and Preussmann R. (1991). Carcinogenic N-nitrosamines in the diet: occurrence, formation, mechanisms and carcinogenic potential. *Mutation Research Genetic Toxicology*, **259**(3), 277–289.

Van Zelm R., Huijbregts M. A. J. and van de Meent D. (2010). Transformation products in the life cycle impact assessment of chemicals. *Environmental Science and Technology*, **44**, 1004–1009.

Verlicchi P. (2017). Hospital Wastewater: Characteristics, Management, Treatment and Environmental Risks. In: The Handbook of Environmental Chemistry Vol. 60, D. Barceló & A. G. Kostianoy (eds). Springer, Berlin.

von Sonntag C. and von Gunten U. (2012). Chemistry of Ozone in Water and Wastewater Treatment. IWA Publishing, London, ISBN 9781843393139.

Wick A., Fink G., Joss A., Siegrist H. and Ternes T. (2009). Fate of betablockers and psycho-active drugs in conventional wastewater treatment. *Water Research*, **43**, 1060–1074.

Wick A., Wagner M. and Ternes T. (2011a). Elucidation of the transformation pathway of the opium alkaloid codeine in biological wastewater treatment. *Environmental Science and Technology*, **45**, 3374–3385.

Wick A., Marincas O., Moldovan Z. and Ternes T (2011b). Sorption of biocides, triazine and phenylurea herbicides, and UV-filters onto secondary sludge. *Water Research*, **45**, 3638–3652.

Wittmer A., Heisele A., McArdell C. S., Böhler M., Longree P. and Siegrist H. (2015). Decreased UV absorbance as an indicator of micropollutant removal efficiency in wastewater treated with ozone. *Water Science and Technology*, **71**(7), 980–985.

Wunderlin P., Mestankova H., Salhi E., Schindler-Wildhaber Y., Schärer M., Schirmer K. and von Gunten U. (2015). Behandelbarkeit von Abwasser mit Ozon _ Testverfahren zur Beurteilung (Treatability of wastewater with ozone – assessment test). *Aqua & Gas*, **7/8**, 25–38.

Wunderlin, P., Joss, A. and Fleiner, J. (2017). Elimination von Spurenstoffen durch granulierte Aktivkohle (GAK) Filtration: Grosstechnische Untersuchungen auf der

ARA Bülach-Furt (Elimination of micropollutants with granulated activated carbon (GAC) filtration: Full-scale experiments on WWTP Bülach-Furt), Interim report (accessed May 2018): https://www.micropoll.ch/fileadmin/user_upload/Redaktion/Dokumente/01_Berichte/02_Technische_Verfahren/03_Aktivkohle/GAK_Bu%CC%88lach_Zwischenbericht_27032017_Final.pdf

Zietzschmann F., Altmann J., Hannemann C. and Jekel M. (2015). Lab-testing, predicting, and modeling multi-stage activated carbon adsorption of organic micro-pollutants from treated wastewater. *Water Research*, **83**, 52–60.

Zimmermann S., Wittenwiler M., Hollender J., Krauss K., Ort C., Siegrist H. and von Gunten U. (2011). Kinetic assessment and modeling of an ozonation step for full-scale municipal wastewater treatment: micropollutant oxidation, by-product formation and disinfection. *Water Research*, **45**, 605–617.

Chapter 7

Anaerobic digestion processes

M. M. Yeshanew[1], G. Esposito[1], D. J. Batstone[2] and P. N. L. Lens[3]

[1]University of Cassino and Southern Lazio, Department of Civil and Mechanical Engineering, via Di Biasio 43, 03043 Cassino, FR, Italy

[2]University of Queensland, Advanced Water Management Centre (AWMC), Level 4, Gehrmann Laboratories, Bldg. 60, Research Road, Brisbane 4072

[3]UNESCO-IHE Institute for Water Education, Westvest 7, 2611 AX Delft, The Netherlands

ABSTRACT

Producing bioenergy cost effectively is a major contribution to improving the living standard of future generations. The anaerobic digestion (AD) technology is a biochemical process which can convert complex organic waste into clean and renewable sources of energy. Various biotechnological developments in the AD technology have enhanced the methane production efficiency and biogas quality of conventional anaerobic digestion. Anaerobic digestion has a broad range of additional applications in new energy carriers such as production of hydrogen, electricity and low-cost chemicals, as well as production of digestate as a low-cost fertilizer. This chapter covers various AD reactor configurations and discusses different aspects for enhancing the AD process such as substrate pretreatment. It also provides an overview of phosphorus recovery during the AD process, and outlines the emerging landscape of novel anaerobic digestion processes.

Editors: Giorgio Mannina, George Ekama, Hallvard Ødegaard and Gustaf Olsson
doi: 10.2166/9781780409719_0261

Keywords: Anaerobic digestion, bioenergy, mathematical modeling, phosphorus recovery, reactor design, substrate pretreatment

7.1 INTRODUCTION

Globally, there is a growing awareness that renewable energy and energy efficiency are vital for both creating new economic opportunities and controlling environmental pollution. As an energy generating process, the anaerobic digestion (AD) technology has been used worldwide and has gained increasing interest since its first application in the 1970s (Van Lier *et al.*, 2015). This technology has the potential of converting organic matter into usable energy in the form of biogas which is renewable and environmentally friendly. The produced biogas can be used directly for cooking and lighting purposes, in combined heat and power gas engines or it can be upgraded to natural gas called biomethane. The methane and the power produced from the AD process can be used to replace the energy derived from oil and gas. Besides, a nutrient rich digestate is produced as a by-product which can be used for land application (Raheem *et al.*, 2016). These advantages have led to process intensification options aiming at improving the energy recovery from the AD process and economic feasibility (Karadag *et al.*, 2015). Over the past decades, there have been substantial improvements and technological advancements in the AD technology, especially better understanding of reactor design and substrate pretreatment methods, which enables the AD process to treat various types of waste.

Utilizing the digestate from the AD process and considering the value of nutrients, such as phosphorus, combined nutrient recovery AD can be more economically advantageous than energy production alone when applying proper recovery methods. Table 7.1 presents some of the physico- chemical characteristics of the digestate (Plana & Noche, 2016). Phosphorus is a critical and non-renewable resource necessary to support global food production (Latif *et al.*, 2017). There are large amounts of phosphorus in the wastewater streams originating from anthropogenic, livestock, natural or industrial sources (Figure 7.1) (Batstone *et al.*, 2015a). During the AD process, except for a minor amount used for cell synthesis, phosphorus is released as inorganic soluble phosphate due to organic matter degradation (Figure 7.2) (De Graaff *et al.*, 2010). The released phosphate tends to bind with the common ions present in the reactor (e.g. Fe^{3+}, Al^{3+}, Mg^{2+}, NH_4^+ and Ca^{2+}) and form insoluble complexes/precipitates such as $FePO_4$, $AlPO_4$, $MgNH_4PO_4$ (struvite) and $Ca_3(PO_4)_2$ (Latif *et al.*, 2015). The majority of these precipitates (around 90%) are separated together with the digestate solids (Kim & Nakhla, 2009). This is an ineffective use of the phosphate, and application of digestate can be expensive due to high transport and application costs, and regulations on the application of digestate on agricultural land (Latif *et al.*, 2017). Hence, an alternative approach is required in order to reduce phosphorus precipitation and increase the soluble phosphate in the digestate following the AD process. Recycling the phosphorus through enhancing its solubility during the AD process is emerging, not only for sustainability reasons but also due to economic drivers that are based on the demand side.

Table 7.1 Physico-chemical characteristics of digestate (Source: Plana & Noche, 2016).

Parameter	Unit	Mean Concentration	Variation Range
pH	–	7.3–9.0	+0.5 to +2
Dry matter (DM)	%	1.5–13.2	–1.5 to –5.5
Organic DM (% DM)	%	63.8–75.0	–5 to –15
Total Carbon (% DM)	%	36.0–45.0	–2 to –3
Total Phosphorus (% DM)	%	0.6–1.7	–
Total Phosphorus	kg/mg FM	0.4–2.6	≈ 0
Water soluble Phosphorus (% of total P)	%	25–45	–20 to –47
Total Nitrogen (% DM)	%	3.1–14.0	–
Total Nitrogen	kg/mg FM	1.2–9.10	≈ 0
Total NH_4^+	kg/mg FM	1.5–6.8	–
C:N ratio	–	3.0–8.5	–3 to –5
Total Potassium (% DM)	%	1.9–4.3	–
Total Potassium	kg/mg FM	1.2–11.5	≈ 0
Total Magnesium	kg/mg FM	0.3–0.7	≈ 0
Total Calcium	kg/mg FM	1.0–2.3	≈ 0

DM = Dry matter; FM = Fresh matter.

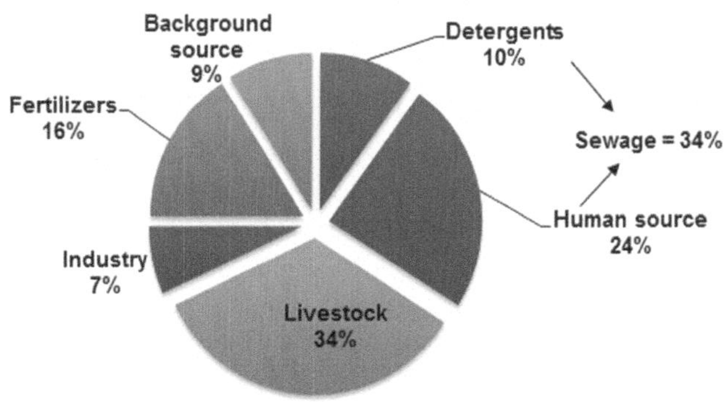

Figure 7.1 Sources of phosphorus in wastewater (source: Rahman *et al.*, 2014).

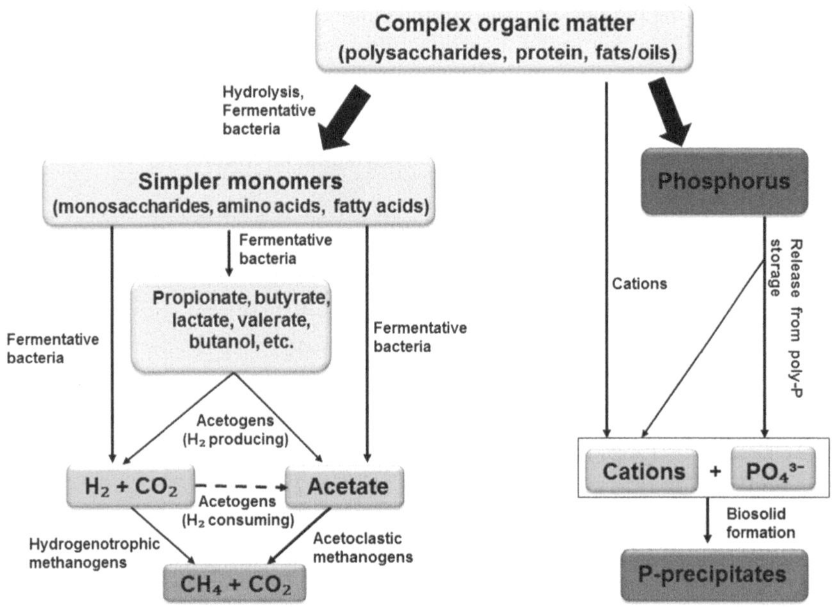

Figure 7.2 Schematic representation of organic matter degradation pathway along with phosphorus precipitation during the AD process at neutral pH conditions (source: Latif *et al.,* 2017).

Starting from the principles of the AD process, this chapter covers the AD process for biogas and biofuel production. Recent developments in the AD technologies, including reactor design, substrate pretreatment methods and operational concerns, are discussed. In addition, different techniques which can reduce phosphorus precipitation in the AD process such as temperature, pH and pressure control are presented.

7.2 PRINCIPLES OF THE ANAEROBIC PROCESSES

The fundamental aspects of the AD process have been well discussed previously (Kothari *et al.*, 2014) and only a brief discussion is presented here. The AD process is a multi-step biological conversion of organic matter to produce mainly biogas and digestate in the absence of oxygen (Ariunbaatar *et al.,* 2014). The organic carbon conversion rate in the AD process is directly proportional to the biogas yield under well-operated conditions. A typical biogas contains 50–70% CH_4, 30–50% CO_2 and traces of other gases.

The AD process consists of four main steps: hydrolysis, acidogenesis, acetogenesis and methanogenesis (Figure 7.2) (Minale & Worku, 2014). In the hydrolysis step, higher molecular organic substrates (such as carbohydrate, protein and lipid) are hydrolyzed by hydrolytic bacteria into soluble organic substrates, e.g. glucose, amino acids and fatty acids (Yeshanew *et al.,* 2016a). The soluble compounds obtained after hydrolysis are subsequently converted in the acidogenesis step into volatile fatty acids (VFAs) along with the generation of by-products, e.g. H_2, CO_2, NH_3 and H_2S. In the third step (acetogenesis), the VFAs are degraded to produce acetate and H_2 by obligate H_2 producing bacteria, the so called acetogens (Kothari *et al.,* 2014). This conversion is highly influenced by the H_2 partial pressure in the reactor. Methanogenesis is the final step of the AD process in which methanogens produce methane from the final products of acetogenesis (i.e. H_2, CO_2 and acetate) as well as hydrogen and CO_2 produced from acidogenesis (Matheri *et al.,* 2017). The process involves two groups of methanogens, namely the acetate consumers and H_2/CO_2 consumers. The former group splits acetate into methane and CO_2, and approximately 70% of the methane is formed through this pathway. The remaining portion of methane (30%) is produced using the H_2/CO_2 pathway, where H_2 serve as electron donor and CO_2 as electron acceptor (Kothari *et al.,* 2014).

Considering the microbial population of the AD process, the first three steps are carried out by the bacterial domain, while the fourth step (methanogenesis) is undertaken by the archaeal domain (Figure 7.2). Bacteria constitute a large domain of prokaryotic microorganisms (Cavinato *et al.,* 2016). The majority of bacteria identified in the AD process mainly contain the phyla of *Firmicutes, Actinobacteria, Spirochaetes, Bacteriodetes* and other *Proteobacteria* (Cavinato *et al.,* 2016). *Clostridia* are a highly polyphyletic class of *Firmicutes,* including the *Clostridium* and *Thermoanaerobacter* genera (Sen *et al.,* 2016). *Acetogens* are another important bacterial community during the AD process, which are generally in obligate syntrophic association with *hydrogenotrophic methanogens* and are responsible for converting propionate, butyrate and lactate into acetate (Cavinato *et al.,* 2016). The methanogenic *Archaea* are a phylogenetically diverse group of strictly anaerobic *Euryarchaeota* with an energy metabolism restricted to the formation of CH_4 from formate, methanol, methylamines, acetate, CO_2 and H_2. *Methanococcales* and *Methanobacteriales* are common hydrogen utilizers, while *Methanosarcinales* are dominant acetate utilizers (Habouzit *et al.,* 2014), made up of the genera *Methanosaeta* (obligate acetate utilizer) and *Methanosarcina* (which has multi-substrate capability). *Methanosaeta* dominates in the majority of mesophilic (30–40°C) digesters, while *Methanosarcina* dominates under adverse conditions, including high ammonia, and thermophilic conditions (Karakashev *et al.,* 2005).

These various AD microbial groups have different optimum environmental conditions (such as pH and temperature, Table 7.2) and a well-designed AD reactor ensures the effective growth of the diverse microbial groups. The process is also affected by operational parameters including organic loading rate (OLR), hydraulic retention time (HRT), total nitrogen and the nature of the substrate (Matheri *et al.,* 2017). For instance, acidogenesis and methanogenesis require a different pH range

Table 7.2 Some examples of anaerobic microbial communities and their respective environmental conditions.

Highest Abundance	Cell Morphology and Physiology	Working Temperature Range	Working pH Range	References
Bacteria Community				
Firmicutes sp.	Round cocci or bacillus, gram-positive	Mesophilic or thermophilic	4.7–6.0	Cavinato *et al.*, 2016
Clostridiales sp.	Rod shaped, spore forming, majority gram-positive, strictly anaerobic	Mesophilic or thermophilic	4.7–6.0	Cavinato *et al.*, 2016
Bacteriodetes sp.	Nonsporulating bacilli, gram-negative, obligate anaerobic	Mesophilic or thermophilic	4.7–6.0	Karakashev *et al.*, 2005
Proteobacteria sp.	Rod shaped, gram-negative	Mesophilic or thermophilic	4.7–6.0	Cavinato *et al.*, 2016
Chloroflexi sp.	Filamentous, green non-sulfur bacteria,	Mesophilic or thermophilic	4.7–6.0	Cavinato *et al.*, 2016
Thermotogales sp.	Rods, thin filaments	Thermophilic or hyphertherophilic	4.7–6.0	Karakashev *et al.*, 2005

Archaea community				
Methanosarcina sp.	Irregular cocci or *Sarcina*-like cells, aceticlastic or hydrogenotrophic	Mesophilic or thermophilic	6.5–7.5	Karakashev *et al.*, 2005
Methanosaeta sp.	Rods or filamentous, strictly aceticlastic	Mesophilic or thermophilic	6.5–7.9	Cavinato *et al.*, 2016
Methanobacteriales	Rods or filamentous, hydrogenotrophic	Mesophilic or thermophilic	6.5–7.9	Karakashev *et al.*, 2005
Methanomicrobiales	Small rods, irregular cocci, flat oval-shaped cells, hydrogenotrophic	Mesophilic	6.5–7.9	Cavinato *et al.*, 2016
Methanobrevibacter sp.	–	Mesophilic or thermophilic	6.5–7.5	Karakashev *et al.*, 2005
Methanococcales	Irregular cocci, hydrogenotrophic	Mesophilic or thermophilic	6.5–7.5	Cavinato *et al.*, 2016
Methanothermobacter	–	Thermophilic	6.5–7.5	Cavinato *et al.*, 2016
Methanothermobacter	Hydrogenotrophic	Thermophilic	6.5–7.5	Karakashev *et al.*, 2005

for their optimal process control. Acidogenic bacteria are less sensitive and require a pH of 5.0–6.5, while methanogenic bacteria are more sensitive and require a pH range of 6.5–7.5 (Cavinato *et al.,* 2016). Therefore, an optimal pH range should be maintained at which both microbial groups can coexist (Sen *et al.,* 2016). The VFAs generated during the AD process lead to a reduction in the pH (Ghimire *et al.,* 2015). This pH change is buffered by weak bases, including ammonia and bicarbonate. Hence, the buffering capacity of the reactor is vital for the process (Yeshanew *et al.,* 2016b). Temperature is another important parameter affecting the overall metabolism and growth rate of the microorganisms of the AD process. The AD microorganisms operate at a broad temperature range, with different operating regimes, i.e. psychrophilic (10–25°C), mesophilic (30–40°C) and thermophilic (50–60°C) conditions (Sen *et al.,* 2016).

7.3 DESIGN AND OPERATION OF AD REACTORS

In the AD process, the design of the reactor is important to improve its efficiency and biogas production. One of the important reactor design considerations in the AD process is to maintain the population of relatively slow growing anaerobic microbes, especially the methanogenic groups, in the reactor. Growth rate must exceed washout rate (or dilution rate), and this is a special consideration for slow growing microbes. This either requires a large digester, or decoupling of the sludge retention time (SRT) from the HRT. Such decoupling maintains a high SRT with a relatively short HRT. High rate AD reactors were developed over conventional technologies to address this (Van Lier *et al.,* 2015), enabling SRTs (100 d+) two orders of magnitude above the HRTs (<1 day). In conventional (flow through) anaerobic digesters, there is a high HRT (20–40 days) to establish a long SRT (=HRT) which leads to a large reactor size (Yeshanew *et al.,* 2016b). High rate reactors (SRT > HRT) are most commonly enabled through biomass immobilization (Habouzit *et al.,* 2014), either on fixed media, settleable media, or through self-supported biofilms called anaerobic granules, with settling of these integrated in the reactor. The retention of biomass inside the reactor provides several further benefits such as good settling characteristics, and high tolerance to inhibitors and operational changes (Escudie *et al.,* 2011). As a result, high rate AD reactors are stable processes and distinctively reduce the required large reactor size and related capital costs, making the AD process more cost effective and attractive for handling different waste streams (Van Lier *et al.,* 2015).

Various high rate AD reactors have been developed during the past two decades and are being implemented at small- and large-scale for treating various kinds of industrial and municipal wastewater (Karadag *et al.,* 2015). Based on the mechanisms of biomass retention, high rate AD reactors are broadly categorized into two classes: biofilm-based such as anaerobic packed bed reactor (APBR), and anaerobic fluidized bed reactor (AFBR) and self-immobilized, including upflow anaerobic sludge blanket (UASB) and expanded granular sludge bed (EGSB) reactor systems (Karadag *et al.*, 2015; Van Lier *et al.*, 2015). A detailed discussion of some of the common conventional and high rate AD reactor types with their design considerations is presented in the following sub-sections.

7.3.1 Covered anaerobic lagoon

The covered anaerobic lagoon is a slow rate AD reactor and an inexpensive option for treating various waste streams with a total solid (TS) content of 0.5–3.0% (Jensen *et al.*, 2014). The covered anaerobic lagoons are designed as earthen pits constructed with impermeable liners (e.g., clay and polyethylene) at the bottom and sides (Figure 7.3a). A typical advantage of the covered lagoon is it does not need either mechanical mixing or external heating. The SRT in a covered lagoon is longer than its HRT. In addition, the HRT can be as long as 3–9 months and the lagoons are sometimes referred to as waste storage units. Because covered lagoons are operated at ambient temperature conditions, their treatment efficiency is tied to the geographical location and climate. They are relatively inefficient compared with other forms of anaerobic digestion, but are low-cost (Jensen *et al.*, 2014).

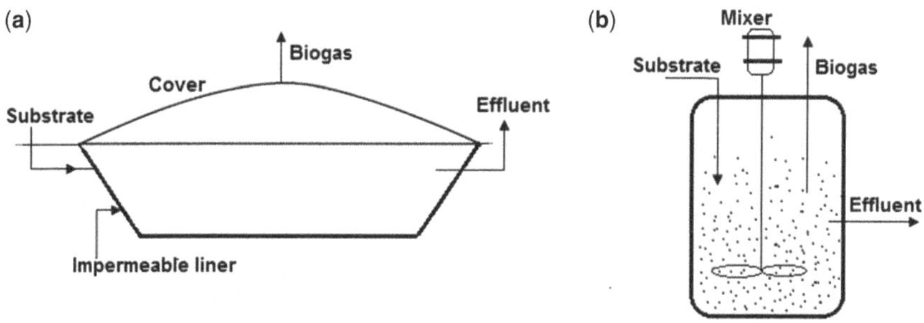

Figure 7.3 Schematic diagrams of (a) covered anaerobic lagoon and (b) CSTR.

7.3.2 Continuous stirred tank reactor (CSTR)

The CSTR is the most commonly used reactor configuration employed for the AD process in treating various wastes with a TS content of 0.5% or higher. The CSTR is maintained at completely mixed conditions by mixing continuously or intermittently using mechanical mixing, biogas sparging or liquid recirculation (Figure 7.3b). The design HRT in the CSTRs ranges from 20 to 40 days (Yeshanew *et al.*, 2016b), with operating HRTs down to 10 d. In a standard CSTR the HRT is equal to the SRT, and the only way to decouple these is through external thickening, and return of solids to the CSTR (recuperative thickening) (Vanyushina *et al.*, 2012). The completely mixed conditions enable rapid dilution of incoming toxic compounds, which allows the CSTRs to handle shock loads and toxic compounds in the waste streams.

7.3.3 Anaerobic packed/fixed bed reactor

The application of the anaerobic packed bed reactor (APBR) was started in the 1960s for industrial wastewater treatment in the USA (Van Lier *et al.,* 2015). It is also commonly known as the anaerobic filter reactor (AFR) or fixed bed reactor. The APBR is filled with an inert and stationary (fixed) carrier material, which provides a support medium for the biomass, and hence allows SRT longer than the HRT (Figure 7.4a) (Mao *et al.,* 2015). The APBR can be applied in up-flow and down-flow mode. In the up-flow mode, wastewater flows upward through the carrier material and the entire filter bed is submerged. A significant portion of the microbial biomass remains entrapped within the interstices or void spaces between the packing medium. The unattached microorganisms form flocs and take a granular shape because of the rolling action of rising gas bubbles, hence contributing significantly to the organic matter removal. In the down-flow mode, the wastewater is spread from the top and the loosely held microbial biomass can be washed out from the bed (Escudie *et al.,* 2011).

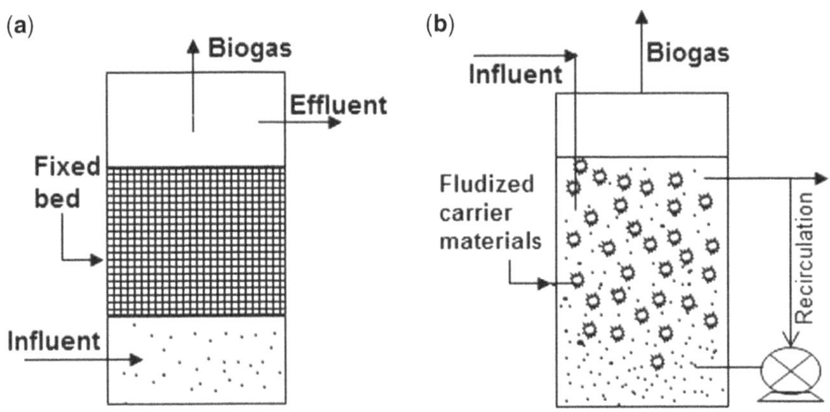

Figure 7.4 Schematic diagrams of (a) APBR and (b) AFBR.

Various types of synthetic packing materials and natural packing materials, such as gravel, coke, straw and bamboo segments, have been used in APBRs. The shape, size, weight, specific surface area, and porosity of the packing material are important aspects in designing the APBR (Van Lier *et al.,* 2015). In APBRs, neither a mechanical agitation nor recirculation equipment is required, hence lowering the construction and operation costs. Besides, advantages such as fast reactor start-up, simplicity of construction and easy operation have been pointed out by several studies (Mao *et al.,* 2015). However, clogging of the carrier material is the major bottleneck for the APBR during long-term operation, in particular for wastewater having a high concentration of suspended solids. Carrier clogging can be tackled by applying a primary settler and/or a pre-acidification step (Escudie *et al.,* 2011).

Intermittent liquid recirculation has also been recommended in order to prevent the occurrence of rapid clogging of the carrier material due to biomass accumulation. In addition, the excess microbial biomass should be periodically removed from the bottom of the reactor in order to minimize clogging phenomena. It is indicated that the hydrodynamic conditions play an important role in biomass retention within the void space (Escudie *et al.,* 2011).

7.3.4 Anaerobic fluidized bed reactor

The anaerobic fluidized bed reactor (AFBR) is a configuration that has been regarded as the second generation modern AD reactor type, since it has a better hydraulic mixing regime compared with the APBR (Van Lier *et al.,* 2015). In the AFBR system, the biofilm develops on the surface of the carrier material, which is kept in suspension or fluidized by the influent flow and/or recirculation of the liquid at velocities sufficient to cause fluidization (Figure 7.4b) (Andalib *et al.,* 2014; Barca *et al.,* 2015). In this reactor configuration, anaerobic microorganisms attach onto the surface of a solid, non-fixed mobile carrier material and form a multi-layer microbial cell, the biofilm (Andalib *et al.,* 2014). The type of carrier material plays a crucial role for the successful development of a matured biofilm and hence reactor performance (Habouzit *et al.,* 2014). Currently, there are several types of available carrier materials having different shapes and sizes, including fine sand (0.1–0.3 mm), basalt, and pumice, or they are made of polyethylene, polypropylene or high-density polyethylene with a typical density slightly less than water (Habouzit *et al.,* 2014; Van Lier *et al.,* 2015).

The bed fluidization favors the contact between the microbial biofilm and the substrate, thus improving mass transfer and treatment efficiencies (Fernandez *et al.,* 2008). Accordingly, several studies indicated its robust performance at higher OLR and shorter HRT (Yeshanew *et al.,* 2016b). In addition, the bed fluidization also eliminates operating problems such as bed clogging and high pressure drop, and permits better hydraulic recirculation and a greater surface area per unit of reactor volume. Hence, the investment costs are much lower due to the reduced reactor volumes with high process efficiency (Yeshanew *et al.,* 2016b). However, this reactor configuration requires more energy for bed fluidization and to maintain a uniform biofilm, e.g. in terms of thickness and density (Mao *et al.,* 2015). AFBRs have been used successfully to treat high strength wastewaters with different biodegradable particulate fractions and chemical oxygen demand (COD) concentrations (Van Lier *et al.,* 2015).

7.3.5 Anaerobic moving bed biofilm reactor

The anaerobic moving bed biofilm reactor (AMBBR) contains carrier materials that can move freely throughout the entire volume of the reactor (Karadag *et al.,* 2015). The moving bed reactor is a very common reactor configuration in aerobic systems (Chu & Wang, 2011), in which the carrier material movement is caused by the agitation set up by the air (Figure 7.5a). In anaerobic conditions, the reactor is often equipped with a submerged pump or mixer in order to enable the carrier material movement (Figure 7.5b) (Gu *et al.*, 2014). Hence, the carrier material should have an optimal density. The main advantages of this reactor configuration are no carrier clogging, efficient mixing, low head loss and good hydrodynamic conditions due to the movement of the biofilm carrier material.

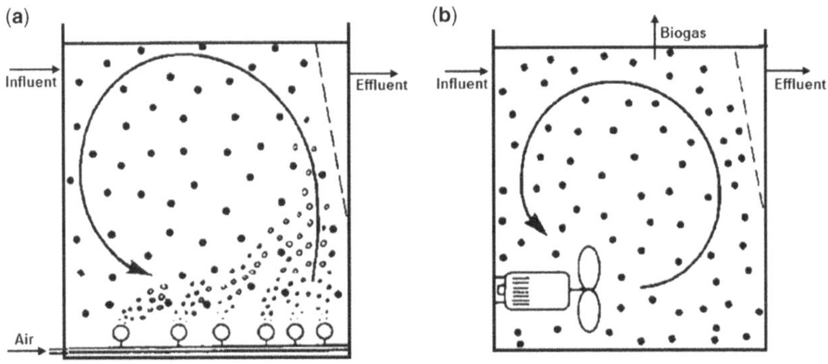

Figure 7.5 Schematic diagrams of (a) aerobic moving bed reactor and (b) AMBBR (Rusten *et al.*, 2006).

7.3.6 Anaerobic sequencing batch biofilm reactor

The anaerobic sequencing batch biofilm reactor (ASBBR) is established by introducing the carrier material for the biofilm development into the conventional anaerobic sequencing batch reactor (ASBR) (Figure 7.6), where the reactor operation consists of four sequential steps: feeding, reaction, settling and effluent withdrawal (Karadag *et al.*, 2015). The first ASBBR was proposed by adding the biofilm carrier material into an ASBR for biomass immobilization. The retention of AD biomass on the carrier materials then represents an important contribution to the improvement of the ASBR performance, since it eliminates the settling step and reduces the washout of active biomass from the reactor (Dutta *et al.*, 2014).

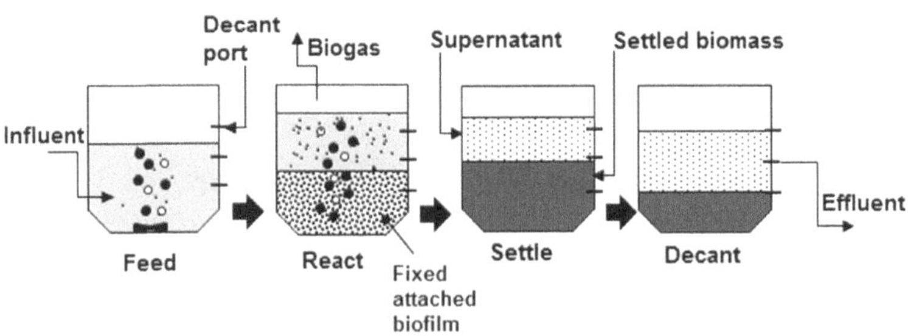

Figure 7.6 Schematic diagram of the ASBBR operation.

7.3.7 Upflow anaerobic sludge blanket (UASB) reactor

The UASB reactor was first applied by Lettinga and his co-workers in 1976 in the Netherlands (Van Lier *et al.*, 2015). The UASB reactor is the most popular industrial anaerobic wastewater process, having a wide application potential, particularly for feeds where the majority of organics are soluble (Figure 7.7a). Many successful performance results have been reported at lab- and pilot-scale applications using the UASB reactor configuration, which has resulted in the establishment of thousands of full-scale reactors worldwide (Van Lier *et al.*, 2015). The performance of UASB reactors relies on the formation of well-settleable sludge aggregates (flocs\granules) and the stability of the UASB granules (Van Lier *et al.*, 2015). The formation of well-settleable sludge aggregates, and internal retention of granules through settling allows the decoupling of the HRT and SRT, so that efficient treatment can be carried out at high OLRs with a significant decrease in reactor size. Furthermore, the natural turbulence caused by the rising of biogas bubbles provides efficient substrate and biomass contact. Thus, mechanical mixing is not required, which significantly reduces the energy demand and its associated cost. Trade of granular sludge is common for re-inoculation or start-up of new AD reactors (Van Lier *et al.*, 2015). Drawbacks include long process start-up, lack of pathogens and nutrient removal, and insufficient organic matter removal in the final effluent. The process is also intolerant to particulate organic and inorganic solids, due to their tendency to interfere with the granulation process (by acting as a physical growth substrate), and their tendency to accumulate in the granular sludge bed.

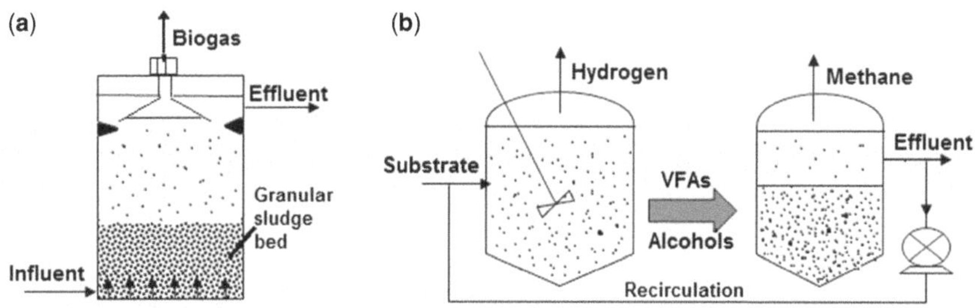

Figure 7.7 Schematic diagram of a (a) UASB and (b) two-stage anaerobic reactor.

7.3.8 Hybrid anaerobic biofilm reactors

In order to improve the treatment performance and overcome the deficiencies of the AD process, a hybrid anaerobic biofilm reactor has been developed by combining the biofilm reactor with another type of reactor in two or more physically separated reactor zones (Karadag *et al.*, 2015). Most of the reported hybrid anaerobic biofilm reactors are a combination of a UASB and an APBR, aiming to improve the biomass loss in the UASB reactor and take advantage of both reactor systems

(Azbar *et al.*, 2009). In recent years, hybrid anaerobic biofilm reactors have been widely applied for the treatment of different types of wastewater.

7.3.9 Two-stage anaerobic reactor

The two-stage AD system was initially proposed as a potential solution to improve the overall AD process efficiency and energy recovery over a single stage reactor (Ariunbaatar *et al.*, 2014). The system is composed of a hydrolysis-acidogenesis step carried out in the first-phase (known as dark fermentation) and an acetogenesis-methanogenesis step carried out in the second phase (Figure 7.7b) (Cavinato *et al.*, 2016). The main advantages of two-stage AD systems are (i) better stabilization of the AD process under two different pH optima, with pH adjustment possible in the first stage, (ii) protection of the methanogenic step from organic overloads, (iii) reduced digestion time and buffering requirements; and (iv) increase of the overall yield of energy recovery (carbon conversion up to 40%) (Ariunbaatar *et al.*, 2014). In the two-stage AD system, hydrogen and methane are produced simultaneously, with the final purpose of mixing the two gasses to achieve biohythane (mixture of H_2 and CH_4) (Yeshanew *et al.*, 2016b). Biohythane is a promising renewable energy carrier which offers a better combustion efficiency and reduced greenhouse gas emissions compared to natural gas only in internal combustion engines (Liu *et al.*, 2013). This gas mixture has been commercially applied as vehicle fuel in some countries such as the USA and India (Cavinato *et al.*, 2016), and is achieving popularity in the transportation sector (Yeshanew *et al.*, 2016b).

7.3.10 Anaerobic membrane bioreactor (AnMBR)

An anaerobic membrane bioreactor (AnMBR) is an advanced and highly compacted AD technology that combines the advantages of anaerobic treatment with membrane separation, making it an attractive approach for the treatment of a broad spectrum of wastewaters (Smith *et al.*, 2012). The reactor type has a dual purpose of organic matter removal and solid separation, thus giving rise to a small footprint and a very high effluent (permeate) quality with no suspended solids (Skouteris *et al.*, 2012). The AnMBR retains all microorganisms in the reactor without washout, irrespective of the HRT. Hence, it demonstrates higher operational stability and efficiency in treating various types of wastewaters under extreme conditions, including high salinity, high suspended solids content and wastewater with poor biomass granulation (Li & Yu, 2016). As a result, AnMBRs are currently of great interest to both researchers and the industrial community, with many laboratory and pilot-scale studies being undertaken. AnMBRs achieve high effluent quality at high OLR (up to 30 kgCOD/(m^3·d), have minimal start-up times, have a short HRT, and are not adversely impacted by input solids, and hence can treat problematic industrial streams (van Zyl *et al.*, 2008; Jensen *et al.*, 2015). The membrane filtration component can exist in three alignments: external cross-flow, internal submerged or external submerged membrane (Ozgun *et al.*, 2013). In the external cross-flow configuration, the membrane is separated from the reactor and the membrane operates under a pressure and thus produces permeate (Figure 7.8b). The suspended anaerobic biomass maintained in the reactor is pumped through the membrane, and hence creates a positive pressure that leads to permeate production. The extracted permeates through these are, generally, at a much higher pressure and operating cost (Judd, 2011). The rejected biomass, also called retentate, is returned to the reactor (Ozgun *et al.*, 2013).

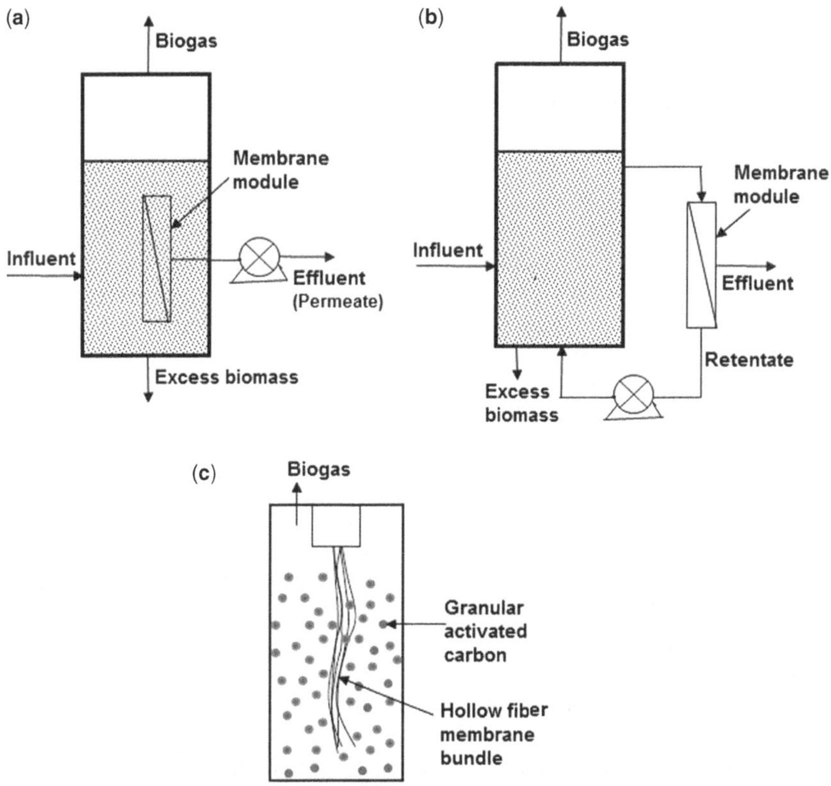

Figure 7.8 Schematics of AnMBR (a) internal submerged membrane, (b) external cross-flow membrane and (c) hollow fiber membrane with GAC.

In the internal submerged AnMBR, the membranes are submerged directly into the suspended biomass in the reactor and permeate is produced by exerting a vacuum on the membrane (Figure 7.8a) (Skouteris *et al.*, 2012). In the external submerged AnMBR, membranes may be located in an external chamber that is separated from the main reactor, however the membranes are still submerged in the suspended biomass and operated under vacuum. In such an external submerged configuration, the suspended biomass from the reactor is pumped to the external chamber, while the retentate is returned to the main reactor. This configuration facilitates membrane cleaning and replacement by allowing isolation of the membrane unit in an external chamber. This separation enables anaerobic conditions to be maintained in the main reactor during membrane cleaning or replacement (Ozgun *et al.,* 2013).

Various membrane configurations such as flat sheet, hollow fiber and tubular membranes are applied in AnMBRs (Smith *et al.,* 2012). Generally, for the submerged AnMBRs flat sheet and hollow fiber membranes are preferred, whereas external cross-flow AnMBRs can be also configured with tubular membranes (Ozgun *et al.,* 2013). Due to capital and operating costs, except where the fluid is extremely problematic, the external and internal submerged membrane units dominate the market (Judd, 2011). Operational costs related to energy requirements for gas/liquid recirculation, membrane fouling control and chemical costs required for membrane cleaning are high, compared with conventional high rate anaerobic digesters.

The level and degree of membrane fouling depends on many factors such as membrane operation, reactor type, membrane configuration, substrate type and sludge characteristics (Ozgun *et al.,* 2013). Cleaning methods can be various depending on the nature of the fouling (Skouteris *et al.,* 2012). Membrane cleaning can either be done physically or chemically. Physical cleaning is closely related to membrane operation such as periodic backwashing, relaxation or short-term increase in shear rate to remove the cake layer accumulated on the membrane surface (Smith *et al.,* 2012). In addition, membranes can also be physically removed from the membrane tank to be cleaned ex-situ, e.g. by applying water jets. Membrane cleaning in place (CIP) can be also performed by using a chemical cleaning procedure with alkali (such as NaOH), oxidants (such as H_2O_2 and NaOCl) or acid (HCl, H_2SO_4 and citric acid) cleaning agents (Ozgun *et al.,* 2013).

In submerged membrane applications, membrane fouling is controlled by operating below critical flux, where fouling cake depth is stable, with aeration by biogas to provide liquid shear (to increase critical flux). To enable higher critical fluxes with minimal gas supply, particulates such as granular activated carbon (GAC) are supplied with the breakthrough being an integration of the AnMBR with an AFBR having the GAC (Figure 7.8c). The GAC acts both as a carrier material and biofilm scrubber from the surface of the membranes (Yoo *et al.,* 2012). Such a configuration allows sustained flux with less frequency of membrane cleaning (Smith *et al.,* 2012). A pilot-scale AnMBR with GAC has been successfully and stably operated for almost two years to treat municipal wastewater, with no chemical cleaning (Li & Yu, 2016). Moreover, membrane acquisition and/or replacement costs have significantly decreased due to a decline in membrane module costs (Skouteris *et al.,* 2012).

7.4 SUBSTRATE PRETREATMENT METHODS FOR ENHANCED AD

In the AD process of complex organic substrates, hydrolysis is the rate-limiting step (Ariunbaatar *et al.*, 2014) because of the complex structure and composition of the substrate (Lee *et al.*, 2014). Various pretreatment methods have been explored to enhance the solubilization of the complex organic substrate and ultimately increase the methane production and solids destruction (Yeshanew *et al.*, 2016a). Pretreatment methods include mechanical, thermal, chemical, and biological (Table 7.3) or a combination of them (Neumann *et al.*, 2016). The degree of hydrolysis and the products from this step directly affect the overall digestion process. The effectiveness of each pretreatment method depends on the characteristics of the substrate, i.e. its nature, composition and structure (Carrère *et al.*, 2016).

7.4.1 Mechanical pretreatment

Mechanical pretreatment reduces the particle size and disrupts the solid particles of the substrates through mechanically/physically breaking down the intermolecular bonds (Lee *et al.*, 2014). The increased surface area allows better contact between the substrates and bacteria, hence facilitating the hydrolysis step. Milling/chopping/grinding, screw press, lysis-centrifugation, sonication, liquid shear/collision and high-pressure homogenization methods are examples of mechanical pretreatments used to improve the physical properties of the substrate (Ariunbaatar *et al.*, 2014). Mechanical pretreatment has no inherent risk of recalcitrant compounds or formation of inhibitors during the course of the pretreatment (Carrère *et al.*, 2016). However, high energy input is required for the mechanical equipment, and performance is relatively modest (Carrère *et al.*, 2016). Full-scale mechanical pretreatment devices are already in use for sewage sludge (Neis *et al.*, 2008) and lignocellulosic substrates (Monch-Tegeder *et al.*, 2014) (Table 7.3).

7.4.2 Thermal pretreatment

Thermal pretreatment was recognized early as a method with a potential to improve the AD process of complex organic substrates (Ariunbaatar *et al.*, 2014). Thermal pretreatment is relatively easy to implement compared with other pretreatment methods, since the biogas produced can be used as an energy source to heat the substrate (Yeshanew *et al.*, 2016a). The main effects of thermal pretreatment are hydrolyzing the cell walls and loosening the intracellular bonds of the substrates, thus allowing the release of soluble organic compounds. Heating temperature and pretreatment time have been identified as the key factors in determining the effectiveness of thermal pretreatment (Rincón *et al.*, 2013). Accordingly, a wide range of temperatures (50–250°C) and treatment times (0.5–12 hours) have been tested (Ariunbaatar *et al.*, 2014).

Table 7.3 Full-scale application of different pretreatment methods for various types of substrate (source: Carrère *et al.*, 2016).

Substrate / Pre-Treatment	Sludge	Manure	Animal by-products	Municipal Solid Waste	Agricultural Residue and Energy Crops	Algae
Mechanical	Sonication High pressure Lysing centrifuge	Grinding Extrusion Maceration	Grinding	Grinding Maceration	Grinding Extrusion	
Thermal	Steam explosion Hydrothermal Thermal hydrolysis	Hydrothermal Low temperature	Hydrothermal Low temperature	Steam explosion		Low temperature
Chemical	[a]FNA		Saponification		Alkali	
Biological	Acid-gas Enzymatic [b]TPAD	Partial composting		Pre-composting	Enzymes Ensiling Composting	

[a]FNA: Free nitrous acid.
[b]TPAD: Temperature phased anaerobic digestion.

Dark shading: Pilot scale.

Light shading: Promising lab-scale results.

High temperature heating is often performed by steam injection to 160°C at autogenous pressure, with the process referred to as thermal hydrolysis. Pressure may be quickly released in a flash tank both to recover steam and further enable cell disruption (Carrère *et al.*, 2016). Higher temperatures result in loss of methane yield due to production of biologically recalcitrant organics, including melanoidins (see below, Carrère *et al.*, 2016). Thermal pretreatment can lead to a positive energy balance due to the increased biogas production, and the required heat can be produced on-site through biogas combustion. It also leads to pathogen destruction, improved dewatering performance and reduced viscosity of the digestate, hence making it easier to handle and dewater the digestate (Carrère *et al.*, 2016). In full-scale applications, thermal pretreatment is widely applied, particularly for treatment of sewage sludge, but also municipal solid waste and animal by-products (Table 7.3).

As an alternative, microwave pretreatment is another potential pretreatment method, which enhances the hydrolysis of the complex organic substrate via thermal and non-thermal effects. The thermal effect is ascribed to the heat generation through the rotation of the dipolar molecules (e.g. water) in the oscillating electromagnetic field. The non-thermal effect is caused by the alignment of the polarized side chains of macromolecules (e.g. protein) with the poles of the electromagnetic field, thus breaking the hydrogen bonds and destabilizing the structure of the molecules (Ariunbaatar *et al.*, 2014). The main disadvantage of thermal hydrolysis is that heat is generated electrically, which is inherently more expensive and inefficient compared with steam derived from biogas.

In spite of the advantages, thermal pretreatment at higher temperatures and microwave pretreatment, in addition to the larger energy input, can lead to the formation of recalcitrant soluble organics. The formed compounds are often inhibitory substances to the AD microbial population (Lee *et al.*, 2014). One of the most well-known phenomena is the Mallaird reaction (Yeshanew *et al.*, 2016a), which occurs between soluble carbohydrates and amino acids, resulting in the formation of complex melanoidin compounds that are difficult to biodegrade by the AD microorganisms (Yeshanew *et al.*, 2016a). Moreover, thermal pretreatments involve heat and also electricity if dewatering is required, as is also required for thermal hydrolysis (Lee *et al.*, 2014).

7.4.3 Chemical pretreatment

Chemical pretreatment methods are employed to hydrolyze the cell wall and cell membrane of the complex organic substrates, as well as chemically denature or destabilize structural proteins. They mainly rely on strong to mild chemical agents to modify the physico-chemical properties of the substrate (Ariunbaatar *et al.*, 2014). The reagents commonly used in chemical pretreatments are acid, alkali, ozone and hydrogen peroxide. Acid pretreatment is known to be efficient to solubilize carbohydrates, while alkali pretreatment is efficient in solubilization of proteins and lignin as well as lipid saponification (Carrère *et al.*, 2016). Chemical treatment can also be performed in combination with high temperatures, which is referred to as a thermo-chemical pretreatment method (Ariunbaatar *et al.*, 2014). Thermochemical pretreatment is usually preferred for highly resistant substrates such as lignocellulosic biomass. An alternative process has recently emerged, which is the application of the free form of nitrous acid (FNA), which has had substantial impacts (20–30% improvement in activated sludge conversion) at relatively low dose rates (1.8 mg HNO_2-N/L), including pathogen destruction (Wei *et al.*, 2018). FNA is mainly effective on activated sludge, and has been found to be ineffective on primary sludge.

Ozonization is another chemical pretreatment option. Ozone is a strong oxidizing agent that can decompose into hydroxyl radicals and react with the organic waste (Lee *et al.*, 2014), leading to solubilization of the complex organic substrates. Nevertheless, the high cost of ozone generation is a deterrent for the full-scale application. From the economic perspective, hydrogen peroxide is a less costly alternative of chemical pretreatment. Nonetheless, it is less reactive with organic substrates as its oxidation potential (1.8 V) is lower than that of ozone (2.1 V) (Lee *et al.*, 2014). Wet oxidation is often used for the pretreatment of lignocellulosic biomass prior to bioethanol fermentation, which is a thermal pretreatment under high pressure with the addition of oxygen.

A significant shortcoming of chemical pretreatment is the need for corrosion-resistant equipment due to the extreme high or low pH values applied. Chemical contamination, risk of formation of recalcitrant compounds (furanic or phenolic compounds) and higher cost of chemicals are some of the disadvantages of chemical and thermochemical pretreatment methods (Carrère *et al.*, 2016).

7.4.4 Biological pretreatment

Biological pretreatment is performed through biological agents such as hydrolytic enzymes, microorganisms (e.g. *Cellulomonas uda, C. biazotea*, *Aspergillus awamori*), mature compost or activated sludge from wastewater treatment plants (Lee *et al.*, 2014). Biological pretreatment is usually a slow process that requires longer retention times and the microorganisms utilize the free and readily available sugars as the main carbon source during the pretreatment step. The application of pure biological enzymes is usually difficult, since they are competing with the indigenous microorganisms during the pretreatment process. Instead, a mixture of enzymes can be used for the hydrolysis of the substrate since different enzymes hydrolyze different components of the complex organic substrate. Biological pretreatments are among the least energy consuming pretreatment methods (Carrère *et al.*, 2016). Nonetheless, pretreatment using enzymes and pure cultures can be costly, thus inexpensive biological agents such as mature compost and activated sludge should be considered in larger scale applications (Lee *et al.*, 2014). Composting and ensiling have been implemented at full scale for manure and energy crops, respectively (Table 7.3).

Alternative biological processes apply elevated temperatures (temperature phased anaerobic digestion – TPAD), or intentionally operate at low pH (mesophilic phased anaerobic digestion – MPAD, or acid-gas) (Carrère *et al.,* 2016). This enables a natural elevated hydrolysis in the first stage, followed by enhanced methanogenesis in the second stage (Ge *et al.,* 2011). TPAD has been found to be more effective than MPAD, with much of the improvement in MPAD being due to the inherent kinetic advantage of a two-stage system over a single stage system, when substrate specific kinetics dominant.

7.5 TECHNIQUES TO ENHANCE PHOSPHORUS RECOVERY DURING AD

Waste streams from different sources contain large amounts of valuable resources such as phosphorus, nitrogen, calcium, magnesium and potassium (Table 7.1), in addition to organic and inorganic carbon (Mehta & Batstone, 2013). Recovery of the valuable resources from the waste, particularly phosphorus, is growing due to the rising prices of phosphorus fertilizer, limited availability of natural phosphorus and its environmental impact on the natural waterways, in addition to operational problems caused by phosphorus precipitation in treatment plant piping (Latif *et al.,* 2015). The phosphorus in the waste stream exists in three different forms: organically bound, soluble (free orthophosphate) and particulate inorganic phosphorus (precipitated) (Bi *et al.,* 2012). The AD process facilitates the release of soluble phosphorus from the biodegradable fraction of the waste (Figure 7.2), with the extent of conversion dependent on the operating conditions employed during the process (De Graaff *et al.,* 2010). When at high enough concentration, the released phosphorus as orthophosphate binds with other cations including Mg^{2+}, Ca^{2+}, Al^{3+} and Fe^{2+} and forms phosphorus precipitates, such that sometimes only 20% of the total phosphorus remains soluble at neutral pH (Latif *et al.,* 2017). The common phosphorus precipitates are struvite (magnesium/potassium ammonium phosphate), aluminum phosphate, iron phosphate and apatite (calcium phosphates) (Latif *et al.*, 2017).

These phosphorus precipitates are a nuisance, forming scale in the reactors and downstream equipment, hence increasing the operational and maintenance costs (Batstone *et al.,* 2015a). Iron and aluminum phosphates are mainly found in biosolids, and reduce the bioavailability of phosphorus when used for agricultural purposes (Kim & Nakhla, 2009). Therefore, it is important to reduce the phosphorus precipitation during the AD process, such that it can be effectively recovered during post-digestion. Reduction of struvite and apatite formation can reduce the maintenance costs of the reactors and removal of iron and aluminum phosphates can improve the digestate quality (Latif *et al.,* 2017).

In addition, to reduce the solids handling costs, digestates are typically dewatered to produce a soluble nutrient-rich (predominantly N and K) side stream (Batstone *et al.,* 2015a). This nutrient-rich side stream can be a feedstock for nutrient extraction/recovery techniques. The remaining particulate-bound nutrients (mainly P) and residual organics are recovered as biosolids, which have value as nutrient amendments for agricultural purposes provided that a suitable biosolids quality can be achieved to match specific application requirements with respect to residual odor, pathogens and heavy metal content (Latif *et al.*, 2017).

Reduced phosphorus precipitation and enhanced phosphorus solubility during the AD process can be achieved through optimizing operational parameters (such as pH, temperature and pressure) or include the addition of chemical additives, such as ethylenediaminetetra acetic acid (EDTA).

7.5.1 Optimizing operational parameters

The phosphorus solubility in the AD process can be enhanced by optimizing the operational parameters such as pH, temperature and pressure (Bi *et al.*, 2012). The pH conditions of the AD reactor affect the phosphorus precipitation in the AD process. Acidic pH conditions, i.e. pH <6, during the AD process can promote phosphorus solubilization. This is due to the higher solubility of struvite and calcium phosphate at lower pH conditions (Latif *et al.*, 2017). Low pH conditions protonate the phosphate ions, thus breaking up the phosphate precipitate and dissolving the phosphorus into the liquid as soluble phosphate ions (Bi *et al.*, 2012). However, lower pH values negatively affect the methanogenic population by accumulating the VFAs and hence reduce the methane production rate (Latif *et al.*, 2017). The process of pH reduction is economically reasonable, mainly due to the very low cost of HCl.

Temperature is another operational parameter that affects the phosphorus precipitation in the AD process (Bi *et al.*, 2012). Operating temperatures below 30°C can enhance the solubility of struvite and thermophilic temperatures could increase the PO_4^{3-} concentration during the AD process. Theoretically, higher phosphorus solubility can be expected for the thermophilic (55°C) than the mesophilic (35°C) AD process. Similarly, high pressure enhances the phosphorus solubility by decreasing the reactor pH through carbon dioxide (CO_2) solubilization into the liquid (Latif *et al.*, 2017). The increased CO_2 solubility may decrease its amount in the reactor headspace, hence increasing the methane concentration of the biogas (Mehta & Batstone, 2013).

7.5.2 Chemical additives

Chemical additives can significantly improve the phosphorus solubility during the AD process. A common chemical additive is EDTA. EDTA is a chelating agent with high affinity for di- (e.g. Ca^{2+} and Mg^{2+}) and tri- (e.g. Fe^{3+} and Al^{3+}) valent metal ions (Zhang *et al.*, 2010). The AD process with added EDTA reduces calcium and magnesium binding with orthophosphate and thereby, reduces formation of precipitants such as struvite and apatite. EDTA is suitable for higher phosphorus release during the AD process without compromising the biogas production. However, it is an expensive chemical and not feasible commercially (Latif *et al.*, 2017).

7.6 BIOFUEL AND BIOENERGY

Fossil resources are presently not regarded as sustainable from the economic, ecological and environmental points of view (Ghimire *et al.*, 2015). The burning of fossil fuels is the main contributor to the increasing level of CO_2 in the atmosphere, which is directly associated with the global warming observed in recent decades (Andalib *et al.*, 2014). In addition, it leads to adverse effects such as toxic gas emission, depletion of petroleum reserves and oil spills (Lee *et al.*, 2014). Therefore, the quest for sustainable and environmentally benign sources of energy for the industrial economies and consumer societies has become urgent in recent years. Consequently, there is renewed interest in the production and use of biofuels from renewable sources (Lee *et al.*, 2014). Other than producing biogas (methane enriched gas), the AD process can also be used for the generation of different forms of bioenergy such as hydrogen and electricity. Direct electricity generation from organic substrate is possible through the use of microbial fuel cells (Wenzel *et al.*, 2017).

7.6.1 Biohydrogen

Hydrogen is considered as the "energy for the future" since it is a clean and renewable energy source with high energy content per mass compared to hydrocarbon fuels (Liu *et al.*, 2013; Carrillo-Reyes *et al.*, 2016). Among the various H_2 production processes, biological methods have shown significant advantages over physico-chemical methods, because they can be less energy intensive when operating using less sophisticated technology (Barca *et al.*, 2015). Hydrogen can be produced biologically by both autotrophic and heterotrophic microorganisms (Ghimire *et al.*, 2015). Autotrophic conversions are mediated by photosynthetic microorganisms (i.e. microalgae, protists and photosynthetic bacteria) utilizing inorganic carbon dioxide as a carbon source, whereas heterotrophs convert the organic carbon sources into simpler compounds with simultaneous production of molecular H_2 (Ghimire *et al.*, 2015). There are two types of heterotrophic conversions: the first one is driven by light energy called photofermentation and the other occurs in the absence of light known as dark fermentation (DF) (Ghimire *et al.*, 2015).

Photofermentation is often performed together with the DF as a two-stage hydrogen production process. This is because the DF not only produces hydrogen but also intermediate metabolites, i.e. VFAs (Yeshanew *et al.*, 2016b). The resulting VFA-rich effluent from the DF can then be used in photofermentation to enhance the overall hydrogen yield (Ghimire *et al.*, 2015). Prior to photofermentation, it is necessary to remove the colloidal particles from the DF effluent to ensure good light penetration. Besides, the DF effluent should have a low ammonium content to avoid the inhibition of the nitrogenase enzyme which catalyzes the photo-hydrogen production (Lee *et al.*, 2014). Dilution has been proposed to reduce the ammonium concentration, but this leads to the requirement of a larger reactor volume (Chen *et al.*, 2010), and the co-process of nitrogen fixation leads to a relatively low light usage efficiency. Thus, the development of more economical ammonia removal techniques is required. Besides, the high energy demand associated with the nitrogenase enzyme, lower solar conversion efficiencies and economic issues of anaerobic photobioreactors are some drawbacks of the photofermentative process (Uyar *et al.*, 2009).

In the photofermentation process, purple non-sulfur bacteria convert organic substrates into hydrogen and CO_2 in the presence of infra-red light (Chen *et al.*, 2010), with additional hydrogen being produced through photo-nitrogen fixation (for growth) (Figure 7.9c). In addition to biohydrogen production, purple non-sulfur bacteria are known to accumulate polyhydroxybutyrate (PHB) biopolymers (Sivagurunathan *et al.*, 2016), and are a potential route for production of this next generation plastic. These microorganisms are receiving increasing attention due to their higher biohydrogen yield potential and better light utilization proficiency, as they are able to absorb and utilize both visible (400–700 nm) and near infrared (700–900 nm) light (Chen *et al.*, 2010).

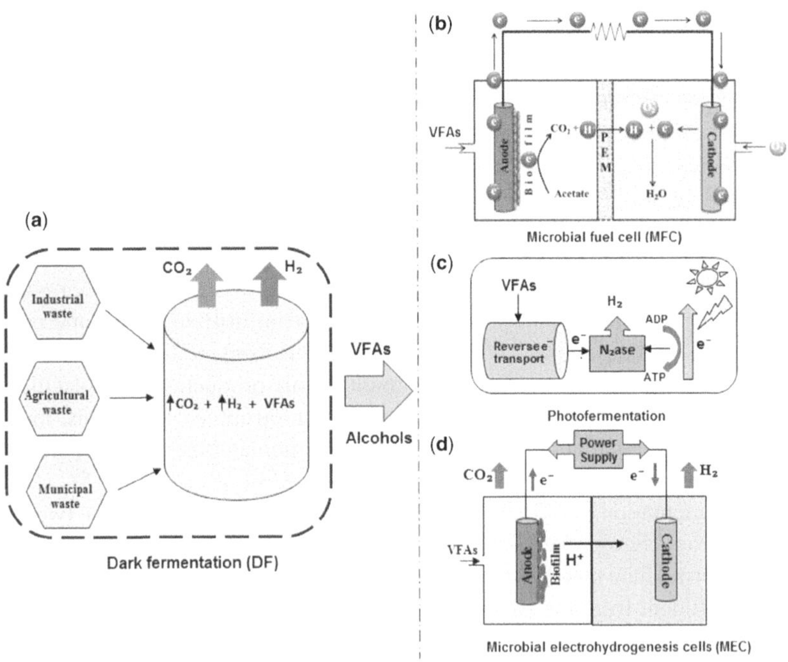

Figure 7.9 Schematic illustration of bioenergy generation or by-product formation using (a) DF, (b) MFC, (c) photofermentation and (d) MEC.

The hydrogen production rate in the photofermentation process is also affected by the type of VFA applied in the photofermentation process (Lee *et al.*, 2014). *Rhodobacter sphaeroides*, a bacterial strain commonly used in photofermentation, could better utilize acetate and propionate for hydrogen production as compared to butyrate (Chen *et al.*, 2010). The maximum hydrogen production rate from acetate and propionate was four times higher than that of butyrate (Uyar *et al.*, 2009). This indicates the need to increase the feed acetate and propionate concentrations for a better hydrogen production performance by *R. sphaeroides* (Uyar *et al.*, 2009).

DF represents cost-effective and key technology for biohydrogen production due its faster conversion efficiencies (Barca *et al.*, 2015). The DF process can utilize a wide range of renewable complex wastes as feedstock (such as crop residue, animal manure, food waste) and produce other valuable platform biochemicals of economic interest (Figure 7.9a) (Lee *et al.*, 2014; Ghimire *et al.*, 2015). Recently, biomass derived starting materials such as plant lignocelluloses as well as microalgae have been identified as a potential next generation feedstock (Sivagurunathan *et al.*, 2016). The process of DF can be achieved by the aid of strict or facultative AD microorganisms in the form of pure cultures, co-cultures or mixed microbial consortia (Ghimire *et al.*, 2015). The latter provide distinct advantages including process flexibility, better tolerance to metabolic stress and higher hydrogen productivity (Guo *et al.*, 2010). Mixed culture systems are favored for industrialization since they can be operated under a non-sterile environment, which reduces the process expenditures and increases the suitability of complex organic substances with different origins (Barca *et al.*, 2015).

The DF technology suffers from some defects regarding scaling up and commercialization. Currently, the major barriers in the scaled-up application of DF for H_2 production are: low H_2 yield (a maximum yield of 4 H_2/mole glucose can be achieved, with acetate as by-product) and the high production costs mainly due to the high cost of feedstock (Ghimire *et al.*, 2015). One of the reasons for insufficient gas yields is the co-formation of soluble metabolite mostly encompassing volatile organic acids and alcohols. This means that a considerable part of the chemical energy stored in the fermentation raw material is diverted to by-product formation and remains unexploited (Lee *et al.*, 2014). Thus, additional efforts are needed recover more energy from the liquid phase residue (fermentation effluent), which would maximize the process conversion efficiencies and valorization of the by-products (Sivagurunathan *et al.*, 2016).

An inherent challenge of the DF process is minimizing the ecological footprint of the process by reducing the water and energy inputs into the process (Guo *et al.*, 2010). Measures involve downstream-associated, two-stage design and may be referred to as integrated systems, when dark fermentation is complemented with either photofermentation, anaerobic digestion or bio-electrochemical systems (Sivagurunathan *et al.*, 2016). Besides multi-step process development, upstream-side strategies – covering adequate start-up, optimization of operational parameters (e.g. pH, temperature, hydraulic retention time, organic loading rate and hydrogen partial pressure), immobilized technology applications as well as bioaugmentation – have been tested and promoted for intensification purposes (Ghimire *et al.*, 2015).

7.6.2 Production of electricity

Bioelectrochemical systems (BESs) combine biological and electrochemical processes to generate electrical energy, hydrogen or other useful chemicals (Yuan & Zhu, 2016). The common examples of BES systems are microbial fuels cells (MFCs) and microbial electrohydrogenesis cells (MECs) (Lee *et al.*, 2014). In both MFCs and MECs, bacteria work under anaerobic conditions at an anode to oxidize various substrates ranging from simple compounds, i.e. sugars and organic acids, to complex organic matter of distinct origin as well as fermentation effluents (Lee *et al.*, 2014). When breaking down the organic material, the bacteria release electrons (negatively charged particles), protons (positively charged hydrogen ions) and carbon dioxide into solution (Wenzel *et al.*, 2017). The anode collects the electrons, which then travel to the cathode via an external circuit, i.e. an electric current. The protons travel through the solution in the cell to the cathode via a membrane. The carbon dioxide can be captured and reused (Figure 7.9b and d).

MFCs are designed to produce electricity and MECs use electricity to drive chemical reactions at the cathode to produce hydrogen and/or other chemicals (Wenzel *et al.*, 2017). In MECs, on oxidizing acetate under standard biological conditions (25°C, 1 bar pressure and pH 7), hydrogen can be produced at the cathode by applying a small circuit voltage, theoretically 0.14 V (Wenzel *et al.*, 2017) (Figure 7.9d). Some exoelectrogenic microorganisms which are capable of electron transfer to an electrode (anode), including the genera *Geobacter*, *Shewanella* and *Pseudomonas sp.* (Rabaey & Rozendal, 2010). One of the challenges of MECs is to suppress the methanogenic activity during electrohydrogenesis with mixed cultures as it negatively affects the hydrogen production rate (Yuan & Zhu, 2016).

In MFCs, organic matter is converted to CO_2, protons and electrons by microorganisms as part of their metabolism using the anode electrode as a final electron acceptor (Lee *et al.*, 2014). The released electrons are transferred to a current collector and migrate through an external resistor to the cathode generating electric current. At the cathode, oxygen reduction to water is catalyzed using the electrons and protons produced on the anode. The development of a biofilm on the anode surface is essential for an improved bio-catalysis and electron transfer to get efficient current production. MFCs have been proposed as a possible technology to harness energy from the remaining organic matter after the DF process for various waste effluents (Wenzel *et al.*, 2017). The fundamental limitation of MFCs is that the technology is expensive, and renewable power is relatively low cost. Therefore, there has been a shift in the field of bio-electrochemical systems away from MFCs, and instead towards using electrical power to drive microbial conversion to upgrade products (Rabaey & Rozendal, 2010).

7.7 MATHEMATICAL MODELING OF ANAEROBIC DIGESTION

Anaerobic processes have been modeled extensively for over 40 years (Pavlostathis & Giraldo-Gomez, 1991) with a focus on (a) characterizing hydrolysis from a rate-limiting perspective, and (b) representing the very good understanding of the underlying biological processes as mathematical dynamic processes (Batstone *et al.*, 2002). Extensive work was done in the 90s on structured biological models, i.e. those which use separate states to represent functional groups of acidogens, acetogens and methanogens (Costello *et al.*, 1991; Siegrist 1993; Kalyuzhnyi & Fedorovich, 1998; Angelidaki *et al.*, 1999) and these models were consolidated into a standardized representation in the IWA Anaerobic Digestion Model No. 1 (Batstone *et al.*, 2002). This model was focused on effective representation of conventional anaerobic processes, including mixed digesters, biofilm systems, and UASB reactors, including simulation of pH through a physico-chemical model. However, it did not address effectively issues such as sulfate reduction, mixed dark fermentation for hydrogen production, or precipitation, mainly because the state of the art for these were not established at the time (Batstone *et al.*, 2006).

As with all IWA biological models, the focus was on biological and chemical kinetics (reaction terms), with transport being implemented by the user. Therefore, the model is equally applicable to multiple modeling domains, including mixed tanks, biofilm and plug flow. Model uptake has been rapid, and many of the initial limitations were addressed through specific extensions (reviewed in Batstone *et al.*, 2006; Batstone *et al.*, 2015b). These include extensive ion chemistry considerations, which include metal ion precipitation and the related ion and sulfur transformations, that were required to address the challenges around phosphorus recovery (Flores-Alsina *et al.*, 2015). Indeed, this is now propagating to plant-wide commonality in representation of physico-chemical processes (Flores-Alsina *et al.*, 2015). This is essential to enable the new generation of resource recovery technologies.

New reactor concepts, such as anaerobic membrane bioreactors are well served by the existing models, including the ADM1 (Spagni *et al.*, 2015), and the main challenge is determining the hydraulic profile and related transport model. Novel processes such as photofermentation, which challenge the traditional concept of anaerobic digestion as a low yield process, can be modeled by introducing new processes such as nitrogen fixation, growth without chemical catabolic processes, and anaerobic polymer storage (Puyol *et al.*, 2017). Another example is bioelectrochemical processes, which include ion migration, and spatially split redox reactions. While these require new approaches, the fundamentals of process modeling that have been established – i.e., a biological basis, reaction balancing, stoichiometry and thermodynamic drivers – are equally applicable, and offer a stable basis on which to continue to enable a rigorous and consistent modeling approach. This will enable model-based approaches to continue to support research through mechanistic analysis of observations, and industrial application through analysis and optimization, as it has benefited conventional anaerobic digestion.

7.8 CONCLUSION AND FUTURE DEVELOPMENTS

AD technology for the production of bioenergy can be optimized by selection of an optimal reactor design and pretreatment strategy. The field, while mature, is still developing, with particularly non-methanogenic anaerobic conversions improving in importance. The area of energy and nutrient recovery is increasing in performance, and likely to drive an increased focus on ways to modify anaerobic digester design to enhance these processes. Practical implementation of the different techniques to reduce phosphorus precipitation (and enable its recovery) during the AD process is limited, and the area of other nutrients (particularly nitrogen) will further increase in importance in the future. Some of these are practically realizable, but require maturity, while others require new concepts driven by innovative research.

7.9 REFERENCES

Andalib M., Elbeshbishy E., Mustafa N., Hafez H., Nakhla G. and Zhu J. (2014). Performance of an anaerobic fluidized bed bioreactor (AnFBR) for digestion of primary municipal wastewater treatment biosolids and bioethanol thin stillage. *Renewable Energy*, **71**, 276–285.

Angelidaki I., Ellegaard L. and Ahring B. K. (1999). A comprehensive model of anaerobic bioconversion of complex substrates to biogas. *Biotechnology and Bioengineering*, **63**(3), 363–372.

Ariunbaatar J., Panico A., Esposito G., Pirozzi F. and Lens P. N. L. (2014). Pretreatment methods to enhance anaerobic digestion of organic solid waste. *Applied Energy*, **123**, 143–156.

Azbar N., Tutuk F. and Keskin T. (2009). Biodegradation performance of an anaerobic hybrid reactor treating olive mill effluent under various organic loading rates. *International Biodeterioration and Biodegradation*, **63**, 690–698.

Barca C., Soric A., Ranava D., Giudici-Orticoni M. T. and Ferrasse J. H. (2015). Anaerobic biofilm reactors for dark fermentative hydrogen production from wastewater: a review. *Bioresource Technology*, **185**, 386–398.

Batstone D. J., Keller J., Angelidaki I., Kalyuzhnyi S. V., Pavlostathis S. G., Rozzi A., Sanders W. T. M., Siegrist H. and Vavilin V. A. (2002). Anaerobic Digestion Model No. 1 (ADM1), IWA Task Group for Mathematical Modelling of Anaerobic Digestion Processes. IWA Publishing, London.

Batstone D. J., Keller J. and Steyer J. P. (2006). A review of ADM1 extensions, applications, and analysis: 2002-2005. *Water Science and Technology*, **54**(4), 1–10.

Batstone D. J., Hülsen T., Mehta C. M. and Keller J. (2015a). Platforms for energy and nutrient recovery from domestic wastewater: a review. *Chemosphere*, **140**, 2–11.

Batstone D. J., Puyol D., Flores-Alsina X. and Rodríguez J. (2015b). Mathematical modelling of anaerobic digestion processes: applications and future needs. *Reviews in Environmental Science and Bio/Technology*, **14**(4), 595–613.

Bi D. S., Guo X. P. and Chen D. H. (2012). Anaerobic digestion of waste active sludge (WAS): the release and recovery of nitrogen and phosphorus. *Fresenius Environmental Bulletin*, **21**(5), 1283–1289.

Carrère H., Antonopoulou G., Affes R., Passos F., Battimelli A., Lyberatos G. and Ferrer I. (2016). Review of feedstock pretreatment strategies for improved anaerobic digestion: from lab-scale research to full-scale application. *Bioresource Technology*, **199**, 386–397.

Carrillo-Reyes J., Trably E., Bernet N., Latrille E. and Razo-Flores E. (2016). High robustness of a simplified microbial consortium producing hydrogen in long term operation of a biofilm fermentative reactor. *International Journal of Hydrogen Energy*, **41**(4), 2367–2376.

Cavinato C., Bolzonella D., Pavan P. and Cecchi F. (2016). Two-phase anaerobic digestion of food wastes for hydrogen and methane production. In: Enriched Methane. The First Step Towards the Hydrogen Economy, M. De Falco and A. Basile (eds), Green Energy and Technology, Switzerland, pp. 75–90.

Chen C., Yeh K., Lo Y., Wang H. and Chang J. (2010). Engineering strategies for the enhanced photo-H_2 production using effluents of dark fermentation processes as substrate. *International Journal of Hydrogen Energy*, **35**, 13356–13364.

Chu L. and Wang J. (2011). Comparison of polyurethane foam & biodegradable polymer as carriers in moving bed biofilm reactor for treating wastewater with a low C/N ratio. *Chemosphere*, **83**, 63–68.

Costello D. J., Greenfield P. F. and Lee P. L. (1991). Dynamic modelling of a single-stage high-rate anaerobic reactor-I. Model development. *Water Research*, **25**(7), 859–871.

De Graaff M. S., Temmink H., Zeeman G. and Buisman C. J. N. (2010). Anaerobic Treatment of concentrated Black Water in a UASB Reactor at a Short HRT. *Water*, **2**(1), 101–119.

Dutta K., Tsai C. Y., Chen W. H. and Lin J. G. (2014). Effect of carriers on the performance of anaerobic sequencing batch biofilm reactor treating synthetic municipal wastewater. *International Biodeterioration and Biodegradation*, **95**, 84–88.

Escudie R., Cresson R., Delgenes J. P. and Bernet N. (2011). Control of start-up and operation of anaerobic biofilm reactors: an overview of 15 years of research. *Water Research*, **45**(1), 1–10.

Fernandez N., Montalvo S., Borja R., Guerrero L., Sanchez E., Cortes I., Colmenarejo M. F., Travieso L. and Raposo F. (2008). Performance evaluation of an anaerobic fluidized bed reactor with natural zeolite as support material when treating high-strength distillery wastewater. *Renewable Energy*, **33**, 2458–2466.

Flores-Alsina X., Kazadi Mbamba C., Solon K., Vrecko D., Tait S., Batstone D. J., Jeppsson U. and Gernaey K. V. (2015). A plant-wide aqueous phase chemistry module describing pH variations and ion speciation/pairing in wastewater treatment process models. *Water Research*, **85**, 255–265.

Ge H., Jensen P. D. and Batstone D. J. (2011). Temperature phased anaerobic digestion increases apparent hydrolysis rate for waste activated sludge. *Water Research*, **45**(4), 1597–1606.

Ghimire A., Frunzo L., Trably E., Escudié R., Pirozzi F., Lens N. L. P. and Esposito G. (2015). A review on dark fermentative biohydrogen production from organic biomass: process parameters and use of by-products. *Applied Energy*, **144**, 73–95.

Gu Q., Sun T., Wu G., Li M. and Qiu W. (2014). Influence of carrier filling ratio on the performance of moving bed biofilm reactor in treating coking wastewater. *Bioresource Technology*, **166**, 72–78.

Guo X. M., Trably E., Latrille E., Carrère H. and Steyer J. P. (2010). Hydrogen production from agricultural waste by dark fermentation: a review. *International Journal of Hydrogen Energy*, **35**, 10660–10673.

Habouzit F., Hamelin J., Santa-Catalina G., Steyer J. P. and Bernet N. (2014). Biofilm development during the start-up period of anaerobic biofilm reactors: the biofilm archaea community is highly dependent on the support material. *Microbial Biotechnology*, **7**, 257–264.

Jensen P. D., Sullivan T., Carney C. and Batstone D. J. (2014). Analysis of the potential to recover energy and nutrient resources from cattle slaughterhouses in Australia by employing anaerobic digestion. *Applied Energy*, **136**, 23–31.

Jensen P. D., Yap S. D., Boyle-Gotla A., Janoschka J., Carney C., Pidou M. and Batstone D. J. (2015). Anaerobic membrane bioreactors enable high rate treatment of slaughterhouse wastewater. *Biochemical Engineering Journal*, **97**, 132–141.

Judd S. (2011). The MBR Book: Principles and Applications of Membrane Bioreactors for Water and Wastewater Treatment. Butterworth-Heinemann, Burlington.

Kalyuzhnyi S. and Fedorovich V. (1998). Mathematical modelling of competition between sulphate reduction and methanogenesis in anaerobic reactors. *Bioresource Technology*, **65**(3), 227–242.

Karadag D., Köroglu O. E., Ozkaya B. and Cakmakci M. (2015). A review on anaerobic biofilm reactors for the treatment of dairy industry wastewater. *Process Biochemistry*, **50**, 262–271.

Karakashev D., Batstone D. J. and Angelidaki I. (2005). Influence of environmental conditions on methanogenic compositions in anaerobic biogas reactors. *Applied and Environmental Microbiology*, **71**(1), 331–338.

Kim M. and Nakhla G. (2009). Phosphorus fractionation in membrane-assisted biological nutrient removal processes. *Chemosphere*, **76**(9), 1283–1287.

Kothari R., Pandey A. K., Kumar S., Tyagi V. V. and Tyagi S. K. (2014). Different aspects of dry anaerobic digestion for bio-energy: an overview. *Renewable and Sustainable Energy Reviews*, **39**, 174–195.

Latif M. A., Mehta C. M. and Batstone D. J. (2015). Low pH anaerobic digestion of waste activated sludge for enhanced phosphorous release. *Water Research*, **81**, 288–293.

Latif M. A., Mehta C. M. and Batstone D. J. (2017). Influence of low pH on continuous anaerobic digestion of waste activated sludge. *Water Research*, **113**, 42–49.

Lee W. S., Chua A. S. M., Yeoh H. K. and Ngoh G. C. (2014). A review of the production and applications of waste-derived volatile fatty acids. *Chemical Engineering Journal*, **235**, 83–99.

Li W. W. and Yu H. Q. (2016). Advances in energy-producing anaerobic biotechnologies for municipal wastewater treatment. *Environmental Protection—Review Engineering*, **2**(4), 438–446.

Liu Z., Zhang C., Lu Y., Wu X., Wang L., Wang L., Han B. and Xing X. H. (2013). States and challenges for high-value biohythane production from waste biomass by dark fermentation technology. *Bioresource Technology*, **135**, 292–303.

Mao C., Feng Y., Wang X. and Ren G. (2015). Review on research achievements of biogas from anaerobic digestion. *Renewable and Sustainable Energy Reviews*, **45**, 540–555.

Matheri A. N., Ndiweni S. N., Belaid M., Muzenda E. and Hubert R. (2017). Optimising biogas production from anaerobic co-digestion of chicken manure and organic fraction of municipal solid waste. *Renewable and Sustainable Energy Reviews*, **80**, 756–764.

Mehta C. M. and Batstone D. J. (2013). Nutrient solubilization and its availability following anaerobic digestion. *Water Science and Technology*, **67**(4), 756–763.

Minale M. and Worku T. (2014). Anaerobic co-digestion of sanitary wastewater and kitchen solid waste for biogas and fertilizer production under ambient temperature: waste generated from condominium house. *International Journal of Environmental Science and Technology*, **11**(2), 509–516.

Monch-Tegeder M., Lemmer A. and Oechsner H. (2014). Enhancement of methane production with horse manure supplement and pretreatment in a full-scale biogas process. *Energy*, **73**, 523–530.

Neis U., Nickel K. and Lunden A. (2008). Improving anaerobic and aerobic degradation by ultrasonic disintegration of biomass. *Journal of Environmental Science and Health. Part A, Toxic/Hazardous Substances & Environmental Engineering*, **43**(13), 1541–1545.

Neumann P., Pesante S., Venegas M. and Vidal G. (2016). Developments in pre-treatment methods to improve anaerobic digestion of sewage sludge. *Reviews in Environmental Science and Bio/Technology*, **15**, 173–211.

Ozgun H., Dereli R. K., Ersahin M. E., Kinaci C., Spanjers H. and Van Lier J. B. (2013). A review of anaerobic membrane bioreactors for municipal wastewater treatment: integration options, limitations and expectations. *Separation and Purification Technology*, **118**, 89–104.

Pavlostathis S. G. and Giraldo-Gomez E. (1991). Kinetics of anaerobic treatment: a critical review. *Critical Reviews in Environmental Control*, **21**(5–6), 411–490.

Plana P. V. and Noche N. (2016). A review of the current digestate distribution models: storage and transport. *WIT Transactions on Ecology and the Environment*, **202**, 345–357.

Puyol D., Barry E. M., Hülsen T. and Batstone D. J. (2017). A mechanistic model for anaerobic phototrophs in domestic wastewater applications: photo-anaerobic model (PAnM). *Water Research*, **116**, 241–253.

Rabaey K. and Rozendal R. A. (2010). Microbial electrosynthesis - revisiting the electrical route for microbial production. *Nature Reviews Microbiology*, **8**, 706–716.

Raheem A., Hassan M. Y. and Shakoor R. (2016). Bioenergy from anaerobic digestion in Pakistan: potential, Development and prospects. *Renewable and Sustainable Energy Reviews*, **59**, 264–275.

Rahman M., Salleh M. A., Rashid U., Ahsan A., Hossain M. M. and Ra C. S. (2014). Production of slow release crystal fertilizer from wastewaters through struvite crystallization - a review. *Arabian Journal of Chemistry*, **7**, 139–155.

Rincón B., Bujalance L., Fermoso F. G., Martín A. and Borja R. (2013). Biochemical methane potential of two-phase olive mill solid waste: influence of thermal pretreatment on the process kinetics. *Bioresource Technology*, **140**, 249–255.

Sen B., Aravind J., Kanmani P. and Lay C.-H. (2016). State of the art and future concept of food waste fermentation to bioenergy. *Renewable and Sustainable Energy Reviews*, **53**, 547–557.

Siegrist H., Renggli D. and Gujer W. (1993). Mathematical Modelling of Anaerobic Mesophilic Sewage Sludge Treatment. *Water Science and Technology*, **27**(2), 25–36.

Sivagurunathan P., Kumar G., Bakonyi P., Kim S., Kobayashi T., Xu K., Lakner G., Tóth G., Nemestóthy N. and Bẻlafi-Bakó K. (2016). A critical review on issues and overcoming strategies for the enhancement of dark fermentative hydrogen production in continuous systems. *International Journal of Hydrogen Energy*, **41**, 3820–3836.

Skouteris G., Hermosilla D., López P., Negro C. and Blanco Á. (2012). Anaerobic membrane bioreactors for wastewater treatment: a review. *Chemical Engineering Journal*, **198–199**, 138–148.

Smith A. L., Stadler L. B., Love N. G., Skerlos S. J. and Raskin L. (2012). Perspectives on anaerobic membrane bioreactor treatment of domestic wastewater. *Bioresource Technology*, **122**, 149–159.

Spagni A., Ferraris M. and Casu S. (2015). Modelling wastewater treatment in a submerged anaerobic membrane bioreactor. *Journal of Environmental Science and Health. Part A, Toxic/Hazardous Substances & Environmental Engineering*, **50**(3), 325–331.

Uyar B., Eroglu I., Yücel M. and Gündüz U. (2009). Photofermentative hydrogen production from volatile fatty acids present in dark fermentation effluents. *International Journal of Hydrogen Energy*, **34**, 4517–4523.

Van Lier J. B., Van der Zee F. P., Frijters C. T. M. J. and Ersahin M. E. (2015). Celebrating 40 years anaerobic sludge bed reactors for industrial wastewater treatment-review paper. *Reviews in Environmental Science and Bio/Technology*, **14**(4), 681–702.

van Zyl P. J., Wentzel M. C., Ekama G. A. and Riedel K.-H. (2008). Design and start-up of a high rate anaerobic membrane bio-reactor for the treatment of a low pH, high strength dissolved organic wastewater. *Water Science and Technology*, **57**(2), 291–295.

Vanyushina A. Y., Agarev A. M., Moyzhes S. I., Nikolaev Y. A., Kevbrina M. V. and Kozlov M. N. (2012). Comparison of different thickening methods for active biomass recycle for anaerobic digestion of wastewater sludge. *Water Science and Technology*, **66**(8), 1787–1793.

Wei W., Wang Q., Zhang L., Laloo A., Duan H., Batstone D. J. and Yuan Z. (2018). Free nitrous acid pre-treatment of waste activated sludge enhances volatile solids destruction and improves sludge dewaterability in continuous anaerobic digestion. *Water Research*, **130**, 13–19.

Wenzel J., Fuentes L., Cabezas A. and Etchebehere C. (2017). Microbial fuel cell coupled to biohydrogen reactor: a feasible technology to increase energy yield from cheese whey. *Bioprocess and Biosystems Engineering*, **40**, 807–819.

Yeshanew M. M., Frunzo L., Lens P. N., Pirozzi F. and Esposito G. (2016a). Mass loss controlled thermal pretreatment system to assess the effects of pretreatment temperature on organic matter solubilization and methane yield from food waste. *Frontiers in Environmental Science*, **4**, 62.

Yeshanew M. M., Frunzo L., Pirozzi F., Lens N. L. P. and Esposito G. (2016b). Production of biohythane from food waste via an integrated system of continuously stirred tank and anaerobic fixed bed reactors. *Bioresource Technology*, **220**, 312–322.

Yoo R., Kim J., McCarty P. L. and Bae J. (2012). Anaerobic treatment of municipal wastewater with a staged anaerobic fluidized membrane bioreactor (SAF-MBR) system. *Bioresource Technology* 120, 133–139.

Yuan H. and Zhu N. (2016). Progress in inhibition mechanisms and process control of intermediates and by-products in sewage sludge anaerobic digestion. *Renewable and Sustainable Energy Reviews*, **58**, 429–438.

Zhang T., Bowes K. E., Harrison J. H. and Chen S. (2010). Releasing phosphorus from calcium for struvite fertilizer production from anaerobically digested dairy effluent. *Water Environment Research*, **82**(1), 34–42.

Chapter 8

Greenhouse gas emissions from membrane bioreactors

G. Mannina[2], M. Capodici[1], A. Cosenza[1], D. Di Trapani[1] and M. C. M. van Loosdrecht[3]

[1]Dipartimento di Ingegneria Civile, Ambientale, Aerospaziale, dei Materiali, Università di Palermo, Viale delle Scienze, Ed. 8, 90128, Palermo, Italy

[2]Department of Earth and Environmental Engineering, Columbia University, 500 West 120th Street, New York, NY 10027, USA

[3]Department of Biotechnology, Delft University of Technology, Julianalaan 67, 2628 BC, Delft, Netherlands
E-mail corresponding author: giorgio.mannina@unipa.it

ABSTRACT

Nowadays, it is widely accepted that wastewater treatment plants (WWTPs) are significantly contributing to anthropogenic sources of greenhouse gas (GHG) emission. Among the GHGs emitted from WWTPs, nitrous oxide (N_2O) has been identified as the potentially major component, due to its high global warming potential (GWP) which is 298 times higher than that of carbon dioxide (CO_2) and also to its capability to react with stratospheric ozone, causing ozone layer depletion. Up until now, most of the experimental investigations aimed at assessing the key mechanisms of N_2O formation and the operational conditions that can enhance its emission, have been carried out on conventional activated sludge (CAS) processes, while the knowledge of N_2O emission from advanced technologies such as membrane bioreactors (MBRs) is still very limited. Moreover, the specific peculiarities of MBRs might hamper the direct transferability of data measured

Editors: Giorgio Mannina, George Ekama, Hallvard Ødegaard and Gustaf Olsson
doi: 10.2166/9781780409719_0293

on CAS systems. Therefore, there is a need to increase the knowledge about N_2O emission from MBRs through experimental and mathematical modelling activities. The present chapter aims to provide some knowledge about GHG emissions from MBR systems, in particular regarding N_2O emissions, by reporting some experimental data carried out on pilot plant systems of different configurations. The effect of specific wastewater features and operational conditions on N_2O production/emission from MBRs are highlighted.

Keywords: Global warming, membrane filtration, nitrous oxide, wastewater treatment

8.1 INTRODUCTION

During the last decade the awareness of the fact that wastewater treatment plants (WWTPs) are responsible for greenhouse gas (GHG) emissions has increased considerably. This is evident from the large number of research papers published on the topic (among others, Kampschreur *et al.,* 2009; Foley *et al.,* 2010; Daelman *et al.,* 2012; Ni *et al.,* 2013; Kim *et al.,* 2015; Mannina *et al.,* 2016a, 2017a, 2018a,b).

The evaluations by the Intergovernmental Panel on Climate Change (IPCC), aimed at identifying the causes, impacts and possible response strategies for mitigating climate change, indicate that the waste and wastewater sector accounts for about 3% of the global GHG emissions (Forster *et al.*, 2007; IPCC Climate Change, 2013). According to the United States Environmental Protection Agency (2013) water and wastewater facilities are amongst the largest energy consumers, thus significantly contributing to the overall GHG emissions, due to pumps, motors and equipment operating 24 hours a day and seven days a week. It is widely accepted in literature that WWTPs emit GHGs through three main sources, i.e., direct, indirect internal and indirect external (Global Water Research Coalition [GWRC], 2011). Direct GHG emissions are due to the biological processes occurring inside the WWTPs and represent the catabolite or obligate intermediates of reactions (CO_2, from microbial respiration; nitrous oxide, N_2O, from nitrogen removal processes, or methane (CH_4) from anaerobic processes). Indirect GHG internal emissions are mainly due to the consumption of electrical or thermal energy. Finally, indirect GHG external emissions are mainly related to sources not directly controlled within the WWTP (e.g., third-party biosolids hauling, production of chemicals and their transportation to the plant, etc.) (Mannina *et al.,* 2016e). Zeng *et al.* (2017) investigating 1079 urban WWTPs across China, have found that indirect GHG emissions could decrease by 32.2% if all plants worked efficiently.

The acquired awareness of "WWTP as source of GHG" has contributed to broaden the traditional operational goal of WWTPs to the minimization of GHG. Indeed, the traditional aim of WWTPs to achieve very stringent effluent limits now includes the GHG in terms of reduction of direct and indirect emissions (Flores-Alsina *et al.,* 2011a). In this regard, during recent years several efforts have been made in order to identify the key issues surrounding the GHG emissions from WWTPs: (i) processes responsible for GHG emissions (Foley *et al.*, 2010; Daelman *et al.*, 2012); (ii) operating factors mainly affecting the GHG emissions (Kampschreur *et al.*, 2009); and (iii) strategies for predicting and reducing GHG emissions (Flores-Alsina *et al.*, 2011b; Corominas *et al.*, 2012; Kim *et al.*, 2015). Several research projects (i.e., SANITAS and ENERWATER, PRIN2012GHG) and a Task Group of the International Water Association have also been developed and promoted in order to better understand the key mechanisms of GHG emissions from WWTPs.

Among the GHGs emitted from WWTPs, N_2O has been identified as the one of greatest interest. The amount of N_2O emitted from WWTPs is considerably lower than that of CO_2 or CH_4. However, due to the high global warming potential (GWP) of N_2O, 298 times higher than that of CO_2, and its capability to react with stratospheric ozone it is required to be investigated (IPCC, 2007). Several studies have been performed with the aim of better understanding the core pathways of its formation (Kampschreur *et al.*, 2009; Law *et al.*, 2013; Ni *et al.*, 2013).

N_2O is mainly produced in the biological nitrogen removal (BNR) processes via nitrification and denitrification both from autotrophic and heterotrophic bacteria (Kampschreur *et al.*, 2009). In some cases, N_2O emissions have been found to contribute to over 80% of the total greenhouse gas footprint emitted from WWTPs (Daelman *et al.*, 2013a,b). Despite the knowledge on GHG emission from WWTPs being quite mature mainly regarding CO_2 and CH_4, for N_2O literature still reveals a huge variability in the N_2O measured data ranging between 0.01–1.8% (N_2O emitted relative to wastewater N-load) (Kampschreur *et al.*, 2009; Ahn *et al.*, 2010; Aboobakar *et al.*, 2013; Ye *et al.*, 2014; Rodriguez-Caballero *et al.*, 2015; Daelman *et al.*, 2015). Literature also shows several difficulties in measuring the real overall N_2O produced by WWTPs; the available literature data case studies are often specific and difficult to transfer to other systems (Marques *et al.*, 2016). This aspect is relevant because most of the studies reported in literature are related to conventional activated sludge (CAS) systems and the knowledge acquired may not be transferred into innovative systems such as membrane bioreactors (MBRs).

MBR systems have attracted increasing attention in recent years, due to several advantages compared with conventional processes (Judd & Judd, 2010). More specifically, MBR systems provide high effluent quality, small footprint and moderate sludge production rates compared to CAS systems (Stephenson *et al.*, 2000). However, MBRs are characterized by some specific peculiarities – biomass selection; absence of secondary clarifier – that can contribute to N_2O production. Moreover, the intensive aeration for fouling mitigation in membrane compartments can promote N_2O stripping. These issues may hamper a direct transferability of the results derived for CAS systems. The main goal of this chapter is to summarize the key operating factors influencing the GHG production/emission from MBR WWTPs. Since the acquired knowledge on CO_2 and CH_4 formation is well developed, attention is primarily focused on N_2O emission here.

8.2 GHG EMISSION MECHANISMS

8.2.1 Direct emissions

8.2.1.1 Carbon dioxide – CO_2

CO_2 is directly produced due to aerobic and anaerobic biological processes. During the aerobic biological processes, the cell growth leads to the oxidation of organic compounds into CO_2. During the anaerobic biological processes, the organic matter is transformed into biogas composed of CO_2 and CH_4. The amount of the fossil CO_2 emissions from WWTPs can vary with the inlet wastewater composition and the plant configuration (Law *et al.*, 2013).

8.2.1.2 Methane – CH_4

CH_4 has a GWP of 34 over a 100-year period (IPCC, 2013). It is mainly produced during the decomposition of a wide range of organic matter in anaerobic conditions. A large fraction of the volatile matter contained in the sludge entering the anaerobic digester is converted into CH_4 (around 60–70% for primary sludge). The process of anaerobic digestion consists of four main sequential steps: hydrolysis, acidogenesis, acetogenesis and methanogenesis (Appels *et al.*, 2008).

Specifically, during the first hydrolysis step, both insoluble organic material and high molecular weight compounds such as lipids, polysaccharides, proteins and nucleic acids, are converted into soluble organic substances (e.g., amino acids and fatty acids). The products of the hydrolysis are further degraded during acidogenesis (second step). During the third step (acetogenesis), the organic acids and alcohols produced by acidogenesis are further digested to produce acetic acid as well as CO_2 and H_2. Finally, in the fourth step (methanogenesis) CH_4 is produced.

WWTPs, where anaerobic processes are implemented, are often a source of CH_4 (California Energy Commission [CEC], 2006; https://www.ncbi.nlm.nih.gov/pubmed/22575155). CH_4, produced during the anaerobic decomposition of organic substrate (activated by methanogenic bacteria), can be released to the atmosphere through the surface of open tanks (Mannina *et al.*, 2016b), or during storage and handling of the digested sludge. This methane emission can easily off-set the reduced fossil CO_2 emission associated with biogas energy production (https://www.ncbi.nlm.nih.gov/pubmed/22575155). Biogas, containing 55–65% of CH_4, can be used as an energy source and hence reduce the energy footprint (and consequently the GHG emissions) of the WWTP. Large amounts of CH_4 can be also produced due to the disposal of untreated sewage sludge to landfill (Czepiel *et al.*, 1993).

8.2.1.3 Nitrous oxide – N_2O

N_2O emissions from WWTPs are mainly related to the BNR processes (Kampschreur *et al.*, 2009; Law *et al.*, 2013), but also to processes that nitrify despite not being designed as nitrification processes. In Figure 8.1 the key pathways related to the N_2O formation during nitrification and denitrification are summarized.

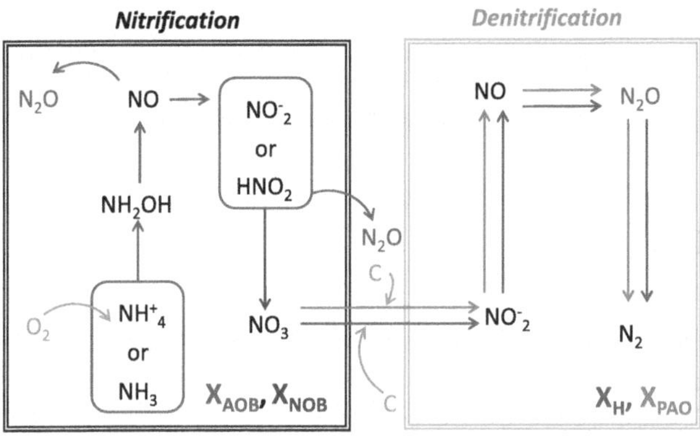

Figure 8.1 N_2O formation pathways during nitrification and denitrification process (X_{AOB} = ammonia oxidizing bacteria; X_{NOB} = nitrite oxidizing bacteria; X_H = heterotrophic bacteria) (Mannina *et al.* 2018c).

The nitrification process is divided into two main steps. In the first step, autotrophic ammonia oxidizing bacteria (X_{AOB}) aerobically oxidize NH_3 or NH_4^+ into NO_2^-. In the second step, autotrophic nitrite oxidizing bacteria (X_{NOB}) aerobically oxidize NO_2^- into NO_3^-. Denitrification leads to the reduction of NO_3^- into N_2 by means of heterotrophic bacteria growth (X_H).

N_2O is produced during the biological nitrogen removal processes due to both X_{AOB} and X_H (Kampschreur *et al.*, 2009; Ni & Yuan, 2015). X_{AOB} can produce N_2O as a product of the hydroxylamine oxidation (NH_2OH), identified as one of the major pathways, or nitrite (NO_2) reduction (X_{AOB} denitrification) (Ni *et al.*, 2013). Heterotrophic biomass produces N_2O as an obligate intermediate during nitrate (NO_3) reduction (Law *et al.*, 2013).

8.2.1.4 Liquid/gas mass transfer

The GHG emissions from WWTPs depend on the gas solubility and on operating and external factors influencing the GHG emission (e.g., stripping effect due to aeration or stirring).

Among the GHG emitted from a WWTP, CO_2 and N_2O are easily soluble in water; at 25°C and 0% salinity, the Henry's law constants of CO_2 and N_2O are 34 and 24 mM atm^{-1}, respectively (Weiss & Price, 1980). The Henry's law constants of CO_2 and N_2O are quite high compared with that of O_2 (1.3 mM atm^{-1}, at 25°C and 0% salinity). Therefore, an accumulation of CO_2 and N_2O in the liquid phase can occur, especially in the absence of external factors (e.g., non aerated tanks).

CH_4 has the same Henry's law constant as O_2, consequently the CH_4 produced inside the WWTP is mostly emitted into the atmosphere.

The main external factors influencing the GHG emission are: (i) temperature of the liquid phase, (ii) aeration, and (iii) stirrer effect. In particular, the temperature influences the coefficients of Henry's law, and the aeration and/or stirrer power promotes the stripping of GHG. Indeed, Ahn *et al.* (2010) have found that N_2O emissions are two to three orders of magnitude higher in aerated zones than in non-aerated ones.

8.2.2 Indirect emissions

8.2.2.1 Energy consumption

CO_2 emissions due to the energy consumption of WWTPs (internal indirect emission) represent important sources of GHG emissions from WWTPs. Pumps, motors, aeration and other equipment, operating 24 hours a day, seven days a week, make the water sector (which includes WWTPs) a large consumer of energy in a community (5–10 W/person) and potentially a large contributor to the community's total GHG emissions (US Environmental Protection Agency, 2013).

The energy (E_l) required for the mechanical devices such as pumps, motors, and aerators can be evaluated by using Equation 8.1 (Singh & Kansal, 2018).

$$E_l = \sum_{i=0}^{n} \frac{P_i \cdot T_i}{Q} \quad \left[\text{kWh m}^{-3}\right] \tag{8.1}$$

Where: P_i is the rated power of *i*th electrical equipment [kW]; T_i is the duration of daily operation of the *i*th electrical equipment [h d^{-1}]; and Q is the daily average sewage inflow to the WWTP [m^3 d^{-1}].

Energy for transporting sludge (E_d) can be evaluated by using Equation 8.2 (Singh & Kansal, 2018).

$$E_d = \frac{\sum_{i=0}^{n} \left(\frac{D_i}{E_i}\right) \cdot CF}{Q} \quad \left[\text{kWh m}^{-3}\right] \tag{8.2}$$

Where: D_i is the total distance travelled daily by *i*th vehicle [km/day]; E_i is the fuel efficiency of the *i*th vehicle [km L^{-1}]; and *CF* is the energy conversion factor for diesel to electricity (15.64 kWh L^{-1}).

The energy consumption due to the chemicals usage (E_{ch}) can be evaluated by using Equation 8.3.

$$E_{ch} = \frac{\left(\sum_{i=0}^{n} W_i \cdot E_{ci}\right)}{Q} \quad \left[\text{kWh m}^{-3}\right] \tag{8.3}$$

Where: E_{ci} is the unit energy consumption values of the *i*th chemical [kWh kg^{-1}]; and W_i is the consumption of the *i*th chemical [kg d^{-1}].

CO_2 emission per unit energy produced is of course related to the type of fossil fuel used and to the energy mix between renewable and non-renewable energy sources. Literature data suggest that the emission factor for the energy usage can vary between 0.45 kgCO$_2$ kWh^{-1} to 1.06 kgCO$_2$ kWh^{-1} (Central Electricity Authority [CEA], 2013).

8.2.2.2 Chemicals usage

Chemicals used during the operation of the WWTPs embody energy which leads to the CO_2 emission (Singh & Kansal, 2018). Thus, knowing the amount of chemicals used during the plant operation, it is possible to evaluate the CO_2 emission due to their usage on the basis of the emission factor. For example, the emission factors due to the used polymer and disinfectant are 13.54 and 1.124 kgCO$_2$ kg^{-1}, respectively (Haas, 2009; Zhang *et al.*, 2010; Singh & Kansal, 2018). Therefore, the total GHG emissions due to chemicals (G_{ch}) can be evaluated by using Equation 8.4.

$$G_{ch} = \frac{\left(\sum_{i=0}^{n} W_i \cdot EF_i\right)}{Q} \qquad \left[\mathrm{kgCO_{2eq}m^{-3}}\right] \tag{8.4}$$

Where: EF_i is the emission factor of the *i*th chemical [kgCO$_2$ kg^{-1}].

8.3 GHG FROM MBR: LITERATURE OVERVIEW

As mentioned above, the issue related to GHG emission from WWTPs has emerged significantly in recent years. Table 8.1 reports some literature data, in terms of N_2O emission, from activated sludge and biofilm systems.

From the observation of Table 8.1, it is possible to notice the huge variability of the N_2O emission factors (expressed as percentage of the influent N loading rate), depending on operational conditions as well as plant configuration.

Concerning the GHG emissions from MBRs specifically, limited information has been reported in literature so far. MBRs are characterized by specific peculiarities (biomass selection independent of settling properties, lack of final settling tank, intensive aeration for fouling mitigation, etc.) that can contribute to GHG production/emission (referring in particular to N_2O). Thus results achieved for conventional activated sludge systems should not be directly transferred to MBR systems. Table 8.2 reports a summary of some results achieved for MBR systems in terms of N_2O emission.

Table 8.1 N_2O emission factor (as % of influent load) in different WWTP configurations.

Reference	N_2O Emission (% of N-Loading)	Type of WWTP	Configuration	Remarks
Zheng *et al.* (1994)	2.3–16%	Nitrifying activated sludge	Lab-scale plant fed with artificial wastewater	N_2O emission increased with decreasing oxygen concentration and SRT
Czepiel *et al.* (1995)	0.04%	Activated sludge	Full-scale plant (11,000 pe)	
Wicht and Beier (1995)	0–14.6% (average: 0.6%)	Activated sludge	25 full-scale plants	N_2O increased with increasing nitrogen load
Sümer *et al.* (1995)	0.00%	Activated sludge	Full-scale plant (60,0000 pe)	N_2O emission increased with nitrite + nitrate concentrations
Osada *et al.* (1995)	1–35%	Oxic-Anoxic SBR	Lab-scale plant fed with real high strength wastewater	N_2O emission higher with longer aeration period in one SBR cycle, probably related to increased nitrite levels
Kimochi *et al.* (1998)	0.01–0.08%	Activated sludge	Full-scale plant (1000 pe)	N_2O emission decreased with proportionally shorter aeration periods
Park *et al.* (2000)	0.2–4.5%	Nitrifying and denitrifying activated sludge	Lab-scale plant fed with real wastewater	N_2O emission decreased after methanol addition
Park *et al.* (2000)	0.2–3%	Aerated biofilm filter	Lab-scale	N_2O emissions increased without substrate addition
Itokawa *et al.* (2001)	0.5–>20%	Oxic-anoxic SBR activated sludge	Lab-scale plant fed with artificial high strength wastewater	N_2O emission increased with C/N decrease

Zeng *et al.* (2003)	90%	Anaerobic-anoxic SBR activated sludge	Lab-scale plant fed with artificial wastewater	N_2O emission increased with increased nitrite concentration – N_2O might be major product of denitrification by GAO
Rezić *et al.* (2007)	3–25%	Rotating tube	Lab-scale	Inverse relationship of N_2O emissions to RPM
Kampschreur *et al.* (2008)	4%	Nitrification stage of activated sludge plant	Full-scale plant (620,000 pe)	
Ahn *et al.* (2010)	0.01–3.3%	Several plant configurations: separate-stage BNR, four-stage Bardenpho, step-feed BNR, step-feed non-BNR, plug-flow, oxidation ditch	Full-scale plants	High nitrite, ammonium, and dissolved oxygen concentrations positively correlated with N_2O emissions from aerobic zones. High nitrite and dissolved oxygen concentrations positively correlated with N_2O emissions from anoxic zones.
Aboobakar *et al.* (2013)	0.01–0.036%	Plug-flow nitrifying activated sludge process	Full-scale plant (210,000 pe)	Emissions were shown to be negatively correlated to dissolved oxygen
Lochmatter *et al.* (2013)	1–9%	Granular sludge	Lab-scale plant fed with synthetic wastewater	Inverse relationship between COD load and N_2O emissions
Kong *et al.* (2013)	2.70%	SBBR	Lab-scale plant fed with synthetic ammonium-rich wastewater	N_2O emissions likely related to intermittently aerated partial nitrification

(*Continued*)

Table 8.1 N_2O emission factor (as % of influent load) in different WWTP configurations. (*Continued*)

Reference	N_2O Emission (% of N-Loading)	Type of WWTP	Configuration	Remarks
Eldyasti *et al.* (2014)	0.6–1.9%	Fluidized bed reactor	Lab-scale plant	N_2O emissions increased with the C/N decrease
Daelman *et al.* (2015)	2.8–11%	Plug flow reactor in series with two parallel Carrousel reactors	Full-scale plant	Seasonal peaks in the nitrite concentration correlated with the N_2O emission. The diurnal variability of N_2O emission depends on the trend of the nitrite and nitrate concentrations in the tank.
Ma *et al.* (2015)	0.5–1.5%	MBBR nitritation-anammox bioreactor	Lab-scale plant	N_2O emissions increased with the increase of inorganic carbon limitation
Todt and Dörsch (2015)	0.7–8.5%	MBBR	Lab-scale plant	N_2O peaks coincided with NO_2 accumulation peaks
Baresel *et al.* (2016)	0.3–0.4%	Activated Sludge Process in the UCT-configuration	Full-scale plant	Measurements of dissolved N_2O required low maintenance and N_2O emission monitoring based on such measurements might therefore be considered as a significant practical improvement
Pan *et al.* (2016)	0.68–3.5%	Step-fed, plug-flow activated sludge reactors	Full-scale plant	The step-feed configuration had a substantial influence on the N_2O emissions

Note: SRT = solids retention time; SBR = sequencing batch reactor; GAO = glycogen-accumulating organisms; COD = chemical oxygen demand; pe = population equivalent; SBBR = sequencing batch biofilm reactor; MBBR = moving bed biofilm reactor; UCT = University of Cape Town

Table 8.2 N_2O emission factor (as percentage of influent load) from MBRs.

Reference	N_2O emission (% of N-loading)	Type of WWTP	Configuration	Remarks
Nuansawan *et al.* (2016)	0.04–0.18%	Two stage anaerobic-aerobic MBR	Lab-scale plant fed with municipal waste leachate	N_2O emissions reduced with sludge recirculation
Mannina *et al.* (2016a)	0.18–5.11%	SB-MBR	Pilot-scale plant, anoxic-aerobic MBR, subjected to salinity increase	N_2O emissions increased with the salinity increase
Mannina *et al.* (2016b)	0–92%	DN-MBR	Pilot-scale plant MBR, in a pre-denitrification scheme, treating saline wastewater contaminated by hydrocarbons	N_2O emissions significantly affected by the joint effect of salt and hydrocarbons
Mannina *et al.* (2017b)	3.82–10.26%	UCT-MBR	Pilot scale plant MBR in a UCT configuration subjected to C/N variation	N_2O emissions increased with the C/N decrease
Mannina *et al.* (2018b)	0.0002–4.01%	UCT-IFAS-MBR	Pilot scale plant IFAS-MBR in a UCT configuration operated at different C/N ratios	Highest N_2O emission related to the presence of free ammonia that caused nitrite accumulation
Mannina *et al.* (2017c)	0.6–38%	UCT-IFAS-MBR	Pilot scale plant IFAS-MBR in a UCT configuration operated at different SRTs	N_2O emissions increased significantly in the MBR compartment with the increase of the air flow rate (for fouling mitigation)
Mannina *et al.* (2018a)	0.6–8.9%	UCT-IFAS-MBR	Pilot scale plant IFAS-MBR in a UCT configuration operated at different HRT/SRTs	N_2O emissions decreased with the decrease of HRT/SRT

Note: SB = sequencing batch; DN = denitrification; IFAS = integrated fixed-film activated sludge; HRT = hydraulic retention time.

Even for MBRs, the literature data show a significant variation for the N_2O emission factor, depending on system configuration, features of feeding wastewater and operational conditions. Due to the increasing recurrence of MBR processes in recent years, it is important to increase the knowledge about GHG production/emission from such systems, by investigating the effect of wastewater features, operational conditions and plant layout on the GHG emissions in ad-hoc experimental campaigns.

8.4 MAIN FACTORS AFFECTING GHG EMISSIONS

The GHG emission from WWTPs has been widely investigated in order to identify its key parameters and its relationship with the operational conditions of the treatment plant.

With reference to N_2O, the emission has been related to the carbon source used during denitrification as well as with the nitrate concentration in the effluent (among others Park *et al.*, 2000; Li *et al.*, 2008; Kampschreur *et al.*, 2009; Yan *et al.*, 2014). The deficiency of dissolved oxygen during nitrification, as well as elevated nitrite concentrations and the incorrect balancing of COD vs nitrogen ratio have been identified as the key factors affecting the N_2O emission (Kampschreur *et al.*, 2009). Hence, the operational conditions play a significant role in N_2O emission (see for instance Figure 8.2).

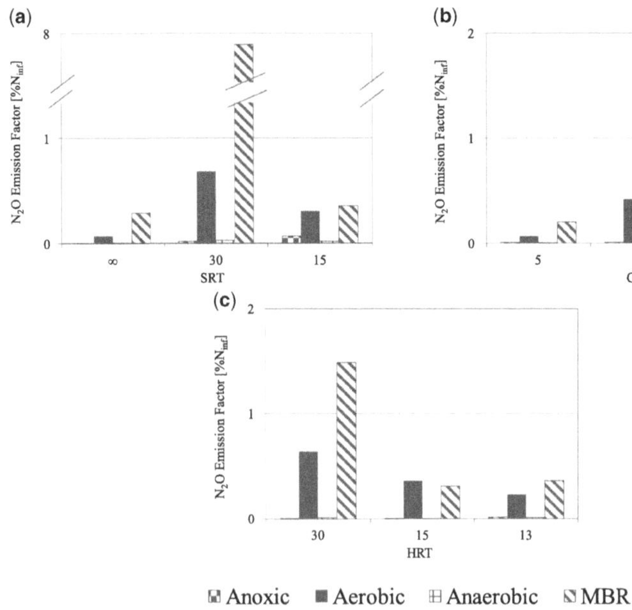

Figure 8.2 N_2O emission factor from a UCT-IFAS-MBR by varying SRT (a); C/N (b); HRT (c) (Mannina *et al.*, 2017a).

Mannina *et al.* (2017a) investigated University of Cape Town (UCT) integrated fixed film activated sludge (IFAS) membrane bioreactor (MBR) (referred to as UCT-IFAS-MBR) pilot plant fed with real wastewater with dosed organic supplements and equipped with funnel shaped covers over all reactors for nitrous oxide collection. The experiments were divided in three phases and aimed at investigating the influence of SRT, influent carbon to nitrogen concentration (C/N) ratio and hydraulic retention time (HRT) operational conditions on the nitrous oxide emission factor. Each experimental phase was divided in three experimental cycles by varying the operational conditions in each phase.

In Figure 8.2a the effect exerted by SRT variation, from indefinite SRT to 15 d is shown; in detail, in each cycle the highest N_2O emission occurred from the aerobic and MBR reactors, thus confirming the crucial role played by the aeration devices in enhancing the emission from WWTPs. In particular, the cycle carried out with SRT = 30 d resulted in the highest emission. The effect of influent C/N ratio, depicted in Figure 8.2b was highest during the third cycle (C/N = 2). During this last cycle, the emission from the aerobic reactor was higher than from the MBR reactor. This is likely ascribable to incomplete denitrification in the anoxic reactor due the organic carbon deficiency that leads to an increase in N_2O production. The N_2O produced in the anoxic reactor was conveyed to the aerobic reactor contributing to the high emission. The effect of combined HRT-SRT variation, depicted in Figure 8.2c, resulted in high emission during the longest HRT (30 h), while no significant differences were found by comparing HRT in the second and third cycles.

The relationship between N_2O and dissolved oxygen concentration was also investigated in Mannina *et al.* (2016a). These authors reported the results of a batch fed membrane bioreactor (SBR-MBR) treating saline wastewater. During the experimentation the authors also investigated the influence exerted by DO concentration on nitrous oxide formation, comparing the results achieved with NaCl concentrations equal to 0 and 10 g L^{-1}. The results are depicted in Figure 8.3.

The DO concentration, together with salinity, significantly affected the N_2O production during the nitrification process. As demonstrated in Figure 8.3a there is a fairly strong exponential correlation, proven by the high value of R^2, between the average DO and N_2O concentration in the aerobic tank. Furthermore, the salt concentration (Figure 8.3c,d) also influenced the pattern of DO and N_2O concentration profiles. Indeed, when no salt was present in the treated wastewater, the nitrification activity lead to an initial decrease of DO promoting an increase in nitrous oxide production. When the nitrification activity was completed, the DO concentration increased and the nitrous oxide concentration reached its lowest value. Conversely, at 10 g NaCl L^{-1}, the DO consumption decreased due to the lower nitrification activity caused by the shock exerted on autotrophic strains by salinity. Such circumstances led to an incomplete nitrification as well as to nitrous oxide formation an order of magnitude higher than that measured during the period carried out with no salt.

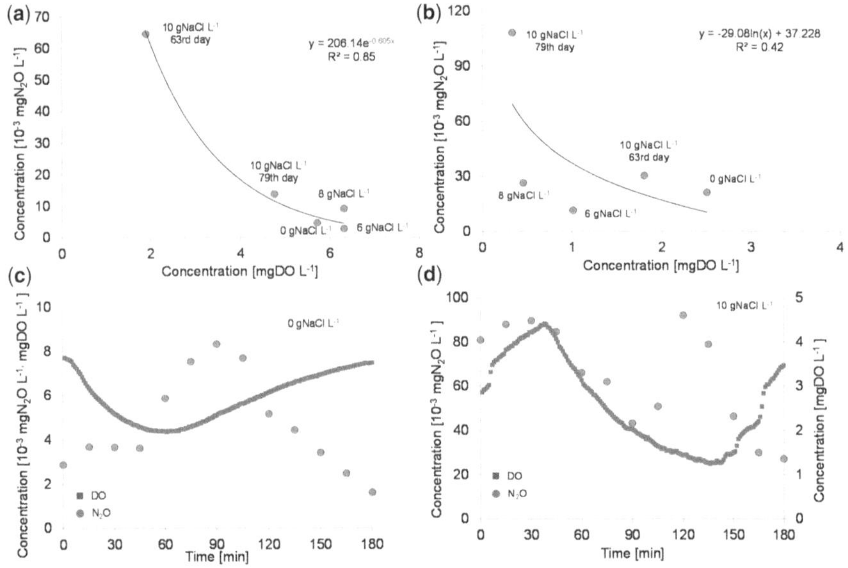

Figure 8.3 Average N_2O concentration in the gas samples withdrawn from the aerobic (a) and anoxic tank (b) in a batch fed membrane bioreactor (SBR-MBR) vs DO concentration; DO and N_2O concentration in the gas samples withdrawn from the aerobic tank during the cycle time at 0 g NaCl L^{-1} (c) and 10 g NaCl L^{-1} (d) (Mannina *et al.*, 2016a).

8.4.1 Direct emissions

With the aim to partially fill the knowledge gap related to GHG emissions from MBR systems, under the Italian research project PRIN2012 entitled "Energy consumption and GreenHouse Gas (GHG) emissions in the wastewater treatment plants: a decision support system for planning and management" a number of experiments have been carried out over 2 years (Mannina *et al.*, 2016a,b, 2017a,b,c, 2018a,b). The main aim was to assess the effect on the N_2O production/emission of different MBR configurations, influent wastewater characteristics (municipal or industrial), operational conditions (SRT, influent C/N ratio, HRT) and type of membrane modules (Mannina, 2017). In the following, a brief description of the pilot plants as well as the main operational features is reported, whilst in Figure 8.4 the schematic lay-out of the pilot plants used is shown.

Pilot plant No.1, referred to as SB-MBR, was designed according to a pre-denitrification scheme in a sequential feeding mode. It consisted of two in-series reactors: one anoxic (volume: 45 L) and one aerobic (volume: 224 L), followed by an MBR compartment (volume: 50 L). The experimental campaign was divided into six phases, each characterized by a different salt concentration in the feeding wastewater. The salt concentration was gradually increased from 0 to 10 g NaCl L^{-1} (Phase I: no salt addition; Phase II: 2 g NaCl L^{-1}; Phase III: 4 g NaCl L^{-1}; Phase IV: 6 g NaCl L^{-1}; Phase V: 8 g NaCl L^{-1}; Phase VI: 10 g NaCl L^{-1}).

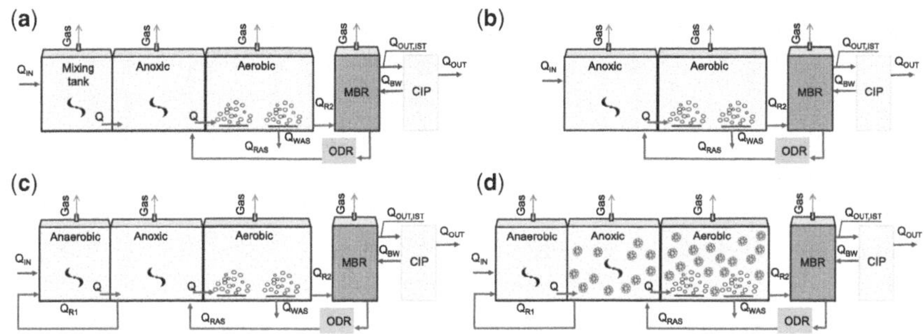

Figure 8.4 Schematic layout of the investigated pilot plants: (a) SB-MBR, (b) pre-denitrification MBR, (c) UCT-MBR, and (d) UCT-IFAS-MBR (Mannina *et al.*, 2016a,b, 2017a,b,c, 2018a,b). (ODR = oxygen depletion reactor; CIP = clean-in-place; RAS = recycle activated sludge; WAS = waste activated sludge).

Pilot plant No.2, referred to as DN-MBR, was characterized by one anoxic (volume: 45 L) and one aerobic (volume: 224L) in-series reactor. The experimental campaign had a duration of 90 days and was divided in two Phases, characterized by different features of the inlet wastewater. In Phase I the salinity of the influent was increased from 10 g NaCl L^{-1} up to 20 g NaCl L^{-1}, while in Phase II the inlet wastewater had a constant salinity (20 g NaCl L^{-1}) and diesel fuel dosage.

Pilot plant No.3, referred to as UCT-MBR was characterized by one anaerobic (volume: 62 L), one anoxic (volume: 102 L), and one aerobic (volume: 211 L) in-series reactor, according to the University of Cape Town (UCT) scheme (Ekama *et al.*, 1983). The UCT-MBR pilot plant was fed with a mixture of real and synthetic wastewater, the latter comprising sodium acetate, glycerol, di-potassium hydrogen phosphate and ammonium chloride. The experimental campaign was divided in two Phases, each characterized by a different value of the influent C/N ratio – in Phase I a C/N of 10 (duration: 41 days) and in Phase II a C/N of 5 (duration: 39 days).

Pilot plant No.4, referred to as UCT-IFAS-MBR, was characterized by one anaerobic (volume: 62 L), one anoxic (volume: 102 L), and one aerobic (volume: 211 L) in-series reactor, according to the University of Cape Town (UCT) scheme (Ekama *et al.*, 1983). The pilot plant was realized according to the IFAS-MBR (integrated fixed-film activated sludge membrane bioreactor) based on the combination of a moving bed biofilm reactor (MBBR) and an activated sludge based membrane bioreactor configuration with the presence of both activated

Table 8.3 Summary of literature studies on N_2O emissions from MBR plants.

Configuration	HRT [h]	Influent COD/N Ratio [mg COD mg^{-1}TN]	SRT [d]	Membrane Typology	Brief Process Description
Sequencing batch operation	20	8	Indefinite (no sludge withdrawal)	Ultrafiltration hollow fibre GE Zenon ZW10®	Evaluation of the effect of a gradual salinity increase in the short term. Experiment divided into six phases (Mannina *et al.*, 2016a)
MBR in a pre-denitrification scheme	16	7	Indefinite (no sludge withdrawal)	Ultrafiltration hollow fibre GE Zenon ZW10®	The joint effect of the salinity and hydrocarbons (20 mg TPH L^{-1}) in the short term was investigated (Mannina *et al.*, 2016b)
MBR coupled to UCT configuration for carbon and nutrient removal	20	10, 5	Indefinite (no sludge withdrawal)	Ultrafiltration hollow fibre PURON®	Evaluation of C/N ratio effect in a BNR process integrated with a membrane module for solid/liquid separation phase (Mannina *et al.*, 2017b)
Hybrid moving bed biofilm membrane bioreactor coupled to UCT configuration for carbon and nutrients removal	30, 15, 13	10, 5, 2	Indefinite (no sludge withdrawal), 30, 15	Ultrafiltration hollow fibre PURON®	Evaluation of the SRT effect on the performance of a UCT-IFAS-MBR pilot plant, evaluating the bio-kinetic activity of the bacterial species (Mannina *et al.*, 2018b)
CAS + MBR	–	–	–	–	Full-scale plant treating domestic and industrial wastewater (Singh & Kansal, 2018)
Activated sludge with sequenced aeration followed by MBR	–	–	–	–	Full-scale plant treating domestic wastewater (GWRC, 2011)
Anaerobic + aerobic pilot plant	5, 2.5	15–17 (anaerobic reactor); 9–16 (aerobic reactor	–	Mitsubishi Rayon, PVDF, 0.4 µm pore size, 0.077 m^2 surface area	Evaluation of the HRT and sludge recirculation effect on CH_4 and N_2O emission from a pilot plant treating fresh leachate mixed with tap water (1:1.5 v/v) (Nuansawan *et al.*, 2016)

sludge and biofilm (Ødegaard, 2017). The suspended plastic carriers for biofilm growth were added to the anoxic and the aerobic reactors, with filling fraction of 15 and 40%, corresponding to specific area of 75 and 200 m^2 m^{-3}, respectively. The experimental campaign had a duration of 340 days and was aimed at investigating the influence of operational variables (namely, SRT, influent C/N ratio and HRT-SRT) on N_2O production and emission.

For further details on the pilot-plant description as well as on experimental campaigns the reader is referred to Mannina *et al.* (2016a, 2017a).

Table 8.3 summarizes the few studies found in literature on N_2O emissions from MBR plants including the MBRs described above operated by Mannina *et al.* (2016a, 2017a), in which a wide range of N_2O emission factors under the different operations were found.

Singh and Kansal (2018) evaluated the total GHG emissions (expressed in terms of equivalent CO_2 emissions) from different WWTPs by adopting a simplified model. Among these plants, four were MBR systems treating domestic wastewater. They found an average GHG emission of 0.85 kg CO_{2eq}/m^3.

Within the work performed by the Global Water Research Coalition (GWRC, 2011) one MBR full-scale plant located in France was monitored in terms of N_2O emissions. An emission factor in terms of N_2O of 0.0011 $kgN_2O\text{-}N\ kgTKN_{influent}^{-1}$ was observed.

Nuansawan *et al.* (2016) investigated an anaerobic – aerobic MBR pilot plant treating leachate under different HRTs (2.5 and 5 days) and with the presence and absence of sludge recirculation from the anaerobic to the aerobic reactor. The main objective of this study was to evaluate the GHG emissions (CH_4 and N_2O) under the two different HRT operation conditions. They found the highest CH_4 (8.26% of the inlet carbon) and N_2O (0.18% of the inlet N) emissions took place from the anaerobic reactor under HRT of 5 days and without sludge recirculation.

By varying the experimental MBR system layout, an intensive campaign (Mannina *et al.*, 2016a,b, 2017a,b,c, 2018a,b) aimed at evaluating the N_2O production was carried out over almost two years (see Figure 8.5). The aim was to create a robust dataset to better understand the N_2O production and emission mechanisms.

Headspace data are representative of the N_2O concentration confined between the liquid surface and the funnel shape cover over the reactors, while the dissolved concentration data are representative of the N_2O concentration in the liquid bulk of each reactor. The data depicted in Figure 8.5 are representative of the aerobic and anoxic reactors where the core part of the nitrogen transformation process occurs.

The data collected reveal a huge variability of N_2O concentrations measured; indeed, the N_2O concentration ranged within six orders of magnitude (from 10^{-1} $\mu g\ N_2O\text{-}N\ L^{-1}$ up to 10^5 $\mu g\ N_2O\text{-}N\ L^{-1}$). This N_2O concentration variability yielded a wide range of the emission factors (EF) measured during the experimentation. In Figure 8.6 the average values of the EF (assessed as percentage of the total nitrogen influent emitted as N_2O according to Tsuneda *et al.* (2005)), are depicted (Mannina *et al.*, 2016a,b, 2017a,b,c, 2018a,b).

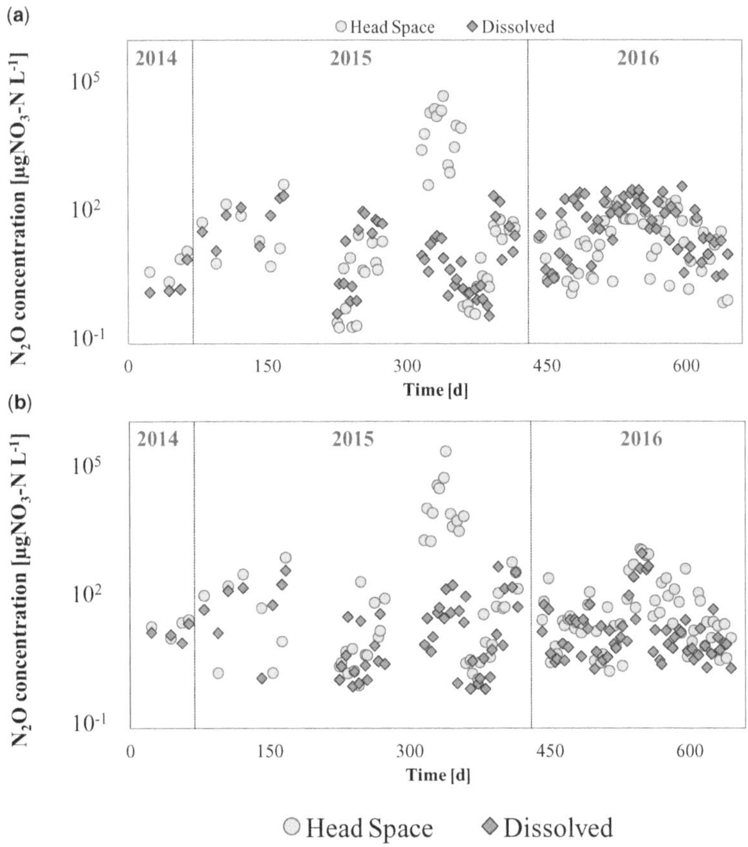

Figure 8.5 Nitrous oxide concentration measured in the head space and in the liquid bulk of aerobic (a) and anoxic (b) reactors over the experimentation (Mannina *et al.*, 2016a,b, 2017a,b,c, 2018a,b).

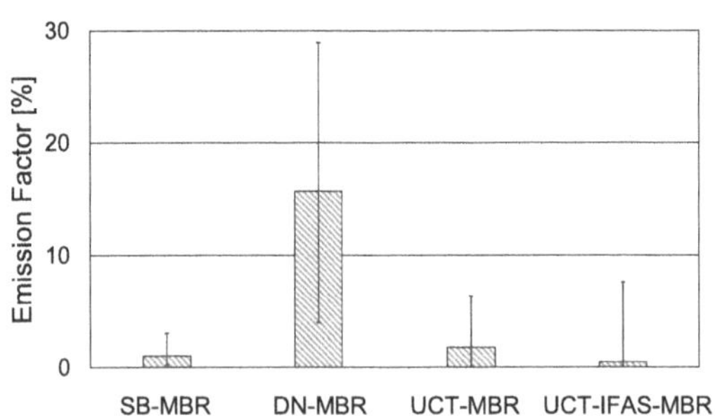

Figure 8.6 N_2O average emission factor measured for each experimental layout. The bars report minimum and maximum value for each configuration.

The data presented in Figure 8.6 demonstrate the influence by type of MBR-plant layout on the N_2O emission. The DN-MBR scheme had the highest emission factor (16% of influent nitrogen on average). It is worth noticing that the influent wastewater composition also played a significant role in the increasing of the N_2O emission. Indeed, the DN-SBR scheme treated an influent wastewater composed also of salt and diesel fuel (Mannina *et al.*, 2016a,b). The joint effect of both salt and hydrocarbon represented a disturbance factor that affected the metabolic activity of the biomass which increased the N_2O production and thus the emission. The role played by the salinity is also noticeable during the SBR-MBR configuration. The stepwise salinity increase resulted in a moderate EF, with the mean EF measured during SB-MBR period equal to 1% of influent nitrogen.

With regard to the UCT-MBR and UCT-IFAS-MBR configuration, the scarcity of carbon availability imposed by the lowest values of influent C/N ratio resulted in an increase of N_2O emission likely due to a carbon limitation for the denitrification process.

To summarize, the configuration that yielded the lowest EF was the UCT-IFAS-MBR, which featured a mean emission of 0.5% of influent nitrogen, but the operational condition of this system did influence the emission factor.

In order to also describe the role played by each reactor in contributing to the total emission, a comparison of mean EF assessed for each reactor during the operation of the UCT-MBR and UCT-IFAS-MBR configurations is given in Figure 8.7.

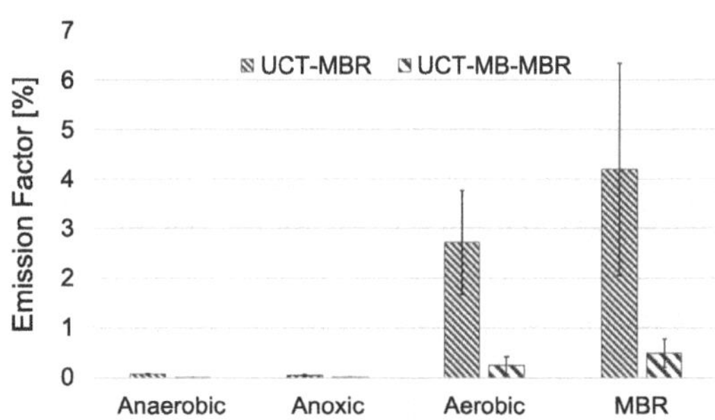

Figure 8.7 Comparison of mean EF measured in each biological reactor during the UCT-MBR and UCT-IFAS-MBR layout (Mannina *et al.*, 2017b,c, 2018a,b).

Figure 8.7 shows the strong reduction in EF that was observed in the UCT-IFAS-MBR layout. This result is likely due to an improvement in biological performance caused by the co-presence of both suspended and attached biomass. The biofilm presence improved the nitrogen removal efficiency thus leading to a lower N_2O emission.

Furthermore, it is worth noting that the highest emissions were measured in the aerated reactor in both MBR layouts, thus confirming the significant role that aeration has on N_2O emission by stripping.

In order to elucidate the influence of the operational condition as well as the layout configuration on the nitrous oxide production, a comparison of mean N_2O dissolved concentration measured in the biological reactors of the UCT-MBR and UCT-IFAS-MBR systems during the period when the influent C/N was varied, is shown in Figure 8.8.

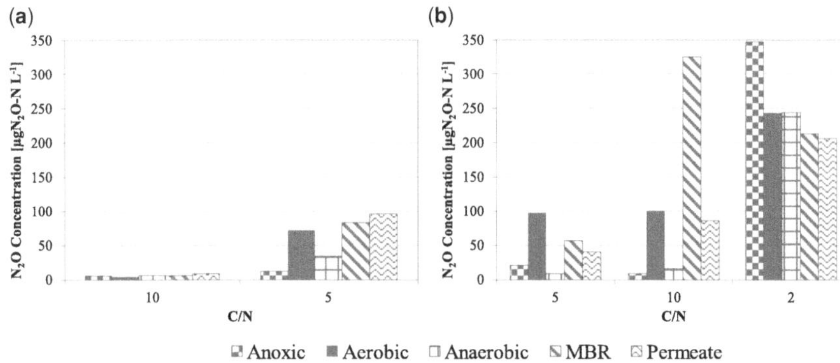

Figure 8.8 Comparison of mean N_2O dissolved concentration measured in the biological reactors in UCT-MBR (a) and UCT-IFAS-MBR (b) during the period when C/N was varied (Mannina *et al.*, 2017b,c, 2018a,b).

Figure 8.8 shows the increase in dissolved N_2O concentration with decreasing influent C/N-ratio. During the UCT-MBR experimental period (Figure 8.8a) the decrease of C/N wastewater fed to the pilot plant resulted in a drastic increase of N_2O dissolved concentration. As an example, the mean N_2O concentration measured in the permeate flow at C/N = 10 was 9.2 μg N_2O-N L^{-1}. Conversely, when the carbon availability was reduced due to the lower C/N, the mean N_2O concentration measured in the permeate flow increased to 95.9 μg N_2O-N L^{-1}, 10 times higher than in the previous period.

The results depicted in Figure 8.8b are similar; however, during the C/N = 2 period, the highest N_2O dissolved concentration was measured in the anoxic reactor. During the C/N = 10 and C/N = 5 periods, the maximum values of N_2O concentration were measured in the aerobic or in the MBR reactor. When C/N = 2 was implemented, the organic carbon deficiency for denitrification lead to a sharp increase in nitrous oxide production which reached an average concentration of 347 μg N_2O-N L^{-1}.

The mean N_2O dissolved concentrations measured in the anoxic reactor during the C/N = 10 and C/N = 5 periods were 9.5 μg N_2O-N $L^{-1}$and 21.5 μg N_2O-N L^{-1}

respectively. As the C/N ratio decreased, the nitrous oxide concentration increased. In a similar way the N_2O concentration measured in the anaerobic reactor during the C/N = 2 period increased up to 244.2 μg N_2O-N L^{-1}, more than 10 times higher than in the other periods. This behaviour is due to the deficiency of organic carbon for the denitrification, which resulted also in an increase in nitrate recycled to the anaerobic reactor. As a consequence, during the C/N = 2 period the anaerobic reactor was overloaded with nitrate thus acting as an anoxic reactor and contributing to this huge N_2O production and a failure of bio P removal (Mannina *et al.*, 2018b).

8.4.2 Indirect emissions

Aeration of MBRs can influence the amount of direct (pollutants discharged into the receiving water body and into the atmosphere) and indirect emissions. Mannina *et al.* (2016d) have recently investigated, for an MBR pilot plant, the effect of the air flow rate variation on the effluent quality index (EQI), the effluent penalty for exceeding concentration limits, and the operational cost (OC) (the exact calculation of the EQI, effluent penalty & OC can be found in Mannina & Cosenza, 2015).

Mannina *et al.* (2016d) found that by increasing the air flow rate, the EQI decreases according to an exponential pattern (with a correlation coefficient, R^2, equal to 0.61) (Figure 8.9a). The decrease of EQI with the increase of the air flow is due to the improvement of the biological processes (carbon removal and ammonia oxidation) with the increase of the dissolved oxygen inside the aerated tanks and due to the decrease of N_2O produced during the nitrification. Indeed, under oxygen limiting conditions, AOB use nitrite as the terminal electron acceptor to save oxygen for the oxygenation reaction of ammonia to hydroxylamine thus contributing to the N_2O production during nitrification (Kampschreur *et al.*, 2009). Mannina *et al.* (2016d) found also that the improvement of the biological processes at high air flow leads to a reduction of the mass of pollutants discharged to the environment with a consequent decrease of the penalty/fee to be paid (Figure 8.9b). Therefore, in terms of both EQI and effluent penalty, the highest air flow (2.3 m^3 h^{-1} in the pilot plant) represented the best operating condition. However, higher air flow increases the OC (Figure 8.9c) and the maximum OC value (3 € m^{-3}) was indeed obtained under the maximum air flow rate condition (2.3 m^3 h^{-1} in the pilot plant).

In Figure 8.10 data related to the relationship between the air flow versus, indirect (Figure 8.10a) and direct emissions (Figure 8.10b) obtained by Mannina *et al.* (2016d) are reported. The total energy required for the mechanical devices, needed to assess the indirect emission, has been quantified according to Singh and Kansal (2018).

An exponential relationship ($R^2 = 0.83$) was obtained between the air flow and the indirect (Figure 8.10a) and direct (Figure 8.10b) GHG emissions.

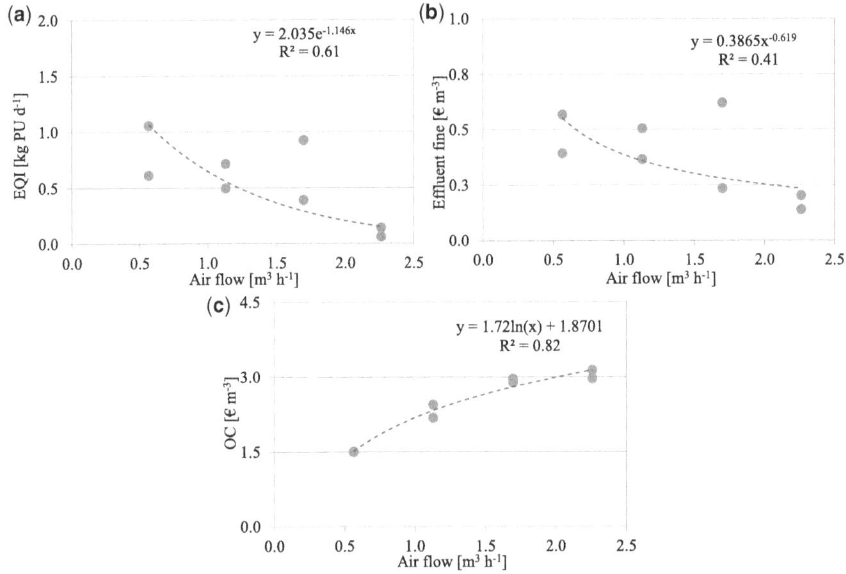

Figure 8.9 Air flow versus effluent quality index (a); effluent fine (b) and operational costs (c) (Mannina *et al.*, 2016d).

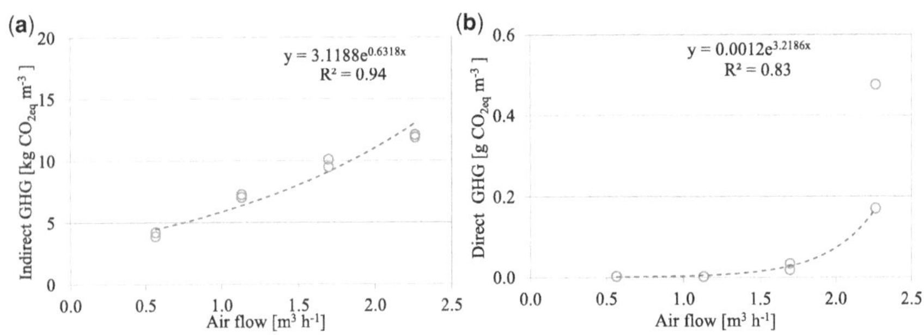

Figure 8.10 Relationship between the air flow and indirect GHG emissions (a); correlation between the air flow rate and direct GHG emissions (b).

In terms of GHG emissions (both direct and indirect) the lowest air flow (0.6 $m^3\,h^{-1}$ in the pilot plant) seems to be more appropriate than the others. However, this result is in conflict with the previous results (i.e. that EQI and effluent penalty have the maximum value under the lowest air flow rate value). This indicates interlinkages between different phenomena involved. Indeed, a "multiple trade-off" has to be performed for identification of the best air flow in order to mitigate GHG emissions and reduce the EQI and OC values.

Singh and Kansal (2018) found that the energy consumption of full-scale MBRs is around 0.42 kWh m^{-3}. This value is about 40% more than that of CAS. The amount of the total GHG emitted from an MBR was found by Singh and Kansal (2018) to be equal to 0.853 $kgCO_2$ eq. m^{-3} (0.564 $kgCO_2$ eq. m^{-3} due to the transport and 0.288 $kgCO_2$ eq. m^{-3} due to the treatment).

8.5 CONCLUSIONS

Wastewater treatment plants represent a significant source of greenhouse gas and this aspect will contribute in the future to reconsideration of the targets that wastewater treatment plants must fulfil. Indeed, besides pollutant removal and operational costs reduction, targets for future WWTPs must include the reduction of greenhouse gas emissions to the environment. In this light, WWTPs will become a sort of "box" from which further knowledge should be acquired, investigating specific topics neglected so far.

The understanding of processes that enhance GHG emission as well as the knowledge of operational variables and conditions that favour their production, represent new and important knowledge acquired by the scientific community over recent years. Indeed, many efforts have been recently devoted to experimental and monitoring activities with the aim to: i. assess the main mechanisms of GHG formation, ii. evaluate the operational conditions that favour their production.

In this context, some issues related to GHG production/emission are still poorly understood and deserve further investigation. For instance, despite the fact that many studies reveal that N_2O formation mostly derives from AOB activity, the conditions that trigger its formation are still not clear. Data obtained by authors highlighted a huge variability of N_2O concentration, demonstrating the influence of plant configuration and operational conditions. In particular, low influent C/N concentration ratios increased the N_2O production significantly.

There is also the need to reconsider the management of WWTPs, by considering GHG emissions. Indeed, if the aim is only limited to respect the imposed effluent quality, at the same time minimizing the operational costs, the GHG emissions might be significant.

As an example, the decrease of the dissolved oxygen set-point inside the nitrification reactor could promote the increase of N_2O production due to incomplete nitrification, despite the reduction of the operational costs.

In light of this, a plant wide mathematical modelling could represent a useful tool for the comparison of different scenarios (in terms of either design or management) for the evaluation of the best system performance, referring to both quality of the liquid effluent, reduction of gaseous emissions and operational costs reduction.

Nevertheless, the complexity of many existing models makes them of little use, except for research purposes. The high number of model parameters coupled to algorithm complexity usually require long computational times that are in contrast with the need of fast response for plant operators.

In view of this, the aim of the scientific community should be the build-up of simplified mathematical tools, derived by complex dynamic mathematical models, to be used as decision support systems able to simulate the quality of gaseous and liquid emissions from WWTPs and to provide useful indications for the optimization of the system management.

8.6 ACKNOWLEDGEMENTS

This work forms part of a research project supported by a grant from the Italian Ministry of Education, University and Research (MIUR) through the Research project of national interest PRIN2012 (D.M. 28 dicembre 2012 n. 957/Ric – Prot. 2012PTZAMC) entitled "Energy consumption and GreenHouse Gas (GHG) emissions in the wastewater treatment plants: a decision support system for planning and management – http://ghgfromwwtp.unipa.it" in which the first author of this paper is the Principal Investigator. Giorgio Mannina is a Fulbright research Fellow at Columbia University, New York, USA.

8.7 REFERENCES

Aboobakar A., Cartmell E., Stephenson T., Jones M., Vale P. and Dotro G. (2013). Nitrous oxide emissions and dissolved oxygen profiling in a full-scale nitrifying activated sludge treatment plant. *Water Research*, **47**, 524–534.

Ahn J. H., Kim S., Park H., Rahm B., Pagilla K. and Chandran K. (2010). N_2O emissions from activated sludge processes, 2008–2009: results of a national monitoring survey in the United States. *Environmental Science and Technology*, **44**, 4505–4511.

Appels L., Baeyens J., Degrève J. and Dewil R. (2008). Principles and potential of the anaerobic digestion of waste-activated sludge. *Progress in Energy and Combustion Science*, **34**, 755–781.

Baresel C., Andersson S., Yang J. and Andersen M. H. (2016). Comparison of nitrous oxide (N_2O) emissions calculations at a Swedish wastewater treatment plant based on water concentrations versus off-gas concentrations. *Advances in Climate Change Research*, **7**(3), 185–191.

CEA (2013). CO_2 Baseline Database for the Indian Power Sector. User Guide Version8.0, Jan. 2013. Government of India, Ministry of Power. Central Electricity Authority. Report by Ministry of Power, Government of India.

CEC – California Energy Commission (2006). Inventory of California Greenhouse Gas Emissions and Sinks: 1990 to 2004. Staff final Report. CEC-600-2006-013-SF. December.

Corominas L., Flores-Alsina X., Snip L. and Vanrolleghem P. A. (2012). Comparison of different modeling approaches to better evaluate greenhouse gas emissions from whole wastewater treatment plants. *Biotechnology and Bioengineering*, **109**(11), 2854–2863.

Czepiel P. M., Crill P. M. and Harris R. C. (1993). Methane emissions from municipal wastewater treatment processes. *Environmental Science and Technology*, **27**, 2472–2477.

Czepiel P., Crill P. and Harriss R. (1995). Nitrous oxide emissions from municipal wastewater treatment. *Environmental Science and Technology*, **29**(9), 2352–2356.

Daelman M. R. J., van Voorthuizen E. M., van Dongen U. G. J. M., Volcke E. I. P. and van Loosdrecht M. C. M. (2012). Methane emission during municipal wastewater treatment. *Water Research*, **46**, 3657–3670.

Daelman M. R. J., De Baets B., van Loosdrecht M. C. M. and Volcke E. I. P. (2013a). Influence of sampling strategies on the estimated nitrous oxide emission from wastewater treatment plants. *Water Research*, **47**, 3120–3130.

Daelman M. R. J., van Voorthuizen E. M., van Dongen L. G. J. M., Volcke E. I. P. and van Loosdrecht M. C. M. (2013b). Methane and nitrous oxide emissions from municipal wastewater treatment e results from a long-term study. *Water Science and Technology*, **67**, 2350–2355.

Daelman M. R. J., van Voorthuizen E. M., van Dongen U. G. J. M., Volcke E. I. P. and van Loosdrecht M. C. M. (2015). Seasonal and diurnal variability of N_2O emissions from a full-scale municipal wastewater treatment plant. *The Science of the Total Environment*, **536**, 1–11.

Ekama G. A., Siebritz I. P. and Marais G. R. (1983). Considerations in the process design of nutrient removal activated sludge processes. *Water Science and Technology*, **15** (3-4), 283–318.

Eldyasti A., Nakhla G. and Zhu J. (2014). Influence of biofilm thickness on nitrous oxide (N_2O) emissions from denitrifying fluidized bed bioreactors (DFBBRs). *Journal of Biotechnology*, **192**(Part A), 281–290.

ENERWATER http://www.enerwater.eu/.

Flores-Alsina X., Arnell M., Amerlinck Y., Corominas L., Gernaey K. V., Guo L., Lindblom E., Nopens I., Porro J., Shaw A., Vanrolleghem P. A. and Jeppsson U. (2011a). A dynamic modelling approach to evaluate GHG emissions from wastewater treatment plants. *Proceedings of World Congress on Water, Climate and Energy*.

Flores-Alsina X., Corominas L., Snip L. and Vanrolleghem P. A. (2011b). Including greenhouse gas emissions during benchmarking of wastewater treatment plant control strategies. *Water Research*, **45**, 4700–4710.

Foley J., de Haas D., Yuan Z. and Lant P. (2010). Nitrous oxide generation in full-scale biological nutrient removal wastewater treatment plants. *Water Research*, **44**(3), 831–844.

Forster P., Ramaswamy V., Artaxo P., Berntsen T., Betts R., Fahey D.W., Haywood J., Lean J., Lowe D.C., Myhre G., Nganga J., Prinn R., Raga G., Schulz M. and Van Dorland R. (2007). Changes in atmospheric constituents and in radiative forcing. In: Climate Change 2007: The Physical Science Basis. Contribution of Working Group I to the Fourth Assessment Report of the Intergovernmental Panel on Climate Change, S. Solomon, D. Qin, M. Manning, Z. Chen, M. Marquis, K.B. Averyt, M. Tignor and H.L. Miller (eds.), Cambridge University Press, Cambridge, United Kingdom and New York, NY, USA, pp. 114–143.

GWRC-Global Water Research Coalition (2011). N_2O and CH_4 Emission from Wastewater Collection and Treatment Systems — State of the Science Report, 2011–29. London, UK.

Haas D. (2009). Energy and Greenhouse Gas Emissions for the SEQ Water Strategy Urban Water Security Research Alliance Technical Report No. 14 (14).

IPCC, Climate Change (2013). The Physical Science Basis. Contribution of Working Group I to the Fifth Assessment Report of the Intergovernmental Panel on Climate Change. 2013. Cambridge University Press, Cambridge, United Kingdom and New York, NY, USA, p. 1535.

Itokawa H., Hanaki K. and Matsuo T. (2001). Nitrous oxide production in high-loading biological nitrogen removal process under low COD/N ratio condition. *Water Research*, **35**(3), 657–664.

Judd S. J. and Judd C. (2010). Principles and Applications of Membrane Bioreactors in Water and Wastewater Treatment, 2nd edn, Elsevier, London, UK.

Kampschreur M. J., van der Star W. R. L., Wielders H. A., Mulder J. W., Jetten M. S. M. and van Loosdrecht M. C. M. (2008). Dynamics of nitric oxide and nitrous oxide emission during full scale reject water treatment. *Water Research*, **42**(3), 812–826.

Kampschreur M. J., Temmink H., Kleerebezem R., Jetten M. S. M. and van Loosdrecht M. C. M. (2009). Nitrous oxide emission during wastewater treatment. *Water Research*, **43**, 4093–4103.

Kim D., Bowen J. D. and Kinnear D. (2015). Comprehensive numerical modeling of greenhouse gas emissions from water resource recovery facilities. *Water Environment Research*, **87**(11), 1955–1969.

Kimochi Y., Inamori Y., Mizuochi M., Xu K.-Q. and Matsumura M. (1998). Nitrogen removal and N_2O emission in a full-scale domestic wastewater treatment plant with intermittent aeration. *Journal of Fermentation and Bioengineering*, **86**(2), 202–206.

Kong Q., Zhang J., Miao M., Tian L., Guo N. and Liang S. (2013). Partial nitrification and nitrous oxide emission in an intermittently aerated sequencing batch biofilm reactor. *Chemical Engineering Journal*, **217**, 435–441.

Law Y., Jacobsen G., Smith A., Yuan Z. and Lant P. (2013). Fossil organic carbon in wastewater and its fate in treatment plants. *Water Research*, **47**(14), 5270–5281.

Li Q. H., Li P., Zhu P. P., Wu J. H. and Liang S. Z. (2008). Effects of exogenous organic carbon substrates on nitrous oxide emissions during the denitrification process of sequence batch reactors. *Environmental Engineering Science*, **25**, 1221–1228.

Lochmatter S., Gonzalez-Gil G. and Holliger C. (2013). Optimized aeration strategies for nitrogen and phosphorus removal with aerobic granular sludge. *Water Research*, **47**, 6187–6197.

Ma Y., Sundar S., Park H. and Chandran K. (2015). The effect of inorganic carbon on microbial interactions in a biofilm nitritation–anammox process. *Water Research*, **70**, 246–254.

Mannina G. (2017). Frontiers in wastewater treatment and modelling-FICWTM 2017. In: Series Lecture Notes in Civil Engineering, G. Mannina (ed.), Springer, ISBN:9783319584201, doi:10.1007/978-3-319-58421-8.

Mannina G. and Cosenza A. (2015). Quantifying sensitivity and uncertainty analysis of a new mathematical model for the evaluation of greenhouse gas emissions from membrane bioreactors. *Journal of Membrane Science*, **475**, 80–90.

Mannina G., Morici C., Cosenza A., Di Trapani D. and Ødegaard H. (2016a). Greenhouse gases from sequential batch membrane bioreactors: a pilot plant case study. *Biochemical Engineering Journal*, **112**, 114–122.

Mannina G., Cosenza A., Di Trapani D., Laudicina V. A., Morici C. and Ødegaard H. (2016b). Nitrous oxide emissions in a membrane bioreactor treating saline wastewater contaminated by hydrocarbons. *Bioresource Technology*, **219**, 289–297.

Mannina G., Cosenza A., Di Trapani D., Capodici M. and Viviani G. (2016c). Membrane bioreactors for treatment of saline wastewater contaminated by hydrocarbons (diesel fuel): an experimental pilot plant case study. *Chemical Engineering Journal*, **291**, 269–278.

Mannina G., Capodici M., Cosenza A. and Di Trapani D. (2016d). Towards a conceptual mathematical tool linking physical and biological processes for a reduction of GHG emissions from an MB-MBR plant. In *Proceedings of X Simposio Internazionale di Ingegneria Sanitaria Ambientale*, 19–23 June 2016, Rome–Italy.

Mannina G., Ekama G., Caniani D., Cosenza A., Esposito G., Gori R., Garrido-Baserba M., Rosso D. and Olsson G. (2016e). Greenhouse gases from wastewater treatment — a review of modelling tools. *Science of the Total Environment*, **551–552**, 254–270.

Mannina G., Capodici M., Cosenza A., Di Trapani D., Laudicina V. A. and Ødegaard H. (2017a). Nitrous oxide from moving bed based integrated fixed film activated sludge membrane bioreactors. *Journal of Environmental Management*, **187**, 96–102.

Mannina G., Capodici M., Cosenza A., Di Trapani D. and van Loosdrecht, M. C. M. (2017b). Nitrous oxide emission in a University of Cape Town membrane bioreactor: the effect of carbon to nitrogen ratio. *Journal of Cleaner Production*, **149**, 180–190.

Mannina G., Capodici M., Cosenza A., Laudicina V. A. and Di Trapani D. (2017c). The influence of solid retention time on IFAS-MBR systems: assessment of nitrous oxide emission. *Journal of Environmental Management*, **203**, 391–399.

Mannina G., Capodici M., Cosenza A. and Di Trapani D. (2018a). Nitrous oxide from integrated fixed-film activated sludge membrane bioreactor: assessing the influence of operational variables. *Bioresource Technology*, **247**, 1221–1227.

Mannina G., Ekama G. A., Capodici M., Cosenza A., Di Trapani D., Ødegaard H. and van Loosdrecht M. C. M. (2018b). Influence of carbon to nitrogen ratio on nitrous oxide emission in an Integrated Fixed Film Activated Sludge Membrane BioReactor plant. *Journal of Cleaner Production*, **176**, 1078–1090.

Mannina G., Cosenza A. and Ekama G. A. (2018c). A comprehensive integrated membrane bioreactor model for greenhouse gas emissions. *Chemical Engineering Journal*, **334**, 1563–1572.

Marques R., Rodriguez-Caballero A., Oehmen A. and Pijuan M. (2016). Assessment of online monitoring strategies for measuring N_2O emissions from full-scale wastewater treatment systems. *Water Research*, **99**, 171–179.

Ni B. J. and Yuan Z. (2015). Recent advances in mathematical modeling of nitrous oxides emissions from wastewater treatment processes. *Water Research*, **87**, 336–346.

Ni B. J., Yuan Z., Chandran K., Vanrolleghem P. A. and Murthy S. (2013). Evaluating four mathematical models for nitrous oxide production by autotrophic ammonia oxidizing bacteria. *Biotechnology and Bioengineering*, **110**, 153–163.

Nuansawan N., Boonnorat J., Chiemchaisri W. and Chiemchaisri C. (2016). Effect of hydraulic retention time and sludge recirculation on greenhouse gas emission and related microbial communities in two-stage membrane bioreactor treating solid waste leachate. *Bioresource Technology*, **210**, 35–42.

Ødegaard H. (2017). New applications for MBBR and IFAS systems. In: Frontiers in wastewater treatment and modelling. Book Series: Lecture Notes in Civil Engineering, G. Mannina (ed.), Vol. 4, Springer Int. Publ. AG, Switzerland, pp. 499–507.

Osada T., Kuroda K. and Yonaga M. (1995). Reducing nitrous oxide gas emissions from fill-and-draw type activated sludge process. *Water Research*, **29**(6), 1607–1608.

Pan Y., van den Akker B., Ye L., Ni B.-J., Watts S., Reid K. and Yuan Z. (2016). Unravelling the spatial variation of nitrous oxide emissions from a step-feed plug-flow full scale wastewater treatment plant. *Scientific Reports*, **6**, Article number: 20792.

Park K. Y., Inamori Y., Mizuochi M. and Ahn K. H. (2000). Emission and control of nitrous oxide from a biological wastewater treatment system with intermittent aeration. *Journal of Bioscience and Bioengineering*, **90**, 247–252.

PRIN2012 – Progetto di Ricerca di Interesse Nazionale (PRIN) (2012). Energy Consumption and GreenHouse Gas (GHG) Emissions in the Wastewater Treatment Plants: A Decision Support System for Planning and Management. http://ghgfromwwtp.unipa.it

Rezić T., Ŝantek B., Novak S. and Marić V. (2007). Heterotrophic cultivation of Paracoccus denitrificans in a horizontal rotating tubular bioreactor. *World Journal of Microbiology and Biotechnology*, **23**, 987–996.

Rodriguez-Caballero A., Aymerich I., Marques R., Poch M. and Pijuan M. (2015). Minimizing N_2O emissions and carbon footprint on a full-scale activated sludge sequencing batch reactor. *Water Research*, **71**, 1–10.

SANITAS Sustainable and Integrated Urban Water System Management. http://www.sanitas-itn.eu/project/overview-and-overall-objectives/.

Singh P. and Kansal A. (2018). Energy and GHG accounting for wastewater infrastructure. *Resources, Conservation and Recycling*, **128**, 499–507.

Sümer E., Weiske A., Benckiser G. and Ottow J. C. G. (1995). Influence of environmental conditions on the amount of N_2O released from activated sludge in a domestic wastewater treatment plant. *Cellular and Molecular Life Sciences*, **51**(4), 419–422.

Todt D. and Dörsch P. (2015). Nitrous oxide emissions in a biofilm loaded with different mixtures of concentrated household wastewater. *International Journal of Environmental Science and Technology*, **12**(11), 3405–3416.

Tsuneda S., Mikami M. and Kimochi Y. (2005). Effect of salinity on nitrous oxide emission in the biological nitrogen removal process for industrial wastewater. *Journal of Hazardous Materials*, **119**, 93–98.

United States Environmental Protection Agency (2006). Global Anthropogenic Non-CO_2 Greenhouse Gas Emissions: 1990 to 2020. US-EPA, Washington, DC.

United States Environmental Protection Agency (2013). Energy Efficiency in Water and Wastewater Facilities. A Guide to Developing and Implementing Greenhouse Gas Reduction Programs.

Weiss R. F. and Price B. A. (1980). Nitrous oxide solubility in water and seawater. *Marine Chemistry*, **8**(4), 347–359.

Wicht H. and Beier M. (1995). N_2O-Emissionen aus nitrifizierenden und denitrifizierenden Kläranlagen (N_2O-emissions from nitrifying and denitrifying sewage treatment plants). *Korrespondenz Abwasser*, **42**(3), 404–406.

Yan X., Li L. and Liu J. (2014). Characteristics of greenhouse gas emission in three full-scale wastewater treatment processes. *Journal of Environmental Sciences*, **26**, 256–263.

Ye L., Ni B. J., Law Y., Byers C. and Yuan Z. (2014). A novel methodology to quantify nitrous oxide emissions from full-scale wastewater treatment systems with surface aerators. *Water Research*, **48**, 257–268.

Zeng R. J., Lemaire R., Yuan Z. and Keller J. (2003). Simultaneous nitrification, denitrification, and phosphorus removal in a lab scale sequencing batch reactor. *Biotechnology and Bioengineering*, **84**(2), 170–178.

Zhang Q. H., Wang X. C., Xiong J. Q., Chen R. and Cao B. (2010). Application of life cycle assessment for an evaluation of wastewater treatment and reuse project – case study of Xi'an, China. *Bioresource Technology*, **101**(5), 1421–1425.

Zheng H., Hanaki K. and Matsuo T. (1994). Production of nitrous oxide gas during nitrification of wastewater. *Water Science and Technology*, **30**(6), 133–141.

Chapter 9

Mixing – new insights and opportunities through computational fluid dynamics

I. Nopens[1] *and U. Rehman*[2]

[1]*BIOMATH, Department of Mathematical Modelling, Statistics and Bio-informatics, Ghent University, Gent, Belgium*

[2]*AM-TEAM, Advanced modelling for process optimisation, Oktrooiplein 1, bus 603, 9000 Gent, Belgium*

ABSTRACT

Mixing is an important aspect in any reactor that is used to intensify a process, be it a bioprocess or a physical-chemical process. However, when modelling for design or operation, often simplified mixing models are used. These simplifying assumptions lead to design flaws resulting in reactors with suboptimal performance. It is demonstrated that more advanced modelling frameworks like computational fluid dynamics (CFD) can provide detailed insight into the behaviour inside reactors. By means of compartmental models derived from these CFD models, better performing reactors can be designed.

Keywords: Bio-kinetic processes, compartmental models, computational fluid dynamics, physical-chemical processes, tanks-in-series

9.1 INTRODUCTION – THE IMPORTANCE OF MIXING

Let's set the scene for this chapter by referring to a quote made by Octave Levenspiel back in 1972 in his book on reactor engineering: "If we know what is happening within the vessel, then we are able to predict the behaviour of the

Editors: Giorgio Mannina, George Ekama, Hallvard Ødegaard and Gustaf Olsson
doi: 10.2166/9781780409719_0321

vessel as a reactor. Though fine in principle, the attendant complexities make it impractical to use this approach". The approach referred to is the modelling framework of computational fluid dynamics (CFD). Indeed, this framework allows detailed insight into what is actually happening in a reactor from the macroscale to the microscale. The latter refers to detailed flow patterns, occurrence of short circuiting, and dead zones. However, to date, the "plain" CFD can be extended with equations for solids and species transport, including process kinetics of both biochemical as well as physical-chemical processes. This further details our knowledge with respect to where process rates are high or low and, hence, where flaws in the design of the reactor are prominent. Hence, these detailed models not only allow us to troubleshoot ill-performing existing reactors, but also allow us to suggest disruptively new reactor designs that solve shortcomings and can outperform the currently used designs which never take this detail of knowledge into account and are based on "crude" rules of thumb from several decennia ago. Hence, it is about time to translate the gathered knowledge in the last decennia into a next generation of reactor design rules and principles. Next to design, also process control is very much affected by this increased knowledge. It has become obvious that large volume reactors are heterogeneous which implies that a sensor reading might be significantly different depending on its exact location. Typically, sensors are placed where maintenance is "easiest", but this might well be in a zone that is not very representative for the measured process variable that is fed to the controller (Rehman *et al.*, 2015).

The complexities referred to by Levenspiel are mainly related to complexity of the model, i.e. the Navier Stokes equations and additional ingredient equations for turbulence, density, rheology, physical and bioprocess kinetics, etc. The equations were more or less in place, but solving them for a full-scale geometry was indeed "impractical" at that time given the computational burden and limited available computing power. This was the premise at that time for the development of alternative models for describing mixing and dispersion in large reactors based on simplifying assumptions.

9.2 "IDEAL MIXING" AND ITS FLAWS AND LIMITATIONS

Two types of "ideal" mixing were defined: plug flow and complete mix (or completely stirred tank reactor – CSTR) (Figure 9.1). In the former, an added pulse or step leaves the system unaltered in shape a certain average time after being fed, i.e. $\bar{t} = \frac{V}{Q}$. In the latter case a pulse will result in an exponential decaying response.

However, in reality these regimes hardly ever exist (apart from small volume reactors or excessively mixed reactors) and "non-ideal" mixing occurs due to the occurrence of short-circuiting, dead zones or internal recirculation (Figure 9.2). It is noteworthy that these "non-idealities" are mainly caused by design flaws (e.g. bad choice of reactor shape or inlet/outlet constructions and locations) or simply because it is impossible to achieve "ideal" mixing in large volume reactors due to the limitation of impeller energy input (to avoid excessive cost) and the effect of liquid viscosity which causes increased energy dissipation.

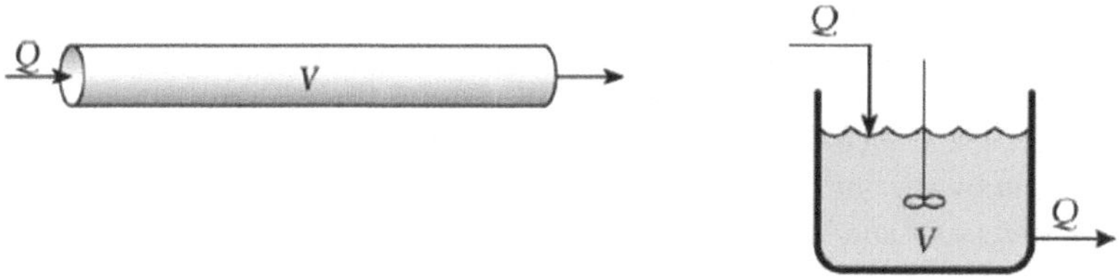

Figure 9.1 Concept of simplified representation of mixing behaviour: plug flow (left) and CSTR (right).

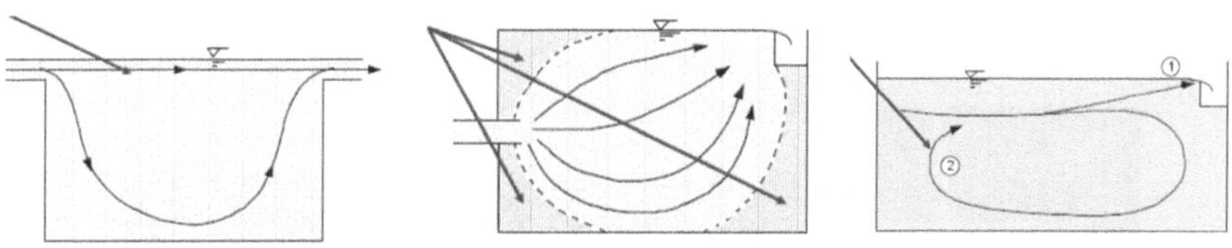

Figure 9.2 Illustration of "non-ideal" mixing behaviour due to short-circuiting (left), dead zones (middle) and internal recirculation (right). Arrows indicate short-circuit stream (left), dead zones (middle) and internal recirculation (right). (1) is the short-circuit stream; (2) is the recirculation stream.

Quantifying the non-ideality is typically performed by means of tracer testing, where an (inert) tracer pulse (or step) is added to the reactor inlet and the tracer concentration in the outlet is measured. This gives rise to the so-called residence time distribution (RTD).

Modelling the non-ideality can be carried out by means of one- or higher dimensional advection-diffusion partial differential equations (PDEs). However, PDEs have been avoided as much as possible as they significantly increase the computational effort. Moreover, diffusion coefficients are not always readily available, though could be quantified from the measured RTD.

A workaround to avoid the use of PDEs is to discretise the spatial component in such a rough way (very coarse grid) that numerical diffusion occurs, which is then is used to mimic the actual diffusion. This approach is better known as the tanks-in-series (TIS) approach. The number of tanks to use can again be determined from the RTD, though this is often an ill-posed problem, i.e. multiple solutions exist. This approach has long been used in the wastewater modelling community. However, given the fact that this is a one-dimensional (1D) approach, it averages the other perpendicular dimensions and assumes homogeneity in that space. Thus, due to the large volumes of the reactors used, this is again a harsh assumption that leads to model structure issues and leads to calibration efforts to compensate for this simplification. Moreover, the actual flow pattern in the reactor is still unknown, leaving us with a "black box" where we only judge an input-output relation.

But what happens in reality? Figure 9.3 illustrates two systems where mixing is not perfect. The coffee cup will eventually get completely mixed due to diffusion or advection by stirring it with a spoon. The activated sludge tank, though, would need an excessive amount of mixing energy to completely mix it. This is mainly due to the large volume. However, what is more important to note is that a TIS approach will never be able to accurately simulate the system behaviour. If the objective is to optimise the system performance, one needs a better model that describes the real behaviour. Only then can one come up with reliable strategies to improve the process. Later in the chapter, we will generalise this observation to physical processes too.

Figure 9.3 Illustrations of non-ideal mixing behaviour in different systems: cup of coffee in which milk is added (left) and an activated sludge reactor dissolved oxygen (DO) concentration (right).

Next to the evidence of observation, another limitation to "averaging" in process models is related to the typical Monod functions that are heavily used in microbial process rate models (e.g. activated sludge model (ASM) family). This is illustrated in Figure 9.4 reproduced from Arnaldos *et al.* (2015). A Monod expression takes a concentration to compute a process rate. In a TIS setting, this is the "average" concentration of the specific CSTR. This process rate is then assumed to be valid in the entire volume of the CSTR. In Figure 9.3 (right) we clearly can observe that this does not always hold. If we assume that this behaviour is reality, the model will severely overestimate the process rate. Often, the modeller will then calibrate the K-value (affinity index) to "artificially" correct for this. However, this is a mere compensation and turns the calibration into a mere fitting exercise leading to a model with a low predictive power.

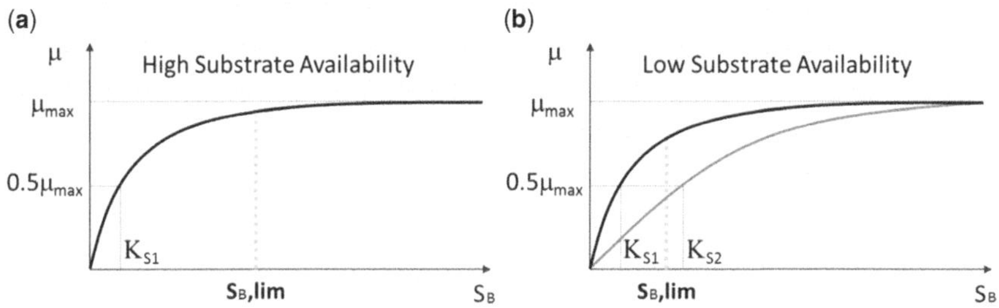

Figure 9.4 Illustration of how spatial averaging leads to errors in bioprocess modelling. (a) when average substrate is high, kinetics saturate, (b) when average substrate is low, kinetics tend to be calibrated.

If we look more closely to the composition of effects that are lumped in the K-value (Figure 9.5), we can conclude that the compensation is mainly trying to correct for the error in the local concentration. Hence, if we can predict that value better, it should avoid calibration and allow us to use the K-value that is only due to transport of the component once it enters the floc.

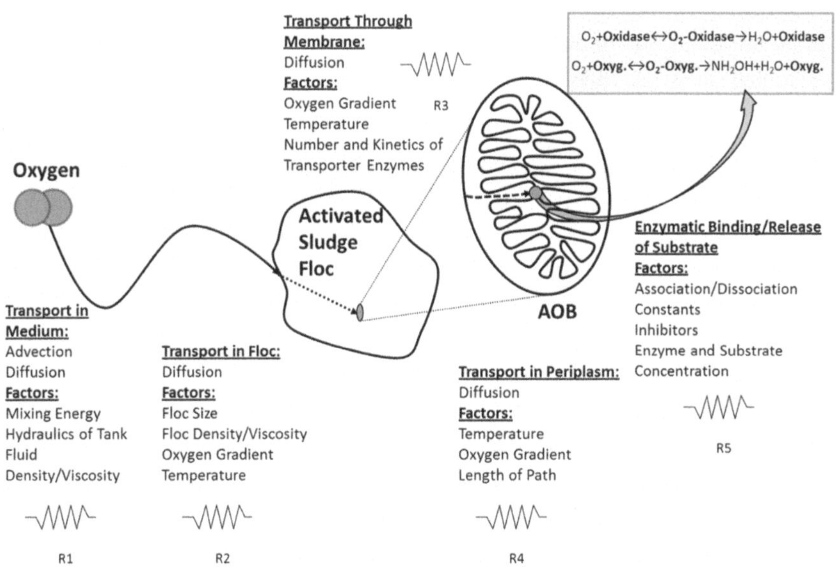

Figure 9.5 Illustration of all lumped mechanisms embodied in the K-value (affinity index). R1-5 refer to resistance introduced by the underlined phenomena.

Furthermore, the poor representation of spatial behaviour makes it problematic to use the model for process control. If we again look at Figure 9.3 (right), we can observe clearly that sensor location might lead to very different readings in the same volume of reactor. Obviously we do not see the coloured representation when we are at the plant. Choosing a different set-point or tuning the controller differently combined with a certain sensor location might cure this, but we then need to have access to this specific spatial behaviour to develop the controller properly. It is obvious that a TIS model will not allow this.

Finally, the true plant overlaid with the coloured spatial concentrations is a powerful tool to illustrate to managers how the system is actually behaving and can convince them to take certain decisions. The question now arises: how do we get to this coloured visualisation of the plant behaviour? The answer is: a more advanced model, i.e. computational fluid dynamics in combination with kinetics (or a population balance model – PBM – for other physical processes).

In the next section, CFD and its application to wastewater is briefly introduced for the unfamiliar reader. This is then followed by a dedicated example of an advanced CFD application and an extension is made to physical-chemical applications. Finally, a critical discussion on how these advanced models can now be used in practice for improved design and operation is given.

9.3 COMPUTATIONAL FLUID DYNAMICS: BRIEF INTRODUCTION

A more thorough way of describing spatial behaviour of a viscous fluid flow is by means of the Navier Stokes equations for conservation of mass and momentum of one or more phases. They are typically solved over a grid or mesh using either finite volume, finite differences or final elements methods. Extra ingredients can be added depending on the modelling objective, such as equations for turbulent motion, species (solute or particle) transport, process kinetics, etc. The study of fluid mechanics using any combination of the aforementioned equations is known as computational fluid dynamics (CFD). CFD has over recent decades become an accepted method for process analysis of fluid flows in many industries (aeronautics, automotive, chemical engineering).

9.3.1 State of the art of CFD in wastewater treatment

CFD has recently also become widely used for analysis of hydraulic problems in water and wastewater treatment plants (WWTPs) but still needs to find wider acceptance for analysis of physical, chemical and biological processes in a WWTP. The application of CFD in this field goes back to the 1980s where it was adopted for settler modelling, mainly as a troubleshooting tool. However, there are substantial financial and risk drivers to conducting CFD for better wastewater design (Wicklein *et al.*, 2016).

Recently, a working group was formed under the wings of the IWA Specialist Group "Modelling and Integrated Assessment" (MIA) that brings together the community active in this field. Laurent *et al.* (2014) described an alternative way of using CFD as a supplement to using simpler models, whereas Wicklein *et al.* (2016) focused on "good CFD modelling practice" for WWTPs and Samstag *et al.* (2016) provided an overview of CFD applications in the unit process of a wastewater treatment train. Nowhere (except possibly disinfection) is CFD used in a widespread or routine way as a design or risk-management tool. This offers clear opportunities to further develop the value of CFD in wastewater process evaluation.

For the people new to CFD (or even the more familiar person), it should be pointed out that the effort and cost of a CFD study has significantly reduced. The effort entails expert manpower (mainly present in academia and consultancy), a software license (unless open source software is used), hardware (multiple cores can speed up solutions) and validation data. The latter is important and can entail 3D velocity patterns (e.g. using acoustic-based techniques), concentration profiles (e.g. sensor grid) and alike. Expertise is present with academia and consultants. Timewise, a CFD model of a fairly easy geometry takes in the order of weeks to be completed and will cost in the €10k–100k range. This means that CFD is now within reach for many design projects that usually have budgets that are an order of magnitude larger.

The next section of this chapter provides some examples of (novel) ways of using CFD and its outcomes to benefit the wastewater sector.

9.3.2 The use of CFD to increase our insight into reactor mixing

9.3.2.1 The case of the bioreactor of the Eindhoven WWTP

A fully integrated CFD-kinetic model was developed for the outer ring (partly oxic and partly anoxic) of a concentric plant design (Eindhoven Water and Resource Recovery Plant (WRRF)) as shown in Figure 9.6 (Rehman *et al.*, 2017). Since coupling to kinetics was required, scalar equations were defined for all species present in the kinetic model. Dissolved species follow the fluid flow and get additional source and sink terms according to the reactions they take part in. These terms can easily be derived from the Gujer matrix of an ASM model. Technically, they are implemented by means of user defined functions (UDF) when the commercial tool Fluent (Ansys) is used.

It was found that significant deviations from completely mixed conditions prevailed due to the specific design (circular) and placement of inlets, outlets, propellers and recirculation flows. A key issue related to the circular design is the fact that the mixed liquor that is internally recycled is pushed to the outer wall when it hits the wall of air bubbles in the bottom aerated zones. This pushes the air plume to the inner wall and results in a significant heterogeneity of species concentrations and, hence, the local activity of microorganisms. The model was validated by means of local measurements of DO by using a construction crane and measuring at 99 different locations. It was further demonstrated that different air and liquid flow rates also significantly changed flow patterns (results not shown).

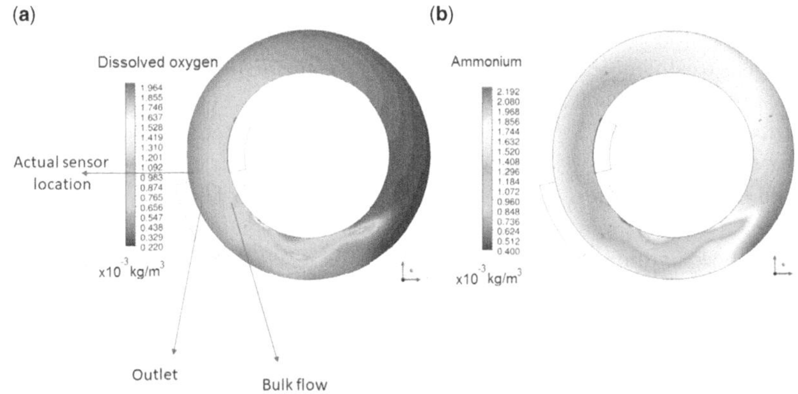

Figure 9.6 Dissolved oxygen (a) and ammonium (b) concentrations at 3.45 m depth in the outer ring of the Eindhoven WRRF.

Given these results it is obvious that a traditional TIS approach will not lead to good results when using it in a flowsheet model due to the fact that it neglects spatial heterogeneities in the 2nd and 3rd dimension.

9.3.3 How can CFD be used to improve current mixing models?

A possible approach is illustrated in Figure 9.7 (Laurent *et al.*, 2014). The rationale is that a more complex model of the system is built and validated and that this gathered knowledge of the process is translated into an intermediate complexity model that can be used in a flowsheet context. How this can be done is demonstrated by Rehman (2016) and briefly outlined in what follows.

The species heterogeneity in any reactor volume can be quantified by introducing cumulative species distributions (CSDs) (Figure 9.8). These can be constructed for each chosen reactor volume by constructing a histogram of a certain species concentration and representing it in a cumulative fashion. Figure 9.8 demonstrates this for different vertical sections of the reactor. CSDs can be interpreted as follows: if they are steep in nature, this means that not a lot of variability in the species concentration occurs in that volume, indicating that it is rather homogeneous and can be represented by a CSTR in a flowsheet model. Vice versa, when the CSD is less steep, this indicates heterogeneity and indicates that this reactor volume would be better represented by more compartments.

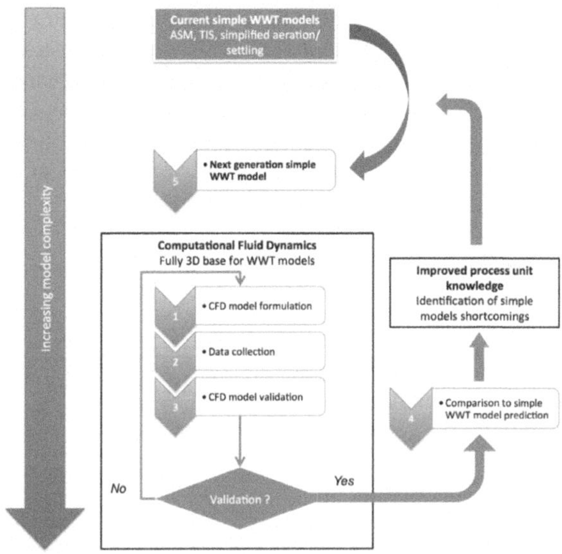

Figure 9.7 Conceptual protocol for the potential use of CFD for wastewater treatment (WWT) process modelling.

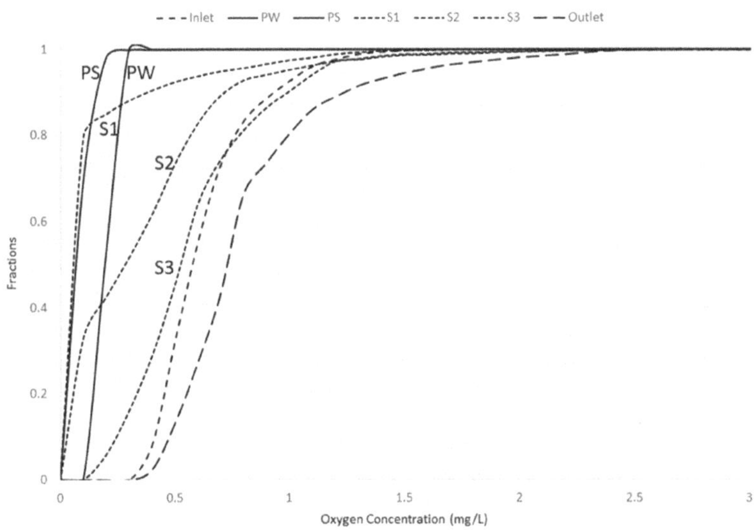

Figure 9.8 Cumulative species distributions of dissolved oxygen at different locations in the outer ring of the Eindhoven WRRF. Steep curves reveal less heterogeneity and vice versa. Codes for the different lines are explained in Figure 9.9.

Hence, CSDs can be adopted further to develop a compartmental model (CM) that can be used instead of a TIS approach. This was demonstrated by Rehman (2016) and is shown for this particular case in Figure 9.9. The CM approach should reduce uncertainty in the mixing model and, hence, unnecessary calibration efforts. We are currently working on demonstrating this, as well as extending the CM to a dynamic CM, i.e. applying the compartmentalisation for different air and liquid flow rate combinations.

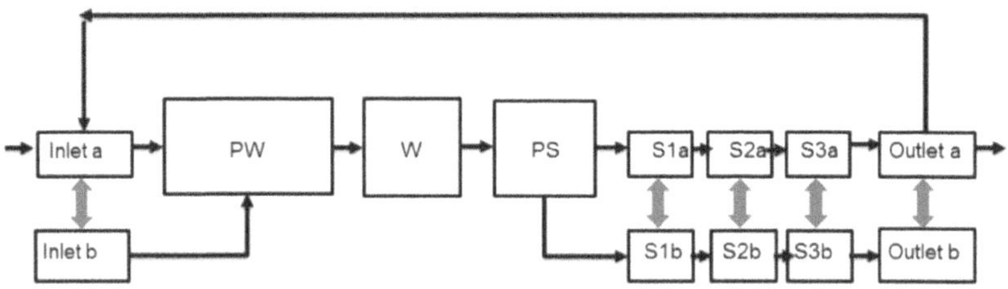

Figure 9.9 Compartmental model of the outer ring of the Eindhoven WRRF. PW: pre winter zone; W: winter zone; PS: pre summer zone; S1-3: Summer zones.

Finally, this integrated CFD-kinetic analysis has clearly shown that the reactor design in this case can be improved. In a follow-up study, the obtained knowledge can now be used to test virtual scenarios that cure the major downsides of this process behaviour by, for example, relocating plate aeration devices, propeller speeds, etc. In this particular case, the CFD model indicated that a high DO zone exists near the inner wall of the outer ring, right at the location of the recycle to the anoxic zone, which is the worst possible location (red zone in Figure 9.6, left). One can now test scenarios to determine how to avoid this high DO zone. In this case, the cause is the propellers in front of the aerated zone and the placement of the aeration plates. A scenario switching of the impellers revealed almost no difference (actually meaning that they are obsolete and could have saved CAPEX) and leads to a reduction of OPEX. Next, relocating aeration plates was tested and indeed led to a reduction of the DO at the recycle location. Several scenarios can easily be tested (note that this does not require remeshing if a smart geometry is implemented, leaving scenarios to be a matter of choosing boundary conditions), which would be very expensive in the real system.

9.3.4 Extending the CFD modelling approach to other unit processes in WWTPs

9.3.4.1 Bioprocesses

There are many examples where the behaviour of a system cannot be described by a systemic model combined with a kinetic model. Some designs seem to behave better than expected, whereas others are underperforming.

Kimura *et al.* (2008) found better performance behaviour of a baffled membrane bioreactor (BMBR) that could not be explained by a CSTR combined with an activated sludge model (ASM). Actually, an innovative design was proposed that induced specific mixing conditions that changed over time due to specific operation of the reactor (Figure 9.10). Through changing the liquid level, the zones between the baffles and the outer wall can either be made aerobic or anoxic. Modelling this with a CSTR will not work. By developing a CFD model, insight into how the system works can be relatively easily obtained. Especially, the transition between aerobic and anoxic conditions is of importance. For optimisation of the design (height of baffles, distance between baffle and outer wall) and operation (air lift, speed of level switch), CFD models are the only option. Furthermore, when scaling up such a design, these items will also be important and a CFD model will prove useful.

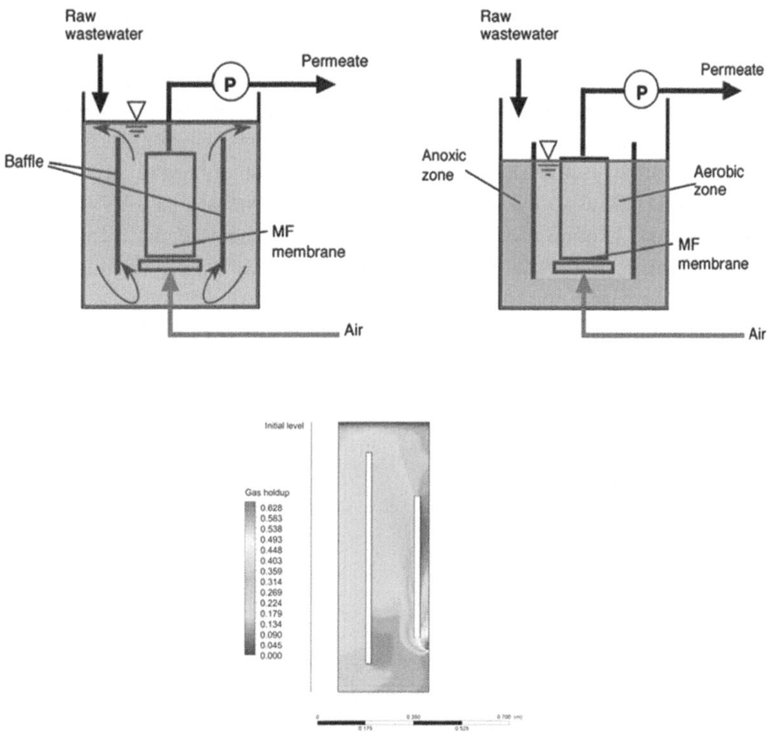

Figure 9.10 Illustration of baffled MBR concept (Kimura *et al.*, 2008) (top) and CFD simulation of the gas holdup (bottom). MF = Microfiltration.

The unexpected appearance of bio-P removal in systems not designed for it is another example (Verrecht *et al.*, 2010; Barnard *et al.*, 2012). However, few authors link this to the specific mixing behaviour in a reactor and try to explain it through other complex theories or calibrating kinetics, whereas it is often caused by the occurrence of dead zones in the corners of ill-designed reactors. This is especially true for sludges with elevated viscosity.

Surface aerated oxidation ditch plants are yet another example of this as demonstrated by Rehman (2016).

Recently, a lot of attention is being given to the mixing behaviour of anaerobic digesters which is pretty much still a black box at full scale mainly due to the difficulty of measuring inside the system. Given the elevated viscosity and the typical limited amount of energy input for mixing, it cannot be expected to behave as completely mixed. Gas bubble production will only locally contribute to mixing, but not to a large extent. But it might be that this type of operation (i.e. not completely mixed) is the key to make the process work. However, it becomes cumbersome to optimise the process without understanding the specifics as to why this is the case. In a sense it seems that reactor design in general is pretty much either business as usual (based on rules of thumb) or trial and error. The time has come to introduce model-based design. It is noteworthy that this is now common practice in, for example, chemical engineering where reactor design is a very hot topic at the moment.

Even at a small scale one can demonstrate that complete mixing can be problematic (Arnaldos *et al.*, 2017). Figure 9.11 shows a CFD-ASM simulation of an unaerated 2 L continuous flow-through respirometer at different flow rates. As can be seen, a high flow rate is needed to achieve almost completely mixed conditions. Compared to an ASM-CSTR model, affinity indices have to be significantly recalibrated to match the outflow concentrations of different species. This emphasises that operating reactors under completely mixed conditions is something we might not want to pursue as this will come at a significant cost. It might be wiser to understand the consequences of incomplete mixing and turn this into a benefit of the system design and operation.

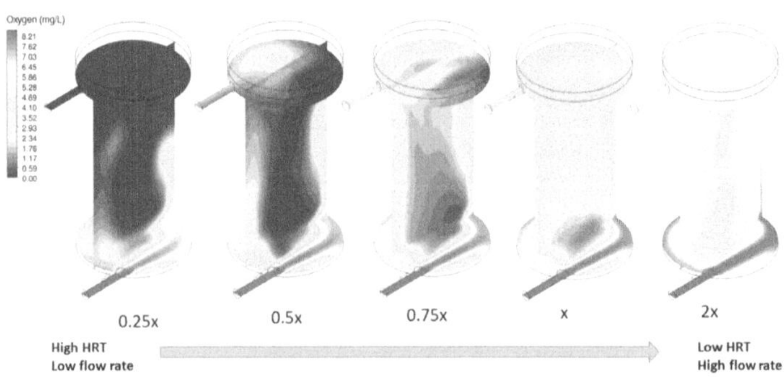

Figure 9.11 Distribution of dissolved oxygen concentration in an unaerated continuous flow-through respirometer.

9.3.4.2 Physical-chemical processes

Thus far focus has been on process kinetics. However, all the above also holds for physical process behaviour. Let us take the example of shear-induced coagulation or flocculation. Here, often the assumption of an average shear (G) is adopted in reactor design. However, pursuing this behaviour is similar to pursuing complete mixing in terms of concentrations. It will be very costly to achieve and the question is whether it is really needed. Coagulation systems are actually designed as merely scaled-up jar tests by means of some rules of thumb based on energy input. However, the energy dissipation will be different at different scales and one actually needs to think again in terms of distributions. Similar to the CSD discussed before, reactors will exhibit a shear distribution. Figure 9.12 demonstrates this for the anoxic section of an A-stage system containing five consecutive mixers intended for promoting flocculation.

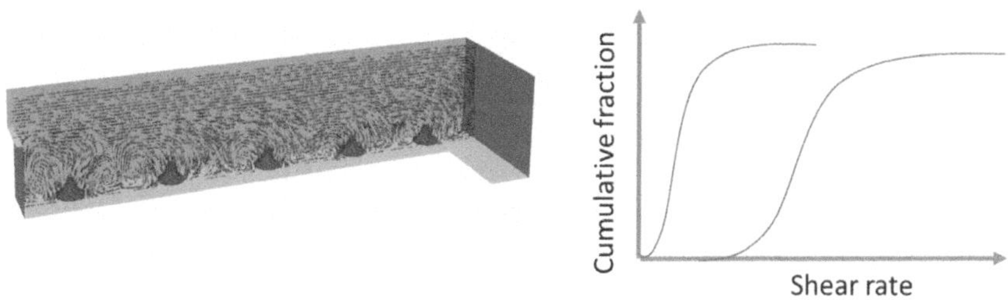

Figure 9.12 Velocity pattern (left) and conceptual cumulative shear distribution (right) of a flocculation reactor.

Depending on the location of a floc or granule it will either be exposed to low shear or high shear. The former will promote flocculation, whereas the latter will promote breakage (depending on floc strength). Obviously, the pattern at which a floc is exposed to different shear environments will be of key importance. This pattern is entirely governed by the mixing pattern and, hence, location of inlet, outlet, inlet flow rate and position and speed of stirring devices. Hence, when scaling up, one should try to achieve a similar shear distribution and recirculation pattern in the reactor. The latter is, however, not straightforward. Again, many studies try to understand why a certain system is not behaving similarly to its small-scale counterparts. The issue is the scale-up rules that are not accounting for the important aspects of mixing. Also here, model-based design based on CFD models coupled with flocculation models is the way to go. Another observable trend is the scale-down of such processes (e.g. tube flocculators) which makes sense as they have a much better controllability. Hence, it might be smarter to design smaller reactors and use numbering-up instead of scaling up. Again, in chemical engineering this is an observable trend in order to enhance reactor performance.

A system that actually combines both impact of kinetics and a physical precipitation process is crystallisation. Indeed, local concentrations are important with respect to supersaturation to promote crystal nucleation and growth. On the other hand, local shear levels promote aggregation and breakage phenomena which drive the crystal size distribution. The latter is important when thinking in a Quality-by-Design (QbD) fashion with respect to reuse of the product. Current reactor designs are definitely not optimal in this respect. Tarrago *et al.* (2016) demonstrated that the CSD could be altered by simply changing the upflow velocity in a struvite crystalliser. However, building knowledge is still required to improve the design in a targeted way. Hence, this is another field where advanced CFD modelling can lead to a large leap in process understanding and optimisation. An illustration of a CFD model simulating the flow field of a crystalliser is given in Figure 9.13. It shows high velocities in the upflow zone and low velocities in the downcomer zones.

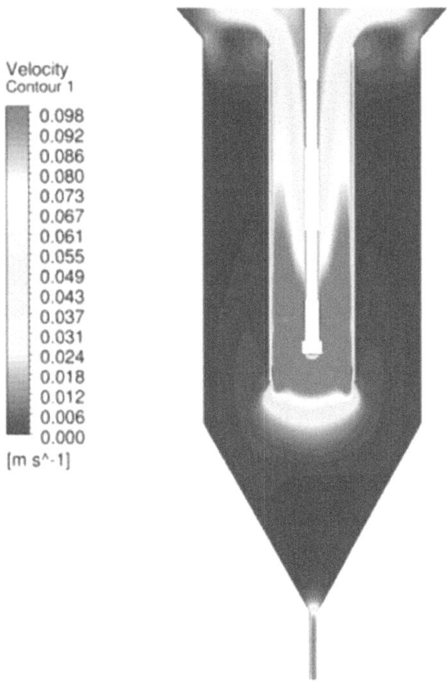

Figure 9.13 Simulated flow field in a crystalliser.

9.4 DISCUSSION

It is clear from the above that mixing behaviour of bioreactors is vastly overlooked and needs more attention. This is important in terms of troubleshooting of ill-designed reactors, which is the classical application of CFD in the wastewater field nowadays. Here, CFD models are used to improve reactor performance by means of introducing baffles, rearranging aeration diffusers, and changing propeller speeds and location. CFD to improve plant performance can extend to other unit operations as well as flow splitter design. Since this is already applied to a large extent, it will not be further discussed.

In a more disruptive way, the know-how gathered from more advanced validated models can and should be adopted in the future for model-based design. We need to build in the performance of a reactor in its design. We need to leave the "era of averaging" as we have the tools at hand now to do a much better job in engineering reactors. This means we need to leave the traditional way reactors are designed. But how can we do this? Before discussing that, some issues need to be highlighted and discussed in more detail: (1) which modelling framework should be used for reactor design (not plant design)? and (2) how do we define optimal mixing?

The former is actually not so straightforward. One would immediately have the reflex of using CFD for this purpose. However, there are a myriad of different design subtleties that one can think of and, hence, the degrees of freedom are enormous. Given the vast computational load of CFD, especially when combined with kinetic and physical models, it is likely not the way to go, at least not in the decades to come. An alternative is to further develop the powerful concept of compartmental models as they are more computationally tractable and, hence, allow the simulation of many more scenarios. A strategy could be to use CMs for optimisation and then link this optimal behaviour to reactor design. Obviously, this is a process that will take time, but is for sure the most promising route at this point. However, some development is needed before we can do so. A proper methodology for constructing CMs needs to be developed. Moreover, at the level of CM some further development is needed. What is the required accuracy? This should be answered by making a thorough comparison of TIS models versus CMs. CMs need to be made dynamic. Note that this was not considered for TIS although we see a significant change in mixing behaviour, even during diurnal fluctuations in dry weather flow. A dynamic CM could be based on simulating several steady state CFD-kinetic scenarios and subsequently interpolating. Which interpolation should be used here? In a next phase, when using CMs for improved process design, one should define an objective function for the optimisation of the compartmental network as well as the exchange fluxes. This methodology needs to be further developed. The next issue that arises is how to translate the optimal CM back into a true reactor design? Here mapping knowledge between CMs and CFD of certain reactor designs is needed. This will require simulations of quite a number of systems with CFD-kinetics at first. However, later on this knowledge should be translated into "next generation design rules" that can be applied without the specific need for an advanced CFD study. In this context, more unconventional reactor types might be better suited for optimising certain processes. Here one needs to think out of the box and discard copying designs that have been used for decades. And finally, why don't we consider building smaller size reactors in order to gain better controllability? Smaller reactors allow much better control of the mixing behaviour, allowing for higher performing reactors at all times. We currently give away this advantage by scaling up, believing that this is the cheaper solution. But was this ever really proven, accounting for the loss in reactor performance? The concept of "numbering up" smaller, better controllable, reactors is used in other fields as well. It might well be that an increased CAPEX for smaller volume reactors pays back relatively shortly in terms of OPEX, but also due to increased performance. The latter should also not be underestimated, as a performance increase of 20–30% might well be possible for certain applications.

As to the question of how to define optimal or "ideal" mixing, this could be an interesting discussion. Currently, ideal mixing patterns are known to be either CSTR or plug flow. However, mostly we are somewhere in between. But that does not mean that these conditions cannot be "ideal" or optimal. We should probably adapt our terminology and connect optimal mixing behaviour to the goal of the system. Optimal mixing can then be very different for different technologies. Striving for complete mixing is most likely not even what is required in many cases. We should use advanced models combining CFD, kinetics and physical processes to pinpoint what degree of mixing is optimal for a certain process and design reactors accordingly. Examples are bioprocesses or physical processes that need alternating conditions to (optimally) work such as simultaneous nitrification/denitrification, mainstream anammox and flocculation processes. In the bioprocesses we typically need to create specific redox conditions. In the case of flocculation we need to create alternate zones of low and high shear to engineer particles with properties such as dense flocs with good settling properties (e.g. Guérin *et al.*, 2017; Moruzzi *et al.*, 2017). In the case of a crystalliser, a combination of optimal conditions of supersaturation for nucleation and crystal growth are needed in combination with good alternating conditions for crystal aggregation and breakage in order to create a custom crystal size distribution, introducing the Quality-by-Design concept that is used in other disciplines where product quality is important. This will become important to establish a resource recovery market where a specific product quality is needed. Figure 9.14 illustrates how one can play with these conditions in the aforementioned crystalliser. Further research in this direction is obviously still needed in order to develop these methodologies.

Now, we need to come back to the question of how we can currently improve reactor design. First, building a CFD model of every unit operation should become the state of the art in every design project. The cost of such a project is not that high and can save significant budget at the end of the project. Once "ideal" reactor designs for unit processes have been designed, the need for these CFD models might decrease and these ideal reactors can be designed off the shelf. It is noteworthy that at academic level, one can further strive for improved processes. In terms of process controllability and sensor location, process dynamics and impact of varying liquid and air flows on reactor hydrodynamics are important for process robustness. More work is needed here. There are several ways this can be dealt with. The straightforward approach is to run the CFD model dynamically, but this will still be computationally cumbersome for some years to come (but at some point it will become feasible). In the mean time we need an approximate solution which could be to decide on several steady state situations the reactor will often be in (e.g. diurnal states; dry versus wet weather), simulate CFD models for these situations, and analyse performance (e.g. hydraulic retention time (HRT), aerobic/anoxic volumes, etc). These steady state "figures" could be stored in a database, a kind of knowledge base, which can be useful to decide on sensor location or sensor set-points based on concentration profiles. Indeed, when one knows how representative a sensor reading is for a reactor, one can dynamically adapt the sensor tuning (or even set-points) to account for this.

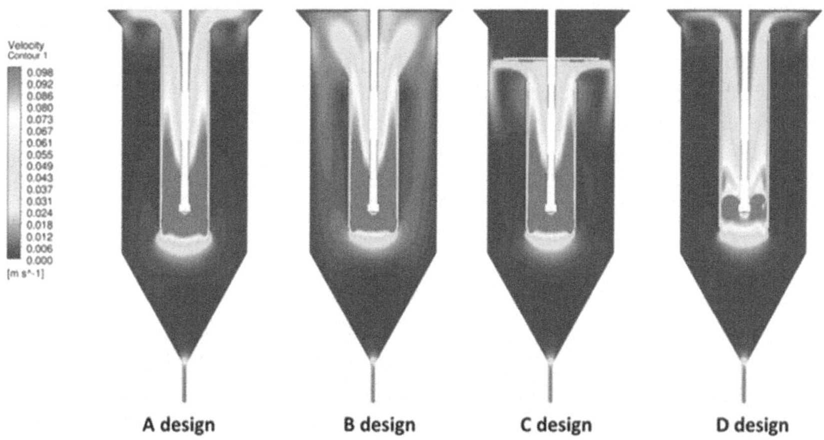

Figure 9.14 Example of how a crystalliser design can be altered to influence the flow pattern, local supersaturation and shear conditions.

It is clear that not all issues have been resolved. Nevertheless, there is already a clear benefit of using CFD as the standard in reactor design. Adopting it will only further extend our process knowledge and that is what engineering should be about. The question "to mix or not to mix" is obviously a rhetorical one as mixing will for sure be required. However, one needs to think about the required level and not blindly accept a design of which the mixing degree is uncertain. Implementing CFD in reactor design will help to answer questions regarding mixing levels. The effort (data requirement, man power, duration) and cost of CFD studies has dropped significantly and justifies its feasibility within any new design project.

9.5 CONCLUSIONS

This chapter clearly indicates that there is much more beyond the traditional way of using CFD for troubleshooting design flaws in wastewater treatment systems. When used in a smart way, it can lead to significant cost savings for existing plants and a higher level of cost-effectiveness for plants yet to be designed, hereby justifying the minor cost and effort of a CFD modelling study in an overall project cost. The methodologies shown are generic and can be extended to different unit processes, both biological as well as physical-chemical. We should also reconsider the terminology of "ideal" mixing with the objective of the process in mind. This will pave the way to better engineered treatment systems, but also to optimal resource recovery processes with quality built into the product.

9.6 REFERENCES

Arnaldos M., Amerlinck Y., Rehman U., Maere T., Van Hoey S., Naessens W. and Nopens I. (2015). From the affinity constant to the half-saturation index: understanding conventional modelling concepts in novel wastewater treatment processes. *Water Research*, **70**, 458–470.

Arnaldos M., Rehman U., Naessens W., Amerlinck Y. and Nopens I. (2017). Understanding the effects of bulk mixing on the determination on the affinity index: consequences on process operation and design. *Water Science and Technology*, **77**(3), 576–588.

Barnard J., Houweling D., Analla H. and Steichen M. (2012). Saving phosphorus removal at the Henderson NV plant. *Water Science and Technology*, **65**(7), 1318–1322.

Guérin L., Coufort-Saudejaud C., Liné A. and Frances C. (2017). Dynamics of aggregate size and shape properties under sequenced flocculation in a turbulent Taylor-Couette reactor. *Journal of Colloid and Interface Science*, **491**, 167–178.

Kimura K., Nishisako R., Miyoshi T., Shimada R. and Watanabe Y. (2008). Baffled membrane bioreactor (BMBR) for efficient nutrient removal from municipal wastewater. *Water Research*, **42**, 625–632.

Laurent J., Samstag R., Ducoste J., Griborio A., Nopens I., Batstone D., Wicks J., Saunders S. and Potier O. (2014). A protocol for the use of computational fluid dynamics as a supportive tool for wastewater treatment plant modelling. *Water Science and Technology*, **70**(10), 1575–1584.

Moruzzi R. B., de Oliveira A. L., da Conceicao F. T., Gregory J. and Campos L. C. (2017). Fractal dimensions of large aggregates under different flocculation conditions. *Science of the Total Environment*, **609**, 807–814.

Rehman U. (2016). Next Generation Bioreactor Models for Wastewater Treatment Systems by Means of Detailed Combined Modelling of Mixing and Biokinetics. PhD thesis, Ghent University, Belgium.

Rehman U., Vesvikar M., Maere T., Guo L., Vanrolleghem P. A. and Nopens I. (2015). Effect of sensor location on controller performance in a wastewater treatment plant. *Water Science and Technology*, **71**(5), 700–708.

Rehman U., Audenaert W., Amerlinck Y., Maere T., Arnaldos M. and Nopens I. (2017). How well-mixed is well mixed? Hydrodynamic-biokinetic model integration in an aerated tank of a full-scale water resource recovery facility. *Water Science and Technology*, **76**(8), 1950–1965.

Samstag R., Ducoste J., Griborio A., Nopens I., Batstone D., Wicks J., Saunders S., Wicklien E., Kenny G. and Laurent J. (2016). CFD for wastewater treatment: an overview. *Water Science and Technology*, **74**(3), 549–563.

Tarrago E., Puig S., Ruscalleda M., Balaguer M. and Colprim J. (2016). Controlling struvite particles' size using the up-flow velocity. *Chemical Engineering Journal*, **302**, 819–827.

Verrecht B., Maere T., Benedetti L., Nopens I. and Judd S. (2010). Model-based energy optimisation of a small-scale decentralised membrane bioreactor for urban reuse. *Water Research*, **44**(14), 4047–4056.

Wicklein E., Batstone D. J., Ducoste J., Laurent J., Griborio A., Wicks J., Saunders S., Samstag R., Potier O. and Nopens I. (2016). Good modelling practice in applying computational fluid dynamics for WWTP modelling. *Water Science and Technology*, **73**(5), 969–982.

Chapter 10

Making water operations smarter

Gustaf Olsson[1] *and Pernille Ingildsen*[2]

[1]*Lund University, Sweden (gustaf.olsson@iea.lth.se)*

[2]*Kalundborg Forsyning A/S, Kalundborg, Denmark (pein@kalfor.dk)*

ABSTRACT

A smart water system operates to produce an acceptable product while keeping energy and resource requirements at a minimum. However, while cost savings have up until recently been our primary aim of making water and wastewater operations more efficient, there is now a more pressing agenda of sustainability driving the smart agenda. The primary aim of developing our utilities into a higher level of smart is hence moving from a singular focus on cost savings to making our water infrastructure adapt and interface seamlessly and sustainably with the natural water systems we depend on and operate inside.

The smart water system ensures the full system's resilience by responding adequately to disturbances and recovering after any upset. An important property of such a system is to be sufficiently transparent so that anybody responsible for the operation – from operators to managers – is sufficiently informed to make rational decisions. A water system includes the whole water cycle: water supply, water distribution, urban drainage and wastewater treatment as well as the customer or demand side. The smart system contains an information flow that we describe by three major ingredients, phrased as the "MAD" principle. It consists of methods and devices for measurements (M), algorithms and software for analysis (A) and procedures and methods for decision (D). The MAD principle can be applied at

Editors: Giorgio Mannina, George Ekama, Hallvard Ødegaard and Gustaf Olsson
doi: 10.2166/9781780409719_0341

any level, from the equipment level to the management and gives the prerequisites for truly smart management and operation.

Keywords: Automation, control, measurements, monitoring, sensors, smart systems

10.1 INTRODUCTION

The word smart is used nowadays as a characteristic of devices, people and phenomena. It may mean clever, sharp or intelligent. Sometimes it is a synonym for elegant, stylish or neat. A smart water system is not the same as an elegant, glamorous or fashionable system. Instead, it operates to fulfil the requirements of producing an acceptable product while keeping the energy and resource requirements at a minimum. Such a system should respond quickly and in an adequate way to disturbances and recover quickly after a major upset. The operation of the system must be sufficiently transparent so that people involved in the operation are well informed to make rational decisions. Furthermore, the complete water cycle should be included: water supply, water distribution, urban drainage, and wastewater treatment (WWT) as well as the customer or demand side.

The primary aim of developing our utilities into a higher level of smart, is now developing from a singular focus on cost savings in operations to a broader view of making the whole water infrastructure adapt and interface seamlessly with the natural water systems we depend on and operate inside in a sustainable and economic fashion. The purpose of this adaption is to ensure catering for both the needs of human society and for our surrounding environment.

This "smart" approach is especially relevant to cities and industries that need to learn to operate in a similar fashion to nature. We ought to try to understand our human infrastructure systems as an integral part of the nature we inhabit. We need to learn to play by the "sustainability play-book". There should be a continuous focus on reducing the damage our water operations do to our nearby surroundings, so that the operation of tomorrow will have a less detrimental impact than that of yesterday.

Due to large amounts of energy and human ingenuity, we have been able to make our living conditions very comfortable and healthy for human consumption. The benefits of the last 100–150 years of "water revolution" have been astonishing and we ought to be more consciously grateful for the great improvement of living conditions most of the world has seen in this period. We understand now, however, that this comfort has come at a cost, a cost that is paid by our nearby environment everywhere. From the natural law of sustainability it follows that what harms our nearby environment will eventually harm our human societies, since we live in a steady symbiosis with the environment. Oblivious to this we have allowed ourselves to squander our water resources.

As we begin to understand the detrimental effects of our modernization and industrialization we need to change and adapt the water infrastructure of our cities and industries. We need to re-educate ourselves about how to operate sustainably moving forward. This is the challenge of this century. However, as long as our water utilities haven't reached a certain level of "smart" we are steering blindfolded; ignorant to our effects and to impending problems from floods and droughts.

So, the challenge is that we need to grow smarter, to sense our environment, our systems and our usage to make sure that the three systems are working in synchronicity, adapting to each other in a sustainable, robust and resilient manner. This is not in conflict with the original aim of "smart" or applying instrumentation, control and automation (ICA) to make operations less expensive. However, it is a much loftier goal to reach true sustainability and the task is much more complex, requiring a deeper understanding of our infrastructure systems, the natural physical and biological systems they interact with and our usage. We need to understand where and when we must be careful, when to save water and when to be ready with emergency plans, i.e. how to be truly sustainable, robust and resilient. Quite possibly we also need to devise new schemes of incentives to make sure that the concept of "optimal operation and management" is aligned with the goal of sustainability.

10.2 TOWARDS SMART OPERATIONS

Water operations can be illustrated by a three-dimensional space. Mass flows form one dimension, indicating that we must satisfy different criteria to make various components fulfil the requirements of the user or the environment. The second dimension describes the flow of resources used and includes energy in terms of electrical, chemical, and thermal energy. Energy is not only a resource that is consumed. The incoming water – raw water for the supply or used water for the treatment – contains both thermal and chemical energy. The used water also contains nutrients that ought to be recovered, such as phosphorus.

Here we will concentrate on the third dimension, the information flow. A key approach to a smarter system is found via the philosophy of ICA. In our book on smart water utilities (Ingildsen & Olsson, 2016) the basis for the information flow is the simple framework that a control engineer defines as a feedback loop. It is here called the "MAD" approach to enable smart water management. The acronym stands for "measure" (M), "analytics" (A) and "decision-making" (D). However, it is emphasized that feedback should be used at all levels, from the simple component to the management and strategic decisions.

To measure is to know: the "M" emphasizes that we need adequate data in space and in time. Measurements include not only signals from sensors and instruments, but also results of laboratory analysis, observations and human communications. Data must be analysed, understood and interpreted – the "A" – in order to become useful information. This is the basis for some decisions – the "D" – manual or automatic. A water operation can only be smart if people in all positions get involved, so that they can see their respective role in the whole plant or system operation.

It is not only the utility employees that should be involved. It is crucial that the water users be engaged. For too long water utilities and operations have been invisible to the users, so we take the "perfect delivery of water" for granted. It should be recognized that most of the energy related to the water cycle is used at home, not within the utility operations.

10.3 MEASUREMENTS

It is obvious that sensors should provide adequate data and sensors and instruments are the very basis of "smart" (Ingildsen & Olsson, 2016, Chapter 3). The measured or observed phenomena form the basis for all feedback. In water systems it is apparent that flow rates and a multitude of concentrations and quality parameters are the foundation for all operations. The instrumentation must be robust, easy to maintain and cost effective. This is even more important in an unmanned process.

For a long time, instrumentation was considered the bottleneck for control and automation in water systems. This has been considerably improved during recent decades. For example an important development of nutrient sensors has taken place in the last two decades, from automated laboratory *ex situ* analysers that had to be protected from the measured system through filtration units to *in situ* sensors that can be placed directly in the liquid to be monitored. There is an interesting development towards more and more "smart" sensors with multiple heads, which can be placed anywhere in the processes. Below we just list some key measurements in water systems and further details of the sensor properties are found both at the vendor websites (search for "water and wastewater sensors" and similar keywords) and in Ingildsen and Olsson (2016).

The introduction of a sensor in a plant or system requires not only confidence in the equipment. If the value of the measurement is not understood, then it is too easy to lose interest in the sensor. Then the performance of the measurement will gradually decrease because of lack of attention or maintenance. Too many high-quality instruments have failed because of this lack of connection with the purpose of the information.

Before we even consider controlling the water quality it must be ensured that the plant equipment is working adequately. If any operating failure is noted, an automatic alarm should be sent to the operating personnel. This includes meters of electric motor speeds, acoustic noise measurements, and indicators that can ensure that pumps and compressors are running. Simple level measurements or weight sensors can confirm that tank levels are within acceptable limits. Gas flow measurements can observe the methane production in a digester, and conductivity sensors can detect changes in the influent composition. The whole idea is *not* to obtain accurate information of the plant state, but to ensure normal operation. Measurements, forming the backbone of the operation in water systems are listed in Table 10.1.

Table 10.1 Key measurements in water systems operations.

Hydraulic measurements	• Flow rates, • Water levels and pressure.
Drinking water quality measurements	• Turbidity, • Free chlorine, • pH, • Content of organic matter, UV254/TOC measurements, • Oxidation/reduction potential, ORP, • Conductivity, • Temperature, • Colour, • Ozone.
Wastewater treatment measurements	• Dissolved oxygen, DO, • Total suspended solids, TSS, • Ammonia, • Nitrite and nitrate, • Phosphate, • Organic content: chemical oxygen demand (COD), biochemical oxygen demand (BOD), • Sludge level, • Respirometry, • Air flow rate in the aeration system.

UV254: UV light 254 nm; TOC: total organic carbon.

It is anticipated that low-cost devices will be available to monitor the transmittance, total suspended solids, particle size and particle size distribution, and turbidity as well as selected organic and inorganic chemical constituents. Developments in biosensors offer a promising basis for providing information about water quality.

There are many interesting sensor developments going on in companies and research laboratories around the world. This will of course increase the chances to make the plant operations smarter in many senses. We mention some developments here (Ingildsen & Olsson, 2016):

- *Microbiology of drinking water*: this is of paramount interest. For example, measurement of *E. coli* or other indicator organisms for faecal contamination in drinking water would be of paramount interest globally. The detection of both parasites like Cryptosporidium and viruses is of primary interest both in rich and in poor countries.
- *Adding processor power to sensors*: making the sensors more complex by adding processor power to the primary sensor will enhance the monitoring capability, see below. Internet of Things (IoT) adds another layer of "smartness" with increasing communication facilities from the sensors directly to the decision.

- *Micropollutants*: detection and measurement of micropollutants in water is an increasing challenge. The first step to be able to remove these contaminants is of course to be able to detect and to measure their concentration.
- *Spectroscopy* offers the ability to measure the whole spectrum online and can give a much more complete picture of the water quality than the measurement of only specific wavelengths.
- *Wireless communication* is part of the IoT development and makes it possible to easier achieve field measurements from distant locations, thus increasing the probability to successfully operate early warning systems.

10.4 MONITORING AND ANALYSIS

Unlike humans, computers are infinitely attentive and can detect abnormal patterns in operating data. *Monitoring* indicates tracking the operational state of a process or a machine via online instrumentation. By making use of vigilant monitoring of key parameters we can reduce the risks for operational problems and growing errors. A consistent monitoring of the "product quality" will help us to avoid problems growing too large, as illustrated in Figure 10.1. By rapidly detecting deviations from "normal" it is possible to minimize the costs of abnormal behaviour. There are numerous applications of monitoring in water and wastewater systems. The textbook Olsson and Newell (1999) contains a comprehensive account of methods for monitoring, detection, diagnosis and control of WWT systems. Monitoring as a basis for early warning systems in WWT is described in literally hundreds of papers. Some overviews are found in Olsson *et al.* (2005), Irizar *et al.* (2008), Hamouda *et al.* (2009), and Olsson *et al.* (2014).

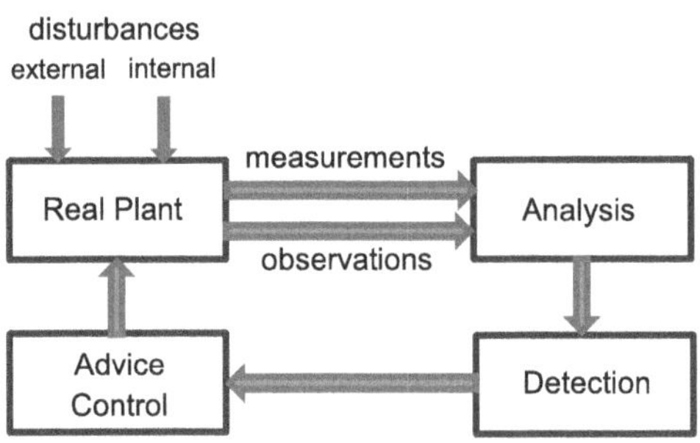

Figure 10.1 A monitoring system is based on analysis of the measurements and observations. A fault or failure can often be detected automatically. Based on the result, some advice can be given and a control action can take place.

As already noted, monitoring the machinery and the plant equipment forms the basic level of diagnosis. Simple indicators can warn the operator if a motor is not running or a pressure is getting too high or too low. Alarms of equipment behaviour or basic physical parameters – like flowrates, pressures, or levels – are essential pieces of information. At the other extreme, biological parameters will change slowly. For example, activated sludge floc settling properties depend, among other things, on the species of organisms. It is important to detect early signs of flow settling changes. Once they are apparent we may have bulking sludge, and it can be too late to make simple corrections. Therefore, early warning systems become critical in biological WWT.

Water and wastewater treatment and transport systems are increasingly being supplied with advanced instrumentation, not only for physical parameters, but also for water quality analysis. This would give a mine of facts if the important and relevant information could be extracted and interpreted, especially in real-time. Any decision and subsequent action taken on a plant relies on an explicit or implicit description of the state of the operation or events that are occurring. The action may be a manual correction or an automatic controller responding to some measurement. The information mostly comes from on-line sensors that are connected to a computer, but often some operator observations are important complements to the on-line sensor information for a decision.

Any monitoring system must determine whether the acquired data are meaningful and correct. So, any data generated in the plant, for local use or for reporting to a remote computer, have to be screened. Before any analysis can be made data rectification or data screening is crucial, so that false conclusions of the data set are avoided. Irrelevant information must be removed, such as noise or extreme data points, so called outliers. It involves at least:

- *Normal range comparisons*: high and low limits are defined;
- *Rate of change*: the trend is usually quite informative and should be averaged over a defined time horizon. Noisy data corrupt the information. Often some sudden changes appear in variables and can be readily detected;
- *Variance*: the variation around some average value may give a useful indication of the properties of the signal. Sometimes a large variance is a sign of poor sensor performance, sometimes of abnormal process behaviour.

A digital *low pass filter* can be applied for measurement data to remove individual spikes and other quick disturbances which do not contain relevant information. Constructing a filter which effectively removes the quick variations and at the same time does not affect the slow variations is always a compromise. The two most popular types of low pass filters are *moving averages* and *exponential smoothing*. Low pass filters in the process industry are almost always implemented with these types of simple filters.

In some instances, it is desirable to detect sudden or fast changes instead of the slow variations. This is done with *high pass filters*. The simplest one is a difference builder that calculates the difference between two consecutive values of a variable. If the difference is higher than a threshold value, then it indicates an undesired fast change.

Process knowledge is an essential condition for adequate diagnosis. A primary observation may lead to a whole chain of examinations of the problem, exactly as in medical diagnosis. For example, a high turbidity of the effluent may be the primary warning. The reason may be a too high influent flowrate, poor sludge settleability, or a high sludge blanket. By back-chaining and asking questions (automatically or manually), the diagnosis will look for the primary cause of the problem.

So far we have considered single signal analysis which can take place in any timescale from seconds to hours, depending on the type of measurement and variable. By combining information from multiple sensors and mathematical models, other operational parameters that are not directly measurable can be estimated: these are known as *soft sensors*. This allows powerful monitoring tools to be built up. Examples include the oxygen uptake rate and the nitrification rate. Massive experience has been gained in monitoring, and numerous advanced algorithms are available for interpreting noisy signals or the detecting out-of-normal variations and extreme values (Olsson & Newell, 1999; Olsson *et al.*, 2005).

Multivariate analysis is a method to detect patterns and correlations in large data sets. It has been used for many years in the chemical process industry, but was only introduced into the wastewater industry in the late 1990s (Rosen & Olsson, 1998). The most well-known method to reduce the dimensionality of the data cloud is Principal Component Analysis (PCA). This technique is simple in the sense that the data can readily be projected onto a smaller dimension. PCA has been used in sequencing batch reactors for monitoring (Villez *et al.*, 2008) and as a basis for control of the phase length (Villez *et al.*, 2010).

PCA methods, however, are insufficient to deal with data that are highly variable, such as influent flow rates and compositions (Rosen *et al.*, 2003). Specifically, the wide range of time constants in a WWT system makes it difficult to look at correlations of data in just one timescale. Various methods to extend the PCA to dynamical data in WWT systems have been developed (Rosen & Lennox, 2001; Lennox & Rosen, 2002). Flores *et al.* (2007) evaluated selected multivariable statistical techniques for various control strategies for plant-wide WWT plant control.

In Corominas *et al.* (2017a,b) many commonly used computer-based methods that can transform data to useful information have been reviewed. Several methods for advanced data analysis are considered. They include methods based on control charts, mass balances, regression models (multi-linear, partial least squares, PLS), self-organizing maps (SOM), principal component analysis (PCA), independent component analysis (ICA, note that ICA denotes two different methodologies), artificial neural networks (ANNs), clustering, fuzzy methods, support vector machines (SVMs), and techniques for the recognition of qualitative features in data series. In addition, computer-based tools for information (Environmental Decision Support Systems, EDSS) and knowledge management (ontologies) are included in the review. It is easy to get lost on how to select the best method for data analysis and there is no simple shortcut for how to find the best one. Mechanistic techniques/models were excluded from this review. A specific review of machine learning techniques applied to the field of water and wastewater is found in Hadjimichael *et al.* (2016).

There is a vast literature on data analysis and there are literally hundreds of books covering various parts of it. Furthermore, several books and articles on elementary data analysis are freely downloadable from the Internet (search for "data analysis", "statistical data analysis", "data mining"). It is important to emphasize that there is a lot of freely available as well as commercial software for data analysis. All the methods described here can be executed on some available software. Software like Excel can be used for a lot of elementary data analysis, while more sophisticated software products such as Matlab (http://se.mathworks.com/) or SAS (http://www.sas.com) contain a multitude of data analysis methods.

10.4.1 Water supply monitoring

Monitoring in water supply systems is an important part of the supplier responsibilities. Water utilities around the world lose between 25% and 50% of treated water to leaks. According to a World Bank report (Kingdom *et al.,* 2006) leakages cause losses of some USD14 billion (10^9) per year. Still many water companies spend far too little effort developing early warning systems to detect leakages. Water is still too cheap in many places. In Europe, the new European Council Directive 98/83/EC is pushing water companies to improve the condition assessment and continuous monitoring of transmission mains to safeguard the security of supply to customers. Traditionally leak detection has been based either on data outside the distribution network or data from sensors brought into the network at certain times. The new development is to establish a permanent set of sensors in the distribution network. More pressure and flow rate sensors can be deployed in water distribution networks and be combined with smart water meetings at the user location (Cardell-Oliver & Peach 2013). A lot of techniques and methods have been developed not only to detect a sudden leakage (burst) or a slow leakage but also to find the location of the leakage, both in a single-pipe system and in a distribution network. Leak detection using pressure and flow-rate sensors is a promising approach. Still, there are many challenges such as the multi-objective problem of trade-offs between leak detection accuracy and the type and placement of sensors. Overviews of current technologies for burst detection and leakage managing are found in Puust *et al.* (2010), Romano *et al.* (2014a,b) and in the keynote by Savić (2017).

Beyond leakage detection awaits closer monitoring of the water quality. Groundwater-based drinking water is for a large part unchlorinated around the world. From a health and sustainability point of view this is a great accomplishment. It bears however a risk of microbial contamination. A risk that is further increased as the infrastructure becomes older and the number of cracks in the system in the pipes increases – in this sense leakage detection or the monitoring of water loss locally in the system can also be used as an indirect indicator of risk of water from the outside of the pipes seeping into the pipes during pressure losses – with the risk of contamination. It has been found that the system is often quite dynamic, and conditions may arise where water can get in. Hence monitoring water quality in such systems will become more and more adapted. As online monitoring of water quality in the distribution systems becomes feasible it may present the opportunity to provide non-chlorinated water from surface water sources. In Kalundborg (Denmark) such a system is currently being constructed. The plant will supply water to industrial processes requiring drinking water quality.

10.4.2 Analysing the user behaviour

Most often the energy used by the water and wastewater operators is discussed while the end-user energy use is forgotten or neglected. The amount of energy used in handling the water in the household is significantly greater than the energy used by the water operators. Water heating will require around 30–50 kWh/m^3 which is an order of magnitude larger than the energy for water delivery and WWT. According to statistics (Reffold *et al.*, 2008) around 90% of the energy related to the water cycle is consumed "at home". Whatever the energy source used for water heating – natural gas, solid waste incineration or electricity – there is an obvious incentive to save user energy use.

It is increasingly common to have smart energy meters installed in the homes and it is apparent that metering will decrease the energy use by 5 to 20% (Sønderlund *et al.*, 2016). The authors also report that smart water meters will have a significant impact on water use anywhere between 2.5 and 29% in various locations. The application of smart meters has been slower in the water industry than in the energy industry. However, with increasing water scarcity it is apparent that the incentives for using metering will increase.

The smart water metering market has emerged in the last decade as water utilities responded to global trends of using detailed and near real time data and analytics to serve their customer better. An advanced metering infrastructure can provide a two-way data link between the utility and the consumers. Thames Water is among the pioneers for this development and has initiated an ambitious smart metering installation programme in London with a target of more than 400,000 smart meters by 2020 (Thames Water, 2014).

Smart measurements at home will not only monitor normal consumption but are able to inform the customer of any abnormal consumption. Leaking faucets or toilets will in the long run increase the consumption. For example, the household water meter can be recorded regularly in periods of a couple of hours to detect abnormal water use or leakages. A low-flow shower head and water saving dishwashers and washing machines can save significant amounts of both water and energy.

10.5 CONTROL AND DECISION

Measurements can be based on sensors, and lab tests as well as on observations. Described data has to be screened and analysed before any decision or control action can be made. This is the letter D in the MAD principle. This is simply what control engineers recognize as a feedback loop. To make a decision there has to be a goal or purpose of the system defined. The decision must be converted to action via an actuator, which is typically a motor, a pump, a valve or a compressor.

The fundamental principle of control is feedback, illustrated by Figure 10.2. The process (for example an aerator, a chemical dosage system, an anaerobic reactor, a water distribution network) is subject to external disturbances all the time. They are mainly caused by the variations in the influent load or water consumption, but can also be caused by internal changes, such as recycles, pumping, etc.

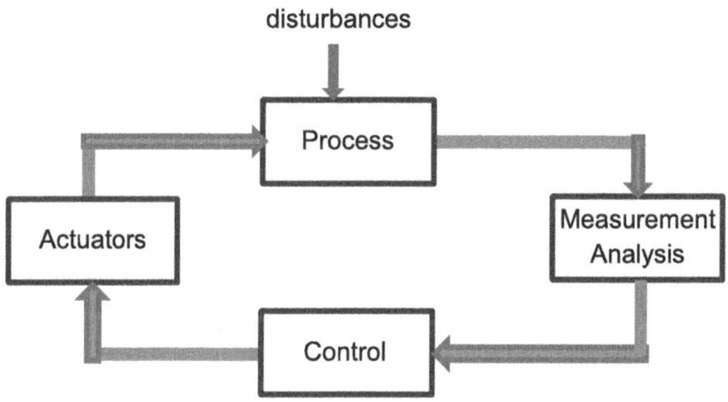

Figure 10.2 Illustration of the feedback principle.

Control theory and practice today can offer "almost anything" in terms of methods and algorithms that the water operator would need. Automatic control is sometimes called the hidden technology. It appears everywhere around us and we do not even think about it. We only notice it when it does *not* work. It is used in a cruise controller in the car, in the temperature control of a room, as well as in the thousands of controllers in water and wastewater operations that make sure that flow rates, pressures, levels and temperatures are right. Our own body contains a huge amount of control systems, for example, to keep our body temperature at 37°C, despite variations in the surrounding environment. Even if control is applied in completely different areas, there is a common theory that is independent of the applications. There are literally hundreds of textbooks on control. A superior textbook has been published recently by Åström and Murray (2014). The authors are world leaders in the development of control theory and engineering. The complete book is available for free on the web and includes the complete text of the book as well as additional examples, exercises, and frequently asked questions.

There are several important "demand pull" driving forces that will encourage an increasing use of ICA (Olsson *et al.*, 2014). They include increasing water use and wastewater load because of growing urbanization and population. The increasing complexity in the function and capability of WWT, including energy and nutrient recovery, more stringent regulations and ever-stronger economic drivers add further driving forces. Climate change and associated extreme weather conditions further add to the complexity of urban water systems.

Several "technology push" forces will make ICA ever more profitable. Today computational power is almost "for free". The instrumentation and actuator development is commented on earlier. The actuator development, commented on below, will also make control more flexible. The education level of operators and process engineers has been raised considerably over recent decades.

Some control systems in the urban water cycle are listed here as typical examples.

10.5.1 Water treatment control

The operation of drinking-water treatment plants is developing from experience-driven to instrument driven. The use of feedforward and feedback control in various units in drinking water treatment enhances the water quality and will also reduce the operational costs (van Schagen *et al.*, 2010). The disturbance is identified and is handled and attenuated via feedback control. For example, a chemical usage reduction of 15% was achieved for the softening treatment step (ibid.).

10.5.2 Water distribution systems

The typical operational problem is the optimization of pump operations, which consists of finding a set of controls that minimizes pumping costs while also meeting the system requirements. The pump optimization problem has been well studied in the literature using different formulations.

Another emerging application of control is variable pressure control. In a traditional water distribution system, the pressure in the main pumping station is kept sufficiently high, so that the most distant customer (the critical point) will have a satisfactory pressure all around the clock. During the day when water consumption is high along the pipe system the pressure will gradually decrease towards the critical point. During the night, however, when consumption is much lower, the pressure drop is smaller. As a result, the pressure at the critical point will rise. Therefore, the pressure variations can be very high, which will cause mechanical stress on the pipe increasing the risk for bursts. Furthermore, with higher-than-necessary pressure the consequences of a leakage will increase.

The obvious goal is to keep the pressure at the critical point just above the minimum pressure all the time. This requires that the pressure along the pipe be adjusted dynamically so that the goal is satisfied. There are two principal ways to achieve this. One is to run a variable speed pump at the head end and reduce the pressure along the pipe with pressure reduction valves. Any such valve, however, will cause a waste of energy. The second and more elegant and energy saving method is to supply the system with several pressure-increasing pumps and avoid the pressure reduction valves. It is crucial to ensure the right balance between the various pressure zones. Systems based on this demand driven distribution principle are in operation (Fantozzi *et al.*, 2017). They are all based on the MAD principle, measuring the critical point pressure, analysing the signals and the information and deciding on the pumping power for the various pressure zones of the system. The web contains a multitude of practical information about pump control: look for "pump control".

10.5.3 Wastewater transport and treatment

Control and automation in sewer networks and WWT plants has been applied for many decades. Most of the conventional control systems are unit-process oriented with the aim to enhance the treatment performance leading to improved effluent quality and reduced energy and chemical consumption. Nowadays there is an increasing interest to consider coupling between different process units with the aim to achieve plant wide control. In some cases also the entire system of sewers and treatment plants has been considered. Using control and automation to improve the capacity of wastewater systems to handle an increasing load has also attracted considerable attention in the past decade.

Unit process control in WWT is mostly a proven technology. This includes:

- Dissolved oxygen (DO) control with a constant or a variable set-point as part of the aerator unit process operation. The calculation of the DO set-point is typically guided by the nitrogen removal performance monitored with ammonia sensors;
- Aeration phase-length control in alternating plants based on nutrient sensors;
- Nitrate recirculation control in a pre-denitrification plant based on nitrate and DO measurements in the aerobic and anoxic zones, respectively;
- Sludge retention time control based on measurements of effluent ammonia concentration and the estimated nitrification capacity;
- Return sludge control based on sludge blanket measurements in the settler;
- The control of anaerobic processes with the goal to stabilize the process and to maximize the biogas production;
- Dosing control in chemical precipitation based on local measurements of pH, alkalinity or phosphate concentration.

Comprehensive reviews of these control systems can be found in Olsson *et al.* (2005, 2014), Åmand *et al.* (2013, 2014) and Olsson (2008, 2012).

Over the years, the function of a WWT plant has evolved from COD removal only, to COD and nutrients (nitrogen and phosphorus) removal, and further to resource (water, energy and nutrients) recovery. Consequently, the complexity of a WWT plant has increased substantially, with strong interactions among different processes units. Such interactions represent both challenges and opportunities for ICA. Interaction between different parts of the treatment plant can cause unintentional disturbances but can also be positively exploited through integrated control or plant-wide control, to derive additional benefits.

A further challenge that future ICA systems ought to address is the mitigation of greenhouse gas emissions, such as N_2O, from biological nitrogen removal plants (Mannina *et al.*, 2016). The control of these parameters at levels that would minimize N_2O emissions and still allow satisfactory nitrogen removal have to be developed and demonstrated.

10.5.4 Integrated control of sewer networks and wastewater treatment plants

The sequential relationship between the sewer, the WWT plant and the receiving water is obvious and the need for control of flow in the sewers for the benefits of WWT plants and the receiving water was recognized early. Integrated control requires both accurate dynamic models of systems and reliable sensors. Following the advent of deterministic models of integrated drainage approximately two decades ago, the issue was researched by several groups, such as Nielsen *et al.* (1996), Pfister *et al.* (1998), Rauch and Harremoës (1999) and Schütze *et al.* (1999). Nearly all investigations focused on semi-virtual cases, utilizing simulation studies. These theoretical studies demonstrated the potential of integrated control, which paved the way to practical implementations.

Following the development of the models, researchers together with operators developed the necessary technologies and experience for reliable online monitoring of sewer systems which complemented the more established monitoring of the treatment plants. Integrated control is now being achieved in real life, for example in The Netherlands (Weijers *et al.*, 2012), in Denmark (Grum *et al.*, 2011), and in Germany (Seggelke *et al.*, 2013).

10.5.5 Computer realizations of control systems

Any control system today is implemented in computer systems from very small PLCs (programmable logic controllers) to larger real-time computers. Often they are named SCADA systems (Supervisory Control And Data Acquisition) and contain computers, networked data communications and graphical user interfaces for process supervisory management. There is a vast literature on SCADA systems and the basic ideas in SCADA are actually explained in several booklets, freely available on the internet. They can describe basic concepts such as the components of a SCADA system, basic computer communication, computer interfaces and human–machine interfaces (HMIs). To find more information about SCADA systems, use search engines and www.youtube.com and look for "SCADA", "PLC" and "real-time computer" process control systems.

10.5.6 Actuators

Actuators are the devices that translate information into "muscles", such as electrical motors, pneumatic devices and hydraulic valves. The performance of these devices is often neglected in many control systems applications, causing them to fail. Plants can be operated more flexibly by assuring adequate controllability of their pumps, motors and valves. The requirements on the actuators often differ for small units compared with large plants. While it is crucial to control the influent flow rate smoothly in a large plant, lower operating costs may be achieved by using on/off control of small pumps or compressors in a small unit. Intermittent aeration can be adequate in a small package plant. The influent flow can also be split in different ways to the process units, depending on the influent water characteristics or the flow rate. However, remote control of actuators in a small plant from a central service group is already a proven technology.

Electric drive systems are vital in water operations, in which any pumping is dependent on an electric motor system. All the major vendors provide a large quantity of information on the internet. Furthermore, there are several short lectures available on www.youtube.com. Search for "electric drive systems".

10.6 TRENDS TOWARDS DECENTRALIZATION

Two developments in parallel will to some extent change the structure of many water supply and WWT systems. The development of renewable energy, primarily solar photovoltaic (PV) and wind power, is already dramatically changing the availability for electrical energy while contributing to decreasing the carbon footprint. Similarly, the interest towards decentralized water systems is increasing, not only in remote areas but also in rapidly growing peri-urban areas (Weber *et al.*, 2007; Bieker *et al.*, 2010; Larsen *et al.*, 2013). These trends will rely on smart solutions using adequate sensors, control and automation.

Available electrical energy is a critical factor to pump or treat water. Access to electrical energy can be enjoyed by 84% of the global population implying that almost 1.2 billion (10^9) people are still without it (International Energy Agency [IEA], 2016). Renewable energy technologies are already making a major contribution to universal access to carbon-free energy. The world now adds more renewable power capacity annually than the capacity it adds (net) from all fossil fuels combined (Renewable Energy Network for the 21st Century [REN21], 2018). In many regions of the world with energy poverty there are abundant renewable energy sources. Particularly in Africa and Asia a lot of rural areas are not connected to any grid infrastructure. Solar and wind power offer huge opportunities in off-grid electricity systems. Small scale off-grid systems have the potential to improve the energy access in rural and peri-urban parts of the developing world, by providing not only lighting and heating but also pumping to gain access to water as well as water re-use and purification using different technologies including desalination, described in detail in Olsson (2018). Already solar power in small-scale installations has provided electrical power and lifted millions of people out of poverty.

Renewable energy offers new possibilities because of its scalability and can deliver energy for all sizes of water operations, from the household level to the village or urban community level. In particular, solar PV is highly modular, from very small lighting systems in remote areas to residential systems and utility-scale projects; from the kW range to several hundred MWs.

During the last few years there has been both a demand pull and a technology push towards more decentralized water supply and WWT. To treat wastewater where it is generated is an attractive option. One of the motivations for decentralized systems is that sewer systems connected to centralized treatment plants are inflexible. Recent decades have witnessed extensive discussions on decentralized systems and source separation. Comparisons between conventional centralized and local solutions have been reported and analysed in detail in the book by Larsen *et al.* (2013).

Increasing water scarcity is a driving force for more water reuse. It is obvious that different qualities of water can be used for different purposes, so all the used water should not necessarily be treated to drinking water standard. Multiple water qualities can be produced and used in the same system, so we need a much more sophisticated water delivery system than is available today. This concept of separate waste streams of highly variable character naturally suggests decentralized solutions: it is in the same spirit as smart water grids. The concept of a smart grid has been embraced in the field of electrical power systems and has inspired its application to other infrastructures.

The ultimate purpose of a smart water grid is to minimize water consumption, ensure an adequate water quality at all times and operate the system as efficiently and consistently as possible. A smart water grid delivers water from suppliers to consumers using a wide range of computer control functions, including sensors, remotely controlled actuators and digital communication. Its measurement and monitoring system keeps track of all the water flowing in the system, i.e. it should monitor not only the consumer behaviour and the demand-side requirements but also any bursts and slow leakages. A smart grid should automatically detect, analyse and respond to malfunctions and optimize asset use. It must be able to resist security attacks and empower and incorporate the consumer, i.e. it must have high levels of situation awareness and predictive capability.

A smart water supply must also offer smart solutions to WWT for various specified standards. Huge differences naturally exist between the decentralized solutions found in many developing countries and the advanced treatments available in wealthier parts of the world. Here the discussion of control and monitoring techniques is primarily focused on high-tech solutions and will consider alternatives to conventional advanced end-of-pipe WWT. To offer a viable alternative to high performance centralized WWT, a decentralized system must provide the same water quality for both small and large discharges, utilize resources efficiently, resist disturbances effectively and be competitive in cost.

10.6.1 ICA in decentralized systems

There are several types of decentralized systems, from small-scale on-site WWT systems to satellite treatment units integrated with centralized systems. Several chapters in Larsen *et al.* (2013) describe various technology solutions. Small, decentralized systems must be easy to operate to be accepted. The user should only have to handle simple routine tasks. Professional service enterprises should be responsible for keeping the plant operating properly and for handling any failures and faults. At the same time, failures of single units must not cause the whole system to collapse. Remote sensing is a proven technology in many other areas and should also be implemented here, both in household size equipment and larger units.

The ICA aspects can look different depending on the scale of the WWT unit (Olsson, 2013). The load to a large treatment plant is mostly less variable, since many waste sources are combined. The requirements on measurement accuracy are likely to be greater for a large plant than for a family-sized unit. However, similar parameters must be tracked. Small systems require some issues to be addressed that differ from those relevant to conventional centralized systems. They include:

- How should the more extreme load variations – both concentrations and flow rates – that may appear in a local plant be handled?
- Does the user have sufficient interest to operate the plant?
- How much process knowledge is needed for the operation and who can provide this knowledge?
- Is the system cost-effective?
- Is the decentralized solution more sustainable than the centralized one?
- What kind of infrastructure is needed?

As already noted a small WWT plant must supply a similar effluent quality or load to the environment as a centralized one and should be cost and energy efficient as well as easy to operate (Larsen *et al.*, 2013). Small plants showing extreme fluctuations in the inflow state and with no professional engineer on site need an operating method that enables them to react efficiently to these fluctuations while at the same time allowing easy maintenance. The design should also be adapted to the available control and operation options.

A large plant with skilled operators requires a different approach to instrumentation than a small package plant. In the first case, more advanced instrumentation can be adequately used and maintained, whereas a small plant requires rugged and low-cost instruments. The accuracy requirements also differ, as do other kinds of requirements, for instance on motors, pumps, compressors and valves. Seen from a small-plant perspective, automated operation will become increasingly important. The plant must operate autonomously, must be resilient against disturbances and must react to sudden changes in load flow rate and composition. This makes detection and early warning systems important. Given the potential of the Internet and the widespread use of network communications, it is obvious that many small units could contact a control centre staffed by professional operators. This is a proven technology and extensive experience has already been acquired in this kind of remote monitoring. Commercial offerings from IoT vendors are already offered on the internet (look for "IoT" and "remote monitoring").

10.6.2 Operator competence

One objection to decentralized systems is their complexity, as the operator may not be sufficiently skilled. However, we should not forget that we are always dealing with highly complex systems in our daily lives. A modern computer is extremely complex. Most car drivers would be quite unable to service their vehicles themselves. Still these complex systems are routinely used by non-specialists. And as decentralized electrical power generation is becoming increasingly widespread, non-professional users frequently utilize, for example, solar photovoltaic systems for power generation. The challenge is to hide the complexity for the unsophisticated user and provide a user interface that is transparent to him or her. Furthermore, available communication and control technology allows many small plants to be centrally supervised.

Experts who remotely supervise many small plants are faced with a diversity of predominantly unstructured data which must be transformed into useable information to characterize and operate a local unit efficiently and safely. Tools must be developed that enable a process specialist located at a remote location to supervise multiple systems and analyse critical operating conditions and at the same time provide the on-site user (the homeowner, hotel or apartment building serviceman etc.) with adequately processed information. This can consist of online maintenance guidance or basic process information.

The constraint today is not the software, hardware or equipment; it is our willingness as customers and users to recognize new possibilities and our ability as water professionals to learn about available possibilities. As Larsen (2011) points out, "soft" issues of acceptance and compliance must be considered at the same time as the more "rational" estimates of cost and performance.

10.7 CONCLUSIONS

There is a great potential to make water systems smarter. Adopting MAD thinking has the following advantages:

- Availability of a huge amount of online information about the operation, from basic equipment and fast timescales to sophisticated water quality analysis in a slower timescale;
- A rich spectrum of analysis methods and computer tools to analyse the raw data to reliable and useful information that can form the basis for decisions;
- Control methods that can handle all possible operational challenges that appear in water supply and WWT systems;
- Flexible actuator hardware that translates decisions to real actions in the plants;

- Communication tools that offer almost unlimited possibilities to transfer information within a water system and to remote locations;
- A remarkable development of decentralized water systems in combination with renewable energy, made possible by autonomous un-manned operations.

As described, this field of ICA is heavy in science and technology and it may seem like a daunting task to get a full overview. However, keeping the MAD principle in mind actually simplifies things. One way of thinking is: what decision do I need to make effectively and efficiently? – for this I need measurements and analysis. What measurements are available? And what is the best analysis I can think of to get to a good decision? Or the other way: with the knowledge from this sensor at this position, what decisions – moving forward – can I make better and more precisely? What does my analysis need to do and what would be the best method?

The principle of MAD is applicable in automation and all the way up to high-level strategic decisions. The framework is always the same, whereas the measurements, the analysis and the decisions are different. Hence, it is of crucial importance that a critical amount of people understand this way of thinking in order for utilities to become smarter, more robust, more resilient, more efficient, more effective and not least more sustainable.

10.8 REFERENCES

Åmand L., Olsson G. and Carlsson B. (2013). Aeration control – a review. *Water Science and Technology*, **67**(11), 2374–2398.

Åmand L., Laurell C., Stark-Fujii K., Thunberg A. and Carlsson B. (2014). Lessons learnt from evaluating full-scale ammonium feedback control in three large wastewater treatment plants. *Water Science and Technology*, **69**(7), 1573–1580.

Åström K. J. and Murray R. M. (2014) Feedback Systems: An Introduction for Scientists and Engineers. The complete book is available for free on the web, called FBSwiki. http://www.cds.caltech.edu/~murray/amwiki/index.php/Main_Page (accessed 15 February 2018).

Bieker S., Cornel P. and Wagner M. (2010). Semicentralised supply and treatment systems: Integrated infrastructure solutions for fast growing urban areas. *Water Science and Technology*, **61**(11), 2905–2913.

Cardell-Oliver R. and Peach G. (2013). Making sense of smart metering data: A data mining approach for discovering water use patterns. *Water: Journal of the Australian Water Association*, **40**(2), 124–128.

Corominas L., Garrido-Baserba M., Villez K., Olsson G., Cortés U. and Poch M. (2017a). Turning passive data into knowledge – a review of wastewater treatment monitoring techniques. *Presented at the IWA Specialized Conference on Instrumentation*, Control and Automation (ICA), Quebec, Canada.

Corominas L., Garrido-Baserba M., Villez K., Olsson G., Cortés U. and Poch M. (2017b). Transforming data into knowledge for improved wastewater treatment operation: A critical review of techniques. *Environmental Modelling and Software*, **106**, 89–103. Available online: http://authors.elsevier.com/sd/article/S1364815217302359.

Fantozzi M., Lambert A., Skovmose Kallesøe C., Hassan A. S., Stærk D., Nielsen A., Bach J. and Riis M. (2017). Pressure Management: An Effective Way to Reduce Non-Revenue Water, Improve Energy Efficiency and Reduce Operation and Maintenance Costs. Grundfos Whitepaper, www.grundfos.com (accessed 9 February 2018).

Flores X., Comas J., Roda I. R., Jiménez L. and Gernaey K. V. (2007). Application of multivariable statistical techniques in plant-wide WWTP control strategies analysis. *Water Science and Technology*, **56**(6), 75–83. DOI:10.2166/wst.2007.586.

Grum M., Thornberg D., Christensen M. L., Shididi S. A. and Thirsing, C. (2011). Full-scale real time control demonstration project in Copenhagen's largest urban drainage catchments. In: *Proceedings of the 12th International Conference on Urban Drainage*, Porto Alegre, Portugal. Available at: www.researchgate.net/publication/266169539_Full-Scale_Real_Time_Control_Demonstration_Project_in_Copenhagen's_Largest_Urban_Drainage_Catchments

Hadjimichael A., Comas J. and Corominas L. (2016). Do machine learning methods used in data mining enhance the potential of decision support systems? A review for the urban water sector. *AI Communications* **29**, 747–756. doi:10.3233/AIC-160714.

Hamouda M. A., Anderson W. B. and Huck P. M. (2009). Decision support systems in water and wastewater treatment process selection and design: A review. *Water Science and Technology*, **60**(7), 1757–1770. DOI:10.2166/wst.2009.538.

IEA (2016). International Energy Agency, *World Energy Outlook* 2016.

Ingildsen P. and Olsson G. (2016). Smart Water Utilities. IWA Publications, London.

Irizar I., Alferes J., Larrea L. and Ayesa, E. (2008). Standard signal processing using enriched sensor information for WWTP monitoring and control. *Water Science and Technology*, **57**(7), 1053–1060. DOI:10.2166/wst.2008.139.

Kingdom B., Liemberger R. and Marin P. (2006). The Challenge of Reducing Non-Revenue Water (NRW) in Developing Countries How the Private Sector Can Help: A Look at Performance-Based Service Contracting. Report for World Bank. https://siteresources.worldbank.org/INTWSS/Resources/WSS8fin4.pdf (accessed 15 February 2018).

Larsen T. A. (2011). Redesigning wastewater infrastructure to improve resource efficiency. *Water Science and Technology*, **63**(11), 2535–2541.

Larsen T. A., Udert K. M. and Lienert J. (eds) (2013). Source Separation and Decentralization for Wastewater Management. IWA Publishing, London. ISBN – 9781843393481.

Lennox J. and Rosen C. (2002). Adaptive multiscale principal components analysis for online monitoring of wastewater treatment. *Water Science and Technology*, **45**(4–5), 227–235.

Mannina G., Ekama G., Caniani D., Cosenza A., Esposito G., Gori R., Garrido-Baserba M., Rosso D. and Olsson G. (2016). Greenhouse gases from wastewater treatment – a review of modelling tools. *Journal Science of the Total Environment*, **551–552**, 254–270.

Nielsen M. K., Carstensen J. and Harremoës P. (1996). Combined control of sewer and treatment plant during rainstorm. *Water Science and Technology*, **34**(3), 181–187.

Olsson G. (2008). Process Control. Chapter 15 In: Biological Wastewater Treatment – Principles, Modelling and Design, M. Henze, M. van Loosdrecht, G. Ekama and D. Brdjanovic (eds), UNESCO, IWA Publishing, London, pp. 393–414.

Olsson G. (2012). ICA and me – a subjective review. *Water Research*, **46**(6), 1585–1624.

Olsson G. (2013). The potential of control and monitoring. Chapter 13 In: Source Separation and Decentralization for Wastewater Management, T. A. Larsen, K. M. Udert and J. Lienert (eds), IWA Publishing, London. ISBN – 9781843393481, pp. 179–192.

Olsson, G. (2018). Clean water using solar and wind: outside the power grid. IWA publications, London.

Olsson G. and Newell B. (1999). Wastewater Treatment Systems. Modelling, Diagnosis and Control. IWA Publishing, London.

Olsson G., Nielsen M., Yuan Z., Lynggaard-Jensen A. and Steyer J. P. (2005). Instrumentation, Control and Automation in Wastewater Systems. IWA Publishing, London.

Olsson G., Carlsson B., Comas J., Copp J., Gernaey K. V., Ingildsen P., Jeppsson U., Kim C., Rieger L., Rodríguez-Roda I., Steyer J.-P., Takács I., Vanrolleghem P. A., Vargas Casillas A., Yuan Z. and Åmand L. (2014). Instrumentation, Control and Automation in wastewater– from London 1973 to Narbonne 2013. *Water Science and Technology*, **69**(7), 1373–1385. doi: 10.2166/wst.2014.057.

Pfister A., Stein A., Schlegel S. and Teichgräber B. (1998). An integrated approach for improving the wastewater discharge and treatment systems. *Water Science and Technology*, **37**(1), 341–346.

Puust R., Kapelan Z., Savic D. A. and Koppel T. (2010). A review of methods for leakage management in pipe networks. *Urban Water Journal*, **7**(1), 25–45.

Rauch W. and Harremoës P. (1999). Genetic algorithms in real time control applied to minimize transient pollution from urban wastewater systems. *Water Research*, **33**(5), 1265–1277.

Reffold E., Leighton F., Choudhury F. and Rayner P. S. (2008). Greenhouse Gas Emissions of Water Supply and Demand Management Options. Environment Agency UK. www.cost.eu/download/5354 (accessed 15 February 2018).

REN21 (2018). Renewable Energy Network for the 21st Century. *Renewables* 2018. Global Status Report. www.ren21.net/status-of-renewables/global-status-report/ (accessed 30 June 2018).

Romano M., Kapelan Z. and Savić D. A. (2014a). Automated detection of pipe bursts and other events in water distribution systems. *Journal of Water Resources Planning and Management*, **140**(4), 457–467.

Romano M., Kapelan Z. and Savić D. A. (2014b). Evolutionary algorithm and expectation maximization strategies for improved detection of pipe bursts and other events in water distribution systems. *Journal of Water Resources Planning and Management*, **140**(5), 572–584.

Rosen C. and Lennox J. A. (2001). Adaptive and multiscale monitoring of wastewater treatment operation. *Water Research*, **35**(14), 3402–3410.

Rosen C. and Olsson G. (1998). Disturbance detection in wastewater treatment systems. *Water Science and Technology*, **37**(12), 197–205.

Rosen C., Röttorp J. and Jeppsson U. (2003). Multivariate on-line monitoring: Challenges and solutions for modern wastewater treatment operation. *Water Science and Technology*, **47**(2), 171–179.

Savić D. (2017). A Smart City without Smart Water is Only a Pipe Dream! Keynote paper, International Association for Hydro-Environment Engineering and Research (IAHR). *World Congress*, 2017, Kuala Lumpur.

Schütze M., Butler D. and Beck M. B. (1999). Optimisation of control strategies for the urban wastewater system—an integrated approach. *Water Science and Technology*, **39**(9), 209–216.

Seggelke K., Löwe R., Beeneken T. and Fuchs L. (2013). Implementation of an integrated real-time control system of sewer system and waste water treatment plant in the city of Wilhelmshaven. *Urban Water Journal*, **10**(5), 330–341.

Sønderlund A. L., Smith J. R., Hutton C. J., Kapelan Z. and Savić, D. (2016). Effectiveness of smart meter-based consumption feedback in curbing household water use: Knowns and unknowns. *Journal of Water Resources Planning and Management*, **142**(12), 04016060.

Thames Water (2014). Final Water Resources Management Plan 2015–2040. https://corporate.thameswater.co.uk/-/media/site-content/thames-water/corporate/aboutus/our-strategies-and-plans/water-resources/our-current-plan-wrmp14/wrmp14_section_0.ashx?la=en (accessed 16 February 2018).

van Schagen K., Rietveld L. C., Veersma A. and Babuška R. (2010). Control-design methodology for drinking-water treatment processes. *Water Science and Technology: Water Supply*, **10**(2), 121–127.

Villez K., Ruiz M., Sin G., Colomer J., Rosen C. and Vanrolleghem P. A. (2008). Combining Multiway Principal Component Analysis (MPCA) and clustering for efficient data mining of historical data sets of SBR processes. *Water Science and Technology*, **57**(10), 1659–1666.

Villez K., Rosen C., D'hooge E. and Vanrolleghem P. A. (2010). Online phase length optimization for a sequencing batch reactor by means of the Hotelling's T2 statistic. *Industrial & Engineering Chemistry Research*, **49**, 180–188.

Weber B., Cornel P. and Wagner M. (2007). Semi-centralised supply and treatment systems for (fast growing) urban areas. *Water Science and Technology*, **55**(1–2), 349–356.

Weijers S. R., De Jonge J., Van Zanten O., Benedetti L., Langeveld J., Menkveld H. W. and Van Nieuwenhuijzen A. F. (2012). KALLISTO: Cost effective and integrated optimization of the urban wastewater system Eindhoven. *Water Practice and Technology*, **7**(2), 1–9.

Chapter 11

Global sensitivity analysis in wastewater treatment modelling

G. Mannina[1,2], A. Cosenza[1], M. B. Neumann[3,4], P. A. Vanrolleghem[5,6]

[1]*Dipartimento di Ingegneria Civile, Ambientale, Aerospaziale, dei Materiali – Università di Palermo, Viale delle Scienze, 90128 Palermo, Italy*

[2]*Department of Earth and Environmental Engineering, Columbia University, 500 West 120th Street, New York, NY 10027, USA*

[3]*Basque Centre for Climate Change – BC3, Leioa, Spain*

[4]*Ikerbasque, Basque Foundation for Science, Bilbao, Spain*

[5]*modelEAU, Département de génie civil et de génie des eaux, Université Laval, 1065 av. de la Médecine, Québec (QC) G1V 0A6, Canada*

[6]*CentrEau, Water Research Centre of Université Laval, Québec, QC, G1V 0A6, Canada (giorgio.mannina@unipa.it)*

ABSTRACT

Global sensitivity analysis (GSA) is a valuable tool to support the use of mathematical models. GSA allows the identification of the effect of model and input factor uncertainty on the model response, also considering the effect due to the interactions among factors.

During recent years, the wastewater modelling field has embraced the use of GSA. Wastewater modellers have tried to transfer the knowledge and experience from other disciplines and other water modelling fields. GSA has been adopted for

Editors: Giorgio Mannina, George Ekama, Hallvard Ødegaard and Gustaf Olsson
doi: 10.2166/9781780409719_0363

several purposes (optimization, calibration, uncertainty assessment, etc.) with the final aim to improve the use of wastewater models. The purpose of this chapter is to provide the key issues surrounding GSA. Specifically, this chapter aims at identifying, for the most popular GSA methods, their potential use, the critical issues to be solved and current limitations.

Keywords: Computational burden, convergence, modelling, numerical methods, sensitivity analysis, water modelling

11.1 INTRODUCTION

Over the past 30 years the knowledge on wastewater treatment has enormously advanced and matured leading to the setting up of detailed principles for describing the key processes (biological, physical, chemical, hydraulic) occurring inside a wastewater treatment plant (WWTP) (Henze *et al.*, 2008). Over the years, these principles have been codified into mathematical models. The term WWTP model is often used to indicate the combination of activated sludge model (ASM), hydraulic model, oxygen transfer model and sedimentation tank model (Gernaey *et al.*, 2004).

The ASMs introduced by Henze *et al.* (2000) represent the most popular models describing the biological (and chemical) reactions taking place inside the activated sludge tanks. The hydraulic models describe the hydraulic conditions inside the tank and the flow rates (internal recycle, return sludge) throughout the different tanks. Regarding the settler tank, the most popular models suggest the description of the settler as an ideal point (without retention time) or as a one-dimensional layered settler (Takacs *et al.*, 1991). WWTP modelling has several scopes (Gernaey *et al.*, 2004): (i) learning, for example the adoption of a mathematical model can improve the knowledge on the process under study; (ii) design, for example the adoption of a mathematical model allows the comparison of different configuration solutions and design alternatives; (iii) process optimization and control, for example a mathematical model allows the prediction of the WWTP behaviour under different influent conditions and provides the best operation conditions to improve the effluent quality. The WWTP models include tens of model parameters (physical, kinetic, stoichiometric, influent fractionation etc.) for which no universal values exist, thus the correct values are required to be identified for each application. With this regard, the model calibration represents a key step in the mathematical modelling. The model calibration consists of the variation of model parameter values in view of identifying the set that provides the best fit between the model response and the set of data obtained from the WWTP under study (Gernaey *et al.*, 2004). The need to calibrate a model depends on the model purpose. For models aimed at educational purposes the default model parameter values can be adopted (Petersen, 2002). Conversely, in the case where the model is aimed at process optimization, an accurate description of the processes under study is required and consequently data collection and model calibration are needed (Petersen, 2002). Since WWTP models are often overparameterized with respect to the available measured data, the parameters that are required to be calibrated (in view of improving the model response) are not unique (Brun *et al.*, 2002). Therefore, the approaches and or procedures aimed at the model calibration often adopt sensitivity analysis (SA) in view of selecting the best parameter subset to be calibrated.

One of the most common definitions of SA has been provided by Saltelli *et al.* (2004) who define SA as: "*The study of how uncertainty in the output of a model (numerical or otherwise) can be apportioned to different sources of uncertainty in the model input*". This means that the SA allows identification of the model parameters and input variables whose values strongly influence the model response (model output). Conversely, it may be of interest to the modeller to see that although some model parameters may not be very well established they do not significantly contribute to output uncertainty.

SA can have several aims such as model calibration, model diagnosis, decision-making support, etc. (Mannina *et al.*, 2011a; Butler *et al.*, 2014; Norton, 2015; Mannina *et al.*, 2017). In the field of mathematical modelling, sensitivity analysis represents a very powerful tool as it provides information about how the variation in the outputs of the model can be apportioned to the variation of the model (input) factors (Saltelli, 2000). This can be of interest, for example to understand which of the uncertain (kinetic, stoichiometric or mass transfer-related) model parameters are responsible for most of the uncertainty in model predictions, for example ammonium effluent concentrations. It may then be possible to conduct dedicated experiments (e.g. batch experiments) to reduce uncertainty in the parameters that were found to be causing much uncertainty in the predictions (Dochain & Vanrolleghem, 2001).

Over the past 15 years, the engineering and scientific communities in the environmental water modelling field have improved knowledge on the use of sensitivity analysis (SA). SA has been identified as constituting a crucial step in any environmental modelling exercise (Jakeman *et al.*, 2006; Norton, 2008).

In the sensitivity analysis literature "Factors" is a term widely used and includes model parameters and model input variables. Saltelli (2000) singles out three main classes of sensitivity analysis methods:

- screening methods;
- local methods;
- global methods.

Screening methods are economical and qualitative methods. Local methods provide a measure of how the model output is affected by factor changes at a specific location in factor space by adopting a one-at-a-time (OAT) approach, indeed other factors are usually set to their nominal or mean values.

Global sensitivity analysis (GSA) on the other hand provides information on how the model outputs are influenced by varying factors across many possible locations over a large range of values (Homma & Saltelli, 1996; Saltelli *et al.* 2004).

GSAs can be conducted for several reasons: (i) to identify factors that mainly influence specific model outputs of interest (*factor prioritization*); (ii) to select which factors interact with other factors (*interacting factors*); and (iii) to identify non-influential factors (*factor fixing*) (Saltelli *et al.*, 2004). Other possible objectives of GSA not explored in this review include *factors mapping* (to search which factors are responsible for producing outputs in a certain region, e.g. above a threshold value) or *variance cutting* (identify a minimal subset of factors to fix, in order to obtain a prescribed reduction of uncertainty in the output).

By means of GSA, modellers are supported to identify critical regions in the factor space, to establish priorities for research and to simplify models (Saltelli *et al.*, 2008).

GSA methods are classified into (Saltelli *et al.*, 2008):

(i) global screening methods, for example Morris screening method (Morris, 1991; Campolongo *et al.*, 2007);
(ii) variance-based methods such as Extended Fourier Amplitude Sensitivity Testing (Extended-FAST) (Saltelli, 2000);
(iii) regression/correlation-based methods such as the standardized regression coefficients (SRCs) method (Saltelli *et al.*, 2008).

In the environmental modelling field, the majority of SA applications have been local and derivative-based due to the fact that these methods are computationally very efficient (Saltelli *et al.*, 2008). However, local sensitivity analysis (LSA) methods can be misleading in the case of highly uncertain input factors and non-linear relationships between model outputs and factors as they provide information only at the "nominal point". The main limits of LSA can be overcome by applying GSA. For a priori unknown model behaviour, the GSA should be the preferred method to apply (Saltelli & Annoni, 2010).

Due to the high computational cost of running simulations for environmental models, the dissemination of GSA applications has been limited (Campolongo *et al.*, 2007). Therefore, modellers have often been reluctant to use GSA methods instead of local methods (Saltelli & Annoni, 2010).

During recent years, modellers have started spending considerable time in understanding the potential of each GSA method applied to complex water engineering models (Mannina *et al.*, 2011b; Sin *et al.*, 2011; Benedetti *et al.*, 2011; Mannina *et al.*, 2014; Ramin *et al.*, 2014). An increasing trend of GSA applications has been found in literature (Figure 11.1). From Figure 11.1 one can observe that GSA applications have quickly spread in the water quality and hydrology fields with 321 and 140 documents in 2017. Moreover, an increasing interest in applications even in the wastewater field has occurred in literature (Figure 11.1).

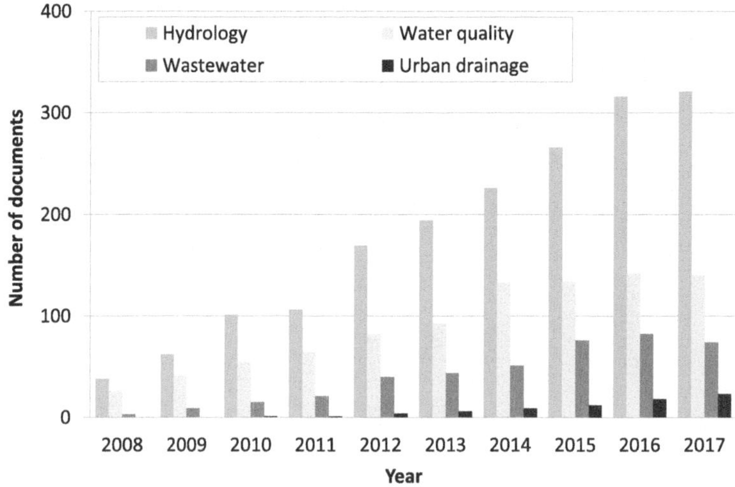

Figure 11.1 Trend of GSA applications found in literature during the last 10 years divided per field of interest (source Scopus, obtained by using the following search words: global sensitivity analysis and hydrology, wastewater, water quality, urban drainage).

Despite the clear advantages of GSA, several questions are still open on GSA: Can the terminology be standardized? Do all model applications require the same numerical settings for GSA? How should modellers test whether the SA has converged?

The purpose of this chapter is to highlight key issues surrounding GSA application in the wastewater modelling field.

Following this introduction, Section 11.2 discusses commonly applied SA methods divided into four main classes: (i) derivative-based LSA; (ii) regression-based; (iii) screening; and (iv) variance-based. Then, the main GSA applications in the wastewater field and some relevant applications in other water modelling fields are provided in Section 11.3. Issues related with the numerical settings in GSA applications (e.g. choice of the criteria for model factor screening in terms of importance or convergence analysis) are presented in Section 11.4. In Section 11.5 details related to the multiple use of GSA methods are presented. Section 11.6 reports a summary and outlook.

11.2 SENSITIVITY ANALYSIS METHODS

In this section a general overview of the features of the most commonly used SA methods is provided. Table 11.1 summarizes the features of each SA method: the ability to cope with interaction among factors and with non-linearity, the computational cost required to employ the analysis, the ability to provide

Table 11.1 Main features of SA methods.

		Features					Sensitivity Measures	
Class	**Method**	**Coping with Interaction**	**Coping with Non-Linearity**	**Factor Fixing**	**Factor Prioritization**	**Cost of Analysis**	**Symbol**	**Definition**
Local SA	Derivative-based	yes	no	no	yes	$2n+1$	$\delta y/\delta x$	Derivative of an output *Y* with respect to an input *X*
Regression-based	SRC	no	(no)	no	yes	500–1000	β_i	Standardized Regression Coefficient
Screening	Morris screening	yes	yes	yes	no	$r\times(n+1)$	μ	Mean of elementary effects
							μ^*	Mean of absolute elementary effects
							σ	Standard deviation of elementary effects
Variance-based	Extended-FAST	yes	yes	yes	yes	$N\times n$	S_i	First-order effect
							S_{Ti}	Total effect
	Sobol'	yes	yes	yes	yes	$N\times(n+2)$	S_i	First-order effect
							S_{Ti}	Total effect

Where: n = number of factors; r = number of trajectories, typically $4\le r\le 10$; N = number of repetitions, typically $500\le N\le 1000$.

information in terms of factors fixing (i.e. how many model factors do not affect the model outputs) and factors prioritization (i.e. how many model factors have a relevant effect on the model outputs variation). The symbols and the definitions in Table 11.1 are in accordance with Saltelli *et al.*, (2008).

In Figure 11.2 are reported the most commonly used sensitivity analysis methods, grouped into qualitative and quantitative methods (Hong & Purucker, 2018).

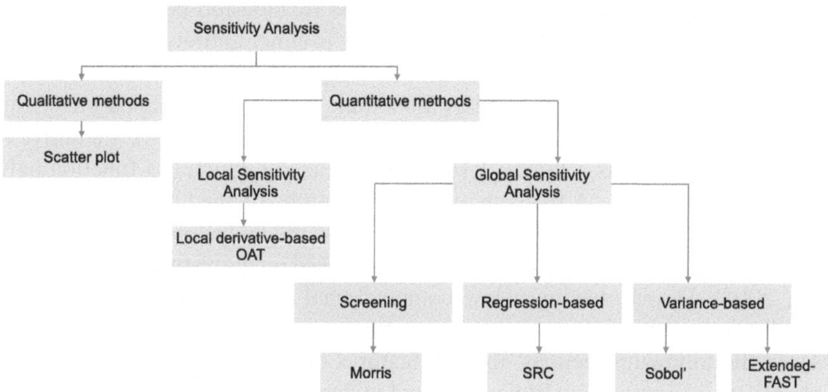

Figure 11.2 Most commonly used sensitivity analysis methods (modified from Hong & Purucker, 2018).

The scatter plot is a qualitative method which allows visual identification of the influence of model factors on model outputs and the possible (non) linear or (non) monotonic dependence between an input and output. Therefore, on the basis of this dependence the scatter plot method is usually adopted as a first step of sensitivity analysis to select the most adequate quantitative method. A disadvantage of the scatter plot method is that interpretation of a scatter plot is subjective (Frey & Patil, 2002).

Quantitative methods are classified into local and global. LSA methods allow identification of the relationship between inputs and outputs by changing the input value (model factor value) within a given point while other factors are usually set to their nominal or mean values by using an OAT approach (Figure 11.2).

GSA methods are model independent and therefore do not require a specific relationship between model factors and model outputs. The proper sampling of the entire parameter space with GSA allows handling of non-linear, non-monotonic and non-additive models. According to Saltelli *et al.* (2004) the application of regression/correlation based methods implies the assumption of model linearity. The application of the variance-based and global screening methods does not rely on special assumptions about the behaviour of the model (such as linearity, monotonicity and additivity of the relationship between input factors and model output) (Saltelli *et al.*, 2008). Specifically, the most common screening method, Morris screening, is mainly effective in identifying important model factors for monotonic models by adopting reasonable computational cost (Table 11.1). The adoption of variance-based methods (e.g. Sobol' or Extended-FAST) is mainly effective for selecting important model factors for all kinds of model structure (linear, non-linear, monotonic and non-monotonic).

For the sake of completeness, it is important to recall the definitions of linear, monotonic and additive models. Specifically, the relationship between Y and an input factor X is monotonic if the curve $Y = f(X)$ is either entirely non-decreasing or entirely non-increasing over all the interval of definition of X. Further, a mathematical model can be defined as monotonic if the rule of monotonicity can be applied for all factors of the model. A model $Y = f(X_1, X_2, \ldots, X_k)$ is additive if f can be decomposed as a sum of k functions f_i, each of which is a function only of the relative factor X_i. A mathematical model can be defined as linear if linear equations connect the model state variables and the model factors.

Most local SA methods used in literature are based on derivatives $\delta y/\delta x$ (Saltelli *et al.*, 2008). Despite having a low computational cost, the derivate-based SA methods can provide information of factor prioritization and interaction (Table 11.1). In the following sections, the key features of each SA method reported in Table 11.1 and Figure 11.2 will be discussed in detail.

11.2.1 Derivative-based

The derivative-based local sensitivity method represents the simplest type of SA. The core concept behind this method is the OAT perturbation of the input factors from their nominal values and the evaluation of the effect on the model outputs. This effect can be visually evaluated by comparing the model responses under different input factors perturbation or evaluating a sensitivity measure (Paton *et al.*, 2013). The sensitivity of the ith input factor on the model output y can be represented by the partial derivative $\delta y/\delta x$ evaluated at the nominal value of the factors. Since input factors may have different units of measurements, the partial derivatives are not comparable. In order to make them comparable they can be re-scaled by adopting a scaling factor. In terms of computational demand, this kind of method requires $2n + 1$ runs, where n represents the number of factors. Setting the size of the perturbation is a critical element in its application, leading to sensitivity to non-linearity or numerical inaccuracy (De Pauw & Vanrolleghem, 2006).

11.2.2 Regression-based

11.2.2.1 Standardized regression coefficients

The rationale of regression/correlation based methods is to perform Monte Carlo (MC) simulations of the model output by using a randomly sampled factor matrix.

Multivariate linear regression according to Equation 11.1 is then used relating model outputs (y) to the factors (x_i) (Saltelli *et al.*, 2008).

$$y = b_0 + \sum_{i=1}^{n} b_i \cdot x_i + \varepsilon \tag{11.1}$$

where n represents the number of factors, b_i are the regression slopes, and ε is the random error of the regression model. For each factor the standardized regression coefficient (SRC = β_i) of the multivariate linear model is calculated according to Equation 11.2.

$$\text{SRC}\ (x_i) = \beta_i = b_i \cdot \frac{\sigma_{xi}}{\sigma_y} \tag{11.2}$$

where σ_{xi} and σ_y represent the factor and the model output standard deviation, respectively. β_i is a valid measure of sensitivity if the coefficient of determination R^2, which indicates the portion of total variance explained by the regression model, is higher than 0.7 (Saltelli *et al.*, 2008). A high absolute value of β_i indicates a relevant effect of the related ith factor on the model output. The sign of β_i indicates the direction of change. More specifically, the positive sign of β_i indicates that increasing the value of the ith factor increases the model output and *vice versa* in the case of a negative sign (Saltelli *et al.*, 2004). In the case of a linear model, the coefficient of determination R^2 is equal to 1 (Saltelli *et al.*, 2008). The SRC method does not provide information about the interaction among factors. It allows identification of the important model factors (*factors prioritization*) if the model is not too far from being linear (Table 11.1).

In terms of computational demand, the regression/correlation based methods are feasible to be used even for complex models with tens of factors. Indeed, the application of these methods requires a limited number of MC simulations, typical numbers found in literature are between 500 and 1000 (Neumann, 2012). However, the regression/correlation-based methods explore only the first-order effects and do not provide any information about the interaction among factors. Thus, these methods can be used only for factor prioritization in cases when the effects of non-linearity are not too strong ($R^2 > 0.7$).

The SRC method has the advantage of being simple to apply. However, it cannot be applied when the relationship between model factors and model output is non-linear or non-monotonic and when interactions are present (Saltelli & Sobol', 1995).

11.2.3 Screening

11.2.3.1 Morris screening

The Morris screening method represents the most used method belonging to the class of global screening methods. It is based on OAT perturbation of the model factors under investigation (Morris, 1991). However, in comparison to the derivative based methods, the OAT perturbation is repeated at different locations within the factors range before being averaged. For each perturbation, the elementary effect (EE) is quantified. The EE represents the relative difference of the model output with, $y(x_1, \ldots, x_{i-1}, x_i + \Delta, x_{i+1}, \ldots, x_n)$, and without, $y(x_1, \ldots, x_n)$, a perturbation Δ of the ith factor (Equation 11.3).

$$\mathrm{EE}_i(x_1, \ldots, x_n, \Delta) = \frac{y(x_1, \ldots, x_{i-1}, x_i + \Delta, x_{i+1}, \ldots, x_n) - y(x_1, \ldots, x_n)}{\Delta} \tag{11.3}$$

The EE is repeatedly computed (r times) at different locations in the factor space (considering a p-level sampling grid). For each of the n factors, the measure of sensitivity is summarized by the mean (μ) and the standard deviation (σ) of the r EEs. The mean μ represents a measure of the importance of the factor in determining model output, whereas the standard deviation σ indicates whether the factor is responsible for introducing non-linearity or interactions (i.e. whether the sensitivity changes for different locations in the factor space) (Table 11.1). A high value of σ indicates that the model output variation is influenced by non-linearity or interactions. Thus, the Morris screening method is able to detect interactions among factors (Table 11.1). In order to avoid the problem that EEs of opposite sign cancel each other out, Campolongo *et al.* (2007) proposed using the mean of the absolute EEs (μ^*). The main objective of the Morris screening method is factors fixing: factors with a low value of μ^* or σ are considered non important and can be fixed anywhere in the factor space.

Further, as suggested originally by Morris (1991), the line $\mu_i^* = +2(\sigma_i/\sqrt{r})$, where $\sigma_i/\sqrt{r}$ represents the standard error of the mean, can be used to discriminate between interacting and non-interacting factors. Factors which lie below the line have a linear effect on the model outputs whereas factors above this line have a non-linear effect or are involved in interactions.

As suggested by Campolongo *et al.* (2007) the required number of simulations for the Morris screening application is equal to $r^*(n + 1)$. Typical numbers are $r = 10$–20 and $p = 4$–8 (Campolongo *et al.*, 2007).

11.2.4 Variance-based

Variance-based methods are founded on the variance decomposition theorem. The main interesting features of variance-based methods are: (i) independence of the model structure; (ii) capability to analyse the influence of each factor within its entire range; (iii) capability to quantify the interaction among factors; and (iv) groups of factors can be considered as single factors. However, the main disadvantage of these methods is their computational cost, as they require a large number of simulations per factor (500–1000 according to Saltelli *et al.*, 2005). In the application of variance-based methods, modellers are often interested in two sensitivity indices: the first-order effect index (S_i) and the total effect index (S_{Ti}) (Table 11.1).

The most frequently used methods are: Extended-FAST and Sobol' (Cukier *et al.*, 1973; Schaibly & Shuler, 1973; Sobol', 1993; Saltelli, 2002).

11.2.4.1 Sobol' indices

The core idea behind the Sobol' indices method is the decomposition of the function $f(x)$, which depends on n input factors $(x_1, \ldots, x_n)$, into summands of increasing dimensionality, namely:

$$f(x_1,\ldots,x_n) = f_0 + \sum_{i=1}^{n} f_i(x_i) + \sum_{i=1}^{n} \sum_{i\neq j}^{n} f_{ij}(x_i, x_j) + \cdots + f_{1,2,\ldots,n}(x_1,\ldots,x_n) \tag{11.4}$$

The integral of every summand of any of its own variables must be zero and all the summands in Equation 11.4 are orthogonal. Thus, the analysis of variance representation of $f(x)$ is based on the satisfaction of the condition of Equation 11.5.

$$\int f_{i_1,\ldots,i_n}(x_{i_1},\ldots,x_{i_n})\,dx_k = 0 \quad \text{for } K = i_1,\ldots,i_n \tag{11.5}$$

The Sobol' indices for a subset of model factors are defined as:

$$S_{i_1,\ldots,i_n} = \frac{D_{i_1,\ldots,i_n}}{D} \tag{11.6}$$

where $D_{i_1,\ldots,i_n}$ is the partial variance of $f_{i_1,\ldots,i_n}(x_{i_1},\ldots,x_{i_n})$ and D is the total variance.

For example, $S_i = D_i/D$ provides the first-order contribution from the ith input parameter to the output variance and $S_{ij} = D_{ij}/D$ is used to compute the second-order contribution from interaction between the ith and jth parameters. Finally, total effect sensitivity indices, which are defined as the sum of all the sensitivity indices as $S_{Ti} = S_i + S_{ij} + \cdots + S_{1\ldots i\ldots n}$ quantify the overall effects of one factor on the model output. The required number of simulations is $N*(n + 2)$; where N is the number of repetitions and n the number of model factors.

11.2.4.2 Extended-FAST

The Extended-FAST method is also based on the variance decomposition. Differently to Sobol' the Extended-FAST evaluates sensitivity independently for each factor using just one simulation because all the terms in a Fourier expansion related to Equation 11.4 are mutually orthogonal.

In the Extended-FAST method S_i, computed according to Equation 11.7, measures how the ith factor contributes to the variance of the model output without taking into account the interactions among factors.

$$S_i = \frac{\mathrm{Var}_{x_i}(E_{x_{-i}}(Y \mid x_i))}{\mathrm{Var}(Y)} \tag{11.7}$$

where E indicates the expectancy operator and Var the variance operator. According to the notation used by Saltelli *et al.* (2004) the subscripts indicate that the operation is either applied "over the ith factor" X_i, or "over all factors except the ith factor" X_{-i}.

The total effect index S_{Ti} is computed according to Equation 11.8 considering all potential interactions with the ith factor.

$$S_{Ti} = 1 - \frac{\mathrm{Var}_{x_{-i}}(E_{x_i}(Y \mid x_{-i}))}{\mathrm{Var}(Y)} \tag{11.8}$$

The difference between S_{Ti} and S_i quantifies the interaction among factors (S_{Si}). Variance-based methods allow identification of important factors (high S_i) (factors prioritization).

Only if S_{Ti} is small can the factor be fixed anywhere within its range of uncertainty (factor fixing). Indeed, if the S_i value is small, it does not necessarily mean that the factor may be fixed because a high S_{Ti} value would indicate that the factor is involved in interactions.

11.3 GSA APPLICATIONS FOR WASTEWATER ENGINEERING

In Table 11.2 the main relevant studies on GSA related to the wastewater field are summarized. Further, Table 11.2 contains key elements of relevant GSA applications in other modelling fields.

In the wastewater modelling field the SRC method has often been adopted for: (i) factors prioritization (among others, Corominas & Neumann, 2014; Wagner *et al.*, 2016; Mannina *et al.*, 2017); and (ii) uncertainty analysis (Benedetti *et al.*, 2011; Sin *et al.*, 2011). Sin *et al.* (2011) applied the SRC method in view of uncertainty analysis of a model of a conventional activated sludge system in which three different scenarios were analysed. In particular, the following scenarios were analysed by Sin *et al.* (2011): scenario 1: stoichiometric, kinetic and influent fractionation parameters are uncertain; scenario 2: hydraulics and mass-transfer factors are uncertain; and scenario 3: stoichiometric, kinetic and influent fractionation parameters, hydraulic and mass-transfer related parameters are uncertain. For each scenario, different model inputs (such as bio-kinetic model parameters, influent fractions, mass-transfer parameters and the like), were considered to be either uncertain or known. The study of Sin *et al.* (2011) was aimed at selecting the most important factors that contribute to the uncertainty of wastewater treatment performance criteria (e.g. effluent quality, sludge production and energy consumption). They found a high ability of the SRC method in identifying the main sources of uncertainty and quantifying their impact on process performance criteria. Conversely, Chen *et al.* (2012) found that for more complex wastewater systems such as membrane bio-reactors the SRC method can

Table 11.2 Literature overview on the main GSA applications.

Method	Convergence Analysis	Number of Factors	Number of Simulation Runs	Threshold Value	Reference	Research Field
SRC	no	26	500	$\|\beta_i\| > 0.1$ important	Sin *et al.* (2011)	Wastewater
SRC		7	500	$\|\beta_i\| > 0.1$ important		
SRC		33	500	$\|\beta_i\| > 0.1$ important		
SRC	yes	15	600–2250	n.a.	Benedetti *et al.* (2011)	
Extended-FAST	no	32	n.a.	$S_i > 0.05$ important	Chen *et al.* (2012)	
Morris Screening	yes	17	90–1260	n.a.	Ruano *et al.* (2012)	
RSA	no	20	10,000	n.a.	Mannina *et al.* (2010)	
SRC	no	79	800	$\|\beta_i\| > 0.1$ important	Mannina *et al.* (2011b); Cosenza *et al.* (2013a)	
Morris Screening			800	$\mu^* > 0.1$ important		
Extended-FAST			39,500	$S_i > 0.01$ important; $S_{Ni} < 0.6$ non-influential		
Extended-FAST	no	79	39,500	$S_i > 0.01$ important; $S_{Ti} < 0.1$ non-influential	Cosenza *et al.* (2014)	
SRC	no	26	500	$\|\beta_i\| > 0.1$ important	Ramin *et al.* (2014)	
Morris Screening			405	$\mu^* > 0.1$ important		
Sobol'	yes	10	5000	n.a.	Cogan and Cellam (2017)	
SRC	yes	33	1200	$\beta_i^2 > 0.1$ important	Mannina *et al.* (2017)	
Morris Screening	yes	14	300	Factors which lay outside the two lines: $\mu_i = \pm 2SEM_i$	Robles *et al.* (2014)	
Morris Screening	no	63	–	$\mu^* > 0.1$ important	Valverde-Pérez *et al.* (2015)	
SRC	no	32	1000	$\|\beta_i\| > 0.1$ important	Wagner *et al.* (2016)	

(*Continued*)

Table 11.2 Literature overview on the main GSA applications. (*Continued*)

Method	Convergence Analysis	Number of Factors	Number of Simulation Runs	Threshold Value	Reference	Research Field
Sobol'	yes	5	36,000	n.a.	Yang (2011)	Relevant applications in other water modelling field
Morris Screening			12,000	n.a.		
Linear regression			3000	n.a.		
RSA			5000	n.a.		
SDP			512	n.a.		
RA	yes	16	3000	n.a.	Gan *et al.* (2014)	
Morris Screening			2800	n.a.		
MARS			3000	n.a.		
FAST			2777	n.a.		
McKay-1			3000	n.a.		
McKay-2			2890	n.a.		
Sobol'			3000	n.a.		
Sobol'	yes	5	420,000	n.a.	Sarrazin *et al.* (2016)	
		13	600,000	$S_{Ti} < -0.0084$ non-influential		
		50	520,000	$S_{Ti} < -0.004$ non-influential		
		5	78,000	n.a.		
Morris Screening		13	560,000	$\mu^* < 0.026$ non-influential		
		50	102,000	$\mu^* < 0.027$ non-influential		
		5	10,000	n.a.		
RSA		13	20,000	n.a.		

		50	30,000	n.a.	
OAT	no	6	85	n.a.	Sun *et al.* (2012)
RSA			10,000	n.a.	
Morris Screening			10,000	n.a.	
SRC	no	42	300	n.a.	Reder *et al.* (2017)
PCC					
Spearman					
RSA	yes	6	30,000	n.a.	Dotto *et al.* (2012)
RSA	no	17	10,000	n.a.	Mannina and Viviani (2010)
RSA	no	28	10,000	n.a.	Freni *et al.* (2009)
SRC	no	17	1000	$\lvert\beta_i\rvert > 0.1$ important	Mannina *et al.* (2012)
Morris Screening			540	$\mu^* > 0.1$ important	
Extended-FAST			8500	$S_i > 0.01$ important; $S_{Ti} - S_i < 0.1$ non-influential	
SRC	no	19	500	$\beta_i^2 > 0.05$ important	Corominas and Neumann (2014)
Morris Screening			200	$\mu^* < 0.1$ non-influential	
SRC			4000	$\lvert\beta_i\rvert > 0.1$ important	
Morris Screening	yes	17	36,000	$\mu^* > 0.1$ important	Vanrolleghem *et al.* (2015)
Extended-FAST			85,000	$S_i > 0.01$ important; $S_{Ti} - S_i < 0.1$ non-influential	

Where: RSA = Regionalized Sensitivity Analysis; RA = Regression analysis; McKay-1 = McKay main effect analysis; McKay-2 = McKay two-way interaction analysis; MARS = Multivariate adaptive regression splines screening; SDP = State-Dependent Parameter; OAT = One-at-A-Time; PCC = Pearson correlation coefficients; SRC = Standardized Regression Coefficient; Extended-FAST = Extended Fourier Amplitude Sensitivity Testing; n.a. not available.

lead to erroneous results, due to non-linearity. Studying a similar system, Cosenza *et al.* (2013a) found that the SRC method provided similar results to the Extended-FAST method in terms of factors prioritization despite the presence of non-linearity. This result has peculiar interest. Indeed, if the modeller knows to deal with a non-linear model and is interested only in factor prioritization he/she can apply the less computationally-demanding method (SRC) without creating errors. Indeed, Mannina *et al.* (2017) have recently successfully applied the SRC method for the factors prioritization of a complex ASM-based model including greenhouse gases as state variable, despite the average R^2 model output being equal to only 0.5. Conversely, Cosenza *et al.* (2013a) found a low similarity in results obtained by applying the Morris screening and Extended-FAST methods, mainly due to a convergence problem.

Despite the fact that the core aim of the Morris screening method is to provide information in terms of factors fixing, it has often been applied in the wastewater field with the aim of factors prioritization thanks to the qualitative knowledge on the factor interactions provided by σ.

Few variance-based method applications are found in the wastewater modelling field mainly due to their higher computational cost compared to other GSA methods. Applications include Cosenza *et al.* (2014) who found significant interactions among the model factors of the ASM2d–SMP model (Cosenza *et al.*, 2013b). Contrary to previous GSA studies for ASMs, Cosenza *et al.* (2014) found the relationship between variables and factors to be non-linear and non-additive.

Some GSA applications derived from other water modelling fields merit being cited here for their peculiarities, which could be of useful interest for future applications in the wastewater field.

Vanrolleghem *et al.* (2015) have applied three GSA methods (SRC, Extended-FAST and Morris screening) to an urban drainage stormwater quality–quantity model. They discussed the issue of sensitivity convergence as further explored in Section 11.5. Vanrolleghem *et al.* (2015) found a considerable deviation of the results of Morris screening from that of the other methods, demonstrating that for their model the best method was the Extended-FAST. However, they underlined the need for using multiple GSA methods and multiple objectives in view of increasing the robustness of the results as previously performed by Corominas and Neumann (2014).

Neumann (2012) compared five SA methods (derivative-based local sensitivity analysis, Morris Screening, SRC, Extended-FAST and an entropy-based method) applied to micropollutants removal in a drinking water model. The author found the same parameter ranking results for the different methods. However, for chemicals leading to high non-linearity, the approximation of first-order effect indices using the local or regression-based methods was poor, leading to different classifications.

Reder *et al.* (2017) have recently applied GSA to a water quality model aimed at evaluating the river water quality in Africa. Specifically, they applied the lowest computationally demanding GSA methods: SRC, Pearson correlation coefficients (PCC) and the Spearman rank correlation coefficient. Despite obtaining reasonable results using a sampling size larger than that recommended in the literature, they pointed out the need to investigate and compare different methodologies. This latter statement embraces all fields of GSA applications since no best GSA method exists and different methods and sampling strategies have to be applied to obtain reasonable results (Pianosi *et al.*, 2016).

Recently Sarrazin *et al.* (2016) presented a research paper focusing on key issues such as the adequate sample size, convergence and the choice of the screening threshold. Sarrazin *et al.* (2016) applied three GSA methods to three hydrological models of increasing complexity (with 5, 13 and 50 model factors). Further, as discussed below, they also proposed a method for a quantitative assessment of the convergence of the sensitivity analysis. They demonstrated the importance of assessing convergence analysis with respect to the objective of the modeller, for example to obtain stable sensitivity indices, a ranking or a screening.

11.4 NUMERICAL SETTINGS

One of the crucial aspects in the GSA applications is the sample size to be adopted in view of obtaining reliable results. Some relevant GSA applications have demonstrated that the sample size suggested in Saltelli *et al.* (2008) can be too low and inadequate to obtain reliable results (Vanrolleghem *et al.*, 2015; Sarrazin *et al.*, 2016).

As recently demonstrated by Sarrazin *et al.* (2016), no relationship exists between the number of model factors and the sample size required to reach convergence. The number of model simulations to be run for carrying out the sensitivity analysis depends on the choice of the sampling repetitions which is therefore crucial. For complex models (large model factors) the sample size can be the bottleneck of the methodology (long calculation time for running the requested model runs).

The sample sizes suggested in literature are often insufficient; therefore, a convergence analysis would be required to ensure that stable results are obtained given the limited sample size. However, a dilemma exists: complex models are likely to be computationally expensive and therefore require sophisticated (again expensive) GSA methods making a convergence analysis unfeasible. In Section 11.4.3 a convergence pursuing numerical method is proposed to obtain reliable results at minimal computational cost.

11.4.1 Open issues

A generally accepted common GSA terminology, permitting ease of comparison between the methods, is still lacking (Mannina *et al.*, 2012; Vanrolleghem *et al.*, 2015).

"Sensitivity indices" provide a measure of the importance of a model factor in determining variation in the model output. When applying sensitivity analysis, one of the most relevant parameters that has to be employed is the so-called "cut-off threshold". The cut-off threshold is a subjective number which allows classification of model factors (i.e. factors classified as being important or non-influential). Therefore, the "cut-off threshold value" of the sensitivity indices for establishing whether a model factor is important or not has a key role in applying GSA (see below Section 11.4.2).

There is missing terminology for comparing results in view of classification (establishing an order of importance of model factors) and for comparing cut-off threshold values for sensitivity indices obtained with different methods.

Another issue surrounding GSA, for which no consensus still exists, is the convergence of the results obtained for different numbers of simulations. The lack of uniformity in the definition of convergence makes it difficult to compare the results obtained for models of different complexities when using different GSA methods (Sarrazin *et al.*, 2016).

In Figure 11.3, a first attempt to standardize the GSA terminology for comparability of results among three classes of methods (i.e. SRC, Morris screening and Extended-FAST) is provided (Vanrolleghem *et al.*, 2015).

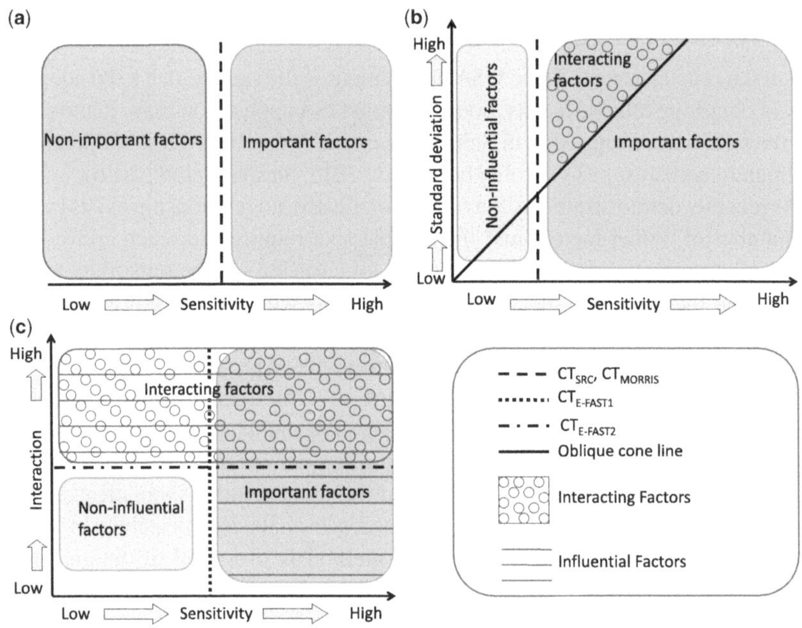

Figure 11.3 Schematic overview of the terminology used for classifying SRC (a), mean (b), first-order effect (c) according to different GSA methods: SRC (a), Morris screening (b) and Extended-FAST (c) (Vanrolleghem *et al.*, 2015).

On the basis of a cut-off threshold (CT) for the sensitivity indices it is possible to classify factors according to their importance (Vanrolleghem *et al.*, 2015). As detailed below, each method provides different sensitivity indices which provide the entity of the importance of each model factor.

The SRC method allows calculation of the standardized regression coefficient (SRC) or slope (β_i) (obtained with respect to the ith model factor and the model output standard deviation – see Section 11.2.2.1) of the multivariate linear regression model. A CT threshold (CT_{SRC} for β_i) has to be defined in view of classifying model factors in terms of importance. Once CT_{SRC} is defined, on the basis of its value, factors can be classified in the following way (Figure 11.3a):

- *important factors*: if SRC coefficient > CT_{SRC};
- *non-important factors*: if SRC coefficient < CT_{SRC}.

In the case of linear models the non-important factors can be fixed anywhere in their space. In the case of non-linear models, non-important factors could be involved in interactions with other factors thus they cannot be fixed (see below).

In the case of the Morris screening method, two indices are calculated. Specifically, the absolute mean (μ*) and the standard deviation of the elementary effect (difference with and without factor variation) computed at different points of the input space. These values provide information about the importance of model factors. A CT threshold (CT_{MORRIS} for μ*) has to be established in view of classifying model factors in terms of importance.

By means of the Morris screening method three different types of factors can be distinguished with respect to the mean and the standard deviation of the elementary effect function (Figure 11.3b):

- *important factors*: if absolute mean sensitivity > CT_{MORRIS};
- *interacting factors*: if absolute mean sensitivity > CT_{MORRIS} and the standard deviation is above a specified cone line;
- *non-influential factors*: if absolute mean sensitivity < CT_{MORRIS}.

In particular, the Morris screening method as modified by Campolongo *et al.* (2007) basically defines a cone whose edges are set by a CT_{MORRIS} and an oblique line called the core line, corresponding to two times the standard error of the sensitivity index (Figure 11.3b) (quantitative characteristics are given below).

In the case of the Extended-FAST method, the ith factor contribution to the variance of the model output with and without taking into account the interactions among factors (S_{Ti} and S_i, respectively) are considered as sensitivity indices. The difference between S_{Ti} and S_i represents the contribution of the ith factor in terms of interaction. A CT for each of the indices ($CT_{E\text{-}FAST1}$ and $CT_{E\text{-}FAST2}$ for S_i and $S_{Ti} - S_i$, respectively) has to be established in view of classifying model factors in terms of importance.

The Extended-FAST distinguishes three classes of factors on the basis of two CTs ($CT_{E\text{-}FAST1}$ and $CT_{E\text{-}FAST2}$) (Figure 11.3c):

- *important factors*: if $S_i > CT_{E\text{-}FAST1}$;
- *interacting factors*: if $S_{Ti} - S_i > CT_{E\text{-}FAST2}$;
- *non-influential factors*: if $S_i < CT_{E\text{-}FAST1}$ and $S_{Ti} - S_i < CT_{E\text{-}FAST2}$.

Non-influential factors which can be identified by both the Morris screening as well as the Extended-FAST method can be fixed anywhere within their variation range without changing the model output variance. With respect to this, it is important to stress once more that non-important does not imply non-influential.

11.4.2 Cut-off criteria for factors classification

In order to classify factors as being important or non-influential or interacting, cut-off thresholds of the sensitivity indices values (CTs) are required. The choice of the cut-off thresholds is arbitrary and depends on the objective of the modeller and on the model structure. For example, in the case where the modeller is interested only in the factors that contribute significantly to the model variability, a high cut-off threshold is selected. A general criterion to select the cut-off thresholds is that the CT value has to be effective in view of differentiation of model factors (e.g. as important or non-influential or interacting). Therefore, a value that classifies all involved factors as either important or non-influential is not suitable. Indeed, this latter case would make the GSA application un-useful, especially when the modellers' aim is to select the factors that need to be calibrated.

In the following, CT-values typically found in literature are reported (see also Table 11.2).

A CT value equal to 0.1 is suggested to select important factors, for the SRC ($|\beta_i|$) and Morris screening (μ^*) methods (CT_{SRC} and CT_{MORRIS}, respectively). This latter value is equivalent to a CT value of 0.01 for S_i obtained in the Extended-FAST method ($CT_{\mathrm{E\text{-}FAST1}}$) (since for linear models $S_i = \beta_i^2$) (Sin *et al.*, 2011; Neumann, 2012; Cosenza *et al.*, 2013a; Vanrolleghem *et al.*, 2015).

For the Extended-FAST method a CT of 0.1 ($CT_{\mathrm{E\text{-}FAST2}}$) was chosen by Vanrolleghem *et al.* (2015) for the threshold value of the interaction (i.e. $S_{Ti} - S_i$). Neumann (2012) chose a CT of 0.05 for S_{Ti} and μ^* in order to identify non-influential factors (factors with S_{Ti} (or μ^*) < 0.05 were considered to be non-influential).

Sarrazin *et al.* (2016) have proposed the method presented below to validate the screening results after applying GSA methods, which could provide support to the modeller to investigate the adequacy of the chosen screening threshold.

11.4.3 GSA applications dealing with convergence analysis

Convergence analysis represents another crucial aspect when dealing with GSA. The high time demand required for convergence analysis is often contrasting with the need of a modeller to achieve results as fast as possible. Therefore, very few quantitative criteria exist for implementing convergence analysis in the wastewater field.

Benedetti *et al.* (2011) proposed a method to reduce the computational cost of Monte Carlo based GSA methods. The authors used two criteria (the model output variability and the stability of the set of factors classified as important as the number of iterations increases) to select the minimum number of simulations to be performed. They found that depending on the analysed variable the results of the convergence analysis vary, highlighting that the achievement of convergence is strongly dependent on the model output considered.

Ruano *et al.* (2012) investigated the convergence of the Morris screening for a wastewater treatment plant model. They proposed a criterion (the position factor) for establishing the achievement of convergence. By increasing the number of replicates (r) of the OAT sampling from 5 to 70, Ruano *et al.* (2012) analysed the average variation of the sum of the rank of the model factors (position factor). Ruano *et al.* (2012) found that the optimal number of replicates was 60–70, which is considerably higher than recommended by Saltelli *et al.* (2008) (namely 4–10 as reported in Table 11.1). The work of Ruano *et al.* (2012) confirmed that the modeller must take care in employing "default" numerical settings proposed in literature especially in the case of complex models.

For the sake of completeness, some quantitative criteria adopted in other water modelling fields are reported below. These criteria could represent the starting point for future convergence analysis applications in the wastewater field.

Recently, Sarrazin *et al.* (2016) have defined quantitative criteria to assess convergence of sensitivity indices, ranking and screening. They propose to assess the maximum width of the confidence intervals of the sensitivity index ($Stat_{\text{indices}}$) as a summary statistic to assess converge (Equation 11.9).

$$Stat_{\text{indices}} = \max(S_i^{ub} - S_i^{lb}), \quad i = 1, \ldots, n \tag{11.9}$$

where S_i^{ub} and S_i^{lb} are the upper and lower bounds of the sensitivity index of ith input factor, respectively; n represents the number of the input factors. The convergence is achieved when the $Stat_{\text{indices}}$ value is close to zero. Regarding the convergence in terms of ranking, Sarrazin *et al.* (2016) have proposed a weighted rank correlation coefficient ($\rho_{s,j,k}$) (Equation 11.10).

$$\rho_{s,j,k} = \sum_{i=1}^{n} \left| R_i^j - R_i^k \right| \frac{\max_{j,k}(S_i^j, S_i^k)^2}{\sum_{i=1}^{M} \max_{j,k}(S_i^j, S_i^k)^2} \tag{11.10}$$

where S_i^j and S_i^k relate to the ith input factor evaluated at the jth and kth resample of the model factors, respectively; while, R_i^j and R_i^k are the related rank positions. Sarrazin *et al.* (2016) suggested that the convergence can be considered reached when the 95% quantile value of the $\rho_{s,j,k}$ distribution (obtained over all possible resamples) is equal to one. Regarding the convergence of the input factor screening, Sarrazin *et al.* (2016) have proposed as a summary statistic ($Stat_{\text{screening}}$) the maximum width of the 95% confidence intervals across the sub-set (X_0) of the input factors that can be considered "low-sensitivity" (having sensitivity index value lower than a fixed threshold) (Equation 11.11).

$$Stat_{\text{screening}} = \max(S_i^{ub} - S_i^{lb}),\ x_i \in X_0 \tag{11.11}$$

The screening convergence can be considered reached when $Stat_{\text{screening}}$ is close to zero.

In the urban wastewater field Vanrolleghem *et al.* (2015) proposed a quantitative method for testing the convergence achievement for SRC, Morris screening and Extended-FAST methods. The core idea of the method, is to evaluate the variability of the sensitivity indices (SC) as the number of simulations increases. In particular, for each model output *j*, the S_{SCj} is computed as the sum of the sensitivity *SC* of all factors normalized with respect to the number of factors (*n*).

$$\mathrm{S}_{\mathrm{SCj}} = \frac{\sum_{i=1}^{n} SC_{i,j}}{n} \tag{11.12}$$

The variability of S_{SCj} is analysed for increasing numbers of Monte Carlo simulations (n_{MC}). More specifically, for the *j*th model output the variability of S_{SCj} is expressed as the percentage of change of S_{SCj} from n_{MCi-1} to n_{MCi} (Equation 11.13).

$$\text{Variability} = \left[\frac{\left(\sum_{i=1}^{n} SC_{i,j}\right)_{n_{MCi-1}} - \left(\sum_{i=1}^{n} SC_{i,j}\right)_{n_{MCi}}}{n}\right] \cdot 100 \tag{11.13}$$

According to this method, the convergence can be considered achieved if Variability (Eq. 13) stays within a selected precision threshold.

In Figure 11.4 an example of a convergence analysis for two different types of models (simplified and complex) is reported (Vanrolleghem *et al.*, 2015). The results are related to the application of SRC, Morris screening and Extended-FAST methods. A variability threshold of the results equal to +/– 3% is considered for evaluating convergence (Figure 11.4) of the two models. The simple model is a rainfall-runoff urban drainage model (Mannina & Viviani, 2010) with five model factors. In Figure 11.4a–c two model outputs are reported (i.e. O1 and O2). The complex model is able to simulate processes related to quantity and quality inside sewer systems and includes 17 model factors. For the sake of conciseness only two model outputs of the complex model are considered in Figure 11.4d–f (i.e. O3 and O4).

From Figure 11.4 one may observe that the sensitivity index of a simple model's outputs with the simple method achieves convergence at a lower number of simulations than with the complex methods. For example, for SRC convergence is achieved with 1000 simulations (Figure 11.4a). Conversely, for the complex model, convergence is achieved at 4000 simulations (Figure 11.4d). Moreover, by analysing Figure 11.4 one may observe that, especially for the complex model, sensitivity indices of the two model outputs do not achieve convergence simultaneously.

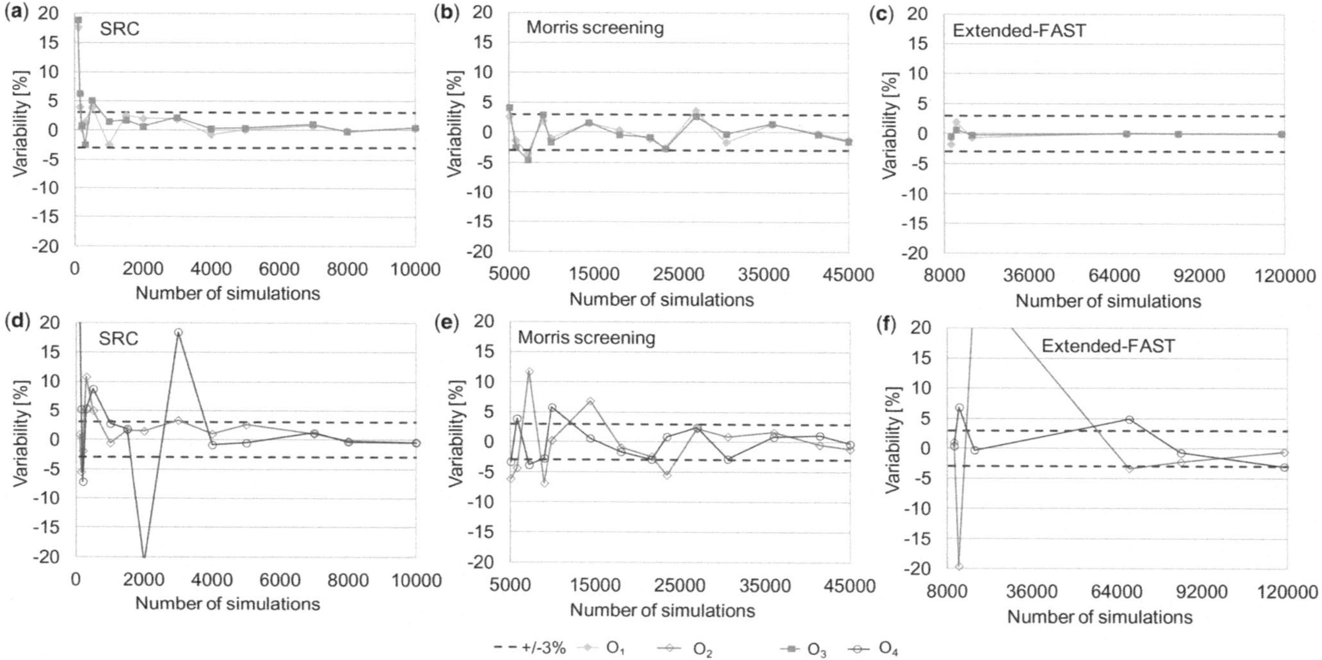

Figure 11.4 Quantitative convergence analysis with increasing sample size (expressed as number of simulations) versus variability (according to Equation 11.7) for different mathematical models and different GSA methods: simple model (a–c) and complex model (d–f) (modified from Vanrolleghem *et al.*, 2015).

11.5 USING MULTIPLE GSA METHODS

11.5.1 Comparison studies

The specific key features of each method compared to the others can be analysed by applying different methods to the same model. Cosenza *et al.* (2013a) have compared the SRC, Morris screening and Extended-FAST methods for a membrane bioreactor wastewater treatment model. They showed a poor similarity between Morris screening and Extended-FAST results in terms of classification into influential/non-influential factors. In terms of identifying important factors Mannina *et al.* (2012) obtained a good agreement applying SRC, Morris screening and Extended-FAST methods to a simple model. In the hydrological field, slight differences of factors ranking were obtained by Yang (2011), for the application of regionalized sensitivity analysis and non-parametric smoothing methods to the HYMOD model. For the regionalized sensitivity analysis method, the different results were debit to the dependence on the choice of the adopted filtering criterion to separate parameter sets into behavioural and non-behavioural. For the non-parametric smoothing method despite the convergence achievement Yang (2011) debited the different results to the low number of simulations performed compared to the other methods.

The results of the comparison studies obtained by Mannina *et al.* (2012) (simple model) and Cosenza *et al.* (2013a) (complex model) are reported in Figure 11.5. More specifically, the Venn diagram of Figure 11.5 shows the important factors selected by applying SRC, Morris screening and Extended-FAST methods to a simple (Figure 11.5a) and a complex model (Figure 11.5b).

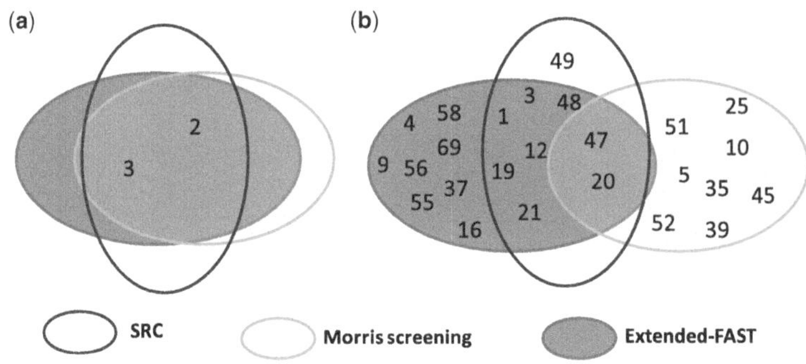

Figure 11.5 Venn diagram related to the comparison of important factors obtained by applying SRC, Morris screening and Extended-FAST methods for a simple model (a) and a complex model (b); numbers represent the factor order index; adapted from Mannina *et al.* (2012).

By analysing Figure 11.5, one may observe that for the case of the simple model the three methods provide the same results in terms of important factors (Figure 11.5a). Conversely, for the complex model a poor similarity is shown in Figure 11.5b.

The findings deduced from one GSA method application are often affected by the subjectivity of the modeller (e.g. cutting threshold for selecting important factors; number of runs; etc.) during the analysis, thus limiting the transferability of the knowledge to other models.

Despite the huge efforts aimed at understanding the peculiarities of each GSA method "*no ideal GSA method exists for all case studies*" (Pianosi *et al.,* 2016).

11.5.2 Sequential use

The Morris screening method is often suggested as a first step to discard non-influential factors and then pursue the GSA by applying a more elaborate method with only the selected potentially influential factors. However, care needs to be taken to assess the convergence for the Morris screening (among others, Cosenza *et al.,* 2013a).

In case of computationally demanding complex models, a general recommendation is that of applying a screening method to discard non-influential factors before applying more demanding methods such as Extended-FAST. Indeed, this is a general recommendation in GSA literature (Saltelli *et al.,* 2008). However, to ensure that this procedure works one must ascertain that no factors are eliminated wrongly in the screening. Examples of such approaches are reported in Sun *et al.* (2012) who suggested using, in the case of a model with a large number of factors, a two-step procedure including first a factors screening step (by using a local method) followed by a GSA step to be applied only to the important factors identified during the first step. For complex models, Gan *et al.* (2014) suggested first using approximate methods (e.g. SRC or local methods) for a rough factor screening and then applying more demanding methods such as Extended-FAST or Sobol's indices.

The problem of the high computational burden can be managed by adopting the meta-model approach, typical of the hydrological field (Song *et al.,* 2015). The core idea behind this approach is the adoption of statistical or experimental design methods as surrogate models (emulators) to assess the relationship between model factors and state variables. However, despite the fact that this approach has often been used in the hydrological field several issues still remain, for example the truthfulness of the meta-models with respect to the real process that has to be described.

11.5.3 Complementary use

The choice of GSA method to apply is not always clear a priori due to unknown model behaviour and the unknown relationship between model outputs and factors. Only by applying different methods simultaneously, can a robust assessment of the impact of model factors on the model outputs and the model's degree of non-linearity be obtained (Neumann, 2012; Corominas & Neumann, 2014).

With this in mind, the complementary use of GSA methods can provide support to the modeller. The complementary use of GSA methods can be performed with two aims:

(1) *Increase robustness of analysis*: if the modeller uses four GSA methods and three are providing similar results and one is not then he/she has to check what is going on;
(2) *Studying multiple objectives*: since not all methods provide information for all objectives, the modeller has to apply complementary methods.

Regarding the multiple objectives, Corominas and Neumann (2014) have applied the SRC and the Morris screening methods to an urban wastewater system model in order to understand how the operational variables of the system (including WWTP) influence the receiving water quality. Specifically, two GSA methods have been applied to identify important operational variables (i.e. SRC) as well as variables that do not influence (i.e. Morris screening) river water quality. The authors have confirmed the recurring statement in the field of water modelling that no ideal GSA method exists and that the combination of several methods can improve the results.

However, despite the advantage of the complementary use of different GSA methods several questions remain to be addressed, such as: Is this really an effective practice for overcoming the computational burden? A way to provide an answer is to test how the results of the sensitivity analysis change after performing a complementary use of GSA methods.

11.6 SUMMARY AND OUTLOOK

This chapter has outlined the application and issues surrounding the use of GSA in the wastewater modelling field. Sometimes applications or criteria from other water modelling fields have been discussed in view of providing the reader with a general overview on GSA. The following summary points can be deduced from this chapter:

(1) The adoption of GSA methods for WWTP modelling has several advantages with respect to local methods. One of the most significant advantages is the capability of GSA methods to take into account interactions among factors. This capability makes the selection of important model factors more precise than local methods. Consequently, the model predictions are more accurate. Moreover, the possibility of identifying the interactions/relationships among the model factors may also reduce the amount of model factors (stoichiometric, kinetic or fractionation) that need to be assessed during the experiments or allow a modeller to better interpret the measured data.

(2) Since the WWTP models often contain tens of model parameters, the sequential use of several GSA methods represents a way for reducing the high computational demand typical of these methods. Therefore, the first adoption of a screening method to discard non-influential factor before applying more demanding methods (such as Extended-FAST) is suggested.
(3) No single GSA method is ideal for all WWTP models. Therefore, the adoption of multiple methods is suggested in view of selecting the best method to be adopted on the basis of the model structure (linear or non-linear, monotonic or non-monotonic). However, a systematic approach to verify the effectiveness of running multiple GSA methods is also required.
(4) Some aspects related to GSA still require further investigations in view of favouring the adoption of GSA in the WWTP modelling field and of improving results of future applications: (i) setting-up of quantitative criteria to assess convergence of GSA results; (ii) establishing a methodology to support modellers for the identification of an adequate screening threshold to be adopted; and (iii) adopting a standardized nomenclature/terminology related to convergence and sensitivity thresholds facilitating comparability among methods.

11.7 ACKNOWLEDGEMENTS

This work forms a part of a research project supported by a grant from the Italian Ministry of Education, University and Research (MIUR) through the Research project of national interest PRIN2012 (D.M. 28 dicembre 2012 n. 957/Ric – Prot. 2012PTZAMC) entitled "Energy consumption and GreenHouse Gas (GHG) emissions in the wastewater treatment plants: a decision support system for planning and management" in which the corresponding author is the Principal Investigator. Marc Neumann acknowledges funding (RYC-2013-13628) from the Spanish Ministry of Economy and Competitiveness (MINECO). This work was also made possible through the financial support of the Natural Sciences and Engineering Research Council of Canada (NSERC) and Primodal Inc. Peter A. Vanrolleghem holds the Canada Research Chair in Water Quality Modelling. Giorgio Mannina is a Fulbright research Fellow at Columbia University, New York, USA.

11.8 REFERENCES

Benedetti L., Claeys F., Nopens I. and Vanrolleghem P. A. (2011). Assessing the convergence of LHS Monte Carlo simulations of wastewater treatment models. *Water Science and Technology*, **63**(10), 2219–2224.

Brun R., Kühni M., Siegrist H., Gujer W. and Reichert P. (2002). Practical identifiability of ASM2d parameters—systematic selection and tuning of parameter subsets. *Water Research*, **36**, 4113–4127.

Butler M. P., Reed P. M., Fisher-Vanden K., Keller K. and Wagener T. (2014). Identifying parametric controls and dependencies in integrated assessment models using global sensitivity analysis. *Environmental Modelling and Software*, **59**, 10–29.

Campolongo F., Carboni J. and Saltelli A. (2007). An effective screening design for sensitivity analysis of large models. *Environmental Modelling and Software*, **22**(10), 1509–1518.

Chen L., Tian Y., Cao C., Zhang S. and Zhang S. (2012). Sensitivity and uncertainty analyses of an extended ASM3–SMP model describing membrane bioreactor operation. *Journal of Membrane Science*, **389**, 99–109.

Cogan N. G. and Chellam S. (2017). Global parametric sensitivity analysis of a model for dead-end microfiltration of bacterial suspensions. *Journal of Membrane Science*, **537**, 119–127.

Corominas L. and Neumann M. B. (2014). Ecosystem-based management of a Mediterranean urban wastewater system: a sensitivity analysis of the operational degrees of freedom. *Journal of Environmental Management*, **143**, 80–87.

Cosenza A., Mannina G., Vanrolleghem P. A. and Neumann M. B. (2013a). Global sensitivity analysis in wastewater applications: a comprehensive comparison of different methods. *Environmental Modelling and Software*, **49**, 40–52.

Cosenza A., Mannina G., Neumann M. B., Viviani G. and Vanrolleghem P. A. (2013b). Biological nitrogen and phosphorus removal in membrane bioreactors: model development and parameter estimation. *Bioprocess and Biosystems Engineering*, **36**(4), 499–514.

Cosenza A., Mannina G., Vanrolleghem P. A. and Neumann M. B. (2014). Variance-based sensitivity analysis for wastewater treatment plant modelling. *Science of the Total Environment*, **470–471**, 1068–1077.

Cukier R. I., Fortuin C. M., Shuler K. E., Petschek A. G. and Schaibly J. H. (1973). Study of the sensitivity of coupled reaction systems to uncertainties in rate coefficients. I Theory. *The Journal of Chemical Physics*, **59**, 3873–3878.

De Pauw D. J. W. and Vanrolleghem P. A. (2006). Practical aspects of sensitivity function approximation for dynamic models. *Mathematical and Computer Modelling of Dynamical Systems*, **12**, 395–414.

Deletic A., Dotto C. B. S., McCarthy D. T., Kleidorfer M., Freni G., Mannina G., Uhl M., Henrichs M., Fletcher T. D., Rauch W., Bertrand-Krajewski J. L. and Tait S. (2012). Assessing uncertainties in urban drainage models. *Physics and Chemistry of the Earth*, Parts A/B/C, **42–44**, 3–10.

Dochain D. and Vanrolleghem P. A. (2001). Dynamical Modelling and Estimation in Wastewater Treatment Processes. IWA Publishing, London, UK.

Dotto C. B. S., Mannina G., Kleidorfer M., Vezzaro L., Henrichs M., McCarthy D. T., Freni G., Rauch W. and Deletic A. (2012). Comparison of different uncertainty techniques in urban stormwater quantity and quality modelling. *Water Research*, **46**(8), 2545–2558.

Freni G., Mannina G. and Viviani G. (2009). Uncertainty assessment of an integrated urban drainage model. *Journal of Hydrology*, **373**, 392–404.

Frey H. and Patil S. (2002). Identification and review of sensitivity analysis methods. *Risk Analysis*, **22**, 553–578.

Gan Y., Duan Q., Gong W., Tong C., Sun Y., Chu W., Ye A., Miao C. and Di Z. (2014). A comprehensive evaluation of various sensitivity analysis methods: a case study with a hydrological model. *Environmental Modelling and Software*, **51**, 269–285.

Gernaey K. V., van Loosdrecht M. C. M., Henze M., Lind M. and Jørgensen S. B. (2004). Activated sludge wastewater treatment plant modelling and simulation: state of the art. *Environmental Modelling and Software*, **19**, 763–783.

Henze M., Grady C. P. L. Jr., Gujer W., Marais G. v. R. and Matsuo T. (2000). Activated Sludge Models ASM1, ASM2, ASM2d and ASM3. Scientific and Technical Report No. 9, IWA Publishing, London, UK.

Henze M., van Loosdrecht M., Ekama G. and Brdjanovic D. (2008). Biological Wastewater Treatment – Principles, Modelling and Design, UNESCO. IWA Publishing, London, UK.

Hong T. and Purucker S. T. (2018). Spatiotemporal sensitivity analysis of vertical transport of pesticides in soil. *Environmental Modelling and Software*, **105**, 24–38.

Homma T. and Saltelli A. (1996). Importance measures in global sensitivity analysis of nonlinear models. *Reliability Engineering and System Safety*, **52**, 1–17.

Jakeman A. J., Letcher R. A. and Norton J. P. (2006). Ten iterative steps in development and evaluation of environmental models. *Environmental Modelling and Software*, **21**(5), 602–614.

Mannina G. and Viviani G. (2010). An urban drainage stormwater quality model: model development and uncertainty quantification. *Journal of Hydrology*, **381**, 248–265.

Mannina G., Cosenza A. and Ekama G. A. (2017). Greenhouse gases from membrane bioreactors: mathematical modelling, sensitivity and uncertainty analysis. *Bioresource Technology*, **239**, 353–367.

Mannina G., Cosenza A., Randrianantoandro M., Anctil F., Neumann M. B. and Vanrolleghem P. A. (2014). Global sensitivity analysis in environmental water quality modelling: Where do we stand? In: *Proceedings of the 7th International Congress on Environmental Modelling and Software*, June 15–19, 2014, San Diego, California, USA, pp. 619–626.

Mannina G., Cosenza A., Vanrolleghem P. A. and Viviani G. (2011a). A practical protocol for calibration of nutrient removal wastewater treatment models. *Journal of Hydroinformatics*, **13**(4), 575–595.

Mannina G., Cosenza A., Viviani G., Vanrolleghem P. A. and Neumann M. B. (2012). Global sensitivity analysis for urban water quality modelling: comparison of different methods. In: *Proceedings of 9th International Conference on Urban Drainage Modelling*, 4–7 September 2012, Belgrade, Serbia.

Mannina G., Di Bella G. and Viviani G. (2010). Uncertainty assessment of a membrane bioreactor model using the GLUE methodology. *Biochemical Engineering Journal*, **52**, 263–275.

Mannina G., Di Bella G. and Viviani G. (2011b). An integrated model for biological and physical process simulation in membrane bioreactors (MBRs). *Journal of Membrane Science*, **376**(1–2), 56–69.

Morris M. D. (1991). Factorial sampling plans for preliminary computational experiments. *Technometrics*, **33**(2), 161–174.

Neumann M. B. (2012). Comparison of sensitivity analysis techniques for modelling micropollutant oxidation in water treatment. *Science of the Total Environment*, **433**(1), 530–537.

Norton J. P. (2008). Algebraic sensitivity analysis of environmental models. *Environmental Modelling and Software*, **23**, 963–972.

Norton J. P. (2015). An introduction to sensitivity assessment of simulation models. *Environmental Modelling and Software*, **69**, 166–174.

Paton F. L., Maier H. R. and Dandy G. C. (2013). Relative magnitudes of sources of uncertainty in assessing climate change impacts on water supply security for the southern Adelaide water supply system. *Water Resources Research*, **49**(3), 1643–1667.

Petersen B., Gernaey K., Henze M. and Vanrolleghem P. A. (2002). Evaluation of an ASM1 model calibration procedure on a municipal – industrial wastewater treatment plant. *Journal of Hydroinformatics*, **4**, 15–38.

Pianosi F., Beven K., Freer J., Hall J. M., Rougier J., Stephenson D. B. and Wagener T. (2016). Sensitivity analysis of environmental models: a systematic review with practical workflow. *Environmental Modelling and Software*, **79**, 214–232.

Ramin E., Flores-Alsina X., Sin G., Gernaey K. V., Jeppsson U., Mikkelsen P. S. and Plósz B. G. (2014). Influence of selecting secondary settling tank sub-models on the calibration of WWTP models – A global sensitivity analysis using BSM2. *Chemical Engineering Journal*, **241**, 28–34.

Reder K., Alcamo J. and Flörke M. (2017). A sensitivity and uncertainty analysis of a continental-scale water quality model of pathogen pollution in African rivers. *Ecological Modelling*, **351**, 129–139.

Robles A., Ruano M. V., Ribes J., Seco A. and Ferrer J. (2014). Global sensitivity analysis of a filtration model for submerged anaerobic membrane bioreactors (AnMBR). *Bioresource Technology*, **158**, 365–373.

Ruano M. V., Ribes J., Seco A. and Ferrer J. (2012). An improved sampling strategy based on trajectory design for application of the Morris method to systems with many input factors. *Environmental Modelling and Software*, **37**(11), 103–109.

Saltelli A. (2000). Sensitivity Analysis. John Wiley, Chichester, UK.

Saltelli A. (2002). Making best use of model valuations to compute sensitivity indices. *Computer Physics Communications*, **145**, 280–297.

Saltelli A. and Annoni P. (2010). How to avoid perfunctory sensitivity analysis. *Environmental Modelling and Software*, **25**(12), 1508–1517.

Saltelli A. and Sobol' M. I. (1995). About the use of rank transformation in sensitivity analysis of model output. *Reliable Engineering and System Safety*, **50**(3), 225–239.

Saltelli A., Tarantola S., Campolongo F. and Ratto M. (2004). Sensitivity analysis in practice. A guide to assessing scientific models. John Wiley & Sons, Ltd., Chichester, UK, ISBN 0-470-87093-1.

Saltelli A., Ratto M., Tarantola S. and Campolongo F. (2005). Sensitivity analysis for chemical models. *Chemical Reviews*, **105**, 2811–2827.

Saltelli A., Ratto M., Andres T., Campolongo F., Cariboni J., Gatelli D., Saisana M. and Tarantola S. (2008). Global Sensitivity Analysis. The Primer. John Wiley, Chichester, UK.

Sarrazin F., Pianosi F. and Wagener T. (2016). Global Sensitivity Analysis of environmental models: convergence and validation. *Environmental Modelling and Software*, **79**, 135–152.

Schaibly J. H. and Shuler K. E. (1973). Study of the sensitivity of coupled reaction systems to uncertainties in rate coefficients. II Applications. *The Journal of Chemical Physics*, **59**, 3879–3888.

Sin G., Gernaey K. V., Neumann M. B., van Loosdrecht M. and Gujer W. (2011). Global sensitivity analysis in wastewater treatment plant model applications: prioritizing sources of uncertainty. *Water Research*, **45**, 639–651.

Sobol' I. M. (1993). Sensitivity analysis for non-linear mathematical models. *Mathematical Modelling and Computational Experiment*, **1**, 407–414.

Song X., Zhang J., Zhan C., Xuan Y., Ye M. and Xu C. (2015). Global sensitivity analysis in hydrological modeling: review of concepts, methods, theoretical framework, and applications. *Journal of Hydrology*, **523**, 739–757.

Sun X. Y., Newham L. T. H., Croke B. F. W. and Norton J. P. (2012). Three complementary methods for sensitivity analysis of a water quality model. *Environmental Modelling and Software*, **37**, 19–29.

Takacs I., Patry G. G. and Nolasco D. (1991). A dynamic model of the clarification-thickening process. *Water Research*, **25**, 1263–1271.

Valverde-Pérez B., Ramin E., Smets B. F. and Plósz B. G. (2015). An innovative enhanced biological nutrient recovery activated sludge system to produce growth medium for green microalgae cultivation. *Water Research*, **68**, 821–830.

Vanrolleghem P. A., Mannina G., Cosenza A. and Neumann M. B. (2015). Global sensitivity analysis for urban water quality modelling: terminology, convergence and comparison of different methods. *Journal of Hydrology*, **522**, 339–352.

Wágner D. S., Valverde-Pérez B., Sæbø M., de la Sotilla M. B., Van Wagenen J., Smets B. F. and Plósz B. G. (2016). Towards a consensus-based biokinetic model for green microalgae: the ASM-A. *Water Research*, **103**, 485–499.

Yang J. (2011). Convergence and uncertainty analyses in Monte-Carlo based sensitivity analysis. *Environmental Modelling and Software*, **26**(4), 444–457.

Index

IWA Publishing's authorised EU representative for General Product Safety Regulations is Diane D'Arras, 15 rue Duret, 75116 Paris, France, e-mail: safety@iwap.co.uk.

Printed and bound by CPI Group (UK) Ltd, Croydon, CR0 4YY
12/05/2026
02108863-0005